Biological Applications
Of Magnetic Resonance

BIOLOGICAL APPLICATIONS OF MAGNETIC RESONANCE

edited by

R. G. SHULMAN
Bell Laboratories
Murray Hill, New Jersey

ACADEMIC PRESS

New York   San Francisco   London

A Subsidiary of Harcourt Brace Jovanovich, Publishers

Academic Press Rapid Manuscript Reproduction

ACADEMIC PRESS, INC.
111 Fifth Avenue, New York, New York 10003

United Kingdom Edition published by
ACADEMIC PRESS, INC. (LONDON) LTD.
24/28 Oval Road, London NW1

Library of Congress Cataloging in Publication Data
Main entry under title:

Biological applications of magnetic resonance.

    1.   Nuclear magnetic resonance.   2.   Biological
chemistry—Technique.   3.   Biology—Technique.
I.   Shulman, Robert Gerson.
QH324.9.N8B56   574.1'9285   79-16020
ISBN 0-12-640750-9

PRINTED IN THE UNITED STATES OF AMERICA

79 80 81 82   9 8 7 6 5 4 3 2 1

CONTENTS

CONTRIBUTORS

Numbers in parentheses indicate the pages on which authors' contributions begin.

IAN M. ARMITAGE (345), Department of Molecular Biophysics and Biochemistry, Yale University, New Haven, Connecticut

AKSEL A. BOTHNER-BY (177), Department of Chemistry, Carnegie–Mellon University, Pittsburgh, Pennsylvania

T. R. BROWN (537), Bell Laboratories, 600 Mountain Avenue, Murray Hill, New Jersey

JAN F. CHLEBOWSKI (345), Department of Molecular Biophysics and Biochemistry, Yale University, New Haven, Connecticut

JOSEPH E. COLEMAN (345), Department of Molecular Biophysics and Biochemistry, Yale University, New Haven, Connecticut

STEVEN K. DOWER (271), Department of Biochemistry, Oxford University, South Parks Road, Oxford, England

RAYMOND A. DWEK (271), Department of Biochemistry, Oxford University, South Parks Road, Oxford, England

D. G. GADIAN (463), Department of Biochemistry, Oxford University, South Parks Road, Oxford, England

C. W. HILBERS (1), Department of Biophysical Chemistry, University of Nijmegen, Nijmegen, The Netherlands

THOMAS R. KRUGH (113), Department of Chemistry, University of Rochester, Rochester, New York

GERD N. LA MAR (305), Department of Chemistry, University of California, Davis, California

JOHN L. MARKLEY (397), Biochemistry Division, Department of Chemistry, Purdue University, West Lafayette, Indiana

W. B. MIMS (221), Bell Laboratories, 600 Mountain Avenue, Murray Hill, New Jersey

MERRILL E. NUSS (113), Department of Chemistry, University of Rochester, Rochester, New York

JAMES D. OTVOS (345), Department of Molecular Biophysics and Biochemistry, Yale University, New Haven, Connecticut

J. PEISACH (221), Department of Molecular Pharmacology and Molecular Biology, Albert Einstein College of Medicine, Yeshiva University, Bronx, New York

G. K. RADDA (463), Department of Biochemistry, Oxford University, South Parks Road, Oxford, England

B. R. REID (45), Department of Biochemistry, University of California, Riverside, California

R. E. RICHARDS (463), Department of Biochemistry, Oxford University, South Parks Road, Oxford, England

G. T. ROBILLARD (45), Department of Physical Chemistry, University of Groningen, Groningen, The Netherlands

ANTONIUS J. M. SCHOOT UITERKAMP (345), Department of Molecular Biophysics and Biochemistry, Yale University, New Haven, Connecticut

P. J. SEELEY (463), Department of Biochemistry, Oxford University, South Parks Road, Oxford, England

R. G. SHULMAN (537), Bell Laboratories, 600 Mountain Avenue, Murray Hill, New Jersey

K. UGURBIL (537), Bell Laboratories, 600 Mountain Avenue, Murray Hill, New Jersey

HYDROGEN-BONDED PROTON EXCHANGE
AND ITS EFFECT ON NMR SPECTRA OF NUCLEIC ACIDS

C. W. Hilbers

Department of Biophysical Chemistry
University of Nijmegen
Nijmegen, The Netherlands

I. INTRODUCTION

Nuclear magnetic resonance spectroscopy is one of the few techniques that permits direct observation of protons situated in hydrogen bonds. This possibility had been realized very early in the application of NMR to structural studies of alcohols (Liddel and Ramsey, 1951). Since then numerous studies have been carried out on small molecules from which it became clear that hydrogen bonding results in downfield shifts of the resonances of protons involved in hydrogen bonds. Kearns *et al.* (1971) were the first to observe hydrogen-bonded proton resonances in nucleic acids. These investigators studied the NMR spectra of transfer RNAs and found that the ring N protons of uridine and guanine, when involved in hydrogen bonding, give rise to resonances shifted far downfield to about 14 ppm from the methyl resonance of the reference 2,2-dimethyl-4-silapentane-1-sulfonate (DSS). These findings provided the experimentalists with "a window" through which important junctions in such biological macromolecules could be observed and therefore triggered a number of experiments in which the structure and stability of nucleic acids were studied. This chapter is primarily concerned with the exchange behavior of the hydrogen-bonded protons and its influence on their NMR spectra. Such studies are important for two reasons. First a good understanding of the exchange phenomena of protons in hydrogen bonds is required in order to be able to derive structural information from the hydrogen-bonded proton resonance spectra. For instance, in so-called melting experiments in which the disappearance of hydrogen-bonded resonances is studied as a function

of temperature, exchange phenomena are to be related to the actual physical disruption of basepaired regions. Second, these
exchange processes are interesting per se. The formation and
disruption of hydrogen bonds of double helical RNA and DNA
structures are key events during a number of biological processes, yet very often they take place far below the thermal
helix to coil transition of these hydrogen-bonded structures.
For instance, during the replication and transcription of DNA,
the two self-complementary strands have to be separated. Physical chemical studies have shown, however, that the spontaneous
opening of double-stranded DNA is a very unlikely event under
physiological conditions. Stretches of basepairs have little
tendency to open up (Gralla and Crothers, 1973; Lukashin *et al.*,
1976). On the other hand, as is known from tritium exchange experiments (Teitelbaum and Englander, 1975a,b) and from chemical
modification methods like formaldehyde and mercury binding (Mc-
Ghee and von Hippel, 1975a,b; Lukashin *et al.*, 1976; Williams
and Crothers, 1975), DNA and also RNA possess a conformational
motility, which results in fluctuational opening of basepairs.
This behavior is likely to play an important role during replication and transcription among other processes. NMR provides a
physical method to approach these problems since the hydrogen-
bonded proton resonances reflect the fluctuational motility in
basepairs well below the thermal melting transition.
    Here the influence of exchange on the hydrogen-bonded proton
spectra will be discussed after some introductory remarks on
the position and assignments of these resonances have been made.

II. LOCATION OF HYDROGEN-BONDED PROTON RESONANCES OF NUCLEIC
    ACIDS

    Protons involved in hydrogen bonds resonate at rather low
fields, well resolved from the bulk of proton resonances from
the molecule under study. While the resonances of most protons
of diamagnetic molecules cover a region of about 10 ppm downfield from the methyl resonance of the reference compound DSS,
resonances from protons participating in hydrogen bonds in nucleic acids are found from about 9 to 16 ppm downfield from the
reference signal, depending on the type of hydrogen bond involved. This downfield shift is mainly the result of polarization of the N-H bond in one molecule caused by the electric
field of atomic charges in the complexing molecule (Berkeley
and Hanna, 1964; Slejko and Drago, 1973; Giessner-Prettre *et
al.*, 1977). In addition, shifts caused by anisotropic diamagnetic susceptibility terms like the ring current shift of the
complexing molecule may also contribute to this downfield shift.

The spectral regions, where hydrogen-bonded ring nitrogen protons and exocyclic aminoprotons participating in basepairs resonate, have been inferred from studies on transfer RNAs and double helical RNA and DNA model systems. In numerous studies, following the discovery of Kearns, Patel, and Shulman, it has been established that resonances from ring N-H protons hydrogen bonded to ring nitrogens, $>$N-H$\cdots$N$<$, are found between 16 to 11 ppm downfield from DSS. This is true for ring N protons in classical Watson-Crick basepairs as well as in nonclassical basepairs. An example is given in Fig. 1, where the spectrum of the double helix formed by oligo A and oligo U is compared with the spectrum of the triple helix formed by oligo A-oligo U-oligo U. One basepair in the triple helix is a normal Watson-Crick combination while the other is a Hoogsteen pair in which the N_3H proton of uridine is complexed to the N_7 ring nitrogen of adenine (see Fig. 1). The ring N protons of AU pairs generally resonate at somewhat lower field values, i.e., between 14.5 and 12 ppm, than the ring N protons of GC pairs, which are found between 13.6 and 11.5 ppm. These numbers should not be considered as exact limiting values, nor are those to be discussed below.

The $C-C^+$ basepair in acid oligo C solutions gives rise to a resonance at 15.5 ppm, which is assigned to the C^+N_3H proton (Kallenbach et al., 1976). Also at low field, 15.2 ppm, the ring proton resonance of inosine in the basepair I-C (Patel, D. J., 1977, personal communication) is found. Hydrogen bonds of the type $>$N-H$\cdots$O=C$<$, i.e., ring N protons participating in hydrogen bonding to a carbonyl oxygen, are found in GU pairs. These protons resonate between 12 to 10.5 ppm. This follows from experiments on yeast tRNA[Asp] (Robillard et al., 1976) and on poly GT (Kearns, 1977). The same type of hydrogen bonds are present in quadrupole complexes of guanosine monophosphates (Pinnavaia et al., 1975) and of oligo I (Kallenbach et al., 1976). These complexes give rise to resonances at 11.1 ppm and 11.8 ppm, respectively.

The resonances of exocyclic aminoprotons hydrogen bonded to carbonyl oxygen are found upfield from 9 ppm and so far have been accessible to detailed study in a few isolated cases (Patel, 1976, 1977). Exocyclic aminoprotons complexed to ring nitrogens resonate around 9 ppm (Steinmetz-Kayne et al., 1977). An example is provided by the acid form of oligo A (see Fig. 2). At low pH, oligo A forms a double helical complex in which one aminoproton is hydrogen bonded to adenine N_7 and the other aminoproton to the phosphate group of the opposite chain. The resonance at 9 ppm was assigned to the first, the resonance at 8 ppm to the second aminoproton (Geerdes, H. A. M., Kremer, A., and Hilbers, C. W., 1977, unpublished results). These data have been collected in Table I.

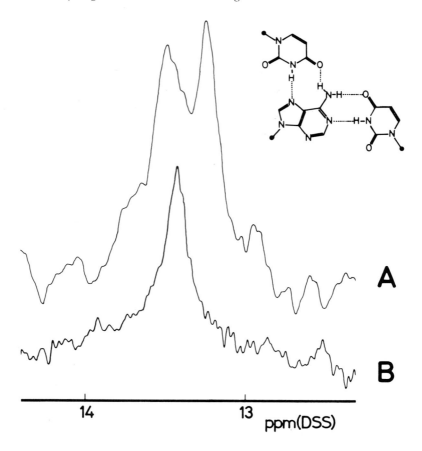

Fig. 1.   The 360 MHz spectra of the oligo A-(oligo U)$_2$
triple helix (upper spectrum) and oligo A-oligo U double helix
(lower spectrum) in a buffer containing 0.12 $\underline{M}$ NaCl, 10 mM so-
dium-cacodylate, 0.5 m$\underline{M}$ EDTA at pH 7.0, recorded at 5°C.   The
concentrations (in monomers) were in the triple helix oligo A
8 m$\underline{M}$, oligo U 16 m$\underline{M}$, and in the double helix both 11 m$\underline{M}$.   A
pairing scheme of the triple combination is given. Note that
the resonances are from the ring N$_3$ protons (Geerdes and Hilbers, 1977, Nucl. Acids Res. 4, 207-221).

In the crystal structure of yeast tRNA[Phe], also ring N
protons have found to be hydrogen bonded to phosphate groups,
for instance, the interaction U$_{33}$N$_3$H-P$_{36}$.   On the basis of an-

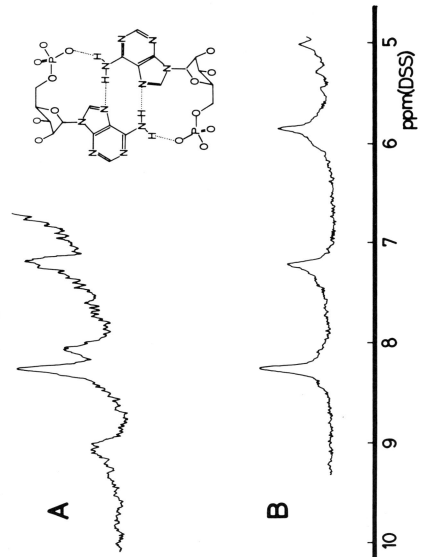

Fig. 2. The 360 MHz spectra of double helical oligo A, approximate chain length 14 nucleotides, recorded in H_2O (A) or D_2O (B) at 25°C, pH 4.5. Concentration oligo A ~ 4 mM. The resonances at 9.1 and 8.1 ppm in (A), not visible in (B), are assigned to the amino protons complexed to AN7 and the phosphate group, respectively.

TABLE I.   *Location of the ring N hydrogen-bonded proton resonances of different basepairs, in ppm downfield from DSS*[a].

Basepair	Type	Location of resonances
	AU Watson–Crick	14.5–12
	GC Watson–Crick	13.5–11.5
	AU Hoogsteen	14.5–12
	s^{4}UA reversed Hoogsteen	15–12
	GU	12–10.5

TABLE I (continued)

Basepair	Type	Location of resonances
	$C-C^+$ acid oligo C	~16
	$I-C$	~15[b]
	m^7G-G	14.5-12.5
	$A-A$ acid oligo A	$N-H\bullet\bullet\bullet N$ ~9 $N-H-OP$ ~8

[a]*Also included are the positions of the amino proton resonances of acid oligo A.*
  [b]*D. J. Patel (private communication, 1977).*

ticodon hairpin studies Kearns (1976) suggests that resonances at ~11.5 ppm common to all tRNA spectra may come from these hydrogen-bonded protons. At present these assignments cannot be considered firmly established, since for yeast tRNA[Phe] this resonance is not affected after nicking the anticodon loop between U_{33}, which provides a hydrogen bond to phosphate P_{36} and G_{34} (Salemink, P. J. M., unpublished results, 1977).

III. RING CURRENT SHIFT CALCULATIONS

Having located the different types of hydrogen-bonded resonances in particular regions, the first problem one encounters when examining NMR spectra is that of assignment. In exchange and structural studies it is necessary that the resonances be assigned to particular protons in the molecule. In general this is a tedious problem, especially for larger nucleic acid structures like tRNA. Two approaches have been taken so far, namely, selective (chemical) modification of the nucleic acid structure (Wong and Kearns, 1974; Reid et al., 1975; Daniel and Cohn, 1975; Salemink et al., 1977) and the application of ring current shift calculations. The latter are used to explain the secondary shifts, i.e., the shifts from the resonance positions that hydrogen-bonded protons have in an "isolated" basepair, one not surrounded by other basepairs. The resonance position in such a basepair is designated as the intrinsic resonance position. Ring current effects are expected to provide the main shift mechanism, since the bases have aromatic character and are therefore expected to be employable to predict resonance positions. The computations used the ring current shift contours provided by Giessner-Prettre and Pullman (1970) and started from the assumption that the RNA or DNA double helical stretches had A'RNA (Shulman et al., 1973) or B-DNA structure (Patel and Tonelli, 1974) in solution, respectively. Only nearest-neighbor contributions were included in these calculations. By studying a number of model systems (Shulman et al., 1973; Patel and Tonelli, 1974), shifts for the different possible basepair combinations were determined and tabulated and used to derive the resonance position of the ring N protons in "isolated" AU and GC basepairs. Calculated and experimental positions of a number of model systems have been collected in Table II.

In general the calculated positions agree within one- or two-tenths of a ppm with the observed positions except for terminal basepairs, which very often yield experimental resonance positions at higher field from the calculated positions. The latter observations can be rationalized by taking fraying ef-

TABLE II. *Observed and calculated positions of the ring N hydrogen bonded proton resonances in different oligonucleotides.[a]*

Oligonucleotide	Basepair	Observed position	Calculated position
1234 d AAAGCTTT[b] .... TTTCGAAT	A.T. (terminal)	13.6	14.3
	A.T. (2 position)	14.2	14.1
	A.T. (3 position)		14.2
	G.C.	13.1	12.9
r AAGCUU[c] ...... UUCGAA	A.U. (terminal)	13.2	13.4
	A.U. (internal)	14.25	14.0
	G.C.	13.5	13.3
d ATGCAT[d] .... TACGTA	A.T. (terminal)	13.15	13.9
	A.T. (internal)	13.8	13.85
	G.C. (central)	12.75	12.74
d CGCG[e] .... GCGC	G.C. (terminal)	13.4	13.3
	G.C. (internal)	13.25	13.0
d CCGG[f] .... GGCC	C.G. (terminal)	13.3	13.4
	C.G. (internal)		13.1

TABLE II (Continued)

Oligonucleotide	Basepair	Observed position	Calculated position
d GGCC[f] • • • • CCGG	G.C. (terminal) C.G. (internal)	13.1	13.3[5] 13.1[5]
d AGAGAG[g] • • • • • • TCTCTC	G.C. (internal) A.T. (internal)	12.6-12.7 13.8-14.0	12.4[5] 14.1
d GCGCGC[g] • • • • • • CGCGCG	G.C. (internal)	13.1-13.2	13.1
d TATATA[g] • • • • • • ATATAT	A.T. (internal)	13.35	13.3

[a]*All calculations were based on a B-DNA structure, using the ring current shift table by Patel and Tonelli (1974) except for r(AAGCUU) for which an A'RNA double helix structure was used (Kan et al., 1975).*
    [b]*Kallenbach et al. (1976).*
    [c]*Kan et al. (1975).*
    [d]*Patel and Hilbers (1975).*
    [e]*Garssen et al. (1977).*
    [f]*Patel (1977).*
    [g]*Patel and Tonelli (1974).*

fects of helix ends into account (see below) and/or by allowing for interstrand stacking interactions. Recently the shift calculations have been refined by including next nearest-neighbor contributions to the ring current shift (Arter and Schmidt, 1976) as well as contributions due to the anisotropy of the diamagnetic susceptibility other than those from ring currents (Giessner-Prettre and Pullman, 1976). Inclusion of these additional effects requires a readjustment of the intrinsic resonance positions. One may legitimately ask whether addition of

these contributions will increase our level of confidence in
these type of calculations. The shift calculations are based
on approximate molecular orbital calculations; in addition,
the structure of the double-stranded RNA or DNA model systems
in solution may be different from the one adopted in the calculations. In a detailed study of the self-complementary
r(AAGCUU) (Borer *et al.*, 1975) the shifts for exchangeable as
well as for the nonexchangeable protons were predicted taking
next nearest-neighbor contributions into account. Much better
agreement was obtained for an A'RNA structure than for a B-DNA
structure. Similar conclusions follow from Patel's work on
model DNA compounds, where preference has to be given to the
B-DNA structure (Patel, 1977, and references cited therein) in
accordance with expectation. These data together with the results in Table II look promising enough to make it worthwhile
pursuing this subject. For the student of RNA and DNA structure in solution, it is very important that the limitations of
these methods theoretically as well as experimentally be better
assessed.

IV. EXCHANGE OF HYDROGEN-BONDED PROTONS

It has been known for a long time that above certain temperatures double helices are disrupted or "melted out." In the
NMR spectra of the hydrogen-bonded ring N protons this melting
becomes manifested by the disappearance of their resonances.
This is interpreted as follows: as a result of the opening of
the double helix the ring N protons become accessible to the
solvent and in aqueous solutions this easily leads to exchange
with water protons. Consequently the hydrogen-bonded proton
will disappear from the spectrum. This process is symbolized
by the following equations:

$$\ce{>N-H\bond{...}N< <=>[k_{-1}][k_1] >NH + N<} \tag{1}$$

$$\ce{>NH + H^{*}OH <=>[k] >NH^{*} + HOH} \tag{2}$$

The latter process is catalyzed by hydroxyl and/or buffer ions
present in solution, i.e.,

$$\ce{>NH + B <=>[k_c] >N^- + BH^+} \tag{3}$$

C. W. Hilbers

where B represents the buffer or hydroxyl ions. The kinetics
of these processes have been described in detail (Eigen, 1964;
Englander *et al.*, 1972; Teitelbaum and Englander, 1975a,b;
Crothers *et al.*, 1973).
  The overall rate constant for transferring a proton from
the hydrogen-bonded state to B is given by

$$\underline{k}_{ex} = \underline{k}_{-1}k_c[B]/(\underline{k}_1 + \underline{k}_c[B]) \qquad k_{ax} = k_{op}\dfrac{k_c \, {}^cB}{k_{cl} + k_c \cdot {}^cB} \qquad (4)$$

The rate constants are defined in the reaction equations above.
It should be realized that in deriving (4) reaction (1) has been
taken to represent the formation of an intramolecular basepair.
  For practical reasons two extremes are usually examined:

(a) $\underline{k}_c[B] \gg \underline{k}_1$ medium Temp $k_c \, {}^cB \gg k_{cl}$ (5)

In this situation the ring N proton reacts with the catalyst
virtually every time the double helix opens up and the rate of
transfer of the ring N proton from the double helix is determined by the dissociation rate of the double helix, i.e.,
$\underline{k}_{ex} = \underline{k}_{-1}$. In other words the reaction is opening limited.

(b) $\underline{k}_c[B] \ll \underline{k}_1$ high Temp $k_c \, {}^cB \ll k_{cl}$ (6)

In this situation the double helix opens and closes many times
before the ring N proton reacts with the catalyst. The proton
transfer rate from the double helical state is now determined
by the preequilibrium describing the dissociation of the basepair in the double helix times the rate at which the proton in
the open state reacts with the catalyst, i.e., $\underline{k}_{ex} = \underline{K}_{diss}\underline{k}_c[B]$,
where $\underline{K}_{diss} = \underline{k}_{-1}/\underline{k}_1$. If the overall rate of proton transfer
from the hydrogen bond is limited by the rate $\underline{k}_c \cdot B$, $\underline{k}_{ex}$ will
depend on pH and the type of buffer present in solution. This
permits an experimental differentiation between the limits (5)
and (6), since for an opening limited reaction such a dependence is not expected.
  The rate constant $\underline{k}_c$, characterizing the transfer of the
ring N proton in the open form to the B ions can be related to
the difference in pK values of the $\geqslant$N-H moiety and the ions.
If the formation of the collision complexes involved in the
proton transfer are diffusion limited:

$$\underline{k}_c = \underline{k}_f \frac{10^{\Delta pK}}{1 + 10^{\Delta pK}} \qquad (7)$$

where $\Delta pK = pK(BH^+) - pK(N\text{-}H)$ (Englander *et al.*, 1972); $\underline{k}_f$ is
the rate constant representing the formation of the collision

complex, which is assumed to be diffusion limited, i.e., of the
order of 10^{10} M^{-1}sec^{-1}. If ΔpK is about two or larger, k_c will
approach the diffusion limit. On the other hand for proton
transfer of the ring NH moiety to HPO$_4^{2-}$ ions one expects
$k_c \approx 5 \times 10^7$ M^{-1}sec^{-1}, since $\Delta pK = -2.3$.

It is well established that exchange reactions as symbo-
lyzed in Eqs. (1)-(3) may result in broadening and/or shifting
of the NMR resonances of the protons involved. In each of the
different chemical environments, i.e., in the double helix, in
the open state, and in the water molecule, the proton is in a
different magnetic environment and will resonate at a different
position in the spectrum. If the exchange is slow the NMR
spectrometer is able to distinguish between the protons in the
different environments. However, with increasing exchange
rates the spectrometer does not know any longer to which environment the proton belongs and the resonance lines will
broaden, collapse, and eventually give rise to one sharp peak,
when the exchange becomes rapid enough. Since the concentration of water is so overwhelmingly high with respect to the
dissolved biological material, the signals will all merge with
the water resonance in the fast exchange limit. The influence
of exchange on the line forms and line positions can be described by the Bloch equations modified for chemical exchange
(McConnell, 1958; Pople et al., 1959). For the present problem these equations have been employed to relate the widths
and positions of the hydrogen-bonded resonances to the lifetime
of the double helix (Crothers et al., 1974).

1. For slow exchange between helix and coil, i.e., under conditions where the reciprocal lifetimes of these two states are
much smaller than the chemical shift difference between the two
corresponding resonances, the excess linewidth at half height,
$\Delta\nu_{\frac{1}{2}ex}$, due to exchange is given by

Helix ~ Coil ~ langsam;
so langsam, daß k_c^rB ≫ k₊₁

$$\tau_{hc}^{-1} = \pi \, \Delta\nu_{\frac{1}{2}ex} \qquad\qquad \Delta\nu_{\frac{1}{2}} = \frac{1}{\pi \, \tau_{hc}} \qquad\qquad (8)$$

where τ_{hc}^{-1} is the probability per unit time that the ring N proton jumps from the helical to the coil state. For the type of
reaction considered Eq. (1), this means that $\tau_{hc}^{-1} = k_{-1}$. The
resonance position coincides with that of the ring N proton in
the double helical state.

2. When the exchange between helix and coil is fast on the NMR
time scale and the fraction in the coil state is small, two
limits equivalent to expressions (5) and (6) above are important.

f_c ~ 0
f_H ~ 1

(a) When exchange with water occurs virtually every time
the helix is open the excess linewidth due to exchange is

$$\overline{\tau_{hc}} = k_{op}$$

$$\tau_{hc}^{-1} = \pi \, \Delta\nu_{\frac{1}{2}ex} = k_{-1} \quad (\text{chem. Austausch} = \text{schnell}) \tag{8a}$$

as for the slow exchange situation. Again the resonance position coincides with that representative of the double helix.

(b) On the other hand when the ring N proton interchanges
many times between the helix and coil states before exchange
with water occurs the exchange induced linebroadening is given
by

a) $\pi\Delta\nu = k_{-1}$

$$\pi \, \Delta\nu_{\frac{1}{2}ex} = f_c \tau_{cw}^{-1}$$

b) $\pi\Delta\nu = K_{diss} \cdot k_c{}^c B$ (9)

while the resonance position is given by

$$\overline{\omega} = f_h \omega_h + f_c \omega_c \tag{10}$$

In these expressions τ_{cw}^{-1} is the probability per unit time the
ring N proton in the coil state is transferred to buffer ions
or water; ω_h and ω_c are the resonance positions of the ring N
proton in the double helix and in the coil, respectively; while
f_h and f_c are the corresponding fractions in these states.

Thus under these circumstances the line broadening is not
determined by the helix lifetime but by the "leakage" of the
ring N proton to buffer or hydroxyl ions, i.e.,

weil Austausch
gesdms.-best

$$\pi \, \Delta\nu_{\frac{1}{2}ex} = f_c \tau_{cw}^{-1} = K_{diss} \, k_c [B] \quad (\text{chem. Austausch} =) \tag{11}$$

langsam

$k_c{}^c B \ll k_{-1}$

V. EXCHANGE OF OLIGONUCLEOTIDES

To probe the structure and dynamics of nucleic acids in solution, a number of NMR studies of oligo- and polynucleotides
have been carried out during the last 5 years. This has been
possible as a result of two developments:

(1) improvement in NMR instrumentation has increased resolution and sensitivity, and
(2) the new techniques available for isolating sufficient
amounts of pure natural occurring RNAs and for synthesizing
model RNA and DNA oligomers.

In particular the investigations on oligomers may serve to
assess the occurrence of the different exchange phenomena and
their relation to the dynamical structure of nucleic acids in
solution. These smaller systems have the well-known advantage

over larger molecules of providing a better resolution. Individual resonances can be followed over a wide range of temperature and solution conditions. Also the assignment of resonances to particular protons in the molecule is less complicated and can be performed to a high degree of certainty. Detailed studies have been performed on the self-complementary hexanucleotides d(ATGCAT) (Patel and Hilbers, 1975; Hilbers and Patel, 1975; Patel, 1975) and r(AAGCUU) Borer *et al.*, 1975; Kan *et al.*, 1975) and the octanucleotide d(AAAGCTTT) (Kallenbach *et al.*, 1976).

The double helical molecule formed by the hexanucleotide d(ATGCAT) gives rise to three hydrogen-bonded ring N proton resonances: one from the terminal AT pairs at 13.15 ppm, one from the internal AT pairs at 13.80 ppm, and one from the central GC pairs at 12.7 ppm (see Fig. 3). These positions were obtained at 3°C. Only three resonances are observed since the double helix contains a twofold axis of symmetry, which makes the terminal basepairs, the internal basepairs, as well as the central basepairs pairwise equivalent. The assignment of these resonances was based on two methods. Assuming that the double helix had a B-DNA conformation the line positions were calculated using the ring current shift rules discussed in the preceding section. This yields 13.9 ppm for the terminal AT pair resonance, 13.85 ppm for the internal AT pair resonance, and 12.73 ppm for the GC pair resonance. These values are to be compared with the experimental positions given above. The resonances at 13.85 and 12.75 ppm were assigned to the internal AT and central GC pair, respectively. Because of its sensitivity to temperature changes the resonance at 13.15 ppm was assigned to the terminal AT pair, resulting in a difference of 0.75 ppm between calculated and experimental positions. This assignment was confirmed by the second approach in which actinomycin-D was bound to the double helical DNA (Patel, 1974). Under the experimental conditions used, the phenoxazone ring system of the oligopeptide intercalates between the two GC pairs. This results in a splitting of the GC and internal AT ring N proton resonances at positions in accordance with the assignments of the ring current calculations. The resonance assigned to the terminal AT pairs was not affected. Also for the other nucleotides mentioned above one resonance turned out to be most sensitive to changes in temperature and was therefore assigned to the terminal basepairs. The resonance positions of the other protons agreed reasonably well with the experimental positions (see Table II).

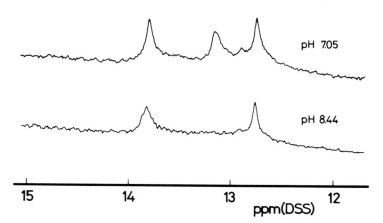

Fig. 3. The 300 MHz hydrogen-bonded proton spectrum of double-stranded d(ATGCAT) recorded at two different pH values both at 3°C. Solution conditions: 14 mM d(ATGCAT) in H_2O, no salt added. The resonances were assigned as follows: AT_{term} at 13.15 ppm, AT_{int} at 13.8 ppm, and GC_{centr} at 12.75 ppm (Patel and Hilbers, 1975, Biochemistry 14, 2651-2656, copyright by the American Chemical Society).

A. Fraying

The temperature-dependent behavior of the terminal basepair resonances has been attributed to so-called fraying processes (Patel and Hilbers, 1975). Fraying reactions consist of rapid opening and closing of the basepairs as a result of which the resonance position of the ring N proton is the weighed average of the fraction of protons situated in the double helical form and the fraction in the open form:

$$\overline{\omega} = f_h\omega_h + f_c\omega_c$$

It is to be expected that at lower temperatures the equilibrium will shift to the base-paired form. In accordance with this expectation lowering of the temperature causes downfield shifts of the terminal basepair resonances in the direction of their calculated position in the double helix (Patel and Hilbers, 1975; Kan et al., 1975; Kallenbach et al., 1976). It turns out that the fraying may even persist down to -25°C. This has been observed for the "terminal AT" resonance of the d(ATGCAT) hexanucleotide, which has not yet completely shifted to its theoretical position at this temperature. At 3°C the terminal AT resonance is not extremely broadened with respect to the other

two resonances.   In view of the fraying model this means that
the closing reaction is so fast that exchange of ring NH protons from the open form to water and buffer ions cannot compete.   In other words the proton transfer is not limited by the
opening of the basepair, but by the rate of transfer from the
open state.   According to Eq. (11) the excess linebroadening is
then given by

$$\pi \, \Delta\nu_{\frac{1}{2}ex} = \underline{K}_{diss} \, \underline{k}_c \, [OH^-]$$

if no buffer is present in solution.   Hence the proton exchange
should be sensitive to base catalysis, which was confirmed by
studying the linewidths as a function of pH (Patel and Hilbers,
1975).
    At higher pH values the terminal AT resonance of double
helical d(ATGCAT) broadens and disappears (see Figs. 3 and 4).
Also the resonance from the internal AT pairs turns out to be
affected, indicating that the proton transfer from this basepair is not determined by the opening rate of the double helical conformation either.   The position of all three resonances
remains unaffected in the pH range 4 to 9, demonstrating that

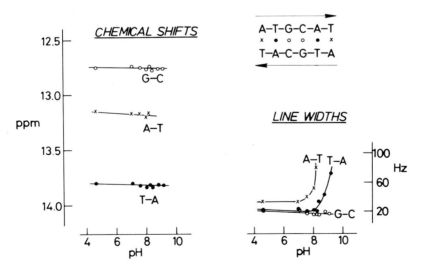

*Fig. 4. Plot of the chemical shifts and linewidths of the
300 MHz hydrogen-bonded proton spectrum of double-stranded
d(ATGCAT) as a function of pH. Solution conditions: 14 mM
d(ATGCAT) in H_2O, no salt added, temperature $3^{\circ}C$ (Patel and
Hilbers, 1975, Biochemistry 14, 2651-2656, copyright by the
American Chemical Society).*

Δ chem. Verschiebung: Maß für d. Änderung des Gleichgewichts zw. Helix-Coil
Δ Linienbreite: Maß f. Beschleunigung des Austauscher bzw. der Helix-Coil
                                                        Umwandlung

the equilibrium between the open and intact state of the basepairs does not change under these conditions (Fig. 4).

The dissociation constant for the equilibrium between these
two states can be calculated from the measured resonance position and the position of the proton in the double helical and
open form. For the terminal basepairs it is given by

$$K_{dT} = (\omega_{TH}-\overline{\omega}_T)/(\overline{\omega}_T-\omega_{TC}) \quad = \quad \frac{\Delta \omega_H}{\Delta \omega_c} \tag{12}$$

where ω_T is the position measured for the terminal AT resonance, ω_{TH} the resonance position in the double helical state,
and ω_{TC} the resonance position in the open state.

The equilibrium constant for the opening of the internal
AT pairs can be calculated in an analogous way. It should be
realized, however, that if opening proceeds from the outside
to the center of the double helix, only those AT pairs open up
for which the terminal pairs are already disrupted. An expression relating K_{dI} to the shifts should take these effects
into account:

$$K_{dI} = (\omega_{TC}-\omega_{TH})(\overline{\omega}_I-\omega_{IH})/(\omega_{IC}-\omega_{IH})(\overline{\omega}_T-\omega_{TH}) = \tag{13}$$

where $\overline{\omega}_I$ is the resonance position measured for the internal
AT ring proton, ω_{IH} its position in the double helix, and ω_{IC}
its position in the open form.

As an example the resonance positions and linebroadenings
of the basepair resonances of d(ATGCAT), dissolved in water
with no salts added, measured as a function of temperature, are
shown in Fig. 5. The equilibrium constants K_{dT} and K_{dI} calculated from these and other measurements are given in Fig. 6.
The corresponding reaction enthalpies have been listed in
Table III. Also included are the reaction enthalpies calculated for the opening and closing of the terminal and internal
AU pairs of r(AAGCUU) (Kan et al., 1975). Whether the differences in enthalpy values are physically significant is difficult to answer. On the one hand, the NMR data are obtained
over a rather small temperature range, limiting the accuracy
of the enthalpies. On the other hand, calculation of the melting temperatures of the AT and AU pairs in d(ATGCAT) and
r(AAGCUU), respectively, yields values that reasonably compare
with average melting temperatures obtained from measurements
of the nonexchangeable protons in these molecules (Patel,
1975; Borer et al., 1975). To this end the equilibrium dissociation constants obtained along the lines indicated above
were extrapolated to unity. As a consequence the experiments
on d(ATGCAT) do suggest that the enthalpy value obtained for
the internal AT pairs is rather sensitive to the counterion

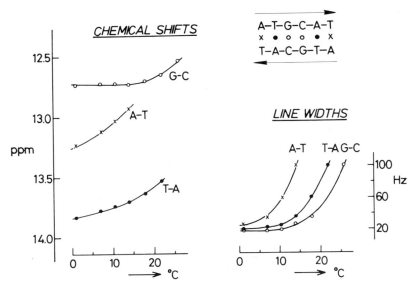

Fig. 5. *Plot of the chemical shifts and linewidths of the 300 MHz hydrogen-bonded proton spectrum of double-stranded d(ATGCAT) as a function of temperature. Solution conditions: 14 mM d(ATGCAT) in H_2O, no salt added, pH 7 at $0°C$ (Patel and Hilbers, 1975, Biochemistry 14, 2651-2656, copyright by the American Chemical Society).*

TABLE III. *Reaction enthalpies in kcal/mol for the disruption of the AT_{term} and AT_{int} basepairs of double helical d(ATGCAT) and of the AU_{term} and AU_{int} basepairs of double helical r(AAGCUU).*

	AT_{term}	AT_{int}	AU_{term}	AU_{int}
Low ionic strength[a]	6.1	11.8		
Mg^{2+} [b]	4.2	4.7		
Medium ionic strength[c]	-	-	8.1	8.2
High ionic strength[d]	4.7	5.1		

[a]*No salt added, pH 7.0.*
[b]*0.1 M NaCl, 0.025 M Mg^{2+}, pH 7.34.*
[c]*0.1 M Cl⁻, 0.01 M phosphate, 0.17 M Na⁺, 10^{-4} M EDTA.* pH ?
[d]*0.3 M NaCl, 0.1 M phosphate, pH 7.*

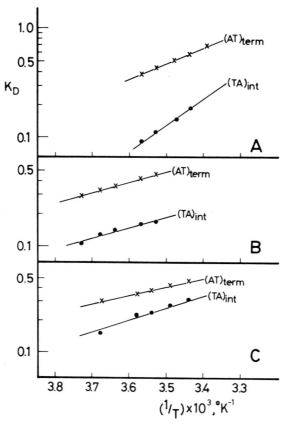

Fig. 6. Plots of the dissociation constants $\underline{K}_d$ of the terminal and internal basepairs in double-stranded d(ATGCAT)
against (1/$\underline{T}$) as a function of solvent conditions.   (A) H_2O, no
salt added, pH 7; (B) 0.3 $\underline{M}$ NaCl, 0.1 $\underline{M}$ phosphate, H_2O, pH 7;
(C) 0.1 $\underline{M}$ NaCl, 0.025 $\underline{M}$ Mg^{2+}, H_2O, pH 7.34 (Patel and Hilbers,
1975, Biochemistry 14, 2651-2656, copyright by the American
Chemical Society).

concentration (see Table III). For other molecules no such data are available. The type of experiments discussed here clearly shows that the melting transition of these short oligonucleotides is not of an all or none character. Intermediate state
concentrations are appreciable during the melting process.

B. *Proton Transfer from the Open State*

The experiments investigating the fraying processes at the ends of double helices demonstrated that the broadening of the hydrogen-bonded proton resonances is strongly influenced by the hydroxyl ions present in solution. Since the excess linewidth under these conditions is given by

$$\pi \, \Delta\nu_{\frac{1}{2}ex} = \underline{K_d} \, \underline{k_{OH}}- [OH^-]$$

the rate constant k_{OH}- characterizing the proton transfer from the open state to $\overline{OH}^-$ can be calculated from the linewidth, line position, and pH. Values obtained for the terminal and internal AT pairs of double-stranded d(ATGCAT) are of the order of 0.5×10^{10} $M^{-1}sec^{-1}$ (Patel and Hilbers, 1975). This shows that the transfer of the ring N_1 proton of thymine in the open state to OH^- is diffusion limited in accordance with predictions from Eq. (7). Values for isolated nucleobases have been published and are found to be in good agreement. The rate of exchange of the ring N proton of purine at 24°C amounts to 3×10^{10} $M^{-1}sec^{-1}$ (Marshall and Grunwald, 1969) and for the ring N proton of uracil it is 1×10^{10} $M^{-1}sec^{-1}$ at 25°C (Eigen, 1964). These results indicate that at lower pH values the ring NH resonances should be observable. Indeed this can be experimentally confirmed; at ~pH 4.0 the exchange rate of the U-N_3H proton in UMP and tetra-U is greatly suppressed and its resonance is visible at 11.3 ppm. Note that at this pH value GMP forms complexes that allow the observation of its ring N_1H proton resonance (see Section II).

One would expect buffer ions, which normally are present in much higher concentration than hydroxyl ions, to significantly enhance the proton exchange from the open state. This has been confirmed by comparing the spectra of double-stranded d(ATGCAT) in the presence and absence of phosphate buffer (Hilbers and Patel, 1975). Computation of the rate constant $k_{HPO_4^{2-}}$, characterizing the rate of proton transfer from the N_1-H moiety of the AT pairs in the open state to HPO_4^{2-}, yields on the average $k_{HPO_4^{2-}} = 0.2 \times 10^6$ $M^{-1}sec^{-1}$. The same results are obtained when these calculations are performed for the AU pairs in double-stranded r(AAGCUU) (Kan *et al.*, 1975). These values are lower than expected on the basis of Eq. (7). Finally it is worth noting that also Mg^{2+} ions tend to enhance the transfer of ring protons from the open state, despite their tendency to stimulate double helix formation (Hilbers and Patel, 1975).

C. Nucleation

The kinetic parameters, characterizing the formation and dissociation of small double helical compounds, as determined by temperature jump experiments, were interpreted in terms of a nucleation process (i.e., the formation of a double helical nucleus consisting of a few contiguous basepairs) followed by a propagation reaction in which the rest of the basepairs is formed much faster (Pörschke and Eigen, 1971; Craig et al., 1971). In order for the double helix to be stable the rate of formation of the basepairs during the propagation reaction must be much faster than the rate of dissociation of the nucleus. Can these processes be distinguished in an NMR experiment?

In the presence of 0.1 M phosphate ions the exchange rate of the ring N protons of the GC pairs of d(ATGCAT) from the open state is so much enhanced that under these conditions Eq. (8a) is valid (see Section IV). Indeed virtually no shift of the GC pairs is observed anymore in temperature-dependent experiments (Hilbers and Patel, 1975). Under these conditions the exchange broadening of the GC hydrogen-bonded ring N proton resonance approaches the dissociation rate of the total double helix, i.e.,

$$\pi \, \Delta \nu_{\frac{1}{2}ex} = k_d = K_{6,2} \tau_{2,0}^{-1} = \left(0.05 \; bis \; 0.1\right) \cdot \tau_{2,0}^{-1} \tag{14}$$

where k_d is the helix dissociation rate constant, $K_{6,2}$ the equilibrium constant defining the dissociation from the intact helix to a particle containing only two GC pairs, and $\tau_{2,0}^{-1}$ the dissociation rate of the two remaining GC pairs. For the hexanucleotide r(AAGCUU) NMR data (Kan et al., 1975) as well as results from temperature jump experiments (Pörschke et al., 1973; Ravetch et al., 1974) are available. The helix dissociation rates obtained by both methods are compared in Fig. 7. The agreement between the results is not perfect, but presumably as good as can be expected in a comparison of experiments performed under different conditions and with different methods. Moreover, the NMR experiments were carried out under conditions where the exchange of the GN_1 protons was not completely opening limited, i.e., the phosphate ion concentration was not high enough to completely validate the use of Eqs. (5) and (8a).

The equilibrium constant $K_{6,2}$ in Eq. (14) can be estimated from the line positions of the internal AT or AU pairs. At 20°C it amounts to about 0.05-0.1 for both molecules, double helical d(ATGCAT) as well as double helical r(AAGCUU). Hence the dissociation rates of the remaining GC pairs are of the order of $6 \times 10^2 sec^{-1}$. From the experiments performed on solutions containing phosphate buffer, the formation rate constant

$\tau_{2,0}^{-1}$

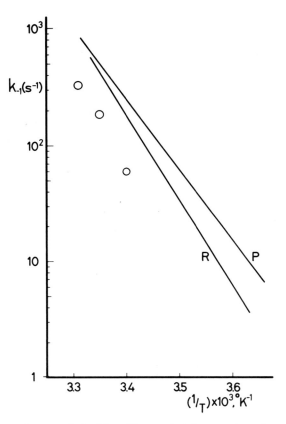

Fig. 7. Plots of helix dissociation rates $\underline{k_{-1}}$ of r(AAGCUU), as a function of $1/\underline{T}$. (P) adapted from Pörschke et al. (1973), Biopolymers 12, 1313-1335; (R) adapted from Ravetch et al. (1974), Nucl. Acids Res. 1, 109-127; (O) calculated from Kan et al. (1975), Biochemistry 14, 4864-4869.

of the AT and AU pairs are estimated as $>> 10^4 sec^{-1}$. Hence the rate of dissociation of the GC pairs is much lower than the rate for formation of the AT pairs, providing direct evidence of the validity of the nucleation propagation model in double helix formation.

D. Influence of Unwinding Protein

   In the foregoing sections we have discussed how changes in temperature and solution conditions can be used to study the dynamical structure of double helical fragments of DNA. Nature

is not free to vary these parameters in order to influence nucleic acid structure and therefore it has developed more subtle
ways. The making and breaking of double helices is influenced
by so-called unwinding proteins. One of these proteins, en-
coded by the filamentous coliphage M13, has a regulating function during synthesis of single-stranded viral DNA from the
double-stranded replicative form (Denhardt, 1975). It has been
shown that the protein is able to lower the melting temperature
of several naturally occurring DNAs by as much as $40^{\circ}C$ (Alberts
et al., 1972). NMR investigations into the mechanism of "unwinding" by this protein have been started (Coleman et al.,
1976; Garssen et al., 1977). Its influence on the ring N proton spectrum of the small helical fragment of DNA formed by the
self-complementary d(pCGCG) is shown in Fig. 8. The thermal
melting of the d(pCGCG) observed as the disappearance of the
hydrogen-bonded protein resonances extends over a temperature
range about $20^{\circ}C$ (Fig. 8A-D). At about $10^{\circ}C$ the resonance from
the terminal GC pairs broadens and disappears, while at about
$25^{\circ}C$ the central GC resonances have broadened beyond detection.
In the experiment of Fig. 8E the ratio of protein to single-
stranded tetranucleotide was chosen as 1:1 and the protein has
melted the double helix even at $0^{\circ}C$. The disappearance of the
hydrogen-bonded resonances in Fig. 8E is not due to linebroadening as a result of an increase of molecular weight of the protein nucleic acid complex with respect to the double helical
tetranucleotide. Under the solution conditions used in the experiment, the protein most likely occurs as a dimer with a molecular weight of 19,400. Even at the low temperatures involved, the linewidths of hydrogen-bonded proton resonances
present in a possible protein double-stranded tetranucleotide
species would amount to about 50 Hz. These resonances would
be easily detectable. The ^{31}P resonances of the phosphates in
the tetranucleotide show that there is fast exchange of the DNA
fragment between the double helical state and the situation
where it is complexed to the gene-5 protein. Combination of
these results with measurements of the hydrogen-bonded proton
resonances of the double helical tetranucleotide in the presence of varying amounts of gene-5 protein permit us to conclude
that the lifetime of the double helix has effectively decreased
under the influence of the gene-5 protein. For a large part,
fluctuations in the double helical DNA structure determine the
unwinding by the protein (Garssen et al., 1977).

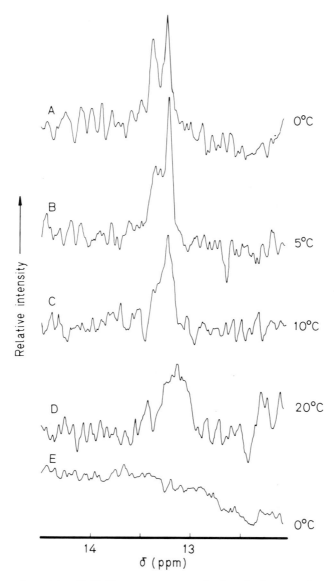

Fig. 8. The 360 MHz spectra of the hydrogen-bonded protons of double-stranded d(pCGCG) as a function of temperature (A-D). The influence of the gene-5 protein is shown in (E). Solution conditions: 2 mM tetranucleotide, 3.25 mM cacodylate, 1 mM EDTA, 0.05 M NaCl, pH 7. In addition, the solution of spectrum E contained 2 mM gene-5 protein. The number of accumulations and the vertical scale are the same for all spectra (A-E). (Garssen et al., 1977, Eur. J. Biochem. 81, 453-463).

VI. EXCHANGE IN TRANSFER RNAS

A. *Melting Studies of tRNAs*

Transfer RNAs are among the smallest natural occurring polynucleotides. Their biological functions have been extensively
described; among these their role in protein synthesis is understood best. The primary sequence of about 90 of these molecules has been determined (Clark, 1977). It turned out that all
of these different molecules can be folded into a common cloverleaf pattern comprised of four double helical regions. An example of this secondary structure of a particular tRNA species,
yeast tRNA^{Phe}, is presented in Fig. 9 together with the nomenclature of the different parts of the molecule. The three-dimensional structure of this tRNA has been clarified by X-ray
crystallography (for reviews see Kim, 1976, and Rich and Rajbhandary, 1976). From this it followed that the DHU loop folds
back onto the TψC loop, stabilized by basepair interactions.
The hydrogen-bonded proton resonance spectra of tRNAs in the
region between 15 and 11 ppm downfield from DSS contain about
26 resonances (see chapter by Robillard and Reid). An example
is provided by the spectrum of yeast tRNA^{Phe} in Fig. 10.

Since the resonances are partly overlapping, the interpretation of these spectra becomes more involved as compared to
the spectra of the model systems discussed above. To overcome
these complications melting studies have been carried out under
solution conditions where sequential melting of the tRNA structure could be expected (Römer *et al.*, 1969; Cole and Crothers,
1972). In principle this would reduce the problem of having
to assign about 26 resonances to one of assigning about five
resonances at a time. Then application of ring current calculations is expected to allow assignments to be made with the
same level of confidence as for the model systems. In addition
such experiments provide information on the solution structure
and also on the stability of the different parts of the tRNA
molecule.

Since the method is of a more general interest in studying
the solution structure of RNA and DNA molecules, problems that
may be encountered during its application are discussed. As
has been mentioned above, resonances in tRNA spectra very often
overlap. Under these circumstances the resonances melting out

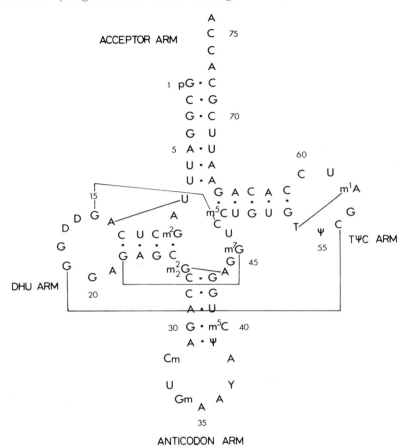

Fig. 9. *Cloverleaf structure of yeast tRNA^{Phe}. Dots indicate the ring N protons of the Watson-Crick basepairs that give rise to resonances below 11 ppm downfield from DSS. Solid lines indicate basepairs involved in the tertiary structure expected to give rise to resonances below 11 ppm.*

C. W. Hilbers

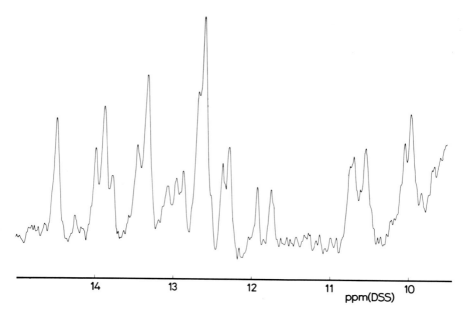

Fig. 10.   The 360 MHz spectrum of the hydrogen-bonded protons in yeast tRNA^{Phe} in the region between 9.5 and 15 ppm
downfield from DSS. Solution conditions: 10 m$\underline{M}$ cacodylate,
10 m$\underline{M}$ Na$_2$S$_2$O$_3$, pH 7.0, temperature 35°C. No Mg^{2+} was added but
the dissolved tRNA may have contained Mg^{2+} ions (Salemink,
P. J. M., and Hilbers, C. W., unpublished).

in a particular temperature range have to be traced back by
taking difference spectra. Interpretation of these difference
spectra very often is only possible if the assumption is made
that the different branches of the molecule melt out in an all
or none process. In other words, fraying effects occurring at
the end of double helical regions during melting are disre-
garded. This may lead to a misinterpretation of the results
if melting transitions cannot be separated well enough on the
temperature axis. However, to what extent these effects are
really important and affect the interpretation can only be
judged for the individual cases studied. For instance, helices
melting at higher temperatures may show fraying effects. This
is clear for the melting of the TψC stem in yeast tRNA^{Asp} (Robillard et al., 1976) and for the anticodon stem in E. coli
tRNA^{Glu} (Hilbers and Shulman, 1974). These arms are expected
to be the most stable parts in these molecules and therefore
the last to melt out. The melting transitions are reasonably
separated on the temperature axis and there is little doubt
that by the combination of ring current calculations and melting

the resonances were correctly assigned to the TψC- and the anticodon stem, respectively. This was confirmed by independent
experiments. For instance, the distribution of the resonances
of the TψC stem of yeast tRNA^{Asp} was compared with that of an
isolated 3' half-molecule. Although the resonances of the TψC
stem in the whole molecule are broadened at 75°C, the corresponding resonances can be observed in the 3' half-molecule
spectrum (Robillard et al., 1976). At lower temperatures (below 35°C) melting has been observed, which results in a decrease of the intensity of the resonances without giving rise
to broadening. In such situations fraying effects are expected
to be absent, which validates the interpretation in terms of an
all or none process (Hilbers et al., 1976; Robillard et al.,
1977).

In general the situation may turn out to be more complicated
and it has proven advantageous to combine the NMR melting experiments with optical melting and temperature jump experiments.
In principle, melting studies of hydrogen-bonded protons can
also be performed in conjunction with melting observed through
the chemical shift changes of nonexchangeable protons, e.g.,
the methyl resonances of modified residues in tRNA (Robillard
et al., 1977). It should be kept in mind, however, that the
nonexchangeable protons may not properly reflect the melting
processes. From the experiments of Borer et al., (1975) and
Patel (1975) it turns out that nonexchangeable protons within the same residue in DNA or RNA fragments exhibit melting
temperatures that may differ by at least 10°C.

B. NMR and Optical "Melting" Experiments

Optical studies measure the double helix disruption by following the increase in UV absorption at about 260 nm. The
double helix to coil transition is usually characterized by a
melting temperature T_m, which is defined as the temperature at
which the measured parameter (in this case the absorption) has
changed halfway from the value characteristic of the double helix to that of the coil form (Bloomfield et al., 1974). Very
often the disappearance of the hydrogen-bonded proton resonances occurs below the optical melting temperature T_m. This
was shown in an early study on the melting of the complex
formed by the pentanucleotides d(AACAA) and d(TTGTT) (Crothers
et al., 1973). The optical melting temperature was found to
be 28°C, while the hydrogen-bonded proton resonances had disappeared already by 9°C. Experiments on the self-complementary
hexanucleotides d(ATGCAT) (Patel and Hilbers, 1975; Patel,
1975) and r(AAGCUU) (Kan et al., 1975) lead to similar conclusions, as do results from the melting of d(AAAGCTTT) (Kallenbach et al., 1976). Also studies of the melting of hairpin

molecules have yielded the same results. For instance, the hydrogen-bonded resonances of the 5' half-molecule of yeast
tRNA^{Phe}, which forms a hairpin, essentially consisting of the
DHU branch of the tRNA molecule, melts around 60°C (Rohrdorf,
1975), while under the same solution conditions the optical
melting temperature is 70°C (Römer et al., 1969).
     In the event that the linewidth is determined by the life-
time of the double helix, i.e., the exchange rate is opening
limited, this is easily understandable (Crothers et al., 1973).
If the lifetime of the double helix at the melting temperature
T_m is much less than 1 msec, i.e., if the linewidth at this
temperature is much larger than 300 Hz, then the hydrogen-
bonded resonances will have broadened beyond detection already
below this melting temperature. Situations that typically can
be encountered in combined optical and NMR studies are sketched
in Fig. 11 (Crothers et al., 1973), where the logarithm of the
dissociation rate has been plotted versus the temperature T.
The shaded region represents the area where severe linebroadening in the NMR spectrum occurs. In going from the low to the
high side of this region the linebroadening increases from 30
to 300 Hz. When the dissociation rate constant is smaller than
the values in the shaded region the resonances will have widths
close to their natural linewidths.
     Comparison of the lines (a)-(d) shows the parameters that
determine how far below the T_m the broadening of the resonances

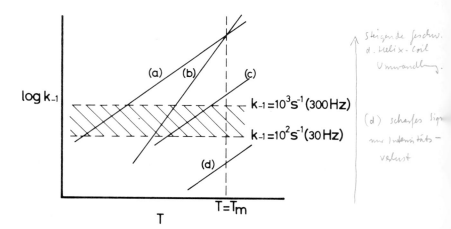

Fig. 11. Plot of the logarithm of the helix dissociation
rate k_{-1} against temperature T. The shaded region indicates
the region where severe linebroadening of the hydrogen bonded
resonances will be observed when the exchange of the ring N
protons is opening limited (Crothers et al., 1973, Proc. Nat.
Acad. Sci. USA 70, 2899-2901).

occurs. The situation we have encountered already in the examples mentioned above is represented by line (a). Line (b) in-
tersects line (a) at the melting temperature, meaning that the
helix lifetime is the same in both situations. However, (b)
has a larger activation energy, meaning that "sharp" resonances
should persist closer to the melting temperature. For very
high activation energies in situation (b) the temperature at
which the NMR resonances broaden should approach T_m. Such situations have been encountered experimentally. For instance, at
60°C the resonances of the acceptor stem of yeast tRNA^{Gly},
though broadened, are still visible in the NMR spectrum, while
the optical melting temperature is at 67°C (Hilbers et al.,
1976). Another example is provided by the melting of the 49
nucleotide 3' end of the 16 S rRNA molecule, which at 60°C shows
very strongly broadened resonances (Baan et al., 1977), while
the optical melting temperature is found to be 60°C (Baan,
1977).

Returning to Fig. 11, it is clear that in situation (c) the
rate of helix dissociation at $T = T_m$ is much smaller than for
(a) and (b), and consequently broadening occurs closer to the
melting temperature. Situation (d) represents a limiting situation in which $k_{-1} < 10^2 sec^{-1}$ at T_m, so that the resonances
should remain sharp but diminish in intensity when one proceeds
through the helix to coil transition. Examples of this situation have also been observed. Optical melting and temperature
jump experiments on yeast tRNA^{Gly} showed a low-temperature transition with k_{-1} values around 25 sec^{-1} at the melting temperature. The corresponding NMR spectra exhibit resonances that
disappeared from the spectra without broadening. Calculation of
the helix-coil equilibrium constants obtained from both methods
are shown in Fig. 12. The differences in absolute values arise
from the fact that the NMR experiments were performed on 0.15 M
KCl solutions and the optical experiments on 0.15 M NaCl solutions. Sodium and potassium ions influence the thermal stability differently in that sodium ions increase the T_m of tertiary
unfolding by 7°C compared to potassium ions (Urbanke et al.,
1975). Since in the low-temperature transition unfolding of
tertiary structure is involved the difference between the NMR
and optical equilibrium constant was considered to be explained
quantitatively. Similar observations have been made recently
in a detailed study of the melting of yeast tRNA^{Phe} (Robillard
et al., 1977). Resonances assigned to tertiary interactions
diminish without broadening between 13 and 26°C in solutions of
30 mM Na$^+$ without Mg^{2+} at pH 6.8. These results are in agreement with temperature jump experiments that yielded 20 msec
lifetimes for the tertiary base-paired regions (Römer et al.,
1969).

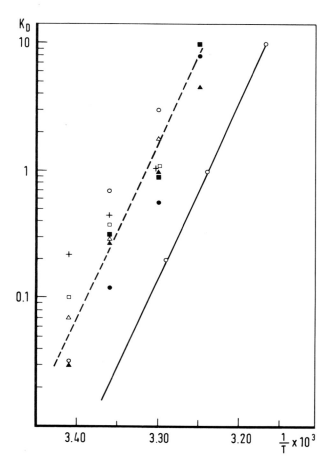

Fig. 12. *Plot of the dissociation constant $\underline{K}_d$ of the low-temperature melting transition of yeast tRNA^{Gly} as a function of temperature. The dashed line represents an average drawn through the values obtained by NMR. Solution conditions: 2 m$\underline{M}$ yeast tRNA^{Gly}, 0.15 $\underline{M}$ KCl, no Mg^{2+}, 0.01 $\underline{M}$ cacodylate, pH 7. The solid line represents the dissociation constants for the same process calculated from the optical differential melting curve. Solution conditions: 0.15 $\underline{M}$ NaCl, 0.01 $\underline{M}$ cacodylate, 10^{-4} $\underline{M}$ EDTA, pH 7 (Hilbers et al., 1976, Biochemistry 15, 1874-1882, copyright by the American Chemical Society).*

C. Relation between NMR Melting and Temperature Jump Experiments

In a temperature jump experiment the sample is raised in temperature by application of a very short (μsec) heating pulse. After the pulse the molecular system (e.g., tRNA) will adapt it-

self to the new temperature and if basepairs are broken during
this process this can be followed by monitoring the change in
the optical absorption at 260 nm. The time dependence of optical absorption changes can be characterized by a relaxation
time τ. The making and breaking of basepairs in tRNAs are intramolecular processes and for this situation the relaxation
time is related to the rate constants of dissociation (k_{-1}) and
formation (k_1) by

$$\tau^{-1} = k_{-1} + k_1$$

At the high-temperature side of a melting transition the dissociation rate constant k_{-1} predominates and determines the
value of τ^{-1}. In addition, at the melting temperature T_m, $\left(k_{-1} = k_1 \right)$
$2\tau = 1/k_{-1}$. This permits one to map the linebroadening meas-
urements in NMR onto the temperature jump results (Crothers *et
al.*, 1974). If one assumes that the linebroadening effects observed in an NMR experiment measure the lifetime of the double
helix, severe linebroadening will be observed in a temperature
range where the helix lifetime is around 5 msec (see Fig. 11).
For a particular melting transition a line drawn through the
relaxation times, determined at the high-temperature side of
the melting and the point 2τ at T_m should intersect the level
at which the helix lifetime is 5 msec in a temperature range
where severe linebroadening is observed in the NMR spectra.
     An example of this procedure is presented in Fig. 13.
Yeast tRNA^{Gly} exhibits a melting transition at 81°C when dissolved in a solution containing 0.01 M sodium cacodylate, 1 M
Na^+, and 10^{-4} M EDTA at pH 7.0. Extrapolation of the high-
temperature relaxation time values to the 5 msec level according to the procedure outlined above yields an intersection at
68°C, while the NMR spectrum starts to broaden at 61°C. On the
basis of this mapping the melting transition is assigned to the
combined melting of the acceptor and the TψC stem. One might
legitimately ask whether the broadening observed in the NMR
spectrum is a direct measure of the lifetime of the base-paired
state. At 68°C the dissociation rate of this structure is
about $0.2 \times 10^3 sec^{-1}$ while the association rate is estimated to
be close to $10^4 sec^{-1}$ (see Fig. 13). Consequently the exchange
between the double helix and coil state does not correspond to
the slow exchange limit. Still the observed linebroadening reflects the helix lifetime since from the difference in pK
values between uracil and cacodylate the rate of disappearance
of the ring N protons from the open state is estimated to be
10^5 to $10^6 sec^{-1}$, which is much larger than the formation rate
of $10^4 sec^{-1}$. Hence the conditions required for Eq. (8a) are
fulfilled.

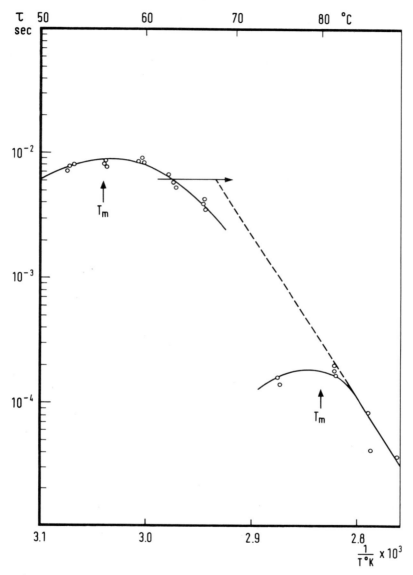

Fig. 13. Plot of the change of the relaxation times of yeast tRNA^{Gly} as a function of temperature determined by temperature jump experiments. The vertical arrows indicate the melting temperatures of the individual transition. The dashed line representing $1/k_{-1}$ has been drawn through the high-temperature τ values and through 2τ at $T = T_m$. It has been extrapolated to the temperature where the helix lifetime is 5 msec.

1. Effect of Hydrogen-Bonded Proton Exchange

formation rate: ~10⁴ s⁻¹ : Mg²⁺ ↑ Stabilität der Doppelhelix [handwritten]

Except for the <u>formation rate of the TψC stem</u>, which may be of the order of 10^5sec^{-1}, <u>the formation rates of the other parts of tRNAs have been found to be of the order of 10^4sec^{-1} or lower in solutions containing 0.1 M Na without Mg^{2+}. There-fore, the linebroadening of these structural elements can be taken as a measure of the lifetime of the corresponding double helical segments, provided the buffer ion concentration is suf-ficiently high.</u> As has been demonstrated (Crothers *et al.*, 1974) even for the disruption of the TψC stem this assumption is not likely to lead to substantial errors.

B ≫ k, [handwritten, left margin]

The occurrence of fraying and other transient phenomena may invalidate this simple approach. However, <u>if the melting of the double helical fragment is not an all or none process</u> the <u>last observable basepair resonances</u> will reflect the lifetime of the <u>total</u> double helix (see section on nucleation). For such a transition to be interpretable it is required that it is separated well enough on the temperature axis.

A few general conclusions can be drawn from the detailed studies carried out so far. It turns out that the tertiary structure forms the weakest part of the tRNA molecule. It is the first structural element to melt out, very <u>often in con-junction with the DHU stem.</u> Once this part of the molecule has been disrupted, the melting of the other stems will follow de-pending on their stability. <u>The latter property is predictable from the rules formulated by Gralla and Crothers</u> (1973) and Borer *et al.* (1974), taking into account the counterion concen-tration. Under physiological conditions the melting of tRNA becomes highly cooperative and the individual transitions co-incide. Yet the tertiary structural elements remain the weakest parts of the structure. This follows from studies on the mech-anism of protein synthesis, where there are strong indications that during this process the interactions between the DHU and TψC loop are disrupted (Richter *et al.*, 1973). This disruption may be triggered by codon-anticodon interaction (Schwarz *et al.*, 1976).

D. T_1 *Relaxation Studies of Yeast* tRNA^{Phe}

The X-ray crystallographic studies of yeast tRNA^{Phe} have shown that all but one of the basepairs involved in the forma-tion of the tertiary structure are non Watson-Crick type pairs.

The bar at the 5 msec level indicates the temperature region in which significant broadening is observed of the resonances of the acceptor and TψC stem of yeast tRNA^{Gly} (Hilbers et al., 1976, Biochemistry 15, 1874-1882, copyright by the American Chemical Society).

It is now well established that the ring N protons of such pairs
involved in hydrogen bonding also give rise to resonances in the
region where the resonances from the normal Watson-Crick pairs
are observed (see Section II and also the chapter by Robillard
and Reid). In determining the structure of RNAs and/or DNAs it
would be very helpful if methods were available by which one
could distinguish secondary from tertiary hydrogen-bonded resonances.

The melting experiments on tRNAs, discussed above, have
shown that the first part of the tRNA structure to melt out is
formed by elements of the tertiary structure, sometimes in conjunction with parts of the secondary structure. This suggests
that the tertiary hydrogen-bonded proton resonances may be distinguished from the secondary ones by means of the different
kinetic properties of their hydrogen bonds. Recently Johnston
and Redfield (1977) as well as Campbell and co-workers (1977)
have made the important observation that the longitudinal relaxation times T_1 of the hydrogen-bonded protons in yeast
tRNA^{Phe} are much longer than their exchange times with water.
A so-called transfer of saturation experiment was performed to
demonstrate this. To this end the resonance of solvent water
is saturated and the hydrogen-bonded resonances from the tRNA
molecule are observed (see Fig. 14). It turns out that by
transfer of magnetization through the proton exchange reaction,
the tRNA resonances become saturated as well, clearly showing
that the normal longitudinal relaxation rate cannot compete
with the exchange process. Hence by measuring the apparent
longitudinal relaxation rate of the hydrogen-bonded protons direct observation of their exchange rates to water is possible.
This permits observation of exchange in a domain where excess
linebroadening as a result of exchange cannot be observed. In
Fig. 15 an example is presented. The longitudinal relaxation
rates of individual resonances of the hydrogen-bonded protons
of yeast tRNA^{Phe} were measured by a saturation recovery method
(Johnston and Redfield, 1977) using the elegant pulse method
developed by Redfield and his co-workers (1975).

Between 15 and 45°C in the presence of 15 mM Mg^{2+} the exchange rates are typically around 9 sec^{-1}, independent of
temperature. In absence of Mg^{2+} two classes of protons can
be distinguished by their difference in exchange rates.
One class essentially behaves as in the presence of 15 mM
Mg^{2+}, i.e., the exchange rates are around 9 sec^{-1} virtually independent of temperature. The second class, consisting of
about nine protons, has higher exchange rates. The resonances
of six of these protons at 14.4, 14.2 (14.4 ppm in the presence
of Mg^{2+} ions), 13.35, 12.95, 12.5, and 11.75 ppm are assigned
to tertiary basepairs, while the remaining three at 14.0, 13.8,
and 11.6 ppm are thought to come from secondary pairs, which
become destabilized during the melting of the tertiary struc-

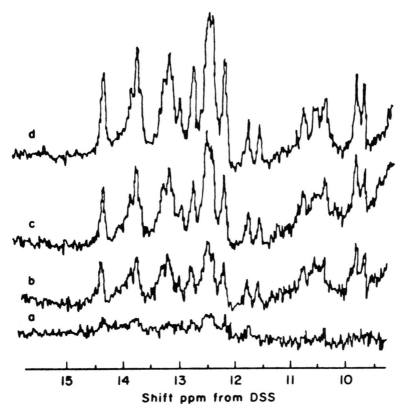

Fig. 14. Demonstration of the transfer of saturation between H_2O and yeast tRNA^{Phe}. (a) The 270 MHz hydrogen-bonded
proton spectrum of yeast tRNA^{Phe} is observed while the water
resonance is saturated by a preirradiation pulse. In spectra
(b) and (c), the preirradiation was applied at 210 and 310 Hz
away from the water resonance. Spectrum (d) was recorded under
conditions where no preirradiation was applied. Solution con-
ditions: 0.1 $\underline{M}$ NaCl, 1.5 m$\underline{M}$ MgCl$_2$, 7 m$\underline{M}$ EDTA, 10 m$\underline{M}$ sodium cacodylate, pH 7.0. Temperature 35°C (Johnston and Redfield,
1977, Nucl. Acids Res. 4, 3599-3616).

ture. Two of the six tertiary resonances have been assigned by
chemical modification methods. Modification of thio U_8 in
E. coli tRNAs led to the conclusion that the U_8 ring N_3 proton
hydrogen-bonded in a reverse Hoogsteen complex to the N_7 atom

of A_{14} must resonate at 14.4 ppm (Wong and Kearns, 1974; Reid et al., 1975; Daniel and Cohn, 1975). By removal of the m^7G_{46} residue it was found that the ring N_1 proton of the m^7G_{46} hydrogen-bonded to N_7 of G_{22} resonates at 12.5 ppm (Salemink et al., 1977). The detailed assignment of the remaining tertiary resonances needs to be confirmed by additional modification experiments. Interestingly the resonance positions of most of the more labile protons detected by the saturation recovery method coincide with the resonances, which are the first to disappear from the spectrum in NMR melting experiments (Hilbers et al., 1973; Römer and Varadi, 1977; Robillard et al., 1977). Above 36°C the exchange rates increase rapidly and it is concluded that in this temperature range they are opening limited. Assuming that the dissociation rate is not strongly dependent on the Na^+ ion concentration, an opening rate of about 100 sec^{-1} is expected at 40°C (Römer and Varadi, 1977), which is in reasonable agreement with the value of 100 to 200 sec^{-1} measured at 42°C in the saturation recovery experiment. At lower temperatures (i.e., below 36°C) it is expected that exchange from the open state k_c [see Eq. (11)] is the rate-determining step (Johnston and Redfield, 1977). Indeed early experiments (Wong et al., 1972) corroborate this notion. Although the spectra of Wong et al. were taken at 220 MHz, having much less resolution compared to what can be achieved at present, it is clear that the exchange of protons responsible for the resonance at 14.4 and 11.7 ppm are subject to base catalysis.

In addition to sorting out tertiary resonances, the saturation recovery method seems promising for studying fluctuations in DNA and RNA structures other than the complete dissociation of base-paired regions. For instance the melting transitions of the individual parts of yeast tRNA^{Phe} in a solution containing 0.1 $\underline{M}$ NaCl and 2 m$\underline{M}$ MgCl$_2$ as determined by temperature jump experiments are highly cooperative and coincide at 70°C (Maass et al., 1971). From these temperature jump measurements one predicts the dissociation rate to vary from about 10^{-8} sec^{-1} at 10°C to 10^{-1} sec^{-1} at 42°C. As mentioned above, in this temperature region Johnston and Redfield (1977) measured exchange rates of 9 sec^{-1} even in the presence of 15 m$\underline{M}$ Mg^{2+}, which is much faster than expected on the basis of the temperature jump experiment. It is not impossible that this effect is due to a direct ring N proton-H_2O nuclear Overhauser effect, since it cannot be ruled out that H_2O molecules reside long enough and close enough to the ring N protons for this to be possible (A. G. Redfield, private communication). On the other hand fluctuations in the base-paired structure of individual basepairs may also contribute to this effect.

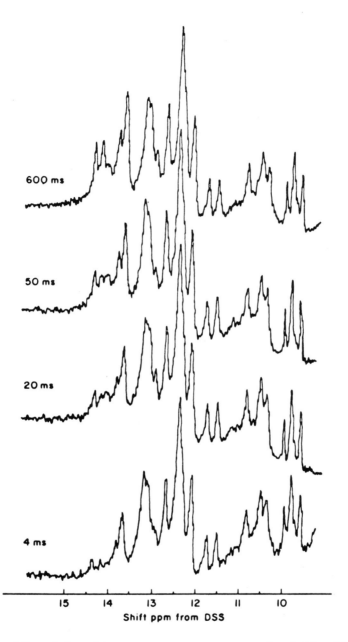

600 ms

50 ms

20 ms

4 ms

15   14   13   12   11   10
Shift ppm from DSS

Fig. 15. Saturation recovery measurement of the 14.4 to
-14.0 ppm spectral region of the yeast tRNA^{Phe} hydrogen-bonded
proton spectrum. The resonances in this spectral region were
saturated by a preirradiation pulse. The recovery to the equilibrium magnetization was recorded by giving observation pulses
after the preirradiation pulse at a time indicated with each
spectrum. Solution conditions: 10 mM EDTA, no Mg^{2+}. Temperature 30°C (Johnston and Redfield, 1977, Nucl. Acids Res. 4,
3599-3616).

VII. CONCLUSIONS

The present chapter has mainly dealt with concepts and methods. The kinetic scheme describing the exchange of the hydrogen-bonded protons and its influence on their NMR spectra is confirmed by experiments of which a number of examples are given. Although the applications have been restricted to nucleic acids, the same principles can be used in studying the hydrogen-bonded proton spectra of proteins. Investigating the influence of exchange on the hydrogen-bonded spectra can yield detailed information on the dynamical structure of biological macromolecules. At physiological conditions where knowledge of the dynamical behavior of the macromolecules is a prerequisite for the understanding of their action this is of particular advantage. In this respect, the development of pulse methods allowing the determination of relaxation times is a welcome addition. One of the main problems seems the unambiguous assignment of the hydrogen-bonded proton resonances in larger nucleic acids, e.g., tRNAs and 5 S rRNAs.

ACKNOWLEDGMENT

I wish to thank Drs. D. J. Patel, G. T. Robillard and A. G. Redfield for making available their results prior to publication. Drs. R. G. Shulman and D. M. Crothers introduced me into the physical chemistry of nucleic acids and I want to acknowledge the excitement of our collaboration.

REFERENCES

Alberts, B., Frey, L., and Delius, H. (1972). *J. Mol. Biol. 68*, 139-152.

Arter, D. B., and Schmidt, P. G. (1976). *Nucl. Acids Res. 3*, 1437-1447.

Baan, R. A. (1977). Ph.D. Thesis, Dept. Biochemistry, Univ. of Leiden, Leiden.

Baan, R. A., Hilbers, C. W., van Charldorp, R., van Leerdam, E., van Knippenberg, P. H., and Bosch, L. (1977). *Proc. Nat. Acad. Sci. USA 74*, 1028-1031.

Berkeley, P. J., and Hanna, M. W. (1964). *J. Am. Chem. Soc. 86*, 2990-2994.

Bloomfield, V. A., Crothers, D. M., and Tinoco, I., Jr. (1974).

"Physical Chemistry of Nucleic Acids." Harper & Row, New York.

Borer, P. N., Dengler, D., Tinoco, I., and Uhlenbeck, O. C. (1974). *J. Mol. Biol.* 86, 843-853.

Borer, P. N., Kan, S. L., and Ts'o, P. O. P. (1975). *Biochemistry 14*, 4847-4863.

Campbell, I. D., Dobson, C. M., and Ratcliffe, R. G. (1977). *J. Mag. Res.* 27, 455-463.

Clark, B. F. C. (1977). *In* "Progress in Nucleic Acid Research and Molecular Biology" (W. E. Cohn, ed.), Vol. 20, pp. 1-19. Academic Press, New York.

Cole, P. E., and Crothers, C. M. (1972). *Biochemistry 11*, 4368-4374.

Coleman, J. E., Anderson, R. A., Ratcliffe, R. G., and Armitage, I. M. (1976). *Biochemistry 15*, 5419-5430.

Craig, M. E., Crothers, D. M., and Doty, P. (1971). *J. Mol. Biol.* 62, 383-401.

Crothers, D. M., Hilbers, C. W., and Shulman, R. G. (1973). *Proc. Nat. Acad. Sci. USA 70*, 2899-2901.

Crothers, D. M., Cole, P. E., Hilbers, C. W., and Shulman, R. G. (1974). *J. Mol. Biol.* 87, 63-88.

Daniel, W. E., Jr., and Cohn, M. (1975). *Proc. Nat. Acad. Sci. USA 72*, 2582-2586.

Denhardt, D. T. (1975). *CRC Crit. Rev. Microbiol.* 4, 161-223.

Eigen, M. 1964). *Angew. Chem. (Int. Ed. Engl.) 3*, 1.

Englander, S. W., Downer, N. W., and Teitelbaum, H. (1972). *In* "Annual Reviews of Biochemistry" (E. E. Snell, P. D. Boyer, A. Meister, and R. L. Sinsheimer, eds.), Vol. 41, pp. 903-924. Annual Review, Inc., Palo Alto, California.

Garssen, G. J., Hilbers, C. W., Schoenmakers, J. G. G., and van Boom, J. H. (1977). *Eur. J. Biochem.* 81, 453-463.

Geerdes, H. A. M., and Hilbers, C. W. (1977). *Nucl. Acids Res.* 4, 207-221.

Giessner-Prettre, C., and Pullman, B. (1970). *J. Theoret. Biol.* 27, 87-95.

Giessner-Prettre, C., and Pullman, B. (1976). *Biochem. Biophys. Res. Commun.* 70, 578-581.

Giessner-Prettre, C., Pullman, B., and Caillet, J. (1977). *Nucl. Acids Res.* 4, 99-116.

Gralla, J., and Crothers, D. M. (1973). *J. Mol. Biol.* 78, 303-319.

Hilbers, C. W., and Patel, D. J. (1975). *Biochemistry 14*, 2656-2660.

Hilbers, C. W., and Shulman, R. G. (1974). *Proc. Nat. Acad. Sci. USA 71*, 3239-3242.

Hilbers, C. W., Shulman, R. G., and Kim, S. H. (1973). *Biochem. Biophys. Res. Commun.* 55, 953-960.

Hilbers, C. W., Robillard, G. T., Shulman, R. G., Blake, R. D., Webb, P. K., Fresco, J., and Riessner, D. (1976). *Biochemistry 15*, 1874-1882.

Johnston, P. D., and Redfield, A. G. (1977). *Nucl. Acids Res.* 4, 3599-3616.

Kallenbach, N. R., Daniel, W. E., and Kaminker, M. A. (1976). *Biochemistry 15*, 1218-1224.

Kan, L. S., Borer, P. N., and Ts'o, P. O. P. (1975). *Biochemistry 14*, 4864-4869.

Kearns, D. R. (1976). *In* "Progress in Nucleic Acid Research and Molecular Biology" (W. E. Cohn, ed.), Vol. 18, pp. 91-149. Academic Press, New York.

Kearns, D. R. (1977). *In* "Annual Review of Biophysics and Bio-engineering" (L. J. Mullins, W. A. Hagens, L. Stryer and C. Newton, eds.), Vol. 6, pp. 477-523. Annual Reviews Inc., Palo Alto, California.

Kearns, D. R., Patel, D. J., and Shulman, R. G. (1971). *Nature 229*, 338-339.

Kim, S. H. (1976). *In* "Progress in Nucleic Acid Research and Molecular Biology" (W. E. Cohn, ed.), Vol. 17, pp. 181-216. Academic Press, New York.

Liddel, U., and Ramsey, N. F. (1951). *J. Chem. Phys. 19*, 1608.

Lukashin, A. V., Vologodskii, A. V., Frank-Kamenetskii, M. D., and Lyubchenko, Y. (1976). *J. Mol. Biol. 108*, 665-682.

Maass, G., Riesner, D., and Römer, R. (1971). *Proc. Int. Congr. Pure Appl. Chem. (Boston)*.

McConnell, H. M. (1958). *J. Chem. Phys. 28*, 430-431.

McGhee, J. D., and von Hippel, P. H. (1975a). *Biochemistry 14*, 1281-1296.

McGhee, J. D., and von Hippel, P. H. (1975b). *Biochemistry 14*, 1297-1303.

Marshall, T. H., and Grunwald, E. (1969). *J. Am. Chem. Soc. 91*, 4541-4544.

Patel, D. J. (1974). *Biochemistry 13*, 2396-2402.

Patel, D. J. (1975). *Biochemistry 14*, 3984-3989.

Patel, D. J. (1976). *Biopolymers 15*, 533-558.

Patel, D. J. (1977). *Biopolymers 16*, 1635/1656.

Patel, D. J., and Hilbers, C. W. (1975). *Biochemistry 14*, 2651-2656.

Patel, D. J., and Tonelli, A. E. (1974). *Biopolymers 13*, 1943-1964.

Pinnavaia, T. J., Miles, H. T., and Becker, E. D. (1975). *J. Am. Chem. Soc. 97*, 7198-7200.

Pörschke, D., and Eigen, M. (1971). *J. Mol. Biol. 62*, 361-381.

Pörschke, D., Uhlenbeck, O. C., and Martin, F. H. (1973). *Bio-polymers 12*, 1313-1335.

Pople, J., Schneider, W. G., and Berstein, H. J. (1959). "High Resolution Nuclear Magnetic Resonance." McGraw Hill, New York.

Ravetch, J., Gralla, J., and Crothers, D. M. (1974). *Nucl. Acids Res. 1*, 109-127.

Redfield, A. G., Kunz, S. D., and Ralph, E. K. (1975). *J. Mag. Res. 19*, 114-117.

Reid, B. R., Ribeiro, N. S., Gould, G., Robillard, G. T., Hilbers, C. W., and Shulman, R. G. (1975). *Proc. Nat. Acad. Sci. USA 72*, 2049-2053.

Rich, A., and Bajbhandary, U. L. (1976). *In* "Annual Reviews of Biochemistry" (E. E. Snell, P. D. Boyer, A. Meister, and C. C. Richardson, eds.), Vol. 45, pp. 805-860. Annual Reviews Inc., Palo Alto, California.

Richter, D., Erdmann, V. A., and Sprinzl, M. (1973). *Nature New Biol. 246*, 132-135.

X Robillard, G. T., Hilbers, C. W., Reid, B. R., Gangloff, J., Dirheimer, G., and Shulman, R. G. (1976). *Biochemistry 15*, 1883-1888.

X Robillard, G. T., Tarr, C. E., Vosman, F., and Reid, B. R. (1977). *Biochemistry 16*, 5261-5273.

Römer, R., and Varadi, V. (1977). *Proc. Nat. Acad. Sci. USA 74*, 1561-1564.

Römer, R., Riesner, D., Maass, G., Wintermeyer, W., Thiebe, R., and Zachau, H. G. (1969). *FEBS Lett. 5*, 15-19.

Rohrdorf, B. F. (1975). Ph.D. Thesis, Dept. of Chemistry, Univ. of California, Riverside.

Salemink, P. J. M., Yamane, T., and Hilbers, C. W. (1977). *Nucl. Acids Res. 4*, 3727-3741.

Schwarz, U., Menzel, H. M., and Gassen, H. G. (1976). *Biochemistry 15*, 2484-2490.

Shulman, R. G., Hilbers, C. W., Kearns, D. R., Reid, B. R., and Wong, Y. P. (1973). *J. Mol. Biol. 78*, 57-69.

Slejko, F. L., and Drago, R. S. (1973). *J. Am. Chem. Soc. 95*, 6935-6944.

Steinmetz-Kayne, M., Benigmo, R., and Kallenbach, N. R. (1977). *Biochemistry 16*, 2064-2073.

X Teitelbaum, H., and Englander, S. W. (1975a). *J. Mol. Biol. 92*, 55-78.

X Teitelbaum, H., and Englander, S. W. (1975b). *J. Mol. Biol. 92*, 79-92.

X Urbanke, C., Römer, R., and Maass, G. (1975). *Eur. J. Biochem. 55*, 439-444.

Williams, M. N., and Crothers, D. M. (1975). *Biochemistry 14*, 1944-1951.

Wong, K. L., and Kearns, D. R. (1974). *Nature 252*, 738-739.

Wong, Y. P., Kearns, D. R., Reid, B. R., and Shulman, R. G. (1972). *J. Mol. Biol. 72*, 725-740.

ELUCIDATION OF NUCLEIC ACID STRUCTURE
BY PROTON NMR

G. T. Robillard

Department of Physical Chemistry
University of Groningen
Groningen, The Netherlands

B. R. Reid

Department of Biochemistry
University of California
Riverside, California

I. INTRODUCTION

The importance of nucleic acids rests in their role in storage and transmission of biological information. Expression of this information involves changes in structure at all levels, secondary, tertiary, and quaternary. Consequently, any technique capable of providing detailed structural information, especially changes in structure, will play a significant role in our understanding of the biological function of these macromolecules.

In the late 1960s Shulman and co-workers (1,2) first applied high-resolution proton NMR to nucleic acids, specifically tRNA. Due to the redundancy of components, however, virtually no resolution could be obtained among aromatic or ribose protons, leaving little hope of extracting structural details from these spectral regions. Prospects for success appeared better from analysis of resonances in the aliphatic and hydrogen-bonded spectral regions. A few resolved resonances were detected between 0 and -4 ppm (from DSS) arising from methylated bases. However, methylation is not uniform in nucleic acids and the paucity of methyl groups limits the usefulness of NMR spectra

in the aliphatic spectral region. Between -11 and -15 ppm at
least 20 protons were observed when H_2O was used as the solvent
(3,4). Their chemical shifts and solvent exchange characteristics indicated that they arose from ring NH protons stabilized
by Watson-Crick type hydrogen bonding in the tRNA. The even
distribution of these hydrogen bonds throughout the molecule
as well as the range of chemical shifts and the reasonably
small number of resonances made them ideal candidates to deliver
a wealth of structural information. In addition, this information is much easier to interpret and potentially more meaningful than that contained in the spectral region from 0 to -9 ppm.
For example, in nucleic acids examination of an abnormally
shifted methyl resonance, or in proteins a histidine resonance
with an abnormal pH dependence of the chemical shift might enable one to draw inferences concerning the presence of aromatic
residues or charged side chains in the immediate vicinity.
Proof of this local structure is a laborious task as the NMR
studies on lysozyme demonstrate (5, and references therein).
In the case of tRNA, however, where hydrogen bonds play an important role in stabilizing the secondary and tertiary structure, making detailed structural correlations is more feasible.
A resonance from an NH proton can be observed in the very low-
field spectral region only if that proton is hydrogen bonded.
Once such a resonance has been sighted and identified, it is
positive proof that the donor and acceptor atoms are within 3
to 4 Å of one another. The observations of up to 40 of these
resonances in a single tRNA spectrum provides us with the opportunity of making detailed NMR studies on (1) the structure and
dynamics of the macromolecule in solution; (2) comparisons of
the solution and x-ray crystallographic structure; (3) alternative structures which may exist in solution. However, two
central problems of spectral interpretation have existed since
the beginning of these studies, which have limited the rate of
progress in this area:

    (1) Spectral integration. Unlike C-H protons, N-H protons
exchange readily with solvent. The ability to observe them by
NMR depends on the rate at which they exchange. Therefore,
their intensity in a given spectrum can vary from 0 to 100% of
a single proton. Hence accurate intensities are particularly
important.

    (2) Spectral assignments. Hydrogen bonding decreases the
solvent-proton exchange rate and makes the protons visible by
NMR. The observed chemical shift of a resonance, however, is
a composite of shifts from hydrogen bonding, exchange processes,
ring currents, diamagnetic anisotropic and electric fields.
Assigning resonances and extracting structural information requires a correct evaluation of these individual effects. In a

classical assignment study of nonexchangeable protons, one
could heat-denature the macromolecule and determine unshifted
resonance positions in the random coil spectrum. Hydrogen-
bonded proton resonances, however, disappear on heating the macromolecule, due to increasing exchange rates with solvent protons. In an investigation of nonexchangeable protons, one
could study resonance positions in model compounds as an aid to
assigning macromolecule NMR spectra. Hydrogen-bonded proton
resonances are almost impossible to observe in small model
compounds because of the rapid proton exchange with solvent.
When they are observed, the system is already large and has
many of the difficulties in interpretation associated with the
macromolecule. In a classical NMR assignment study one could
monitor the pH dependence of chemical shifts and extrapolate
assignment data and possibly structural information. Hydrogen-
bonded proton exchange rates, however, are both acid and base
catalyzed. While their resonances are observable at neutral
pH, deviation of more than 1-2 pH units in either direction
leads to exchange broadening and eventual loss of information.

Clearly the study of hydrogen-bonded protons by NMR is a
complicated matter in which many standard NMR procedures are
not applicable. As a result, the past 5 years has been a developmental phase, which has seen new attempts and approaches
to integrating and assigning hydrogen-bonded proton spectra.
Several reviews have been written during the past few years on
NMR studies of tRNA in which the contributions of various groups
have been recounted (6,7). Rather than simply repeat this process we feel that the topic could be better served by synthesis
of data from a number of independent studies. Our intention is
to try to forge a consensus as to what, at the present time,
can be regarded with reasonable certainty and what problems remain to be solved.

II. HYDROGEN-BONDED PROTONS

A. *Spectral Integration*

Three elements are required for accurate integration:
(1) well-resolved spectra, (2) reasonably high signal/noise,
(3) a correct standard. Instrumental developments over the past
7 years have resulted in an increase of more than 1.5 in spectral resolution, and more than 4 in signal/noise. Equally as
important as these technical developments has been the progress
in large-scale purification procedures, which enable reasonably
large quantities of tRNA to be isolated in homogeneous form with
high specific activity. Our experience has been that the reso-

nance linewidth bears a sort of inverse relationship with the
tRNA's specific activity. Broader resonances lead to greater
overlap and problems with integrating individual regions of the
spectrum. Correct integration is also a question of choosing
a proper standard, either internal or external. Early integrations on 220 MHz spectra of limited resolution were based on
external standards. Resonances from about 20 protons were
found between -11 and -15 ppm[1] in the spectra of yeast tRNA[Phe]
and E. coli tRNA[Met] using a heme methyl resonance of cyano-
methyl myoglobin[f]as the standard (3,4). As spectral resolution
improved with higher frequency instrumentation, internal standards were more often used. Wong et al. (8) integrated the
300 MHz spectrum of yeast tRNA[Phe] in Fig. 1 by setting the
-14.4 ppm resonance equal to one proton or the -13.7 ppm peak
equal to three protons. This integration gave a total of 18
to 19 protons in the spectrum whereas the previous integrations
with an external standard resulted in approximately 20 protons
(3,4). In principle, integration via an internal standard is
more accurate than via an external standard since it works
within a single spectrum, avoiding the difficulties of normaliz-
ing sample concentration, NMR tubes, and instrument settings.
The danger, however, is that one may set a hydrogen-bonded proton resonance to an integral value when its intensity is subin-
tegral because of exchange processes, solvent conditions, etc.
In spite of such difficulties, internal standards continued to
be used and continued to indicate that only the 20 or so secondary structure Watson-Crick interactions from the cloverleaf
contributed resonances in the region between -15 and -11 ppm
(3,4,6,9-12).

The 270 MHz spectrum of E. coli tRNA[Val]$_1$ in Fig. 2 gave the
first indication that more than just secondary structure resonances existed below -11 ppm. Sixteen resolved peaks were observed, several of which were clearly more than one proton in
intensity. On the basis of the number of resonances alone it
was obvious that more than 20 protons contributed to the spectral intensity. Assuming that any one of the lowest intensity
resonances was one proton in intensity, 26 ± 1 protons were
found below -11 ppm (13). Subsequent reexamination of the spectrum of yeast tRNA[Phe] under different solvent conditions and
higher frequency also gave a more highly resolved spectrum,
where 14 resonances could be seen (Fig. 3), seven of which were
unit proton intensity. As in the case of E. coli tRNA[Val]$_1$ this
spectrum integrated to 26 ± 1 protons (14). These integrations
have been challenged by Bolton et al. (15) on the basis of much

---

[1]Throughout this review the spectral region downfield of
DSS is denoted by negative numbers (i.e., -11 ppm).

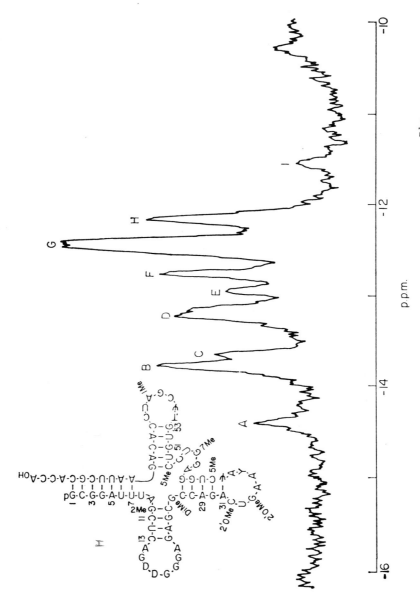

Fig. 1. The 300 MHz proton NMR spectrum of yeast tRNA^{Phe} (83).

E. coli tRNA$_1^{Val}$

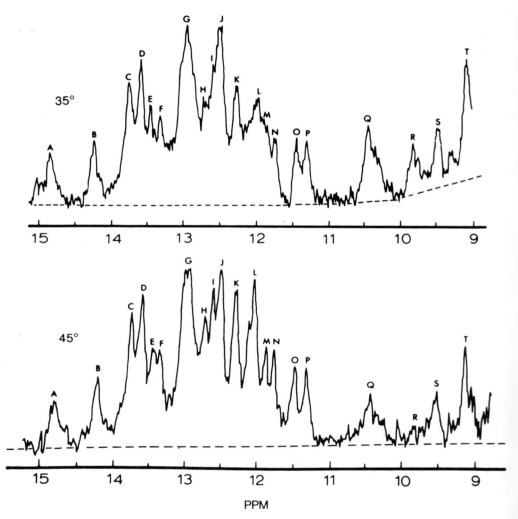

Fig. 2.   The 270 MHz proton NMR spectrum of E. coli
tRNA$_1^{Val}$ (13).

less well resolved spectra.   The standard used was the intensity of the aromatic region in D$_2$O, which contains the resonances
of the methine protons of the four bases.   With such a standard
they find a total of 23 $\pm$ 1 resonances in the same spectral re-

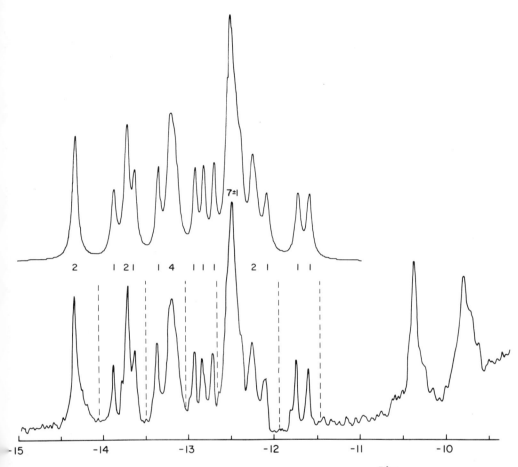

*Fig. 3.   The 360-MHz spectrum of yeast tRNA^{Phe}.   A water-
dialyzed, lyophilized sample was redissolved to 1.1 m$\underline{M}$ concentration in 10 m$\underline{M}$ sodium cacodylate-10m$\underline{M}$ EDTA (pH 7.0).   Peak
intensities are indicated on the experimental spectrum (lower).
The computer-simulated spectrum (upper) contains 26 Lorentzian
lines of 25 Hz full linewidth at half-height (16).*

gion where we have observed 26 $\pm$ 1.   Given the lack of spectral
resolution and the consequent necessity to reach outside the
low-field region for an intensity standard, it is not likely
that the spectrum can be integrated with such accuracy.   In
contrast to the data of Bolton *et al.* (15), the spectra in
Fig. 4 of Reid and co-workers (16) conclusively demonstrate
that, with very pure tRNA samples, good resolution, and reasonably high signal/noise, it is not necessary to go outside the

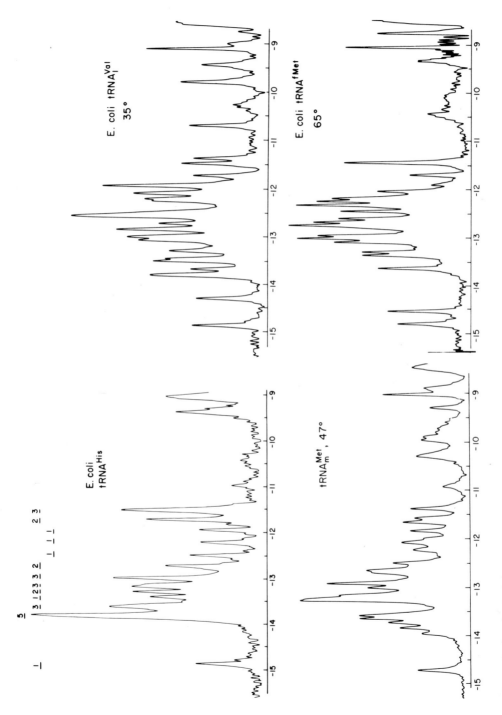

E. coli tRNA$^{Val}_1$
35°

E. coli tRNA^{fMet}
65°

E. coli
tRNA^{His}

tRNA$^{Met}_m$, 47°

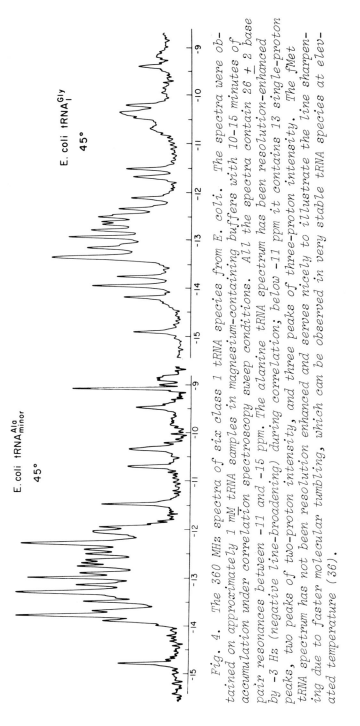

Fig. 4. The 360 MHz spectra of six class 1 tRNA species from E. coli. The spectra were obtained on approximately 1 mM tRNA samples in magnesium-containing buffers with 10–15 minutes of accumulation under correlation spectroscopy sweep conditions. All the spectra contain 26 ± 2 base pair resonances between –11 and –15 ppm. The alanine tRNA spectrum has been resolution-enhanced by –3 Hz (negative line-broadening) during correlation; below –11 ppm it contains 13 single-proton peaks, two peaks of two-proton intensity, and three peaks of three-proton intensity. The fMet tRNA spectrum has not been resolution enhanced and serves nicely to illustrate the line sharpening due to faster molecular tumbling, which can be observed in very stable tRNA species at elevated temperature (36).

low-field region for accurate integrations. The six class 1
tRNA[2] species in Fig. 4 all have 19 or 20 secondary structure
Watson-Crick base pairs each contributing one proton resonance
in this spectral region. Choosing any of the several lowest
intensity resonances as one proton, all of the spectra were
found to contain 26 to 28 resonances in the region below -11
ppm (16).

In conclusion, all class 1 tRNA species mentioned above
show resonances from 19-20 secondary structure hydrogen bonds
and an additional 6-7 tertiary structure hydrogen bonds in the
region below -11 ppm. The low values found in earlier studies
and in the more recent studies of Bolton *et al.* (15) appear to
have arisen from poor resolution, which led to the choice of
an incorrect standard and/or solvents containing insufficient
quantities of cations to completely stabilize the tertiary
structure interactions.

In addition to the resonances below -11 ppm, spectra of
tRNA in H_2O consistently show resonances from exchangeable protons between -9 and -11 ppm. Their numbers vary substantially
as can be seen in Fig. 4. Their origin will be discussed later.

B. *Origin of Resonances between -10 and -15 ppm*

Katz and Penman demonstrated that hydrogen bond formation
between G and C resulted in a downfield shift of the guanine
N1 imino proton resonance (18). While this shift was only ~1
ppm in polar aprotic solvents, resonances were observed considerably further downfield by Crothers *et al.* (19) and other
investigators studying base pairing of short oligonucleotides
in H_2O (20-24). From such studies and also by comparison of
NMR spectra of different species of tRNA (25) it has been established that the resonances between -10 and -15 ppm origi-
nated, in large part, from hydrogen-bonded imino protons of
Watson-Crick GC and AU base pairs. The x-ray crystallographic
studies of yeast tRNA[Phe] have also revealed several types of
ring-NH hydrogen bonding in tertiary structure interactions
other than the expected Watson-Crick type. These interactions
are presented in Fig. 5 (26). The five interactions containing
ring-NH...ring N hydrogen bonding can be expected to generate
resonances in the spectral region below -11 ppm. The low-field
position is caused by deshielding effects both from hydrogen

[2]*"Class 1" tRNA are those species of tRNA containing four
base pairs in the D helix and a variable loop containing five
bases. An alternative nomenclature is D4V5. N.B. E. coli
tRNA[Gly][1] is a D4V4 species.*

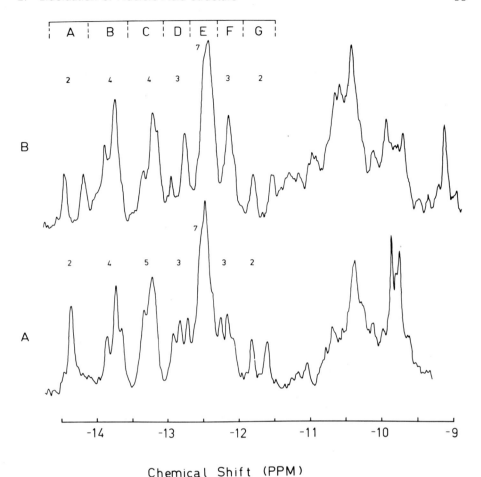

Chemical Shift (PPM)

*Fig. 5.   The 360 MHz spectra of yeast tRNA^{Phe}, which show
the difference in integrated intensity and resonance positions
in the presence (A) and absence (B) of Mg^{2+}.*

bonding and from the ring currents of the hydrogen bond donor
and acceptor bases. The resonances upfield of -11 ppm probably
originate from ring-NH...exocyclic oxygen, ring-NH...phosphate,
or exocyclic -NH...ring N type bonding. Part of the reason
that they are further upfield is that they experience only one
strongly deshielding ring current instead of two as in the case
of the ring-NH...ring N interactions.

C. Observation of Tertiary Structure Resonances

Assuming that the various functional states of tRNA are
controlled by small changes in tertiary structure, rather than
gross unfolding of secondary structure, it is reasonable to expect the changes to be reported only by the few hydrogen-bonded
proton resonances involved in tertiary structure interactions.
The bulk of the protons will be insensitive to these changes
since they are buried within helices. In order to use these
resonances as probes of the state of tertiary structure it is
necessary first to identify the tertiary structure resonances
and, once identified, to assign them.

The various experiments designed to identify resonances
arising from tertiary structure interactions all rely on the
observations that tertiary structure can be specifically perturbed by altering monovalent and divalent cation concentrations (27-30).

The first resonance that we can link with tertiary structure occurs around -13.2 ppm in the yeast tRNA^{Phe} spectrum.
This resonance is present in Mg^{2+} solutions but simply disappears when Mg^{2+} is removed and the intensity of the complex peak
at -13.2 ppm in Fig. 6 changes from five to four protons (31-
33). In low Na^+ and in the absence of Mg^{2+} the remaining tertiary structure resonances can be seen up to 25°, but this one
resonance is still absent even at 5°C. The resonance appears
to arise either from a tertiary structure interaction that is
very sensitive to the presence of Mg^{2+}, or to a weak, Mg^{2+}-sen-
sitive, secondary structure interaction.

Examination of the sequence data of E. coli and yeast tRNAs
shows that many bases involved in tertiary structure interactions in yeast tRNA^{Phe} are conserved in corresponding positions
in the sequences of other class 1 tRNAs (34-35). This suggests
that the tertiary structure interactions are conserved within a
class of tRNA. Bolton and Kearns (36) reasoned that, in unfractionated tRNA, common tertiary interactions would generate resonances at similar field positions, while resonances from secondary structure interactions, whose base composition varied substantially, would not reinforce one another, but would be
smeared out in the spectrum. Therefore, they expected to affect
only tertiary resonances by perturbing the structure. The NMR
spectrum of unfractionated E. coli tRNA was measured in the
presence of Mg^{2+}. After addition of EDTA to remove the Mg^{2+}
the measurement was repeated. The two spectra and their difference are shown in Fig. 7. As expected the removal of Mg^{2+} appears to weaken the tertiary structure and to cause a loss of
some hydrogen bond proton resonances. Four positive peaks are

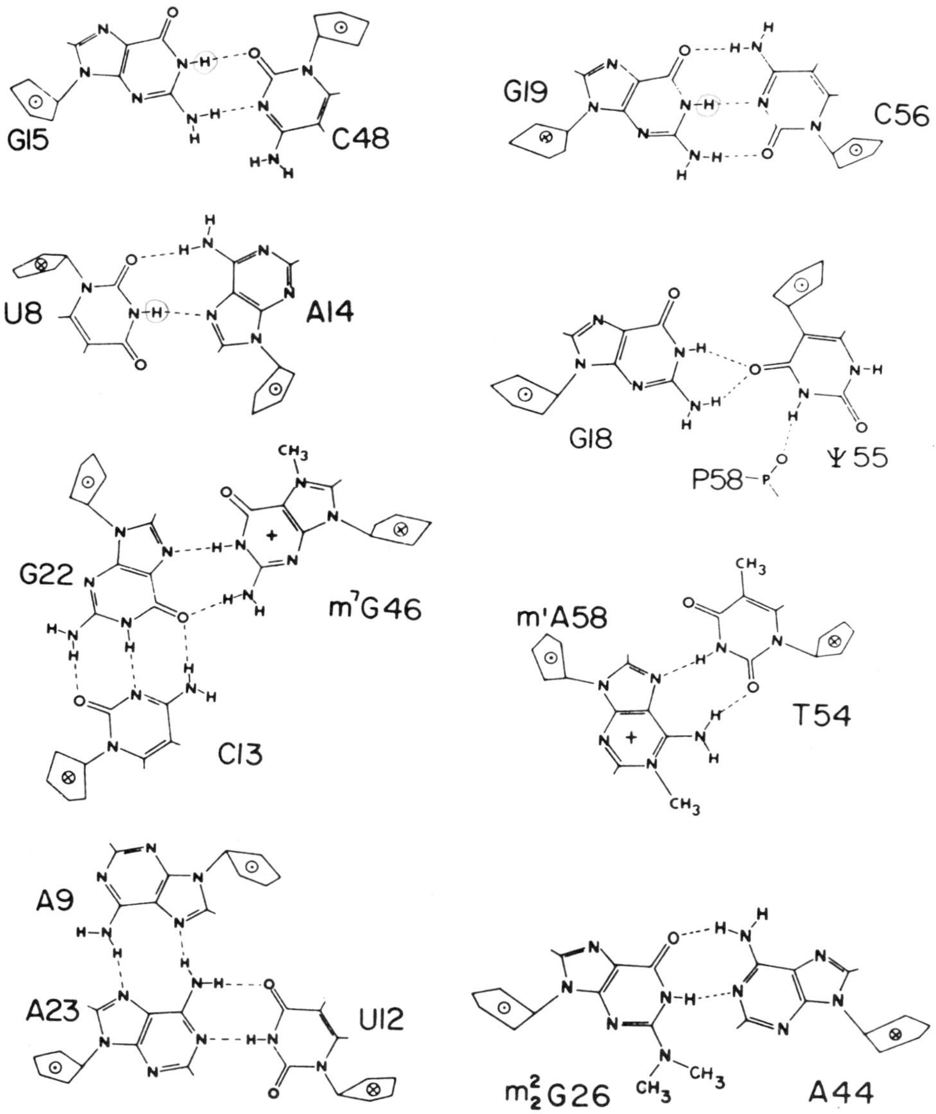

Fig. 6. The crystallographically observed tertiary structure hydrogen bonds in yeast tRNA^{Phe} which involve ring-NH
protons (26).

seen in the difference spectrum at −14.9, −13.8, −13.0, and
−11.5 ppm. The same experiment on unfractionated yeast tRNA
gave losses at −13.8, −13.0, and −11.5 ppm but none at −14.9
ppm. Based on the design of these experiments the resonances
at these positions were attributed to common tertiary structure
interactions. The specific assignments will be discussed in
the following section.

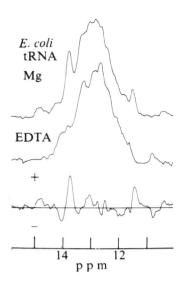

Fig. 7. The 300 MHz spectra of unpurified E. coli tRNA.
The upper spectrum is taken in the presence of Mg^{2+}. The
middle spectrum contains EDTA to remove the Mg^{2+}. The lower
spectrum is the difference between the plus and minus Mg^{2+}
spectra (31).

The most important factor influencing the chemical shift
position of a hydrogen-bonded proton resonance in these tRNA
NMR spectra is the ring currents arising from the neighboring
bases in the helix. The ring current shift contribution from
a single base can be as large as 1 ppm in the case of the purines and as small as 0.1 ppm in case of the pyrimidines. It
is surprising, therefore, that the experiments using unfractionated tRNA worked as well as they did. Even though there
are common tertiary structure interactions within a given tRNA
class, the neighboring bases are not conserved and, therefore
the ring current shifts from neighboring bases should vary considerably. This does not appear to be the case for at least
some of the resonances in Fig. 7. However, it is possible that
some tertiary structure resonances, for example, the peak at
-14.9 ppm, do not show up in Fig. 7 for just this reason.

Another problem arises from the use of unfractionated tRNA.
It is not possible to integrate a difference spectrum. If a
resonance experiences 0.5 ppm shift in 60% of the species and
1.5 ppm in the other 40% then two resonances for one base pair
may be observed at different positions with different intensities. At best this approach can give only qualitative information.

Robillard *et al.* (31) and Romer and Varadi (37) have applied the same perturbation technique but used a single purified tRNA species instead of unfractionated tRNA. Figure 8
shows a superposition of the 13 and 22° spectra of yeast
tRNA^{Phe} in 30 mM Na$^+$ after being stripped of Mg^{2+} by dialysis
against EDTA (31). The first thermal transition occurring under these conditions is the disruption of tertiary structure.
The blackened areas indicate the region where intensity is
lost. Romer and Varadi (37) used K$^+$ and Cs$^+$ to destabilize
tertiary structure preferentially and found the loss of resonances below -11 ppm in the same regions as shown in Fig. 8.
If we compare these results with those of Bolton and Kearns
(36) we find that a resonance is lost in each region where Bolton and Kearns observed a loss. Using pure tRNA there were,
as expected, more resonances lost below -11 ppm. We interpret
the additional intensity lost above -11 ppm as resonances from
tertiary structure interactions in which the proton is bonded
to only one ring nitrogen. A similar approach involving temperature-dependent intensity losses under magnesium-limiting

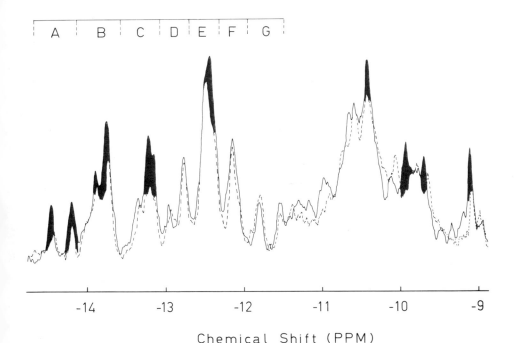

Chemical Shift (PPM)

*Fig. 8. The 360 MHz spectra demonstrating the positions
of major intensity loss during the first melting transition
between 13° (solid line) and 22° (dashed line) for yeast
tRNA^{Phe} in 30 mM Na$^+$ pH 6.8 (102).*

conditions has been used by Reid and co-workers to identify the tertiary resonances in *E. coli* tRNA$_1^{Val}$ (38). The presence of magnesium raises the transition temperature at which the first losses are observed at -14.9, -14.3, -13.4, -12.9, -12.2, and -11.6 ppm. Tertiary resonances were also observed at similar positions in yeast tRNA^{Phe} except for the -14.9 ppm resonance, which was now observed at -14.3 ppm (38). At this point we should emphasize that an underlying assumption in this approach is that the tertiary structure interactions will always be the weakest base pairs, in terms of helical lifetimes, in the molecule; this is only true for tRNA species in which the four cloverleaf helices are relatively stable.

A promising new approach to distinguishing resonances from tertiary structure as opposed to secondary base pairs has been developed recently by Johnston and Redfield (32). They apply a long pulse (39,40), which excites the tRNA protons resonating in the low-field region without exciting solvent and thus can measure the hydrogen-bonded proton spectrum in the Fourier transform mode without interference from the solvent peaks. Figure 9 presents a selective T_1 saturation recovery experiment. An initial saturating pulse is applied at -14.2 ppm and the spectrum is measured as a function of the waiting time between the saturation and observation pulse. Since the protons generating the low-field resonances are in fast exchange with water relative to their intrinsic T_1, the rate of saturation recovery is a direct measure of the water-exchange rate. By irradiating each peak individually, while varying temperature, they obtain the temperature dependence of the exchange rate for each proton. As a result of these observations Johnston and Redfield find that the water-exchange rates for nine protons are preferentially accelerated in the absence of Mg^{2+}. They are the resonances in Table I whose exchange rates are greater than 15 sec^{-1} at 42°. Because of their rapid exchange rates and sensitivity to Mg^{2+} they have been attributed to protons of tertiary base pair interactions or secondary interactions whose melting is strongly coupled to tertiary structure.

Figure 10 correlates the resonances below -11 ppm, which have been associated with protons from tertiary structure interactions in the various studies cited. Assuming that each secondary structure Watson-Crick hydrogen bond generates one resonance below -11 ppm, 20 of the approximately 26 resonances in this region of the yeast tRNA^{Phe} spectrum would be attributed to secondary structure, leaving approximately six resonances for the tertiary structure interactions. The task of choosing six resonances from those indicated in Fig. 10 and assigning them to specific tertiary interactions will be the subject of the following section.

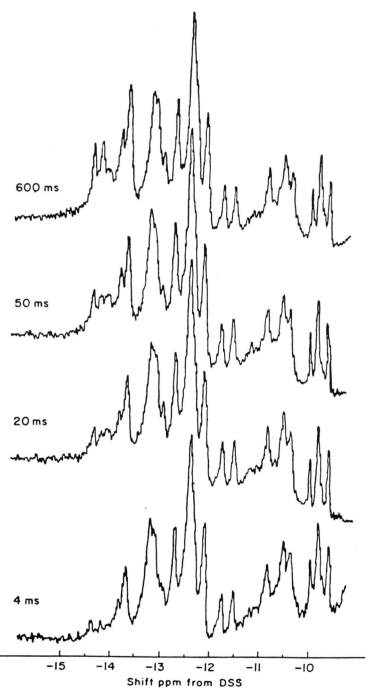

600 ms

50 ms

20 ms

4 ms

Shift ppm from DSS

Fig. 9. The 270 MHz saturation recovery measurements on yeast tRNA^{Phe} in 10 mM EDTA, 10 mM sodium cacodylate pH 7. A selective saturation pulse is applied at −14.2 ppm followed by a waiting period listed at the left of the spectrum and then an observation pulse (32).

Table I[a]

Resonance intensity	Peak ppm from DSS	Exchange rates (sec⁻¹) 10 mM EDTA temperature (°C)				
		10	30	36	39	42
1	-14.4	4	15	19	29	125
1	-14.2	5	16	20	26	65
1	-14.0	7	33	24	33	--
1	-13.9	-	12	18	--	38
2	-13.8	8	5	8	--	7
1	-13.35	-	24	37	--	125
3	-13.2	6	10	4/12	--	12
1	-12.95	-	7	16	--	100
2	-12.7	7	6	12	--	13
7	-12.5	5	3/8	4/12	--	15/29
3	-12.2	4	8	10	--	12
1	-11.75	-	7	28	--	100
1	-11.6	-	8	15	--	125

[a]From Johnston and Redfield (32).

D. Specific Assignments of Tertiary Structure Hydrogen Bond Resonances

1 s⁴U8-A14 (U8-A14)

Sequence comparisons show that many class 1 tRNAs isolated from *E. coli* contain 4-thiouridine (s^4U) in place of uridine at position 8. Treatment of s^4U with cyanogen bromide replaces the sulfur by an oxygen, thereby generating uridine (41). When this procedure was carried out on a series of *E. coli* class 1 tRNAs containing s^4U at position 8, the removal of the sulfur resulted in a shift of a resonance from -14.9 ppm in the thiolated species (13,42). Photo-oxidation of the s4U to U in *E. coli* tRNA$^{Met}_{fl}$ caused a similar upfield shift of the resonance at -14.8 ppm (17). Another approach used by Daniel and Cohn (17) alkylated the thio group of s^4U8 with a bromoacetamide spin label. The resulting s-alkyl enol form (no ring NH) contained no resonance at -14.8 ppm.

On the basis of chemical modification data therefore, one is compelled to conclude that, in spectra of *E. coli* tRNA carrying s^4U at position 8, the resonance near -14.8 ppm should be assigned to the s^4U8-A14 hydrogen/bonded proton (i.e., Fig. 4).

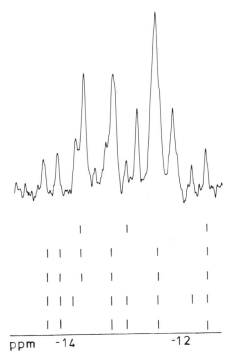

Fig. 10. Tabulation of positions attributed to tertiary structure resonances. (Top) Bolton, Kearns (15); Romer, Varadi (37); Robillard et al. (31); Johnston, Redfield (32); Reid et al. (38) (Bottom).

Recent results of Reid and co-workers (43) have shown that several large variable loop class 3 tRNAs (D3VN, where N is 13-20 nucleotides in the variable loop) do not contain the low-field resonance at -14.8 ppm although they do contain s^4U8 and A14. While it is tempting to speculate that the large variable loop is responsible for the absence of the s^4U8-A14 resonance, this is incorrect; the small variable-loop class 2 *E. coli* tRNA^{Gln} (D3V4) contains s^4U8 and A14, yet has no resonance below $^{-}14.25$ ppm. Thus it appears that tRNAs containing a three-base-paired D stem, in which GC 13 is absent, do not form the 8-14 tertiary base pair. This observation might also suggest that the 15-48 interaction discussed later may not be formed in such species.

Returning to chemical modification studies, it is apparent that most macromolecules rarely survive such modification without some side effects, which are reflected in the NMR spectrum and complicate interpretation. This was also the case for the

CNBr treatments just mentioned (13,42). Fortunately, nature
occasionally provides tRNA species that are only partially thiolated. Class 1 tRNAs containing a single s^4U have an absorption peak at 340 nm with an intensity of 2% relative to the intensity at 260 nm. A preparation of *E. coli* tRNA$_3^{Gly}$ isolated
by Hurd and Reid was found to contain some s^4U at position 8
but the intensity at 340 nm relative to that at 260 nm was only
30 to 40% of the expected value, indicating that in this tRNA
U8 is only 30 to 40% thiolated. The NMR spectrum in Fig. 11
shows a number of equally intense resonances between -11 and
-12 ppm, which are equivalent to one proton intensity. On

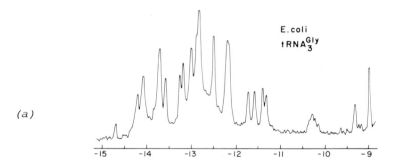

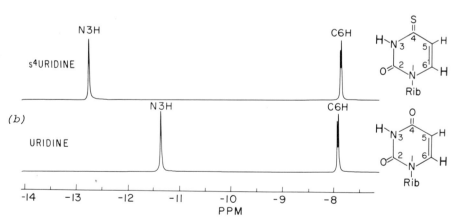

Fig. 11. (a) (upper) The 360 MHz spectrum of E. coli
tRNA$_3^{Gly}$ at 45° in 15 mM magnesium chloride, 0.1 M NaCl, 20 mM
sodium cacodylate, pH 7.0. The absorbance of this sample at
340 nm indicated that it contained only 0.3 moles s⁴U8 per mole
tRNA. (b) (lower) The 360 MHz FT NMR spectra of 4-thiouridine
and uridine in dry DMSO. Each spectrum was collected using
four quadrature pulses of 13.5 μsec separated by 20 sec at a
sample concentration of 0.025 M nucleoside.

this basis the resonance near -14.7 ppm is only 30% of a single
proton in intensity, while that at -14.2 ppm is ~70%. The only
conclusion that can be drawn is that the s^4U8-A14 tertiary
structure hydrogen-bonded proton resonates in the vicinity of
-14.8 ppm while the nonthiolated U8-A14 proton resonance occurs
in the region of -14.3 ppm. The spectrum in the lower part of
Fig. 11 shows that the N3 proton of s^4U resonates 1.4 ppm further downfield than the N3 proton of uridine when the spectra
are measured in very dry DMSO (44). Thus the origin of the
shift differences between s^4U-A and U-A hydrogen-bonded proton
resonances appears to lie in the stronger deshielding experienced by the N3 proton in s^4U compared to U.

2. m^7G46-G22

Procedures have been established by Lawley (45), Wintermeyer and Zachau (46), and Simsek et al.(47) whereby 7-methyl
guanosine can be converted to a N-ribosyl-amino pyrimidine
derivative by mild alkali treatment. Subsequent mild acid
treatment in the presence of nucleophilic amino bases causes
removal of the resulting pyrimidine derivative and cleavage of
the ribose backbone; in the absence of nucleophilic amines removal occurs without chain cleavage.

Since the method is specific for m^7 guanosine, it can be
used to perturb the m^7G46-G22 hydrogen-bonded proton resonance.
Removal of the base should result in the loss of this resonance.
Figure 12 compares spectra of three class 1 tRNAs before and
after removal of m^7G by the procedure of Simsek et al. (47) in
the absence of aniline. In each case the removal was monitored
by the loss of m^7GMP in the nucleotides produced by complete digestion of the tRNA. In the case of yeast tRNA^{Phe} and E. coli
tRNA^{Val} a number of resonances shift and some broaden upon removal of the m^7G; however, in both species only one resonance
is lost at approximately -13.35 ppm (48). Thus, the resonance
at -13.35 ppm is the most likely assignment for the m^7G46-G22
hydrogen bond. In the case of E. coli tRNA$^{Met}_{f1}$, however, a
resonance at approximately -14.6 ppm is lost and no loss occurs
in the region of -13 to -14 ppm. This shift to lower field is
precisely what is expected on the basis of the stacking interactions observed crystallographically. The m^7G46-G22-C13

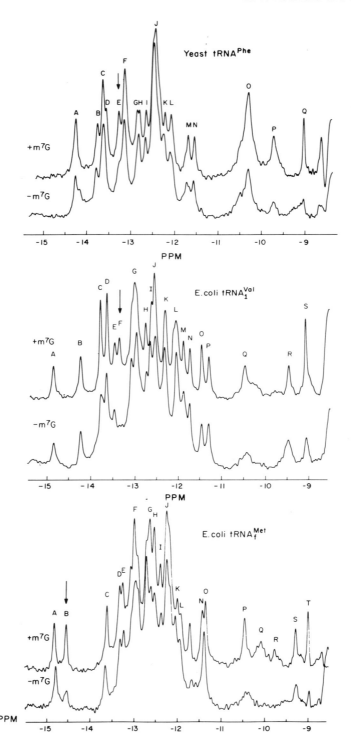

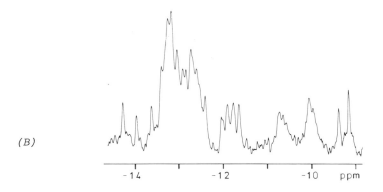

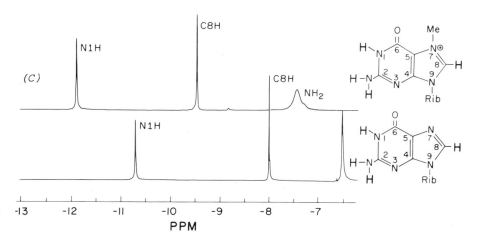

Fig. 12.   (A) The 360 MHz NMR spectra of matched pairs of yeast tRNA^{Phe}, E. coli tRNA$^{Val}_1$, and E. coli tRNA$^{Met}_f$, each with and without 7-methyl guanosine. Duplicate samples of each tRNA were prepared, one of which was subjected to m^{7}G removal according to a modification of the procedure of Simsek et al. (47) in which the aniline was omitted from the reaction mixture. The spectra were accumulated for 10-15 minutes under correlation spectroscopy conditions in the same buffer described in Fig. 11a.   The extent of m^{7}G removal was ~70%; in addition to the loss of the hydrogen-bonded resonance (indicated by an arrow), a narrow resonance at -9.1 ppm from the aromatic C8H of m^{7}G is also lost.   (B) The 360 MHz spectrum of yeast tRNA^{Met} in 10 m$\underline{M}$ sodium cacodylate, 15 m$\underline{M}$ MgCl$_a$, 0.1 $\underline{M}$ NaCl, pH 7, 36°C (tRNA provided by P. B. Sigler).   (C) The 360 MHz spectra of guanosine and 7-methyl guanosine in dry DMSO. The spectra were collected under the same pulse FT conditions described in Fig. 11b.   The N1H, C8H, and amino protons in m^{7}G are all strongly deshielded by the delocalized positive charge.

triple interaction is stacked on top of the A9-A23-U12 triple
such that the m^7G46-G22 N1 hydrogen-bonded proton is located
over the six-membered ring of A9 as marked by the x in Fig.
13A. Based on the atomic coordinates of the x-ray crystal
structure, our ring current calculations show that A9 contributes a 1 ppm upfield shift to the resonance position of the
proton in yeast tRNA^{Phe} (14). In E. coli tRNA$_{f1}^{Met}$, however, the
A9-A23-U12 is replaced by G9-C24-G12. Figure 13B shows that,
in order to make these hydrogen bonding interactions, G9 must
rotate from the "syn" to the "anti" configuration (the dotted
ring). As a result the m^7G46-G22 proton is no longer over the
plane of the G9 ring and should feel virtually no shift at all
from G9. In the absence of the 1 ppm upfield shift from base
9, the m^7G N1H proton would move approximately 1 ppm downfield
and thus explain its occurrence at -14.5 ppm. Thermus thermo-
philus tRNA$_f^{Met}$ has the same sequence in the region around the
m^7G46-G hydrogen bond as does E. coli tRNA$_{f1}^{Met}$ and its NMR spectrum shows the same two resonances below -14 ppm found in
E. coli tRNA$_f^{Met}$ (49). Another test of the validity of this explanation is available. If m^1G instead of G occurs at position
9, the methyl group would make the proposed syn to anti flip
unfavorable since no hydrogen bond to the N1 could form. As a
result, the m^1G9 would remain in the same approximate configuration as A9, and contribute a sizable upfield shift (although
not as great as A9) to the m^7G46-G22 hydrogen bond proton;
hence the m^7G46-G22 resonance should not be seen below -14 ppm
in such cases. Yeast tRNA$_f^{Met}$ is a class 1 tRNA that should
possess the m^7G46-G22 interaction and it does have m^1G at posi-

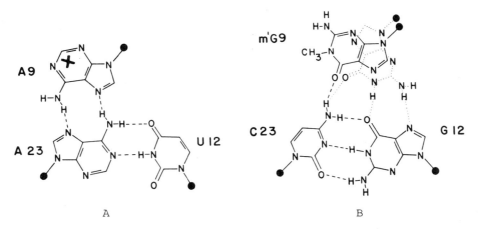

Fig. 13. (A) Crystallographically observed hydrogen bonding scheme of the A9-A23-U12 interaction. (B) Proposed hydrogen bonding scheme for G9-G12-C23 (34). The x in (A) is the
author's insertion.

tion 9. Consistent with this proposal, the spectrum of yeast tRNA$_f^{Met}$ in Fig. 12B contains only one resonance below -14 ppm belonging to the U8-A14 interaction. The m^{7}G46-G22 resonance is shifted upfield into the region between -13 and -14 ppm.

Rather than chemically removing the m^{7}G, another procedure for assigning the m^{7}G46-G22 hydrogen-bonded proton resonance would be by examining the NMR spectra of E. coli tRNA$_{f3}^{Met}$, where m^{7}G has been replaced by an A. Since adenine does not contain a ring N proton, $_{f1}^{Met}$-$_{f3}^{Met}$ difference spectrum should show the loss of only one resonance and possibly the shifting of other resonances.

Salemink et al. (50) isolated E. coli tRNA$_{f3}^{Met}$ and compared the spectrum with that of tRNA$_{f1}^{Met}$ (Fig. 14A-C). One resonance was missing at -13.5 ppm and one was shifted approximately 0.1 ppm upfield to -13.7 ppm. The -13.6 ppm resonance was attributed to the m^{7}G46-G22 hydrogen-bonded proton. The results of Salemink et al. (50) on E. coli tRNA^{Met} are similar but not identical to those of Daniel and Cohn (51).

Comparing E. coli tRNA$_{f1}^{Met}$ and E. coli tRNA$_{f3}^{Met}$ spin labeled on s^{4}U8, a difference of one resonance at approximately -13.4 ppm was found, which they attributed to the m^{7}G46-G22 hydrogen-bonded proton (51). The position is still not in complete agreement with Salemink et al. but the discrepancy has been attributed to different buffer conditions used in the measurements (50).

Salemink et al. (50) also carried out m^{7}G removal experiments on yeast tRNA^{Phe} similar to those of Hurd and Reid (48) but arrived at a completely different conclusion. They used the procedure of Wintermeyer and Zachau (46) to remove the m^{7}G and monitored the removal by the decrease in the intensity of the 292 nm absorption (52-54). Their NMR spectra before and after removal (Fig. 14D-F) show a loss of a resonance at -12.5 ppm. There was no loss at 13.3 ppm, where the loss was observed by Hurd and Reid (48).

Lastly, to further complicate an already confused topic Kearns and Bolton claim that the m^{7}G46-G22 hydrogen-bonded proton resonance does not occur at all in the low-field spectral region since they observed no resonance lost upon m^{7}G removal from unfractionated tRNA (7).

The negative result of Bolton and Kearns (7) is no proof of the absence of this resonance. As we have stressed earlier this type of result on impure tRNA indicates, at most, that the proton experiences differing shifts in the various species.

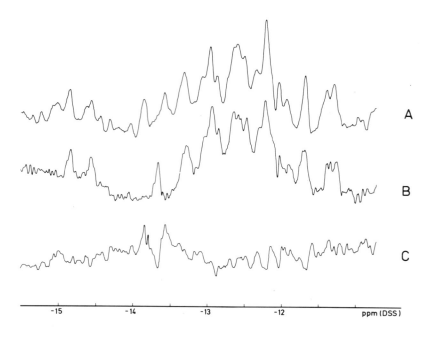

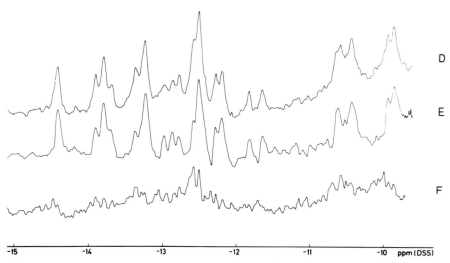

Fig. 14. The 360 MHz spectra of (A) E. coli tRNA$_{f1}^{Met}$, (B) E. coli tRNA$_{f3}^{Met}$, and (C) the difference (A)-(B). (D) Yeast tRNA^{Phe} untreated; (E) yeast tRNA^{Phe} treated to remove m^{7}G; (F) difference spectrum (D)-(E) (50).

It is worth noting that Hurd and Reid observed an additional loss of an extremely sharp resonance located at -9.1 ppm when m^7G was removed from tRNA. They assigned this resonance to C8H of m^7G based on the observations that (a) this resonance is not exchangable with D_2O over short time periods indicating that it is an aromatic proton attached to carbon: (b) this resonance disappears to an extent that parallels the extent of m^7G removal; (c) the aromatic C8H of m^7G is ca 1.5 ppm downfield from the guanosine C8H resonance at ~-7.5 to -8 ppm (see Fig. 12C) (48). Hence one can monitor the removal of m^7G directly from the spectrum if the sweep range is extended up to -9 ppm. Unfortunately the spectra of Salemink et al. did not extend this far upfield to establish to what extent they had removed the m^7G.

A further, independent difficulty occurs in rationalizing an m^7G N1H assignment at -12.5 ppm in yeast tRNA[Phe]. We expect the m^7G-G^0 resonance position to be approximately 1.6 to 1.8 ppm to lower field than the G-C^0 position of about -13.5 ppm. This arises from an inherent 1.4 ppm lower field position for the m^7G N1H resonance than the G N1H (see Fig. 12C) plus an additional 0.2 to 0.4 ppm deshielding ring current from the hydrogen bond acceptor G base. The only way to generate an m^7G46-G22 resonance at -12.5 ppm in the yeast tRNA[Phe] spectrum is through a 2.5 to 3.0 ppm upfield ring current shift from neighboring bases. Yet three independent estimates of the net upfield shift on the m^7G46-N1H (14, 55, 54) indicate values between 1.6 and 1.8 ppm. These shifts are consistent with a final resonance position of -13.4 ppm found by Hurd and Reid (48) but not with that of -12.5 ppm from Salemink et al. (50).

At the present time therefore, the most self-consistent set of results points to the -13.5 ppm resonance as the position of the m^7G46-G22 hydrogen-bonded proton in the spectrum of yeast tRNA[Phe].

3. T54-A58

The hydrogen-bonded proton NMR spectra of yeast tRNA[Phe] and *E. coli* tRNA[Val] both contain two resonances below -14 ppm. One has already been assigned to the (s^4) U8-A14 reversed Hoogsteen interaction. The removal of m^7G46 does not cause any loss of intensity in this spectral region (see Fig. 12A) and hence the second resonance cannot be assigned to the m^7G46-G22 tertiary hydrogen bond. On the bases of our ring current calculations (14,57) and comparative spectroscopy of several tRNA species (58) we suggested that the second resonance could also be assigned to the reversed Hoogsteen interaction T54-A58. The intrinsic starting position of the hydrogen bond proton resonance

was expected to be at a lower field position than in the case
of a Watson-Crick type AU interaction due to the difference in
the type of hydrogen bonding (14,57). Unfortunately there is
no single experiment that unambiguously assigns the T54-A58 hydrogen bond resonance. Instead we must combine a series of observations. One of the strongest arguments is the spectrum of
the TψC fragment containing bases 47-76, from $E.$ $coli$ tRNA$_1^{Val}$
shown in Fig. 15. Calculations based on the x-ray crystal
structure coordinates (see Fig. 18) show that the resonances
between -13.3 and -12.2 ppm are assigned to base pairs 53, 51,
52, and 49, while GU 50, a "wobble" base pair with two ring NH
hydrogen bonds, probably accounts for the two sharp resonances
at -11.9 and -10.8 ppm. All the expected secondary structure
Watson-Crick hydrogen bonds in this helix are GC interactions
and their resonances occur above -13.6 ppm. Only two resonances occur at lower fields, -14.3 ppm and -13.8 ppm, and both
resonances "melt" early if the temperature is raised. As
shown in the inset of Fig. 15, an extra AU interaction could
arise in the fragment spectrum by the base pairing of U47 to
A66. This interaction plus the T54-A58 hydrogen bond account
for the two low-field resonances observed. There is, nevertheless, an ambiguity concerning which of the two resonances

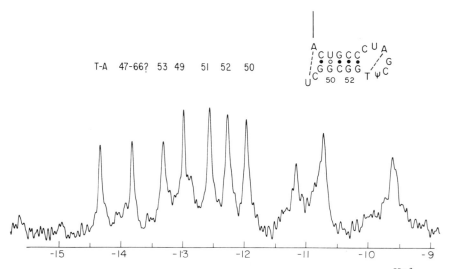

Fig. 15. The 360 MHz spectrum of the E. coli tRNA$_1^{Val}$ fragments 47 to 76 containing the rT helix base pairs. The tRNA
was cleaved at m^{7}G46 according to Simsek et al. (47) and the
30-residue 3' fragment isolated according to Reid et al. (38).
The dialyzed, lyophilized fragment was dissolved in 100 mM NaCl,
5 mM Na phosphate pH 7.0, and the spectrum taken at 25° under
correlation sweep conditions.

arises from T54-A58. In the case of yeast tRNA[Phe] (31,32,59)
both resonances below -14 ppm are tertiary structure interactions. They melt in the first transition as shown in Fig. 8
and they have rapid exchange rates as shown by the saturation
transfer experiments of Johnston and Redfield (32) (see Fig.
10 and Table I). There are no other AU or AT crystallographically observed tertiary structure interactions that are candidates for this resonance; therefore, by default it must be assigned to the T54-A58 tertiary.

Assigning the T54-A58 resonance in other class 1 tRNA species is more difficult. Of the six class 1 tRNAs whose spectra
are shown in Fig. 4, three do not have extra resonances below
-14 ppm that could be attributed to the AT even though they
contain enough resonances below 11 ppm to include this tertiary
base pair (16). Ring current calculations based on the x-ray
crystal structure coordinates indicate that the AT hydrogen-
bonded proton in yeast tRNA[Phe] and *E. coli* tRNA$_1^{Val}$ receives a
total of 0.4 to 0.5 ppm upfield ring current shifts from bases
50 to 62 in the TψC loop and helix. With one exception the only
systematic difference between either yeast tRNA[Phe] or *E. coli*
tRNA[Val] and the other tRNA species that do not have an AT resonance below -14 ppm is the replacement of U59 by a purine. Normal helical stacking of a purine in position 59 should contribute a substantial upfield shift to the T54-A58 hydrogen-bonded
proton resonance and could account for shifting a resonance
from -14.3 to -13.6 ppm. From the x-ray structure of yeast
tRNA[Phe], however, we learn that U59 does not stack in a helical
fashion. In order to allow both the GC53 and AT hydrogen bonds
to form, the segment U59,C60 bulges outward. In this conformation the plane of U59 is approximately 13 Å from the AT hydrogen-bonded proton. At such a distance neither a purine nor pyrimidine would contribute a noticeable ring current shift to the
AT hydrogen-bonded proton resonance.

Clearly it appears that the crystallographic data on yeast
tRNA[Phe] do not coincide with the NMR data from these other tRNA
species. Two explanations are possible. First, upon replacement of a pyrimidine with a purine at position 59 there may be
a change in the nature of the stacking of residue 59 on base
pair 54-58 so that the shift from base 59 is stronger than expected for a pyrimidine in position 59. Second, the T54-A58
interaction could change from a reversed Hoogsteen to a normal
Watson-Crick type interaction. Since these have less deshielded
intrinsic positions (-14.3 ppm) (14,57) the resonance would be
shifted to -13.8 ppm even without a substantial shift from
A59.

The studies of Bolton and Kearns (36) (see Figs. 7 and 10)

showed that one of the common resonances lost upon removal of Mg^{2+} occurred at -13.8 ppm and they assigned it to the common tertiary T54-A58. On the basis of this assignment they concluded that the extra resonance below -14 ppm in yeast tRNA^{Phe} and E. coli tRNA$_1^{Val}$ must arise from a secondary base pair. It appears, from comparing spectra of many different tRNAs, that their assignment of T54-A58 was probably correct in unfractionated tRNA but the other conclusion was wrong. At least half of all the E. coli tRNA NMR spectra have only one resonance, assigned to the 8-14 base pair, below -14 ppm. Thus, in most species, the AT base pair resonance occurs around -13.8 ppm. However, in the case of E. coli tRNA$_1^{Val}$ and yeast tRNA^{Phe}, this resonance occurs below -14 ppm. This observation should point out the danger of trying to make detailed assignment using a mixture of 60 molecular species. Furthermore it reinforces our earlier observation that there is no reason to expect common tertiary resonances from all tRNAs to have common chemical shifts. While some tertiary interactions involve bonding to the D helix, and the sequence GC10-GC11-UA12-CG13 is the most common sequence for this helix (60), one can see why some tertiary resonances may have similar positions; for other tertiary resonances and in other tRNAs one can expect a variety of chemical shifts.

4. G19-C56

This interaction is the only tertiary structure base pair involving Watson-Crick type hydrogen bonding. Thus, in this case, we have a reasonably accurate estimate of the unshifted starting position upon which to base chemical shift estimates. Ring current calculations, based on the original atomic coordinates of yeast tRNA^{Phe} (62), estimated a 0.86 ppm upfield shift from neighboring bases, which put G19-C56 at -12.68 ppm (14). However no resonance is lost in the yeast tRNA^{Phe} NMR spectrum at this position during the tertiary structure melting transition but a partial loss in intensity does occur at -12.95 ppm (31). This is the same position at which Johnston and Redfield detect a resonance with a temperature-sensitive exchange rate (see Table I), which they attribute to a tertiary structure interaction (32). When the ring current calculations were repeated on refined atomic coordinates (63) the calculated shift experienced by the G19-C56 proton was found to be 0.63 ppm instead of the 0.86 ppm obtained from the less accurate coordinates. This puts the calculated value of the G19-C56 resonance at -12.91 ppm, essentially where it is observed in exchange rate measurements of Johnston and Redfield (32) and by Mg^{2+} and temperature sensitivity studies by Reid et al. (38). It is not clear why no intensity loss was observed by Romer and Varadi (37) or why we only observe a partial loss at this position (31).

However, one possibility is that all regions of tertiary structure may not unfold exactly together in the absence of Mg^{2+}.

The reliability of the assignments made to the four tertiary structure interactions, U8-A14, T54-A58, m^7G46-G22, and G19-C56, is quite high. For the first three there is experimental evidence in the form of the data presented in Fig. 10, where all of the studies on pure yeast tRNA^{Phe} indicated tertiary structure resonances at the positions assigned. The chemical modification and fragment studies have removed most ambiguity in the specific assignments. The reliability of the G19-C56 assignment is more uncertain since we lack agreement between the melting studies (31,37) and the exchange rate studies of Johnston and Redfield (32). However, there is agreement between Johnston and Redfield's observations and the ring current calculations on refined coordinates, which do lend support to the current G19-C56 assignment around -12.95 ppm. For the remaining tertiary structure interactions involving ring NH protons, m_2^2G26-A44, G15-C48, and G18-ψ55, there are still uncertainties. In the following sections we shall present the various experiments and discuss the claims for certain assignments.

5. m_2^2G26-$A44$

This anomalous base pairing between two purine residues at the top of the anticodon helix has been independently observed in all of the crystallographic studies and the electron density, indicating a ring NH...ring N hydrogen bond, is unambiguous. Thus a resonance from the Pu26-Pu44 interaction is certainly to be expected in the -11 to -15 ppm region of the spectra of yeast tRNA^{Phe} and E. coli tRNA$_1^{Val}$ (where these positions are occupied by A26 and G44). The spectra of both tRNAs contain resonances at ~-12.4 ± 0.1 ppm that are sensitive to perturbations of tertiary structure (see Fig. 10). It is possible that the Pu26-Pu44 hydrogen-bonded proton resonates at this position, but unfortunately this interaction involves relatively common bases (A or G or methylated G) with no exceptional reactivity properties, so that selective chemical modification is not a viable approach for assignment. Nevertheless Reid and colleagues have used comparative analysis of a considerable number of tRNA species to address this question indirectly (58, 64). Yeast tRNA^{Phe} and the E. coli species tRNA$_1^{Val}$, tRNA$_f^{Met}$, tRNA$_m^{Met}$, and tRNA$_3^{Gly}$ contain Pu26 and Pu44, and all contain a tertiary resonance at -12.3 ± 0.2 ppm in their NMR spectra.

On the other hand E. coli tRNA$_1^{Arg}$ contains A26 and C44, both of which lack a ring NH proton and the corresponding positions in E. coli tRNA^{His} are occupied by C and U, respectively. In

the spectra of these two tRNAs there does not appear to be a
tertiary resonance near -12.3 ppm in that all resonances between -12.0 and -12.8 ppm are satisfactorily attributed to
secondary base pairs, on the assumption of ideal helical stacking, in the respective cloverleaf sequences. In *E. coli*
tRNA^{Lys} the corresponding positions are occupied by A26 and
U44; in the spectrum of this tRNA the region around -12.4 ppm
appears to be devoid of tertiary resonances but a new extra
resonance appears 1 ppm to lower field (61). This type of approach suggests that the Pu26-Pu44 resonance, in those species
that contain it, is found in the neighborhood of -12.4 ppm and
varies by up to 0.2 ppm in different species. Further support
for this interpretation comes from studies of tRNA fragments
containing the combined DHU helix and anticodon helix; the 26-
44 interaction is located at the junction of these two helices
and such fragments contain an additional resonance at -12.4 ppm
(38). Since the procedures of Reid *et al.* (38) as well as of
Robillard *et al.* (14) both assign the Pu26-Pu44 hydrogen-bonded
proton resonance by default (i.e., looking for a resonance not
assignable to any other interaction) they are subject to the in-
accuracy of their respective predictive methods. Reid *et al.*
(38) try to minimize this source of error by examining a large
number of homologous tRNA NMR spectra. Nevertheless, their
predicted assignments for each spectrum still assume ideal A
RNA helical stacking. On the other hand, Robillard *et al.* (14,
57) are not forced to assume ideal stacking but are limited to
structures that have been solved crystallographically and are
subject to the inaccuracies of the available atomic coordinates.
They try to minimize this source of error by repeating these
calculations on newly refined coordinates whenever they become
available. The contrast between the two methods is only noticeable in the assignment of resonances from those regions of
the crystallographic structure that deviate substantially from
ideal stacking. For instance, when calculated from the x-ray
crystallographic coordinates (12,14,59) the shift experienced
by GC28 and AU52 are ~0.5 ppm larger than those estimated by
Reid *et al.* (38), while the shift on GC13 is ~0.7 to 1 ppm less
than the estimates of Reid *et al.* These differences led Reid
et al. (38) to assign AU52 to region B (-13.8 ppm) in Fig. 10
while Robillard *et al.* (14) put it in region C and had a vacan-
cy in region B to which the 26-44 tertiary was assigned. In
another case Reid *et al.* (38) put GC13 around -11.75 ppm, while
Robillard *et al.* (14) put it around -12.5 ppm but GC28 was calculated between -11.5 and -12 ppm.

In the final assignments Reid *et al.* (38) find the Pu26-
Pu44 interaction at -12.5 ppm. However Robillard *et al.* (14)

could place it at -13.8 ppm, and the -12.5 ppm resonance would
be attributed to GC13, which is sensitive to tertiary structure
perturbation and shifts upfield to -11.75 when tertiary structure breaks (31) (see Section III).

6. G15-C48 and G18-ψ55

The remaining two crystallographic hydrogen-bonded ring NH
protons to be found in the low-field spectrum are the reverse
Watson-Crick G15-C48 and G18-ψ55 (see Fig. 5). Both of these
hydrogen bonds involve ring NH...carbonyl oxygen bonding (65);
we have previously discussed the reasons why such hydrogen-
bonded protons should be expected at the upfield end of the
spectrum (66). Hence these are prime candidates for the unassigned resonances in the region from -12 to -11 ppm, or even
higher.

In all tRNAs the 15-48 interaction involves normal unmodi-
fied bases that do not permit assignment by selective chemical
modification. When one compares the spectra of $E.$ $coli$ tRNA$_3^{Gly}$
with $E.$ $coli$ tRNA$_1^{Gly}$ (which can not have a 15-48 ring NH hydrogen bond since G15 is replaced by A15) one observes four protons between -11.7 and -11.3 ppm and at least one proton at
~-10.4 ppm in tRNA$_3^{Gly}$, while tRNA$_1^{Gly}$ contains only two protons
between -11.7 and -11.3 ppm and at least three protons at
~-10.4 ppm (Reid et $al.$, unpublished data). In addition to
G15-C48 and A15-C48 one can also study this interaction in the
form A15-U48, which occurs in $E.$ $coli$ tRNA^{Trp}. The spectrum of
tRNA^{Trp} reveals a single proton at -11.5 ppm and two protons at
~-11.8 ppm; between -10 and -11 ppm there is a weak partial resonance at -10.2 ppm and two protons at -10.8 ppm (10,67). However this type of analysis is seriously complicated by the fact
that, as shown in Fig. 15, the two imino...carbonyl protons of
GU base pairs resonate in the -10 to -12 ppm region (38,68).
Although the UN3H is expected at lower field than the GN1H, the
position of these two resonances varies appreciably as a result
of ring current shifts from the particular nearest-neighbor sequence surrounding a given GU pair. Since tRNA$_1^{Gly}$ and tRNA$_3^{Gly}$
contain a GU pair in different environments, and tRNA^{Trp} contains two GU pairs, the interpretation of which resonances in
the -10 to -12 ppm are derived from 15-48 and/or 18-55 is quite
difficult. The comparative spectroscopic evidence can be interpreted to suggest that the 15-48 proton and the 18-55 proton
are both present in the -10 to -12 ppm region of the spectrum
and the balance of the somewhat meager data on this point
slightly favors assigning 15-48 between -10 and -11 ppm with
18-55 at ~-11.5 ppm.

Lastly we should mention the claims of Kearns (7) that a

resonance at ~-11.5 ppm arises from the non-base-paired "buried"
N3H of U33. This assignment is unlikely from the observed temperature stability of this resonance in intact tRNAs; their argument is based on analysis of a poorly defined fragment of *E.
coli* tRNA$_f^{Met}$ (12) and will be discussed at greater length in a
later section on tRNA fragment spectra.

III. ASSIGNMENTS VIA CALCULATIONS

A. *Contributions to Chemical Shifts*

    Assigning tertiary resonances via the chemical modification
approach discussed in the preceding section is relatively laborious and not applicable to all tRNA species. Similarly, the
assignment of secondary resonances from helical tRNA fragments
as described later is exceedingly laborious and not without pitfalls in interpretation. An alternative approach that has been
developed during the past 5 years is to calculate the resonance
positions of the protons by summing the shift contributions
arising from neighboring atoms and rings.

    The diagram in Fig. 16 illustrates two types of protons that
can be readily monitored in NMR spectra of tRNA, namely, NH protons in hydrogen bonds and CH_3 protons. The various factors
that contribute to the chemical shift of the resonances from
these protons are

    (1) ring currents arising from delocalized electrons of the
aromatic rings;
    (2) polarization effects due to an enhancement or depletion
of charge via protonation or hydrogen bonding;
    (3) the anisotropy of diamagnetic susceptibility of various
proximal groups such as carbonyls, etc.

To calculate a proton resonance position the estimated contributions from these individual shift effects are summed and applied to the intrinsic unshifted NH or CH_3 resonance position
(if known). This is a straightforward task in the case of
methyl protons since the unshifted resonance position is readily
obtainable from the NMR spectrum of dilute solutions of the free
methylated nucleoside as will be discussed later. It is impossible, however, to make the corresponding determination for NH
protons with acceptable accuracy. One cannot define the unshifted resonance position of the ring NH protons of donor bases
in H_2O, or even the unshifted resonance positions for these protons when hydrogen bonded in AU or GC complexes, because they
undergo rapid exchange with solvent protons. The exchange can

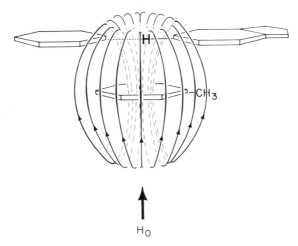

*Fig. 16. A diagrammatic sketch of the induced local mag-
netic field generated by the ring current of a nucleotide in a
RNA helix and the resulting field effect on proximal protons.
The local induced field at the CH_3 position will augment the
primary field $\underline{H}_0$ (downfield shift) whereas it will oppose the
primary field at the hydrogen bond of the upper base pair (upfield shift). Reciprocal effects of the ring currents from the
upper two nucleotides will be exerted on the ring NH and CH_3
protons of the lower nucleotide and similar effects will occur
from the neighboring base pair below.*

be eliminated by using aprotic solvents but then quantitative
extrapolation back to the resonance position in H_2O is not possible due to the differing solvent diamagnetic properties (69-
72). An alternative approach, first used by Crothers *et al.*
(19), measured the resonance positions of NH protons in complexes of complementary oligonucleotides such as d(AACAA) with
d(TTGTT). In theory the helical bimolecular complexes should
be sufficiently stable that the solvent-exchange rates of the
hydrogen-bonded protons are slow. This allows the NH proton
resonances to be observed in the NMR spectrum. The positions
obtained, however, are no longer unshifted positions. As seen
in Fig. 16, the presence of adjacent bases causes the hydrogen-
bonded protons to experience shifts from induced local magnetic
fields of the ring currents. Thus, intrinsic resonance positions must be estimated by subtracting the values of the shift
contributions from other rings in the helix. These shift
values, in turn, are dependent on the helix geometry, i.e., the
position of the bases, which itself is dependent on the ionic
strength and pH of the solvent. Furthermore, even knowing the

position of the bases, the estimates of the shift contributions
are dependent on the accepted values for the ring current of
each base; these values have changed as the methods used to calculate them undergo continued refinement (73,74). As a result
of these approximations it is difficult to see how one can
reach an intrinsic position with an accuracy of better than
±0.3 ppm.

The original studies of Crothers *et al.* on the oligodeoxy-
ribonucleotide complex d($\frac{AACAA}{TTGTT}$) showed resonances in two regions
of the NMR spectrum with an intensity ratio of 1:4. The smaller
peak was upfield at -12.4 ppm and indicated that GC hydrogen-
bonded protons resonate at higher field than their AT counter-
part (19). In similar studies, Patel and co-workers (20-22) observed three resonances for the three different hydrogen bonds
in the duplex d($\frac{ATGCAT}{TACGTA}$). Intrinsic resonance positions of -14.6
± 0.2 and -13.6 ± 0.1 ppm were calculated for the AT and GC protons respectively, when the observed chemical shifts were corrected for ring current shifts from neighboring bases assuming a
B-DNA helical geometry (75). Because of C2' endo versus C3' endo
puckering, oligoribonucleotide complexes are expected to have
helical geometries that are not the same as their deoxy counter-
parts. Nevertheless, studies of base pairing in complexes such
as r($\frac{CCGG}{GGCC}$) (23) or r($\frac{AAGCUU}{UUCGAA}$) yielded intrinsic resonance positions for the GC and AU protons similar to those cited above
when ring current corrections were made based on either A or A'
helical geometry (76).

Once estimates of intrinsic resonance positions were established, Shulman and colleagues used these values together
with tables of ring current shifts in order to interpret hydrogen-bonded proton NMR spectra of various tRNAs (25). The tables
were generated from projected overlaps of different base combinations in the 12-fold A' RNA geometry compiled by Arnott (77)
and the ring current isoshielding contours of Giessner-Prettre
and Pullman (78). Using these tables one can calculate a resonance position for a Watson-Crick hydrogen-bonded proton when
the adjacent bases in the helix are known. The initial fits
between the observed and calculated spectra based on the published ring current shifts (78), were not very satisfactory.
A substantial improvement was obtained, however, when the magnitude of the adenine and guanine ring current shifts were increased by 20% and the intrinsic unshifted position was moved
further downfield. The resulting agreement between observed
and calculated resonance positions for six different tRNAs was
±0.2 ppm (25). These correlations between observed resonances
and predicted chemical shifts were carried out on the assumption
that no tertiary resonances were present. Hence all resonances
were attributed to the secondary structure base pairs seen in

the cloverleaf. Since -14.8 ppm was the lowest resonance position in any of the early spectra (25) it was selected as the intrinsic resonance position for the hydrogen-bonded proton in the
Watson-Crick AU base pair. The choice of such a low-field intrinsic position and the need to predict resonances as far upfield as -11.5 ppm are apparently the factors that forced Shulman *et al.* (25) to increase the A and C ring current shifts 20%
above published values. As shown above, however, the spectra
contain several tertiary resonances and x-ray and chemical modification studies indicate that resonances below -14.3 ppm are
derived from reversed Hoogsteen or AT hydrogen bonds from tertiary interactions. Consequently these procedures for resonance assignment must be reexamined.

B. *Ring Current Calculations and the X-Ray Crystal Structure*

    Three sets of ring current calculations based on the x-ray
crystal structure coordinates of yeast tRNA^{Phe} have appeared in
the past two years (14,55,56). Since the results and the procedures used differ substantially we shall consider them separately.

    Kan and Ts'o (56), working with the refined coordinates of
Rich *et al.*, summed the ring current shifts arising from each
base within 10 Å of every hydrogen-bonded proton. The magnitude
of the shifts were estimated from isoshielding contours (74).
The resulting chemical shift positions of the individual resonances were obtained by subtracting the net ring current shifts
from the intrinsic starting positions for the various types of
hydrogen-bonded protons (79). The most disturbing feature of
these results is that the two resonances at -14.4 ppm in yeast
tRNA^{Phe} are assigned to the secondary structure hydrogen bonds
AU 6 and AU 12. In evaluating their results, they claim to
have, within a tolerance of ±0.1 ppm, only four resonances out
of 25 incorrectly calculated and thus conclude that their ring
current procedure and their intrinsic chemical shifts are correct. Unfortunately they ignore available data indicating that
their assignments are incorrect. As we have already demonstrated, experimental data (some of them three years old) indicate
that the two resonances below -14 ppm originate from tertiary
structure interactions. Thus the AU 6, AU 12, T54-A58, and U8-
A14 assignments are each in error by at least 0.5 ppm. The errors originate in incorrect choice of intrinsic positions. If,
however, one changes their estimated intrinsic position then
all the other AU secondary structure resonance assignments would
also deviate substantially more than Kan and Ts'o have reported.
The goal of this approach should be not simply to predict a
spectrum that looks like the experimental spectrum, but rather

to predict a spectrum in which specific resonances occur where
they are experimentally assigned.

Geerdes and Hilbers (55) have given more attention to intrinsic positions and comparisons with experimental assignments
in analyzing the results of their calculations. In attempting
to determine the intrinsic resonance positions of the Watson-
Crick and the reversed Hoogsteen AU hydrogen-bonded protons
they followed the formation of oligo-A:oligo-U duplexes and
oligo-A:(oligo-U)$_2$ triplexes (80). In the latter both Watson-
Crick and reversed Hoogsteen hydrogen-bonded resonances were
expected to be observed and, in fact, they were observed as
separate resonances. However the choice in assigning these two
resonances was arbitrary and gave intrinsic positions, when
corrected for estimated neighboring ring current shifts, of
-14.5 ppm for both the reversed Hoogsteen and Watson-Crick AU
hydrogen-bonded protons. Geerdes and Hilbers do point out,
however, that the opposite choice is equally probable and would
result in values of -14.8 and -14.3 ppm for the reversed Hoogsteen and Watson-Crick AU resonances, respectively. We shall
come back to these values shortly. Having chosen intrinsic positions, they carried out ring current calculations on yeast
tRNA^{Phe} using published coordinates (62,81) according to the
method of Giessner-Prettre and Pullman (78). With two different sets of coordinates they calculated no resonances at
-14.4 ppm but predicted three protons between -14.15 and -14.0
ppm, including the two tertiary AU-type resonances and a secondary resonance (AU 12).

Initially we observed similar discrepancies between experimental resonance positions and our own calculated positions but
felt that the problem rested in one or more incorrect input parameters employed in the calculations and not in a legitimate
discrepancy between the crystal structure and the solution
structure (14). As indicated earlier, the estimation of intrinsic starting positions from experimental data is only accurate
to ±0.3 ppm and could be a major source of error. Therefore,
on the assumption that there was a strong similarity between
the crystal structure and the solution structure, we attempted
an empirical optimization of ring currents and intrinsic positions using the following procedure. The ring current chemical
shift of each hydrogen-bonded proton was calculated (14,82,59)
using atomic coordinates (62) and varying one ring current at a
time between 0 and 200% of its published values while the other
three ring currents were held constant at their published values
(73). For every new ring current value we had a complete set of
chemical shifts for each hydrogen-bonded proton resonance. Four
files were generated in this manner, one for each ring current
that was varied. The files containing the A and G ring current

variations were then combined into a two-dimensional grid in a computer. A third dimension was added for varying the AU^O and GC^O value. Each intersecting point in the grid contained a complete set of calculated resonance positions. A computer search was carried out comparing each set of values in the grid with the experimentally observed resonance positions by a RMS (root mean square) error test. The A and G ring currents that resulted in the closest similarity between the calculated spectrum and the observed spectrum were then used to generate a new grid, this time varying the C and U ring currents and the AU^O and GC^O values. A search of this grid gave new C and U ring currents. These new C and U ring currents were used to generate another A,G grid. After searching this grid, the final optimized ring current values for A and G and the AU^O and GC^O values were obtained. The optimization process converged at this point and further searching did not provide better values of ring current shifts.

The calculated spectrum resulting from the optimized values for ring currents, AU^O and GC^O is compared with the observed spectrum in Fig. 17. The most important feature of the calculated spectrum is that there is agreement with the experimentally determined tertiary resonance positions for four of the six tertiary structure resonances. Furthermore the agreement for the secondary structure resonances, when compared with fragment spectra (83), is better than with earlier predictive methods (25). The success of these calculations is due predominantly to altering the value for AU^O in a Watson-Crick pair. We have seen experimentally that this position is around -14.5 ± 0.3 ppm. Our value is within this range at -14.35 ppm. The reversed Hoogsteen AU^O we find at -14.9 ppm. We should recall at this point that these new AU^O values are very close to the equally probable alternative set of values that Geerdes and Hilbers could have derived from their experimental data on the oligo A-(oligo U)$_2$ triplex (55). The change in ring currents and intrinsic resonance positions resulting from the refinement is shown in Table II.

In a recent review Kearns (6) has criticized the approach presented above, claiming that a different set of atomic coordinates would have led to different conclusions. This is certainly true, and for this reason we have carried out the same iterative refinement of ring currents and AU^O and GC^O intrinsic positions with four different sets of partially refined atomic coordinates obtained from different laboratories (62,84,85,86). The coordinates of Sussman and Kim (62) gave the lowest RMS error between the observed and calculated spectrum with the smal-

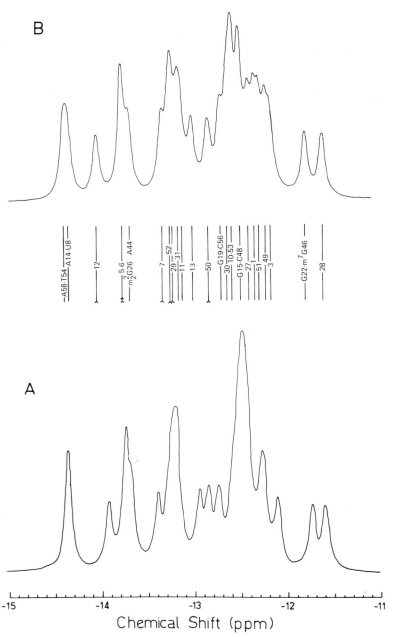

Fig. 17. *Comparison of the ring current calculated spectrum (B) with a computer simulation of the observed spectrum (A) of yeast tRNA^Phe. The assignment predictions for the secondary structure interactions result from the ring current calculations*

Table II

Nucleotide	Starting ring current (73)	Optimized ring current (14)
Adenine		
Hexagonal ring	0.88	0.76
Pentagonal ring	0.67	0.58
Guanine		
Hexagonal ring	0.25	0.28
Pentagonal ring	0.63	0.72
Cytosine	0.27	0.21
Uracil	0.08	0.10

Base pair	Starting intrinsic position (ppm)	Optimized intrinsic position (14) (ppm)
Watson-Crick AU	-14.5 ± 0.3	-14.35
Watson-Crick GC	-13.6 ± 0.1	-13.54
Reversed Hoogsteen		
AU or AT	-14.5 ± 0.3	-14.9

lest alterations in the ring currents and intrinsic positions, and for these reasons they were chosen as the standards for predicting assignments (57).

C. Reliability of Assignment Predictions Based on These Optimized Ring Currents and Intrinsic Positions

While the optimized ring currents and intrinsic positions in Table II yielded an attractive fit between the observed spectrum and that calculated from the atomic coordinates (Fig. 17), this was no guarantee that the resulting assignments were correct. Corroboration was obtained in two ways: first, by successfully predicting the experimental spectrum of a new tRNA with a completely different sequence (E. coli $tRNA_1^{Val}$) (57), and second, by experimentally verifying the assignments in these two spectra by melting studies (31,38) and hairpin fragment studies (38).

1. Calculation of the NMR Spectrum of E. coli $tRNA_1^{Val}$

No atomic coordinates are available for the structure of other tRNAs with which to repeat these calculations. The cloverleaf structures of yeast $tRNA^{Phe}$ and E. coli $tRNA_1^{Val}$, however, indicate that these structures are homologous. The mole-

cules are composed of the same number of nucleotides and are
stabilized by a similar number of tertiary hydrogen bonds (38).
Furthermore all additional base pairs participating in tertiary
structure are conserved with respect to sequence position and
chemical shift position (38,57,59). Assuming homologous structures, hypothetical atomic coordinates of *E. coli* tRNA$_1^{Val}$ were
generated using the yeast tRNA^{Phe} ribose-phosphate atomic coordinates (62) and substituting new bases at the appropriate
glycosidic bonds. After idealization to minimize unfavorable
stereochemistry and Van der Waal's interactions, the resulting
coordinates were assumed to represent the atomic structure of
E. coli tRNA$_1^{Val}$. When these coordinates were used, along with
the previously optimized ring currents and intrinsic positions
(Table II) to calculate the NMR spectrum of *E. coli* tRNA$_1^{Val}$ the
lower spectrum in Fig. 18 was obtained (57). There is clearly
a striking, although not perfect, resemblance to the experimental spectrum above it. Considering the method whereby the
atomic coordinates were generated the agreement between the two
spectra is excellent and lends support to the correctness of
the optimized ring currents and intrinsic positions as well as
the computational procedures employed. The residual disagreement between the observed and calculated spectra most likely
originates from one or more of the assumptions that made these
calculations possible at this stage [i.e. (1) that the structures of yeast tRNA^{Phe} and *E. coli* tRNA$_1^{Val}$ are completely homologous; (2) that the ring current is the only important parameter determining the resonance positions.] In the future these
calculations will have to be repeated on other crystallographically determined structures using highly refined coordinates.
Furthermore the computations themselves must be refined to include smaller chemical shift contributions from diamagnetic
anisotropy, electric fields, and the dielectric properties of
the solvents.

    Recently Arter and Schmidt (87) computed the ring current
shift contribution that should be experienced by each proton
in various RNA helices using Giessner-Prettre ring currents and
eliptical integral calculations (73,75). Their values for the
chemical shifts of hydrogen-bonded ring NH...ring N protons of
Watson-Crick base pairs in 11-fold A-RNA helices are essentially
the same as those we have calculated for tRNA using the optimized ring currents and the Haigh and Mallion computations (82).
The major helices in tRNA closely approximate the A-RNA 11-fold
screw pitch. Using the shift tables of Arter and Schmidt (87)
and AU^O and GC^O values for Watson-Crick type interactions of
-14.35 and -13.45 ppm, respectively, Reid *et al.* (38) have been
able to predict class 1 tRNA NMR spectra for a large series of
tRNAs with better agreement than ever before observed. Some of

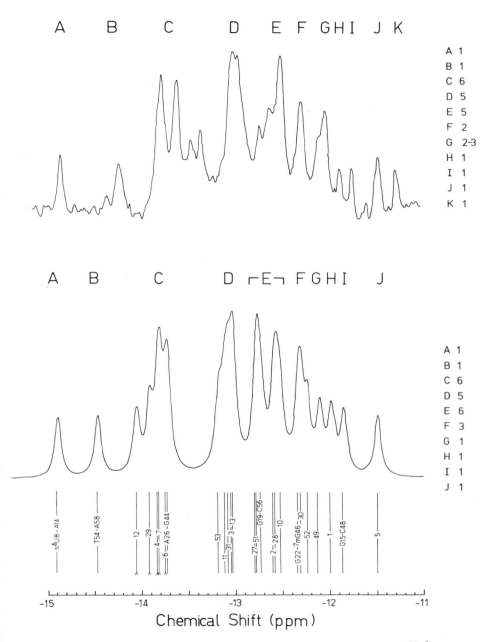

Fig. 18. Upper: 360 MHz spectrum of E. coli tRNA$_1^{Val}$; lower: ring current calculated spectrum using coordinates for E. coli tRNA$_1^{Val}$ and procedures discussed in the text (14).

these spectra and the predicted resonance positions are presented in Fig. 19.   Although the approach involves the uncertainties concerning terminal base pairs and the assumption of
ideal stacking for all bases, as we have already discussed, the
correspondence between predicted and experimental spectra sup-
ports the optimized intrinsic positions and ring currents discussed above.

2. Experimental Assignment of the Yeast tRNA^{Phe} NMR Spectrum by Melting Studies

Hilbers and Shulman proposed that hydrogen-bonded proton
NMR spectra could be assigned by correlating the disappearance
of blocks of resonances at elevated temperatures with the thermally induced melting of specific helices in the cloverleaf
(88).   In the cases where T-jump data were available, the kinetics of a given melting transition could also be defined (89,
90).   The "melting" patterns of the resonances in the low-field
NMR spectrum of yeast tRNA^{Phe} are very sensitive to the mode of
sample preparation, the type and ionic strength of the buffer,
and most importantly, the presence of divalent metal ions.
Thus, depending on the precise conditions employed varying
melting profiles have been obtained (91,92).   In order to verify the correctness of predicted hydrogen bond resonance assignments it is necessary first to define the sequence of thermal
transitions independently of the hydrogen bond resonance spectra.   Methyl groups of the modified bases that generate resonances in the high-field spectral region are very useful for
defining these transitions as shown in the following section.

a. Methyl and methylene resonances.   Smith and co-workers
(1,2) observed broad signals from tRNA in the spectral region
0 to -4 ppm.   On the basis of chemical shifts, they were tentatively attributed to specific methyl and methylene protons.
More positive assignments of the methyl resonances to specific
methylated bases were made on E. coli tRNA^{Tyr} by Koehler and
Schmidt (93), by compiling NMR information on isolated methylated bases and the spectra of an homologous tRNA E. coli
tRNA$^{Tyr}_{su^{+}_{III}}$, lacking certain of the methylated bases.   The investigations of Kan et al. (94) on yeast tRNA^{Phe} used these resonances to monitor structural features.   As can be seen from the
cloverleaf in Fig. 1, this tRNA contains 12 methylated bases
evenly distributed over the DHU, anticodon, and TψC loops and
helices; these aliphatic moieties function as built-in reporter
groups.   Spectral assignments of the high-temperature denatured
species were made using data from the free monomeric nucleosides
and isolated fragments (94).   Subsequent studies between 40 and
90° demonstrated that certain resonances reported specific
structural changes by selectively shifting position, giving a

rough correlation between melting monitored in this manner and optical melting studies similar to that observed in *E. coli* tRNA^{Ala} by Smith *et al.* (2).

In more recent investigations Kan *et al.* (95) employed an incremental procedure to assign the anticodon loop hexanucleotide 34-39 via spectra of the di-, tetra-, and pentanucleotide subunits. We have used the assignments from these two studies (94,95) to monitor the melting of yeast tRNA^{Phe} (31). In contrast to the kinetic "melting" reported by hydrogen bond resonances, the transitions monitored via methyl group resonances are directly comparable to optically detected transitions. Base unstacking, detected optically as increases in absorption, is detected by NMR as changes in chemical shift with the same T_m.

A clear picture of the thermally induced unfolding of yeast tRNA^{Phe} has arisen out of extensive differential melting as well as temperature-jump, optical, and fluorescence measurements (27-29,96). Table III lists the characteristics of these transitions for two solvents. The spectra in Fig. 20 show that the individual methyl resonances are readily resolvable using the resolving power and signal/noise of current superconducting spectrometers. In accordance with the data in Table III, the structure in the presence of Mg^{2+} is very stable up to 65°. Above this temperature it undergoes a sharp transition to a nearly completely denatured state. When Mg^{2+} is removed and Na$^+$ decreased to 30 m$\underline{M}$, the melting pattern changes. The spectrum at 15° in Fig. 21A shows, upon comparison with Fig. 20, that the structure is basically the same in these two solvents. The thermal stability, however, is substantially altered.

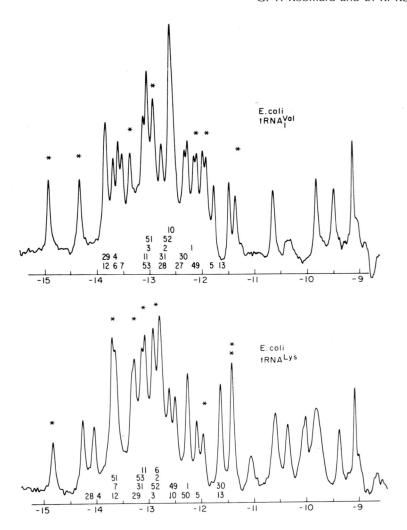

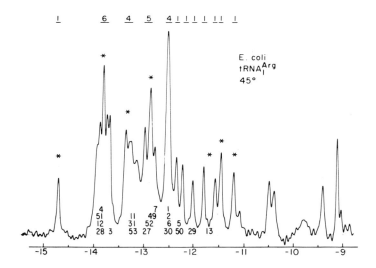

Fig. 19. The 360 MHz NMR spectra of the E. coli species tRNA$_1^{Val}$, tRNA^{Lys}, and tRNA$_1^{Arg}$ together with the predicted positions of the secondary cloverleaf base pair resonances. The nearest-neighbor and next-to-nearest-neighbor upfield shifts were generated from the cloverleaf sequence and the ring current shifts on the ring NH proton in 11-fold helix geometry described by Arter and Schmidt (87); these shifts were added to starting unshifted positions of AU^O = -14.35 ppm and GC^O = -13.45 ppm. The uncertain shift values for base pairs at helix termini (1, 13, 31, and 53) were estimated from the crystal structure stacking diagram of yeast tRNA^{Phe}. Noncloverleaf tertiary resonances are designated by asterisks. The lower two spectra were obtained in the presence of excess magnesium and the top spectrum is in the presence of limiting magnesium ion. The spectra were obtained on 1 mM tRNA samples using fast-sweep correlation for 10-15 minutes of signal-averaging (16).

Table III

Transitions Observed during the Thermal Denaturation of yeast tRNA[Phe]

Solvent	Transition	Structure involved	T_m	Relaxation time	Reference
30 mM Na⁺ pH 6.8	1	Tertiary structure	25	Slow: 10 msec	(27)
		Residual tertiary structure			
	2, 3	Acceptor stem Anticodon helix	35–40	Slow: 2–23 and 17–475 msec	(96)
	4	TψC helix	45–50	Fast: 20–100 μsec	(103)
	5	D helix	60–65	Fast: 20–100 μsec	(103)
2 mM Mg²⁺	1	Complete structure	70		(27)

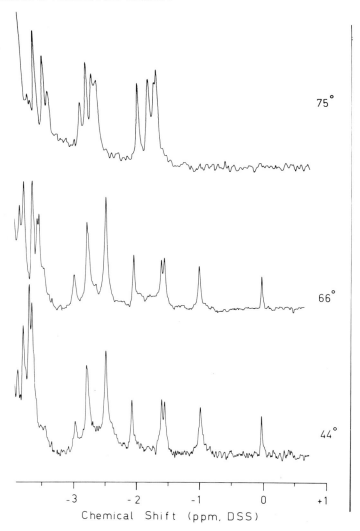

Fig. 20. The 360 MHz spectra of yeast tRNA^Phe in 15 mM MgCl₂, 0.1 M NaCl, pH 7.0, m D₂O. The spectra were obtained using correlation spectroscopy (57).

The transitions observed as a function of temperature in Fig. 21 are as follows. First, between 20 and 40° a large structural change occurs, reflected by decreasing intensities of some resonances and the appearance of others (see the thymine T resonance, for example). Second, between 40 and 60° the thymine experiences a shift to lower field. Third, between 50 and 80° the resonances of the dihydrouridine and m₂²G26 also

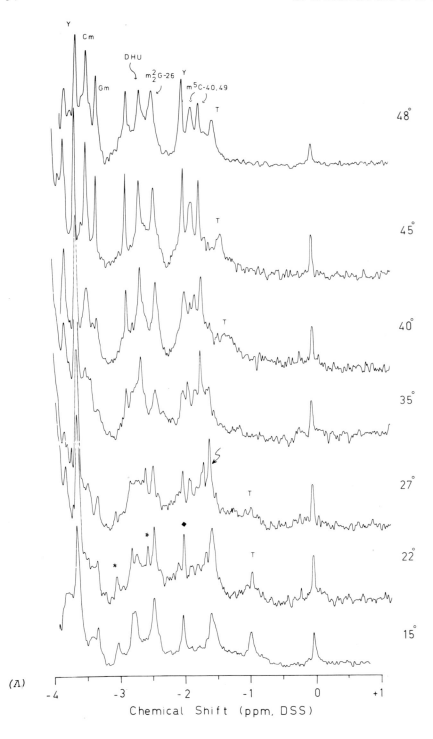

(A)

Chemical Shift (ppm, DSS)

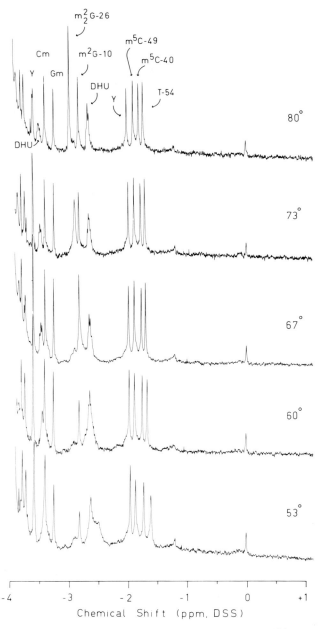

Fig. 21. *The 360 MHz spectra of yeast tRNA^{Phe} in 10 mM sodium phosphate, 20 mM NaCl, pH 7.0, D$_2$O (31). Mg^{2+} was removed by dialysis V.S. EDTA. (A) 15-48°; (B) 53-80°. The assignments in the 80° spectrum are those of Kan et al. (94).*

shift to lower field. The assignments in the 80° spectrum
(Fig. 22) are taken from Kan *et al.* (94). The plot of chemical
shift vs. temperature in Fig. 22 shows that uncertainty in the
assignment of the m_2^2G26 and dihydrouridine resonance at 45° occurs due to the merging of the two resonances at 60°. This ambiguity was removed by irradiating the dihydrouridine C6H multiplet at -3.45 ppm in the 45° spectrum. A sharpening of the
resonances at -2.6 ppm resulted, demonstrating that this is the
DHU C5H resonance. On the basis of only chemical shift data
from the intact tRNA and the 1-45 fragment, Kan *et al.* (94,97)
chose the incorrect assignments at 40°, attributing the -2.47
ppm resonance to dihydrouridine and the -2.6 ppm resonance to
m_2^2G26. In this case, however, the fragment study does not
solve the ambiguity since these two resonances cross over each
other as the temperature is raised in the fragment spectra as
well (97). The optical transitions listed in Table III correlate perfectly with the transitions observed in Figs. 20, 21,
and 22. The D helix is the last to melt, having an optical T_m
of 60 to 65°. By NMR we observe that both the dihydrouridine
methylene resonances, coming from the loop region of the D helix, and the m_2^2G26 methyl resonances at the internal terminus
of the D helix undergo the same transition with a T_m of 60 to
65°. They must be reporting the melting of the D helix. The
transition occurring at the next lowest temperature in Table III
is the melting of the TψC helix with a T_m of 45-50°. This corresponds perfectly with the second transition reported by NMR
as the shifting of the thymine methyl resonance, also with a
T_m of 45-50°. Both high-temperature transitions are fast on
the NMR time scale and are seen as a smooth shifting of resonances.

The remaining low-temperature transitions are slow on the
NMR time scale. As the temperature decreases below 48° the thymine resonance broadens and those of the m^5C residues, the Y
base, the dihydrouridines, and thymine all decrease in intensity while new resonances appear. Although such a transition
causes substantial difficulties in assignments, since the con-
tinuity of a shifting resonance is lost, by following the sequence of transitions from the high-temperature side one can
still define which portions of the molecule undergo rearrangements. Proceeding in such a manner, the spectra between 45 and
15° in Fig. 22 show that both the anticodon helix and tertiary
structure are underoing transitions in this range as reflected

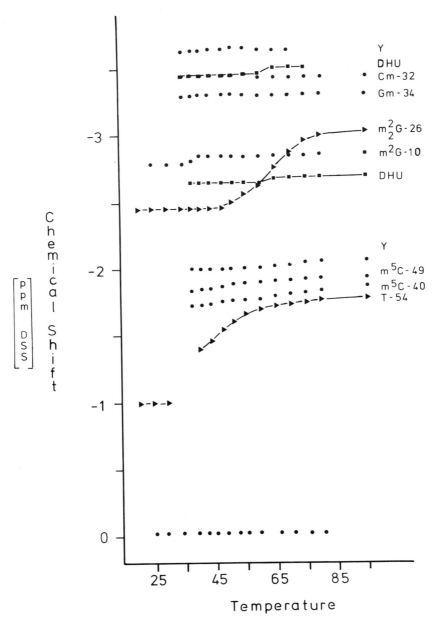

Fig. 22. Plot of the chemical shifts of the resonances in Fig. 21 vs. temperature (31).

by shifts of methyl resonances from bases in the anticodon loop
and helix as well as bases from regions involved in tertiary
structure. The T_m and T-jump data in Table III are in complete
agreement, showing slow transitions for the acceptor and anticodon helix and tertiary structure in this temperature range.

One can readily see, therefore, that when monitored under
similar solvent conditions the aliphatic NMR melting sequence
and the optical T_m values will be the same. Knowing that the
same sequence occurs at NMR concentrations and optical concentrations allows one to predict the sequence of melting and the
T_m values of the individual steps that should be observed when
following the temperature dependence of the structure by hydrogen-bonded proton NMR spectra.

    b. *Hydrogen-bonded proton spectral region.* The procedure
for correlating optical and NMR melting monitored by hydrogen-
bonded proton resonances is subject to a different set of constraints than that used for mapping the same melting followed
by methyl resonances. Since hydrogen-bonded protons exchange
with solvent at a rate dependent on the coil-helix transition
rate, the resonances can broaden and/or disappear at temperatures well below the T_m for base unstacking (89,90). This matter is considered in detail by Hilbers in another chapter of
this volume. Suffice it to say that when the relaxation time
of a transition is longer than 5 msec, meltings as monitored by
hydrogen-bonded proton resonances will be reported as a loss of
resonance intensity without line broadening at the T_m of the
optically observed transition. If, however, the relaxation
time is shorter than 5 msec, the hydrogen-bonded proton resonances will report the transition by broadening and decreasing
in intensity at a temperature below the T_m of the optically observed transition. Thus, for two cases such as pictured in
Fig. 21 the following sequence of events will be observed during
melting in the hydrogen-bonded proton NMR spectra:

    (1) At about 25° the resonances from tertiary structure in-
teractions will melt by decreasing in intensity without broadening.
    (2) The acceptor and anticodon helices will be the next to
melt in the same manner as the tertiary structure resonances
but at ~35°.
    (3) The TψC and D helices will melt by broadening and de-
creasing intensity. The T_m for these events will be 40 and 45°
for the TψC and D helix, respectively.

    To check the accuracy of the calculated ring current shifts
and the resulting assignment predictions in Fig. 17 we now
simply need to compare the resonances that melt in a given tran-

sition with the predicted positions. Analysis of the melting
in Fig. 23 provided the following results (31, 59):

   (1) During the melting of tertiary structure between 13°
and 26°, resonances melt from four of the six regions where tertiary structure resonances were predicted. Note especially the
loss of both resonances below -14 ppm.
   (2) Concomitant melting of the acceptor and anticodon helices is accompanied by the loss of intensity, by 35°, for nine
of the eleven predicted acceptor and anticodon hydrogen bond
resonances.
   (3) TψC helix melting occurs with the loss of four out of
five of the predicted resonances.
   (4) The remaining D helix spectrum contains four resonances,
which are the same as those in the D helix fragment spectrum.
Three of these were correctly predicted. The fourth, we suggest, shifted upon breakdown of tertiary structure, as discussed earlier (see arrows in Fig. 23). When the U8-A14 interaction breaks it appears that A14 rotates back under GC13 and
shifts the GC13 resonance upfield (31). Similar observations
of the appearance of new resonances at intermediate temperatures
have been made in the case of yeast tRNA^{Gly} (90).

   The comparison of the predicted resonance positions with the
resonances lost upon melting show good although not perfect
agreement. Some of the inaccuracies could be due to the ring
currents and calculation procedures, the atomic coordinates used
in the calculations, or the procedure for checking the predicted
positions by monitoring melting. Structural changes that occur
during melting can alter resonance positions. When these altered resonances eventually melt they may not agree with positions predicted from the native structure (31).

IV. LOW-FIELD RESONANCE ASSIGNMENT VIA NUCLEOTIDE FRAGMENTS

   In favorable cases spectra of tRNA fragments may assist in
assigning the spectrum of the intact molecule. This technically
difficult approach was pioneered by Lightfoot et al. (83) to
check assignments and theoretical ring current shifts estimates.
They observed resonances in fragments that generally agreed to
within 0.2 ppm with where they had been predicted by the first
ring current estimates of Shulman et al. (25). This is not too
unexpected since the calibration and alteration of the ring
current values included fragment data. Furthermore, the resonance positions in fragment spectra occurred where resonances
were observed in the intact molecule. Fragment studies can,
however, be complicated if the fragment in question can adopt a
different and more thermodynamically favorable base pairing

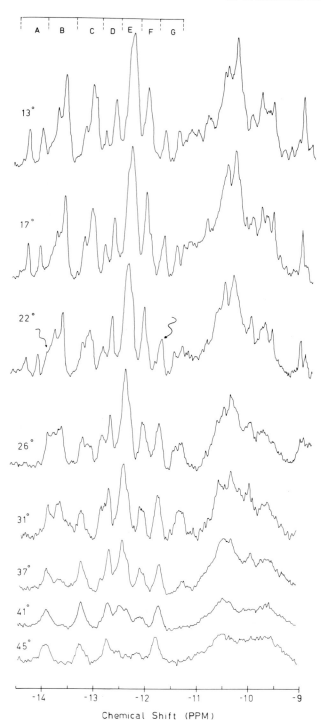

scheme than that in the intact molecule. For these reasons it
is desirable to monitor a series of tRNA fragments containing
a given helical structure to determine if the resonances are
the same in the resulting spectra. In the case of the anticodon helix of yeast tRNA^{Phe}, for example, Lightfoot et al. (83)
observed the same resonances in the spectra of the 21-57 fragment as in the anticodon hairpin (bases 27-45). In a similar
study we observed the same resonances from the 3' half-molecule
of yeast tRNA^{Asp} and the TψC hairpin fragment alone, demonstrating that only the TψC helix is formed in the 3' half-molecule
(98).

The original intention of using fragment spectra to determine the accuracy of ring current predictions has been fulfilled. With current procedures (59,87) predictions usually
come within 0.1 ppm of the observed resonance position. Satis-
fying though this is, it is hardly astonishing. Ring current
shift tables have been computed for idealized A-RNA helices
(59,87). If such helices exist anywhere, it would most likely
be in these isolated fragments where there should be few, if
any, strong tertiary structure forces causing deviations from
ideality. In the future, therefore, it will probably become
unreasonable to go through the trouble of fragment isolation,
purification, and characterization followed by NMR studies just
to assign secondary structure Watson-Crick hydrogen-bonded proton resonances. They can now be predicted with equal accuracy
and much less effort.

NMR studies on tRNA fragments can still provide important
insights into tertiary structure folding in several ways. First,
as in the case of the 47-76 fragment of E. coli tRNA$_1^{Val}$ (Fig. 15)
extra resonances from tertiary structure interactions within a
single loop may be directly observed. Kearns has attempted to
assign the -11.5 ppm resonance in yeast tRNA^{Phe} using NMR spectra of anticodon fragments from E. coli tRNA$_f^{Met}$ and yeast tRNA$_{UUG}^{Leu}$
(7). The spectra of both of these fragments contain a resonance
in the region of -11.7 to -11.3 ppm that cannot be assigned to
any of the secondary structure hydrogen-bonded protons. Since
this resonance is present only at low temperature, they attributed it to the N3H of the constant U on the 5' side of the anticodon (U33), which at low temperature is protected from exchange
with solvent. These data cannot be extrapolated to yeast tRNA^{Phe}
as Kearns has done, for two reasons. First, temperature-de-

Fig. 23. Temperature dependence of yeast tRNA^{Phe} in 30 m$\underline{M}$
Na, no Mg^{2+}, pH 6.8. The spectra are taken at 360 MHz. The
arrows in the 22° spectrum indicate positions of intensity increase during melting (31).

pendent studies on the anticodon fragment containing base pairs
20-57 of yeast tRNA^{Phe} show no resonance between -11 and -12
ppm down to 25° in the presence of 10 mM MgCl$_2$, 0.1 M NaCl.
At lower temperature a resonance shows up only at -11.1 ppm but
it is accompanied by large intensity changes throughout the entire spectrum, indicating that the new resonances can be attributed to aggregation. Second, the protected U33 resonance observed in fragments by Kearns (7) "melts" at low temperature
even in the presence of Mg^{2+} long before the secondary structure
resonances begin melting. However, in the case of intact yeast
tRNA^{Phe} the -11.5 ppm resonance is quite stable up to moderate-
ly high temperatures in the presence of Mg^{2+} (92).

The observation of elements of tertiary structure in fragments, when detected, can provide the most direct method of
tertiary structure resonance assignments. However, if elements
of tertiary structure are missing in fragments, one can observe
terminal secondary resonances at positions different from their
resonance positions in the intact molecule. The alternative approach of melting the intact tRNA is not without assignment ambiguities either, due to the shifting of stable resonances during unfolding. If a resonance at a given position begins to
melt at the same time that another resonance shifts into this
position the net intensity change at this position will be zero
even though a resonance has been lost. This will obviously lead
to erroneous assignments.

V. THE -11 ppm TO -9 ppm SPECTRAL REGION

All tRNA spectra examined to date contain some proton resonances in the region between -9 and -11 ppm. The distribution
of resonances in this region is characteristic for individual
tRNA species and the intensity varies greatly as a function of
temperature or solvent for a given tRNA (see Fig. 4, and compare Fig. 3 with Fig. 8). We have previously suggested (13,66)
that amino protons, downfield shifted by atypical hydrogen
bonding to a ring N, were a likely source of resonances in this
region. Several such amino...ring N hydrogen bonds, e.g., the
A9-A23-U12 triple, have been observed in the crystal structure
of yeast tRNA^{Phe}. The pH-dependent studies of Steinmetz-Kayne
et al. (92) on oligo A and tRNA lend support to this hypothesis.
Other possible candidates for resonances in this neglected region of the NMR spectrum include slowly exchanging non-hydro-
gen-bonded ring NH protons (especially at low temperature), up-
field-shifted ring NH...carbonyl hydrogen bonds, e.g., the 15-
48 reverse Watson-Crick tertiary base pair, and the secondary
GU base pairs found in most tRNAs, which will be discussed in
the next section. Finally, we should mention the surprising

observation of Hurd and Reid (48) that one of the protons in
this region, namely, the sharp resonance invariably observed
between -9.0 and -9.1 ppm, is the nonexchangeable aromatic C8H
proton of m^7G that is strongly deshielded by a delocalized positive charge (see Fig. 12c).

A. GU Base Pairing in tRNA

Careful analysis of the spectra of tRNA helical fragments
containing internal GU base pairs (see Fig. 15) reveals the
presence of more than one resonance in the region between -10
and -12 ppm, which can only be attributed to GU hydrogen bonding. Such GU resonances have also been observed in hairpin
helices derived from 16 S rRNA (68) and strongly indicate that
resonances in this region are to be expected in the spectra of
tRNA species containing GU base pairs. We have previously observed that the intensity in the general region between -10 and
-11.1 ppm is correlated with the GU content of the tRNA (13,
98). The UN3H is inherently 0.7 ppm lower field than the GN1H;
in GU pairs the separation should be even greater since the
UN3H will experience a downfield shift from the in-plane ring
current of G, whereas the GN1H will experience much less of an
effect due to the relatively weak ring current of the in-plane
uridine. The observation of the ring NH protons of both U and
G, and their spectral position, directly indicate that they
form a "wobble" base pair involving two ring NH...carbonyl hydrogen bonds. For reasons described above, we expect the intrinsic U^O position to be found ~1 ppm or more to lower field
than the intrinsic G^O position in a GU pair. Since both of the
GU protons occur in stacked helical sequences in the secondary
structure, they are subject to the same differential upfield
ring current shifts from whichever nucleotides occupy the
nearest-neighbor base pair positions. Unfortunately there are
several uncertainties that currently prevent us from predicting
the two resonances of a GU pair. In order to assume wobble
geometry the bases must move laterally in the GU plane so that
the ring NH can bond to the carbonyl oxygen. We have no way
of knowing whether the G moves to optimize a vertical stack, or
the U moves, or if both bases move. In the case of yeast
tRNA^{Phe} the coordinates of N3H of U69 (expected to be the lower
field of the two GU protons) reveal that it experiences a large
upfield shift from A5 (the base below on the 5' strand). G4,
on the other hand, is not strongly shifted (62). Hence in
yeast tRNA^{Phe} there appear to be no GU protons among the 26
resonances in the -11 to -15 ppm region and we would assign both
ring NH protons of the GU pair to the ~3 proton peak at -10.5
ppm. In E. coli tRNA$_1^{Val}$ the rT helix fragment spectrum reveals
a sharp resonance from GU50 at -11.9 ppm and there is an unas-

signed resonance at this position in the 27 resonance spectrum
of this tRNA. Hence in this case it appears that one of the
two GU protons (U64 N3H) is quite far downfield and is observed
in the -11 to -15 ppm region. Such variations are to be expected since the GU ring NH protons are susceptible to the same
0.2-1.8 ppm range of nearest-neighbor ring current shifts as
AU or GC pairs. Thus the uridine proton of GU4 in yeast
tRNA^{Phe} is stacked with GC above and AU below and resonates at
10.5 ppm, whereas the uridine proton of GU50 in *E. coli* tRNA$^{Val}_1$
is stacked with GC above and CG below and appears to resonate
at -11.9 ppm. We tentatively deduce that the G proton of GU
pairs resonates between -10 and -11 ppm in both cases. Although
some aspects of GU pairing are beginning to emerge from such
studies, a detailed analysis of the resonance positions of GU
pairs will require a much more extensive series of model hairpin
helices in which the GU pair can be observed in a large variety
of possible nearest-neighbor shift environments.

B. Structural Changes and/or Multiple Conformations

It is not improbable, considering the number of catalytic
events in which tRNA participates, that the tRNA undergoes conformational changes that control specificity in these different
reactions. Such possibilities are, therefore, never far from
an investigator's mind when analyzing data. At the present
time, however, there is little if any firm NMR evidence for
functionally significant multiple conformations, or conformational changes.

In their recent study, Kan *et al.* (58) observed that in
MgCl$_2$ the intensity of the thymine 54 methyl protons of yeast
tRNA^{Phe} was only 60% of the expected value, and they suggested
that the other 40% was in a second peak under the resonance at
-1.55 ppm. From these results they concluded that, in the native state, the thymine experiences two different conformations.
Our results on yeast tRNA^{Phe} show clearly that the thymine resonance at -1 ppm has the same intensity as the other three methyl
resonances between -1.5 and -2 ppm (see Fig. 21). We find no
evidence for two conformational states reported by the thymine.

Kastrup and Schmidt (100) have carried out similar NMR studies on methyl resonances of *E. coli* tRNA$^{Val}_1$ at sodium ion concentrations close to 0.25 M. The limited number of methylated
bases in this tRNA produces a far less complicated high-field
spectrum than yeast tRNA^{Phe}. However, at physiological temperature the spectrum shows several sub-integral peaks between -1.2
and -1.9 ppm; this intensity appears at the expense of the lowered thymine intensity at -1.10 ppm. Our own data on this tRNA

(66,101) also reveal these partial resonances, especially in
the presence of Mg^{2+}. Similar partial intensity was observed
around -1.5 ppm by Bolton and Kearns (102). These abnormal intensities are not artifacts of the data collection process.
They are observed both in correctly executed pulse experiments
as well as in spectra collected by rapid sweep correlation procedures. An even more interesting aspect of the foregoing
spectra is that the m^6A methyl resonance at -2.55 ppm was observed to contain less intensity than the DHU methylene resonance at -2.8 ppm, although it should contain 1.5 times more
intensity. Chemical analysis and high-temperature spectra of
the random coil reveal that incomplete methylation of m^6A is
not responsible for this phenomenon and the most likely explanation is the existence of alternative conformations of the
tRNA with different environments for the anticodon m^6A and probably also the thymine residue. The linewidths would indicate
that these putative alternative states of the molecule have
relatively long lifetimes (several milliseconds) and are slowly
interchanging on the NMR time scale.

Lastly, significant binding of the oligonucleotide CGAA to
yeast tRNA^{Phe} was observed when this tRNA was complexed to
$(pU)_8$, a codon analog (104). This was interpreted as evidence
that the codon-anticodon interaction triggers a break between
the D and T loops, resulting in binding of CGAA to the TψCG sequence (104). Geerdes *et al.* (105) monitored both the high-
and low-field proton NMR spectral regions of yeast tRNA^{Phe} in
the presence of the codon UUCA, in an effort to observe the
breaking of the D loop-T loop interactions. No changes were
found that could be attributed to the breaking of these interactions or a change in structure.

VI. HAVE WE LEARNED ANYTHING WE DIDN'T ALREADY KNOW?

Ultimately, a given technique or experimental approach must
be evaluated in terms of what it has taught us about the subject
matter we are studying. Thus we must ask whether NMR studies
have increased our understanding of tRNA structural and functional properties.

A. *Structure*

Even though the x-ray crystallographic structure of yeast
tRNA^{Phe} has been solved in the past few years, it should be re-
membered that NMR had been applied to tRNA and delivered important information concerning the folding of the molecule before determination of the crystal structure. The observation of

Watson-Crick hydrogen-bonded proton resonances in a series of
tRNAs (25,28) supported the prediction of Holley (99) that the
tRNA was probably folded into the secondary structure of a cloverleaf and stabilized by Watson-Crick type hydrogen bonding.
The details of the three-dimensional base pairing in the crystal structure emerged during the 1974-1975 period, but even
without this information, the errors in the integrated low-field
intensity measurements from the previous 4 years would have been
exposed and extra base pair resonances detected from tertiary
structure (13,17,66). This development was largely a matter of
increased resolution, improved magnetic field homogeneity, and
highly purified tRNA. Considering, however, the difficulties
and disagreements relative to the assignment of these tertiary
resonances discussed in previous sections, it is apparent that
these tertiary resonances probably could not have been assigned
without the aid of crystallographic information.

The principal question that NMR has been able to answer
concerns the solution structure of the molecule. It could be
concluded from the NMR spectra of the hydrogen-bonded protons
that the secondary and tertiary structure in solution is similar to the crystal structure simply because we can observe resonances for the protons that are involved in crystallographically
observed hydrogen bonds. It should, however, be stated explicitly that using NMR, like any other single physical technique,
is like looking in a convex mirror. One sees only a little bit
of reality in enormous detail while all else shrinks away on
the periphery. It is likely that, using NMR alone, we could
not detect a different or a second solution structure if that
structure were present at a level of 10% and/or contained the
same interactions found in the crystal structure.

B. Kinetics

The functional significance of conformational states is related to the stability of the individual structural elements
within that state. Thus, the stability of the various loops
and helices in tRNA determine the rigidity or flexibility of a
given conformation. Temperature jump relaxation techniques have
been employed to unravel the kinetics of folding and unfolding
of the individual tRNA helices. However, the identification of
the structural element related to each kinetically resolved
transition is laborious and indirect (27-29). High-resolution
proton NMR has made the task of identification simpler and more
accurate. From a knowledge of the relaxation rates and T_m of a
particular transition one can predict both the "melting" temperature and the characteristics of a given melting transition
to be observed in the NMR spectrum. From the number and chemi-

cal shift of the resonances affected by a given transition one can identify the structural element related to this kinetic event. Thus, NMR has been able to provide a direct link between structural and kinetic studies.

X-ray crystallography has laid a foundation from which we can develop an understanding of the structure-function relationships of tRNA. The dynamics of the molecule, however, are not provided by x-ray; yet they cannot be forgotten. The initial NMR studies comparing solution characteristics with the crystal structure have been limited, necessarily, to a <u>static</u> solution model. The future development of our understanding of the functional aspects of tRNA structure will depend even more heavily on correlations between the dynamic properties of tRNA in solution and the static crystallographic structure. The kinetic and thermodynamic properties of the various structural elements and conformational states will be provided by techniques capable of monitoring them in solution. It can be expected, therefore, that the standard NMR applications as discussed in this review as well as the more esoteric NMR techniques (i.e., nuclear overhauser enhancement and saturation transfer measurements, proton observation of N^{15} enriched nucleic acids) will play a critical role in advancing our knowledge of nucleic acid structural and functional properties.

ACKNOWLEDGMENTS

The 360 MHz NMR spectra shown in this review were obtained on Bruker HXS 360 spectrometers at either the Stanford Magnetic Resonance Laboratory, which is supported by NSF Grant No. GR 23522 and NIH Grant No. RR00711, or the University of Groningen, which is supported by the ZWO (Netherlands Foundation for the Advancement of Pure Research). G.T.R. acknowledges the support of the ZWO during the writing of this review.

B.R.R. gratefully acknowledges the support of the USPHS (NCI Grant No. CA 11697), the American Cancer Society (Grant No. NP 191), and the National Science Foundation (Grant No. PCM 73-01675) for support during this research. The extremely pure *E. coli* tRNA species analyzed during these studies were purified to homogeneity by Susan Ribeiro and Lillian McCollum at University of California, Riverside, and their superb skills are gratefully acknowledged. We thank Ralph Hurd, Ed Azhderian, and Joe Abbate of the Biochemistry Department, University of California, Riverside, who participated in much of the laborious NMR data collection; special thanks are due to Ralph Hurd for helpful comments, stimulating discussion, and permission to use some

of the data from his dissertation research. Samples of yeast
tRNA$_f^{Met}$ were prepared and supplied by G. Ackerman and P. Sigler.

Finally we would like to thank those who provided preprints
of their work before publication.

REFERENCES

1. Smith, I. C. P., Yamane, T., and Shulman, R. G. (1968).
   Science 159, 1360.
2. Smith, I. C. P., Yamane, T., and Shulman, R. G. (1969). *Can.
   J. Biochem. 47*, 480.
3. Kearns, D. R., Patel, D. J., and Shulman, R. G. (1971).
   Nature 229, 338.
4. Kearns, D. R., Patel, D. J., Shulman, R. G., and Yamane, T.
   (1971). *J. Mol. Biol. 61*, 265.
5. Dobson, C. M. (1977). *In* "NMR in Biology" (Dwek, Campbell,
   Richards, and Williams, eds.), pp. 63-93. Academic Press,
   London.
6. Kearns, D. R. (1977). *In* "Annual Review of Biophysics and
   Bioengineering" (L. J. Mullins, W. A. Hagins, L. Stryer,
   and C. Newton, eds.), Vol. 6, pp. 477-503. Annual Reviews,
   Inc., Palo Alto, California.
7. Kearns, D. R. (1976). *Progr. Nucl. Acids Res. Mol. Biol.
   18*, 91-149.
8. Wong, Y. P., Kearns, D. R., Reid, B. R., and Shulman, R. G.
   (1972). *J. Mol. Biol. 72*, 725.
9. Jones, C. R., and Kearns, D. R. (1975). *Biochemistry 14*,
   2660.
10. Jones, C. R., Kearns, D. R., and Muench, K. H. (1976). *J.
    Mol. Biol. 103*, 747.
11. Wong, K. L., Kearns, D. R., Wintermeyer, W., and Zachau,
    H. G. (1975). *Biochim. Biophys. Acta 395*, 1.
12. Wong, K. L., Wong, Y. P., and Kearns, D. R. (1975). *Biopolymers 14*, 749.
13. Reid, B. R., Ribeiro, N. S., Gould, G., Robillard, G. T.,
    Hilbers, C. W., and Shulman, R. G. (1975). *Proc. Nat. Acad.
    Sci. USA 72*, 2049.
14. Robillard, G. T., Tarr, C. E., Vosman, F., and Berendsen,
    H. J. C. (1976). *Nature 262*, 363.
15. Bolton, P. H., Jones, C. R., Bastedo-Lerner, D., Wong, K.
    L., and Kearns, D. R. (1976). *Biochemistry 15*, 4370.
16. Reid, B. R., Ribeiro, N. S., McCollum, L., Abbate, J., and
    Hurd, R. E. (1977). *Biochemistry 16*, 2086.
17. Daniel, W. E., and Cohn, M. (1975). *Proc. Nat. Acad. Sci.
    USA 72*, 2582.
18. Katz, L., and Penman, S. (1966). *J. Mol. Biol. 15*, 220.

19. Crothers, D. M., Hilbers, C. W., and Shulman, R. G. (1973). *Proc. Nat. Acad. Sci. USA 70*, 2899.
20. Patel, D. J., and Tonelli, A. E. (1974). *Proc. Nat. Acad. Sci. USA 71*, 1945.
21. Hilbers, C. W., and Patel, D. J. (1975). *Biochemistry 14*, 2656.
22. Patel, D. J., and Hilbers, C. W. (1975). *Biochemistry 14*, 2651.
23. Arter, D. B., Walker, G. C., Uhlenbeck, O. C., and Schmidt, P. G. (1974). *Biochem. Biophys. Res. Commun. 61*, 1089.
24. Kan, L. S., Borer, P. N., and Ts'o, P. O. P. (1975). *Biochemistry 14*, 4864.
25. Shulman, R. G., Hilbers, C. W., Kearns, D. R., Reid, B. R., and Wong, Y. P. (1973). *J. Mol. Biol. 78*, 57.
26. Rich, A., and RajBhandary (1976). *Ann. Rev. Biochem. 45*, 805.
27. Romer, R., Riesner, D., and Maass, G. (1970). *FEBS Lett. 10*, 352.
28. Romer, R., and Hach, R. (1975). *Eur. J. Biochem. 55*, 271.
29. Urbanke, C., Romer, R., and Maass, G. (1975). *Eur. J. Biochem. 55*, 439.
30. Holbrook, S. R., Sussman, J. L., Warrant, R. W., Church, G. M., and Kim, S. H. (1977). *Nucleic Acid Res. 4*, 2811.
31. Robillard, G. T., Tarr, C. E., Vosman, F., and Reid, B. R. (1977). *Biochemistry 16*,5261.
32. Johnston, P. D., and Redfield, A. G. (1977). *Nucleic Acid Res. 4*, 3599.
33. Bolton, P. H., and Kearns, D. R. (1977). *Biochim. Biophys. Acta 47*, 10.
34. Kim, S. H., Sussman, J. L., Suddath, F. L., Quigley, G. J., McPherson, A., Wang, A. H. J., Seeman, N. C., and Rich, A. (1974). *Proc. Nat. Acad. Sci. USA 71*, 4970.
35. Klug, A., Ladner, J., and Robertus, J. D. (1974). *J. Mol. Biol. 89*, 511.
36. Bolton, P. H., and Kearns, D. R. (1975). *Nature 255*, 347.
37. Romer, R., and Varadi, V. (1977). *Proc. Nat. Acad. Sci. USA 74*, 1561.
38. Reid, B. R., McCollum, L. Ribeiro, N. S., Abbate, J., and Hurd, R. E. (1978). *Biochemistry* (in press).
39. Redfield, A. G., Kunz, S. D., and Ralph, E. K. (1975). *J. Mag. Resonance 19*, 114.
40. Redfield, A. G. (1976). *In* "NMR: Basic Principles and Progress" (Diehl, P., Fluck, E., and Kosfeld, R., eds.), Vol. 13, p. 137. Springer, Göttingen.
41. Walker, R. T., and RajBhandary, U. L. (1972). *J. Biol. Chem. 247*, 4879.
42. Wong, K. L., Bolton, P. H., and Kearns, D. R. (1975). *Biochim. Biophys. Acta 383*, 446.
43. Hurd, R. E., Robillard, G. T., and Reid, B. R. (1977).

Biochemistry 16, 2095.

44. Hurd, R. E., and Reid, B. R. (1978). *Biochemistry* (in press).
45. Lawley, P. D. (1966). *Progr. Nucleic Acid Res. Mol. Biol. 5,* 89.
46. Wintermeyer, W., and Zachau, H. (1970). *FEBS Lett. 11,* 160.
47. Simsek, M., Petrissant, G., and RajBhandary, U. L. (1973). *Proc. Nat. Acad. Sci. USA 70,* 2600.
48. Hurd, R. E., and Reid, B. R. (1978). *Biochemistry* (in press).
49. Kyogoku, Y., Inubushi, T., Morishima, I., Watanabe, K., Oshima, T., and Nishimura, S. (1977). *Nucleic Acid Res. 4,* 585.
50. Salemink, P. J. M., Yamane, T., and Hilbers, C. W. (1977). *Nucleic Acid Res. 4,* 3727.
51. Daniel, W. E., and Cohn, M. (1976). *Biochemistry 15,* 3917.
52. Cantoni, G. L., and Davies, D. R. (1971). *In* "Procedures in Nucleic Acid Research," Vol. 2, pp. 524-564. Harper & Row, New York.
53. Ogasawara, N., Watanabee, Y., and Inoue, Y. (1975). *J. Am. Chem. Soc. 97,* 6571.
54. Lawley, P., and Brookes, P. (1963). *Biochem. J. 89,* 127.
55. Geerdes, H. A. M., and Hilbers, C. W. (1977). *Nucleic Acid Res. 4,* 207.
56. Kan, L. S., and Ts'o, P. O. P. (1977). *Nucleic Acid Res. 4,* 1633.
57. Robillard, G. T., Tarr, C. E., Vosman, F., and Sussman, J. L. (1977). *Biophys. Chem. 6,* 291.
58. Reid, B. R., and Hurd, R. (1977). *Acc. Chem. Res.* (in press).
59. Robillard, G. T. (1977). *In* "NMR in Biology: (R. A. Dwek, I. D. Campbell, R. E. Richards, and R. J. P. Williams, eds.), pp. 201-230. Academic Press, London.
60. Barrell, B. G., and Clark, B. F. C. (1974). "Handbook of Nucleic Acid Sequences." Joynson-Bruvvers Ltd.
61. Abbate, J., and Reid, B. R. Personal communication.
62. Sussman, J. L., and Kim, S. H. (1975). *Biochem. Biophys. Res. Commun. 68,* 89.
63. Sussman, J. L., and Kim, S. H. Personal communication.
64. Reid, B. R., and Hurd, R. E. (1978). *Biochemistry* (in press).
65. Ladner, J. E., Jack, A., Robertus, J. D., Brown, R. S., Rhodes, D., Clark, B. F. C., and Klug, A. (1975). *Proc. Nat. Acad. Sci. USA 72,* 4414.
66. Reid, B. R., and Robillard, G. T. (1975). *Nature 257,* 287.
67. Azhderian, E., and Reid, B. R. Personal communication.
68. Baan, R. A., Hilbers, C. W., Van Charldorp, R., Van Leerdam, E., Van Knippenberg, P. H., and Bosch, L. (1977). *Proc. Nat. Acad. Sci. USA 74,* 1028.

69. Shoup, R. R., Miles, H. T., and Becker, E. D. (1966). *Biochem. Biophys. Res. Commun. 23,* 194.
70. Newmark, R. A., and Cantor, C. R. (1968). *J. Am. Chem. Soc. 90,* 5010.
71. Hurd, R. E., and Reid, B. R. (1977). *Nucleic Acid Res. 4,* 2747.
72. Katz, L. (1969). *J. Mol. Biol. 44,* 279.
73. Giessner-Prettre, C., and Pullman, B. (1965). *CRHS Acad. Sci. Paris 261,* 2521.
74. Giessner-Prettre, C., Pullman, B., Borer, P. N., Kan, L. S., and Ts'o, P. O. P. (1976). *Biopolymers 15,* 2277.
75. Arnott, S., and Hukins, D. W. L. (1972). *Biochem. Biophys. Res. Commun. 47,* 1504.
76. Arnott, S., Hukins, D. W. L. (1972). *Biochem. Biophys. Res. Commun. 48,* 1392.
77. Arnott, S. (1971). *Progr. Biophys. Mol. Biol. 22,* 181.
78. Giessner-Prettre, C., and Pullman, B. (1970). *J. Theor. Biol. 27,* 87.
79. Kearns, D. R., and Shulman, R. G. (1974). *Acc. Chem. Res. 7,* 33.
80. Bloomfield, V. A., Crothers, D. M., and Tinoco, I. (1974). *In* "Physical Chemistry of Nucleic Acids," pp. 322-335. Harper & Row, New York.
81. Kim, S. H., Suddath, F. L., Quigley, G. J., McPherson, A., Sussman, J. S., Wang, A. H. J., Seeman, N. C., and Rich, A. (1974). *Science 185,* 435.
82. Haigh, C. W., and Mallion, R. B. (1971). *Mol. Phys. 22,* 955.
83. Lightfoot, D. R., Wong, K. L., Kearns, D. R., Reid, B. R., and Shulman, R. G. (1973). *J. Mol. Biol. 78,* 71.
84. Quigley, G. J., Seeman, N. C., Wong, A. H., Suddath, F. L., and Rich, A. (1975). *Nucleic Acid Res. 2,* 2329.
85. Stout, C. D., Mizuno, H., Rubin, J., Brennan, T., Rao, S. T., and Sundaralingam, M. (1976). *Nucleic Acid Res. 3,* 1111.
86. Klug, A. Personal communication.
87. Arter, D. B., and Schmidt, P. G. (1976). *Nucleic Acid Res. 3,* 1437.
88. Hilbers, C. W., and Shulman, R. G. (1974). *Proc. Nat. Acad. Sci. USA 71,* 3239.
89. Crothers, D. M., Cole, P. E., Hilbers, C. W., and Shulman, R. G. (1974). *J. Mol. Biol. 87,* 63.
90. Hilbers, C. W., Robillard, G. T., Shulman, R. G., Blake, R. D., Webb, P. K., Fresco, R., and Riesner, D. (1976). *Biochemistry 15,* 1874.
91. Hilbers, C. W., Shulman, R. G., and Kim, S. H. (1973). *Biochem. Biophys. Res. Commun. 55,* 953.
92. Steinmetz-Kayne, M., Benigno, R., and Kallenbach, N. R. (1977). *Biochemistry 16,* 2064.
93. Koehler, K. M., and Schmidt, P. G. (1973). *Biochem. Biophys.*

Res. Commun. 50, 370.

94. Kan, L. S., Ts'o, P. O. P., V.d. Haar, F., Sprinzl, M., and Cramer, F. (1974). *Biochem. Biophys. Res. Commun. 59,* 22.

95. Kan, L. S., Ts'o, P. O. P., V.d. Haar, F., Sprinzl, M., and Cramer, F. (1975). *Biochemistry 14,* 3278.

96. Coutts, S. M., Riesner, D., Romer, R., Rabl, C. R., and Maass, G. (1975). *Biophys. Chem. 3,* 275.

97. Kan, L. S., Ts'o, P. O. P., Sprinzl, M., V.d. Haar, F., and Cramer, F. (1977). *Biochemistry 16,* 3143.

98. Robillard, G. T., Hilbers, C. W., Reid, B. R., Gangloff, J., Dirheimer, G., and Shulman, R. G. (1976). *Biochemistry 15,* 1883.

99. Holley, R. W., Apgar, J., Everett, G. A., Madison, J. T., Marguisee, N., Merrill, S. H., Penswick, J. R., and Amir, A. (1966). *Science 147,* 1462.

100. Kastrup, R. V., and Schmidt, P. G. (1975). *Biochemistry 14,* 3612.

101. Reid, B. R. (1977). *In* "Nucleic Acid Protein Recognition" (H. J. Vogel, ed.), pp. 375-389. Academic Press, New York.

102. Bolton, P. H., and Kearns, D. R. (1976). *Nature 262,* 423.

103. Romer, R., Riesner, D., Maass, G., Wintermeyer, W., Thiebe, R., and Zachau, H. G. (1969). *FEBS Lett. 5,* 15.

104. Schwarz, U., Menzel, H. M., and Gassen, H. G. (1976). *Biochemistry 15,* 2484.

105. Geerdes, H. A. N., van Boom, J. H., and Hilbers, C. W. (1978). *FEBS Lett. 88,* 27.

NUCLEAR MAGNETIC RESONANCE STUDIES
OF DRUG-NUCLEIC ACID COMPLEXES

Thomas R. Krugh
Merrill E. Nuss

Department of Chemistry
University of Rochester
Rochester, New York

I. INTRODUCTION

The interaction of drugs with nucleic acids has been studied by a wide range of spectroscopic techniques. In this chapter we shall concentrate on the application of magnetic resonance techniques to elucidate the structure and dynamics of drug-nucleic acid complexes. However, even this topic is too broad to review thoroughly in this chapter. We have therefore selected to outline the principles and provide examples of the proper use and interpretation of nuclear magnetic resonance data, with special emphasis on the nuclear magnetic resonance studies of drugs that intercalate into DNA. The word in-tercalation comes from the Greek and literally means to "insert between." Thus an intercalating drug is one in which the planar portion of the drug molecule is inserted between the adjacent base pairs of a double-stranded region of a DNA or RNA helix, although the word intercalation is also used for situations in which an aromatic molecule is inserted between adjacent bases on a single-stranded nucleic acid. The intercalation mode of binding was first proposed by Lerman (1961) for the binding of the aminoacridines to DNA. The physiological activity of intercalating drugs is generally attributed to their ability either to produce mutations or to interfere with DNA or RNA polymerase. Some of the most promising anticancer drugs are intercalators. There is also a wide range of compounds in which only a portion of the molecule may intercalate (for example, the intercalation of aromatic amino acids is probably one of the modes involved in protein-nucleic acid recognition). It is therefore important that we obtain a detailed molecular, kinetic, and thermodynamic understanding of the intercalation process. A number of drugs bind to the outside of DNA or RNA; not only can magnetic resonance discriminate between outside binding and intercalative binding, but it may also be used to provide details on the nature of this type of binding.

Although this chapter will concentrate on magnetic resonance results, we shall also discuss selected experiments using fluorescence, circular dichroism, and visible absorption spectroscopies, because we wish to illustrate the advantages of combining the optical spectroscopic data with magnetic resonance data. In brief, not only do the optical spectroscopies provide an independent means of obtaining kinetic, thermodynamic, and equilibrium data, but they also provide an important correlation between the results obtained for the binding of the drugs to oligonucleotides (i.e., model systems) and polynucleotides (the systems of biological interest). X-ray crystallographic studies on drug-oligonucleotide complexes have provided detailed stereochemical information on selected complexes, and we shall compare and contrast the information available from the single crystal x-ray diffraction studies and the high resolution NMR studies of the solution complexes.

The drugs that will be used as primary examples are actinomycin D, ethidium bromide, daunorubicin (or adriamycin), and 9-aminoacridine; the chemical structures of these molecules are shown in Figs. 1 and 13. Actinomycin D has been extensively studied over the last 20 years, and as we shall show later in this chapter, the interaction of actinomycin D with deoxynucleotides is still providing valuable information toward the goal of obtaining a molecular and thermodynamic basis for the binding of the drug to DNA. There are a number of reviews on the history, synthesis, binding studies, and description of models for the binding of the actinomycins to DNA (e.g., see Meienhofer and Atherton, 1977; Hollstein, 1974; Lackner, 1975; Sobell, 1973; Wells, 1971; and the many references therein). Likewise, the readers are referred to the reviews by Henry (1976) and Arcamone *et al.* (1972) for background references on daunorubicin and adriamycin. Reinhardt and Krugh (1977, 1978) and Le Pecq (1971) provide background references on ethidium bromide. It has been well documented that the primary mode of binding of actinomycin D and ethidium bromide is by intercalation of the phenoxazone and the phenanthridinium rings, respectively, into DNA. It is generally accepted that the planar portion of the daunorubicin molecule intercalates into DNA.

Both nuclear magnetic resonance and electron spin resonance have been used to study the complexes of intercalating drugs with nucleic acids. Since ^{1}H NMR has been the most extensively utilized, it has contributed the most to our knowledge of drug-nucleic acid complexes to date. However, electron spin reso-

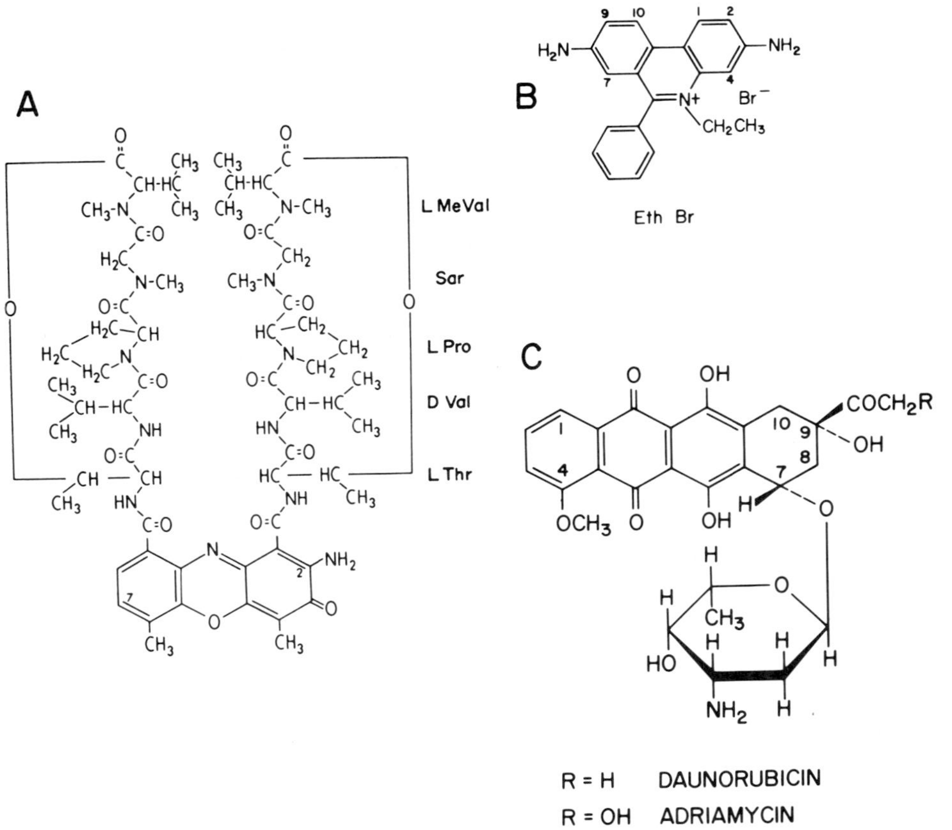

Fig. 1. The chemical structure of (A) actinomycin D; (B) ethidium bromide; and (C) daunorubicin (also frequently called daunomycin) and adriamycin. Reproduced with permission from Krugh and Young (1977).

nance experiments, using either paramagnetic metal ions as probes or spin labeled drug or nucleic acid molecules, have received increasing attention over the last few years (e.g., see Dugas, 1977; Piette, 1974; Reuben and Gabbay, 1975; Hong and Piette, 1976; Sinha et al., 1976; Chiao and Krugh, 1977; Reuben et al., 1976).

II. MAGNETIC RESONANCE PARAMETERS

The most useful parameters in nuclear magnetic resonance studies in solution are the chemical shifts, the indirect spin-spin coupling constants, the lineshape of the resonance (in those cases where chemical exchange phenomena are observed), and the relaxation times (especially the spin-lattice relaxation time, T_1). Each of these parameters may provide molecular or thermodynamic information on the drug-nucleic acid complexes as discussed below.

III. CHEMICAL SHIFTS

The chemical shifts of the drugs, the nucleic acids, and the drug-nucleic acid complexes are the parameters most frequently used to obtain information on the structure of the complexes. First, of course, one must assign the resonances of the drug and the nucleic acids used. This can be a challenging task in itself, especially for complicated molecules like acti-nomycin D (Fig. 1). The reader is referred to the reviews cited above and the articles cited below for a discussion of the methods used for the assignment of the resonances.

The change in the chemical shifts of the drug and nucleic acid resonances that are observed when they form a complex (e.g., as illustrated by the equilibrium expression below) is the important experimental parameter.

$$\text{drug} + \text{nucleic acid} \underset{\longleftarrow}{\overset{K}{\longrightarrow}} \text{drug:nucleic acid complex}$$

The portion of the drug molecule that intercalates is sandwiched between the aromatic base pairs, and therefore the chemical shifts of the drug resonances are influenced by the ring current effects (e.g., see the texts by Emsley et al., 1965; Swek, 1973; Ts'o, 1974a,b; James, 1975 for a discussion of ring current effects). A great simplification in the interpretation of the NMR spectra of drug-nucleic acid complexes is obtained if one observes large ring current shifts which arise from the formation of an intercalated complex because these large ring current shift will almost certainly account for all or most of the observed changes in the chemical shifts of the drug resonances. The chemical shifts of the nucleotide resonances will also be affecte as a result of the intercalation of the planar chromophore. To familiarize the reader with these concepts, we present in Fig. 2 the isoshielding contours for the guanine base that were recently published by Giessner-Prettre and Pullman (1976); however, the reader is also referred to

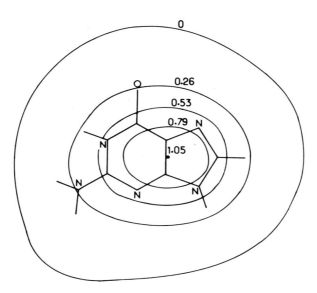

*Fig. 2.   Intermolecular shielding values (Δδ in ppm) due
to the sum of the contributions of the ring current and of the
atomic diamagnetic susceptibility anisotropies in guanine
(in a plane 3.4 Å distant from the molecular surface).   Reproduced with permission from Giessner-Prettre and Pullman (1976).*

Giessner-Prettre et al. (1976) for a more generalized presentation of the effects of the ring currents of the nucleic acid
bases on the chemical shifts of nuclei in their vicinity.   The
isoshielding contours in Fig. 2 show that drug nuclei that are
in a plane 3.4 Å above or below the plane of the guanine base
will experience a ring-current-induced shift equal to the
values given in the figure.   With the measurement of several
ring current shifts one can construct models for the geometry
of the drug-nucleic acid complex as will be illustrated in a
later section.   It is important to use as much data as possible
when interpreting the changes in the chemical shifts observed
upon complex formation for the following reasons.   The isoshielding contours are based upon theoretical calculations that
necessarily contain certain simplifying assumptions.   The shapes
of the isoshielding contours of the nucleic acid bases have been
refined over the past few years, but the general agreement between the experimental data and the refined theoretical calculations has provided a firm basis for this type of approach (e.g.,
see chapter by Hilbers; as well as Kearns, 1977; Kallenbach and
Berman, 1977; Giessner-Prettre et al., 1976; Borer et al.,
1975a).   There are several factors (in addition to the question

of the accuracy of the theoretical calculations) that should
also be considered when trying to extract detailed geometrical
information from the chemical shift changes ($\Delta\delta$) that accompany either drug-nucleic acid complex formation, or the helix
to coil transition of nucleic acids (in either the presence or
absence of added drugs).

A. Solvent Effects

Solvent effects are also referred to as solvent exclusion
effects or solvent transfer effects. When a drug molecule
forms an intercalated complex with a nucleic acid, the protons
are transferred from an aqueous solvent to the hydrophobic environment of the nucleic acid bases: This effective solvent
change can influence the chemical shifts of the drug nuclei.
In fact, Arison and Hoogsteen (1970) and Angerman et al. (1972)
took advantage of the solvent effects exhibited by certain actinomycin D resonances as an aid in the complete assignment of
the proton NMR spectrum of the drug. It is difficult to assess
the exact magnitude of this effect for the various protons of
the drugs under the situation of interest (i.e., the formation
of a drug-nucleic acid complex) but they may be on the order of
0.1 ppm (or possibly less). Of course, the same effects may
complicate the interpretation of the changes in the chemical
shifts of the nucleotide resonances during the helix to coil
transition.

B. Hydrogen Bonding Effects

The formation of hydrogen bonds between the drug and the
nucleic acid bases or the sugar phosphate backbone may result
in a perturbation of the electronic structures of the drug or
the nucleotide bases, which may in turn result in additional
changes in the chemical shifts of the resonances from the
bases. This effect is not expected to be very important for
exocyclic hydrogen bonds (such as with the amino groups of
ethidium bromide (Figure 1) or the nucleotide bases (e.g., the
2-amino-group of guanine) but the formation or disruption of
hydrogen bonds involving donor or acceptor atoms in the aromatic
bases of either the nucleic acid or the drug may lead to changes
in their chemical shifts and/or ring currents. Borer et al.,
(1975) have estimated an upper limit of 0.1 ppm for this effect
for the helix to coil transition of the oligonucleotides.

C. *Conformational Averaging*

In solution studies, the complex (or molecule) of interest
may exist in a family of energetically allowed conformations,
and it is very important to remember that there is a nonlinear
relationship between conformational parameters and chemical
shifts. This is analogous to the more familiar problems encountered in the interpretation of lanthanide induced shifts
or paramagnetic induced nuclear relaxation (e.g., see La Mar
et al., 1973; Krugh, 1976). One could imagine unusual situations in which the conformational averaging would lead to unreasonable "average" geometries. Performing the experiments
at different temperatures and/or ionic strengths would help to
sort out this complication, unless the different conformations
were nearly energetically equivalent.

D. *Other Effects*

The influence of local diamagnetic anisotropy effects, polarization effects, and the introduction of a positively
charged chromophore (such as ethidium bromide) may also influence the induced chemical shifts ($\Delta\delta$) and/or the ring currents
of the nucleotide bases.

The reader is also referred to an excellent discussion in
Borer *et al.* (1975) for a more detailed presentation of the
complicating factors in the interpretation of chemical shift
data for the determination of the structure of oligonucleotide
double helices. Fortunately, the ring current shift is generally
much larger than the other effects discussed above. In addition,
we reiterate that, in several test cases, the agreement between
the induced shifts calculated on the basis of ring currents and
the experimentally observed values provides a firm basis for the
use of this approach.

An example of the possible complications that may be encountered wherein the observed shifts are not completely determined by ring current shifts can be illustrated from the
literature on the magnetic resonance spectra of actinomycin D
and the complexes of actinomycin D with mono-, di-, and oligonucleotides. The 100 MHz ^{1}H NMR spectra of actinomycin D, a
solution of actinomycin D plus dGMP (in a 1:2 ratio), and a solution of actinomycin D with GMP (1:3 molar ratio) are shown in
Fig. 3. Note that the H7 and H8 resonances of actinomycin D
have approximately the same chemical shift (at 100 MHz) and
thus only a single resonance is observed. When 5'-dGMP is added, the two protons experience the same changes in their chemical shifts and thus they continue to exhibit a single reso-

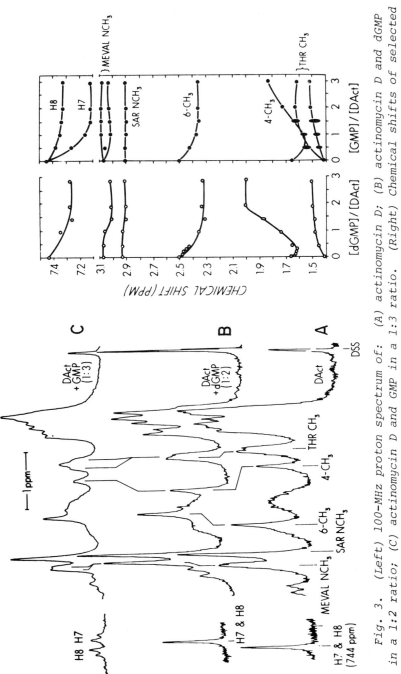

Fig. 3. (Left) 100-MHz proton spectrum of: (A) actinomycin D; (B) actinomycin D and dGMP in a 1:2 ratio; (C) actinomycin D and GMP in a 1:3 ratio. (Right) Chemical shifts of selected actinomycin D groups plotted as a function of the nucleotide/drug ratio for deoxyguanosine 5'-monophosphate and guanosine 5'-monophosphate. Reproduced with permission from Krugh and Neely (1973a).

nance. In the actinomycin D complex with GMP, the H7 and H8 resonances of the phenoxazone ring exhibit a well-resolved AB pattern (the doublet of doublets in the top spectrum shown in Fig. 3). The chemical shifts were monitored as a function of the nucleotide to drug ratio for the addition of both dGMP and GMP and the ^{1}H NMR "titration" curves are plotted on the right of Fig. 3. If we interpret these data only in terms of ring current shifts, then we would incorrectly conclude that the stacking of the guanine base on the phenoxazone ring of actinomycin D was markedly different in the dGMP and GMP complexes. However, Krugh and Neely (1973a) and Krugh and Chen (1975) used several lines of reasoning to conclude that the local diamagnetic anisotropy of the 2'-OH group in GMP is responsible for the differential shifts observed for the H7 and H8 protons in the actinomycin D complex with GMP, and that the orientation of the guanine base in the vicinity of the 6-CH$_3$, the H7, and the H8 protons is similar (but not identical) in both of these complexes.

IV. EFFECTS OF THE SELF-AGGREGATION OF DRUGS ON THE CHEMICAL
    SHIFTS

It is important to remember that both the drugs and the nucleotides may form aggregates in the millimolar concentration range generally used in the proton magnetic resonance studies of drug-nucleic acid complexes. For example, ^{1}H NMR spectroscopy has been used for many years to study the aggregation of nucleic acid bases, nucleotides, and oligonucleotides (e.g., see the texts by Ts'o, 1974a,b) as well as the aggregation of planar aromatic chromophores such as acridine orange (e.g., Blears and Danyluk, 1967). In the millimolar concentration range actinomycin D aggregates to form dimers, and even higher degrees of aggregation may occur above ~5 mM (e.g., see Müller and Emme, 1965; Crothers et al., 1968; Angerman et al., 1972; Auer et al., 1977). The concentration dependence of the ^{1}H NMR chemical shifts of actinomycin D provides a good probe for studying the aggregation of actinomycin D because it also provides the opportunity of determining the geometry of the dimer. Angerman et al. (1972) showed that actinomycin D aggregation can be satisfactorily represented by a dimerization equi-

librium with a temperature-dependent equilibrium constant in
the range of $1 \times 10^3 \ M^{-1}$ to $3 \times 10^3 \ M^{-1}$. Moreover, the chemical shift data showed that the dimer is formed by an interaction
between the phenoxazone rings of actinomycin. From the direction and the relative magnitudes of the chemical shift trends
of the phenoxazone ring protons (e.g., see the data in Fig. 4),
it was concluded that the predominant form of the dimer is one
in which one chromophore is inverted with respect to the other
(Angerman et al., 1972). These results were verified by Krugh
and Neely (1973a) and Krugh and Chen (1975). An interesting
aspect of these studies is the observation that the calculated
values of the chemical shifts of the actinomycin D dimer are
temperature dependent, which implies that the structure of the
dimer is temperature dependent. This observation has been
recently confirmed from an analysis of the concentration and
temperature dependence of the circular dichroism spectrum of
actinomycin D (Auer et al., 1977).

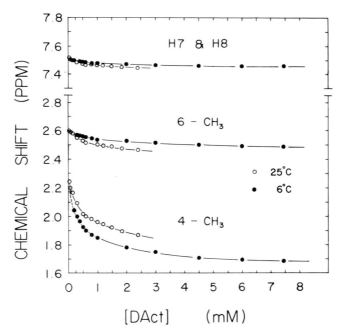

Fig. 4. Concentration dependence of important groups of
actinomycin D in D_2O-5 mM phosphate buffer (pD 7.4) at 6 and
$25^{\circ}C$ (no salt). The threonine methyl, sarcosine N-methyl, and
methylvaline N-methyl resonances do not change much during the
dilution and are not shown for clarity. Reproduced with permission from Krugh and Chen (1975).

As a result of the self-aggregation of drug molecules, the observed changes in the chemical shifts in the presence of nucleotides will be a sum of the chemical shift changes that result from complex formation and the chemical shift change that accompanies the disruption of the actinomycin D dimers. Krugh and Chen (1975) illustrated this point by comparing the data from an experiment in which an increasing amoung of dGMP was added to different concentrations of actinomycin D (12 and 0.29 mM), as shown in Fig. 5. The 0.29 mM actinomycin D spectra were recorded at 25°C because at this temperature the actinomycin D resonances have a much narrower linewidth at 25°C than at 6°C. The narrower linewidths provided the sensitivity required for the submillimolar experiments. The 4-CH$_3$ resonance in Fig. 5 shows the most dramatic change as a function of the initial drug concentration, as anticipated from the large shielding of this resonance when a dimer is formed. Both experiments give rise to nearly identical final limiting chemical shifts for the actinomycin D resonances in the presence of excess 5'-dGMP. The difference between these values and the infinite dilution chemical shifts are used to calculate the "induced chemical shifts" ($\Delta\delta$), which result from complex formation of an actinomycin D monomer with (two) nucleotides. In the first published report on the actinomycin D complex with dGMP (only a 1:1 ratio was studied), Danyluk and Victor (1970) did not consider the influence of the actinomycin D dimerization on the complexation shifts and thus proposed an incorrect geometry for the complex. Arison and Hoogsteen (1970) avoided the problem of actinomycin D aggregation in aqueous solutions by the addition of 2 mole-% dimethylformamide and they correctly interpreted their ^{1}H NMR data on a 2:1 5'-dGMP:actinomycin D solution in terms of a complex in which the guanine bases were located above and below the phenoxazone chromophore of actinomycin D. It is an interesting scientific anecdote that this important observation (i.e., the formation of a 2:1 nucleotide:drug complex) appears only in the abstract of the paper, because all referees insisted that Arison and Hoogsteen's interpretation of their data in terms of the formation of a 2:1 complex was not warranted (B.H. Arison, personal communication).

V.   CHEMICAL SHIFTS IN CARBON-13 SPECTRA

It has been well-established that ^{13}C chemical shifts are much more sensitive to solvent, electronic, and geometrical influences than are proton shifts (e.g., see Stothers, 1972; Levy and Nelson, 1972). In addition, the inherently lower sensitivi-

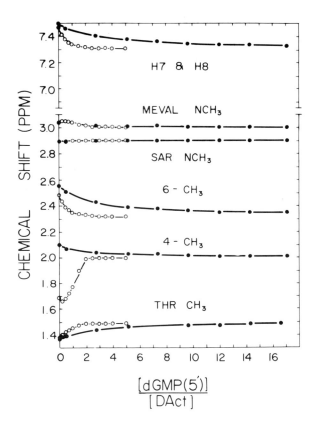

Fig. 5. Chemical shifts of selected protons of Actinomy-cin D plotted as a function of the nucleotide/drug ratio for dGMP (5') at 6°C, 12 mM actinomycin D (O) and at 25°C, 0.29 mM actinomycin D (●). Reproduced with permission from Krugh and Chen (1975).

ty in ^{13}C spectroscopy makes it much more difficult to deter-mine the infinite dilution chemical shifts that are needed to evaluate the induced chemical shifts ($\Delta\delta$ values) to be used in the ring current analysis. An example of the problems that may be encountered and the caution required in the interpretation of the ^{13}C $\Delta\delta$ values comes from a comparison of the ^{13}C NMR studies of the 2:1 dGMP:actinomycin D complex by Patel (1974a) and the subsequent reinvestigation and extension of the data by Krugh and Chen (1975). Patel (1974a) used a two-point ex-trapolation to estimate the infinite dilution chemical shifts of 5'-dGMP that were then used to calculate the induced shift ($\Delta\delta$) values of the ^{13}C resonances of guanine for complex for-mation with actinomycin D. Some selected $\Delta\delta$ values that were reported by Patel (1974a) are: $\Delta\delta$(C6) = +1.16 ppm; $\Delta\delta$(C5) = +0.21 ppm; $\Delta\delta$(C2) = -0.10 ppm; $\Delta\delta$(C4') = -1.21 ppm.   Patel focused on the unusually large value for the C6 carbon and suggested the formation of the complexes shown in Fig. 6B as additional complexes to those shown in Fig. 6A.   The existence of a significant fraction of the complexes in the geometry illustrated by the complex on the right of Fig. 6B would result in a $\Delta\delta$ value in the range of 1.0 to 1.5 ppm for the C6 resonance of dGMP; however, this same complex should give rise to a $\Delta\delta$ value in the range of 0.7 to 1.0 ppm for the C5 carbon resonance of dGMP, in contrast to the observed value ($\Delta\delta$(C5) = +0.21 ppm) (Patel, 1974a).   In addition, none of the complexes in either Fig. 6A or Fig. 6B are consistent with the $\Delta\delta$(C2) = -0.10 value reported by Patel (1974a).   Krugh and Chen (1975) reinvestigated the concentration dependence of the ^{13}C resonances of dGMP and used a nine point graphical extrapolation to obtain more reliable estimates of the infinite dilution

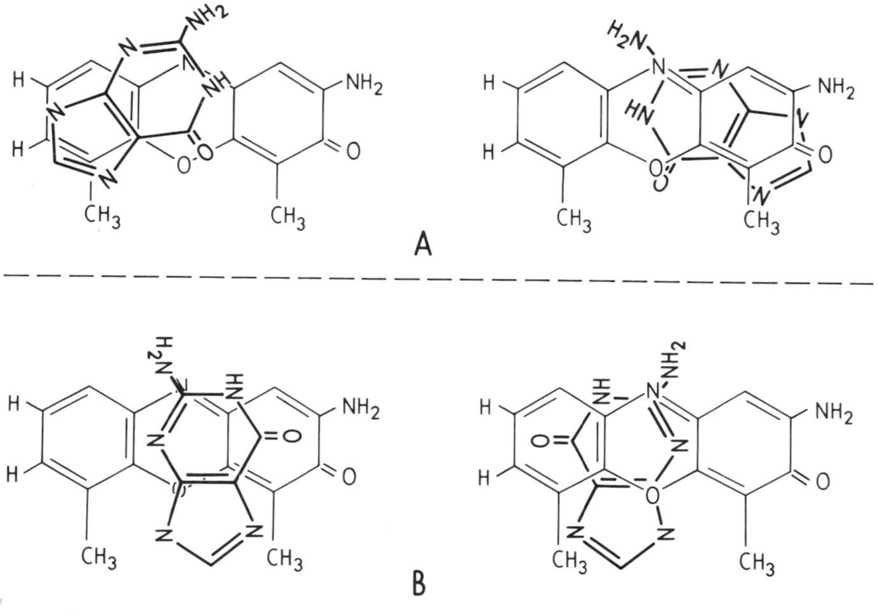

Fig. 6.   (A) Stacking patterns observed in the crystal structure of actinomycin D and deoxyguanosine (Jain and Sobell, 1972).   It should be noted that these structures represent projections of the guanine ring onto the phenoxazone ring and thus the apparent overlap of the rings will depend upon the projection plane.   (B) The alternative geometries of the complex as proposed by Patel (1974a).   Reproduced with permission from Krugh and Chen (1975).

shifts. Using these data, Krugh and Chen (1975) calculated revised values for the induced chemical shifts of the dGMP carbons upon complex formation with actinomycin D ($\Delta\delta$(C6) = +1.40 ppm; $\Delta\delta$(C5) = +0.34 ppm; $\Delta\delta$(C2) = +0.08 ppm; $\Delta\delta$(C4') = -1.01 ppm). These revised values make the ^{13}C NMR results even more difficult to rationalize using only ring current arguments. In fact, there is no combination of the four complexes shown in Fig. 6 that is consistent with the large $\Delta\delta$ value for the C6 resonance, while at the same time predicting only relatively small values for the C2 and C5 carbon resonances. It should be noted that the ^{13}C data must also be reconciled with the proton NMR data on both the same complex (Arison and Hoogsteen, 1970; Krugh and Neely, 1973a; Patel, 1974a) as well as related complexes (Krugh and Neely, 1973b) Patel, 1974b). In summary, the ^{13}C chemical shifts of the guanine resonances in the 2:1 dGMP:actinomycin D complex present an internally inconsistent pattern when an analysis is attempted using only ring current effects, and the interpretation of the data as evidence for the formation of the complexes shown in Fig. 6B is not supported by either the proton magnetic resonance data or the optical titration data (Krugh and Chen, 1975). We conclude that other factors (*vide supra*) are influencing the ^{13}C chemical shifts and we advise that caution must be exercised when interpreting changes in the ^{13}C chemical shifts in terms of the geometry of the complexes formed.

VI. CHEMICAL SHIFTS IN ^{31}P SPECTRA OF DRUG-NUCLEIC ACID COMPLEXES

^{31}P NMR has been extensively utilized to study biological systems both *in vivo* and *in vitro* during the past few years (e.g., see chapters by Ugurbil *et al.*, 1979 and Gzdico *et al.*, 1979; and the references below). The assignment of the phosphorus resonances of nucleoside mono-, di-, and triphosphates is relatively straightforward. The chemical shifts and coupling constants for a number of mononucleotides and dinucleotides have been reported (e.g., Cozzone and Jardetzky, 1976a,b; Reinhardt and Krugh, 1977). However, the assignment of the phosphorus resonances in oligonucleotides ($n \geq 3$) has not been reported to date. The chemical shift of the terminal phosphate group in each nucleotide is sensitive to the pH of the solution in the pH = 5 to pH = 8 range, which provides a convenient probe for measuring the pH of intact cells (e.g., see chapters by Ugurbil *et al.*, 1979 and Gzdico *et al.*, 1979), and simultaneously, an unreliable probe for monitoring drug-nucleic acid complex formation.

The interesting questions to be addressed with ^{31}P NMR
studies of nucleic acids concern the formation of the sugar-
phosphate backbone and the changes in the conformation of the
backbone that accompany intercalation, double-strand to single-
strand transitions, etc. The ^{31}P NMR spectra of the actinomycin D complex with pdGpdC was first reported by Patel (1974b)
where the ^{31}P resonance of the internucleotide phosphate (i.e.,
the phosphodiester group in pdGpdC) shifted downfield 1.7 ppm
upon complex formation. In an accompanying paper, Patel
(1974c) showed that when actinomycin D was intercalated into
the double-stranded form of the self-complementary hexanucleotide d-ApTpGpCpApT, three ^{31}P resonances exhibited substantial
downfield complexation shifts (Fig. 7). The two resonances
that are furthest downfield were assigned to the phosphorus
nuclei at the intercalation site and Patel (1974c) suggested
that the third downfield shifted resonance could arise from
weak surface binding of actinomycin D to the hexanucleotide
double helix, but we prefer the assignment of this resonance
to a phosphorus at a site adjacent to the intercalation site.
These initial results and the previous ^{31}P NMR spectra of tRNA
(Guéron, 1971) suggested that ^{31}P NMR would be a sensitive
probe of nucleic acid conformational changes. These data did
not provide any quantitative information on the conformational
changes of the nucleic acid that result from intercalation because there is no well documented detailed correlation between
the ^{31}P chemical shifts of nucleic acids and the geometry of
the phosphodiester backbone of the double helix [although Go-

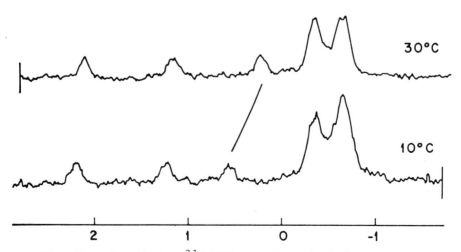

Fig. 7.   The 40-MHz ^{31}P NMR spectra of >1:2 Act-D:
d-ApTpGpCpApT in 0.1 M NaCl-D_2O and 0.01 M EDTA (pH 7) at 30
and 10°.   Reproduced with permission from Patel (1974c).

renstein (1975) and Gorenstein and Kar (1975) have reported
CNDO calculations and Gorenstein *et al.* (1976) and Patel and
Canuel (1976a) and Patel (1976a,b, 1977a) have measured the
temperature dependence of the ^{31}P NMR chemical shifts of dinucleotides and polynucleotides in an initial attempt to provide
the necessary correlation]. For example, Patel (1976a, 1977a)
has concluded that the downfield chemical shifts that are observed for the ^{31}P NMR resonances of the oligonucleotides during the helix to coil transition (e.g., see Figs. 8 and 9) result from changes in the torsional angles (ω, ω') on conversion
from stacked double helical strands (in a *gauche*, *gauche* conformation) to unstacked strands in the coil state (in a *gauche*,
trans conformation). This correlation, or conclusion, is based
upon previous experimental and theoretical studies that suggested that the helix and coil states differ primarily in the
phosphodiester torsional angles (e.g., Sundaralingam, 1973;
Tewari *et al.*, 1974; Olson, 1975; Yathindra and Sundaralingam,
1974; Calascibetta *et al.*, 1975) as well as the experimental
observation that the phosphorus resonances move downfield during the helix to coil transition (e.g., Guéron and Shulman,
1975). However, an unusual feature that is evident in the temperature dependence of the ^{31}P NMR chemical shifts of the C-
and G-self-complementary deoxytetranucleotides (Figs. 8 and 9)

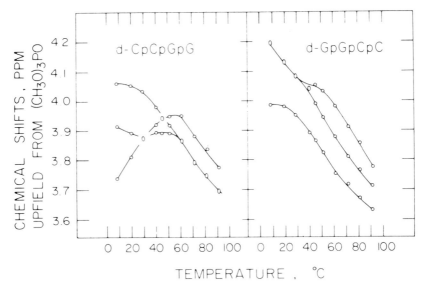

*Fig. 8. The temperature dependence (5-95°C) of the three
backbone phosphate chemical shifts of 20 mM (strand concentration)d-CpCpGpG and d-GpGpCpC in 0.1 M phosphate, D_2O, pH 6.75.
Reproduced with permission from Patel (1977a).*

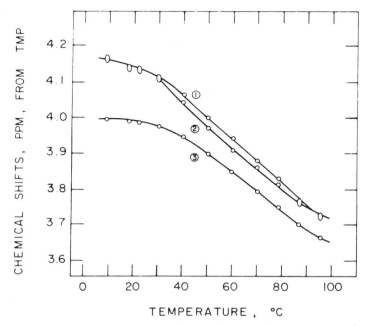

Fig. 9. The temperature dependence of the three internu-cleotide phosphates of d-CpGpCpG in 0.1 M cacodylate, 0.01 M EDTA, D$_2$O, pH 6.5. Reproduced with permission from Patel (1976a).

is that one does not usually observe the characteristic sigmoidal melting curves that might be anticipated by analogy to the chemical shift versus temperature profiles for the nonexchangeable base protons (and the UV absorbance data).

As a further example of the use of ^{31}P NMR spectroscopy in the study of drug-nucleic acid complexes, we present in Fig. 10 the ^{31}P NMR spectra of d-(CpGpCpG) and a 1:2 complex of actinomycin D with this deoxytetranucleotide (Patel, 1976a). It is interesting to note that there are two downfield shifted resonances, which are assigned to the phosphorus nuclei at the intercalation site, and that the remaining four phosphates have only slightly different chemical shifts (demonstrating the asymmetry of the actinomycin D complex with nucleic acids which results primarily from the asymmetry of the phenoxazone ring of the drug (e.g., see Patel, 1976a; Reinhardt and Krugh, 1977; and references therein).

After the initial results on the ^{31}P NMR studies of actinomycin D complexes with pdGpdC (Patel, 1974b) and with d-ApTpGpCpApT (Patel, 1974c) the authors' laboratory initiated a program to determine the effect of other intercalating drugs on

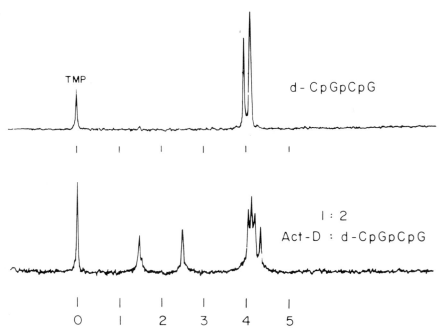

Fig. 10. The 145.7-MHz Fourier transform ^{31}P NMR spectrum (proton noise decoupled at 360 MHz) of d-CpGpCpG (top) and 1:2 Act-D:d-CpGpCpG (bottom) in 0.1 M cacodylate, 0.01 M EDTA, D_2O, pH 6.5 at 30°C. Reproduced with permission from Patel (1976a).

the ^{31}P chemical shifts, and to determine the influence of the nucleotide sequence at the actinomycin D intercalation site on the ^{31}P chemical shifts. Reinhardt and Krugh (1977) showed that neither ethidium bromide nor 9-aminoacridine intercalation complexes with dinucleotides exhibited the large (>1 ppm) down-field shifted ^{31}P resonances that were observed with actinomycin D. Optical and proton magnetic resonance studies on the solutions used for the ^{31}P NMR experiments verified that these model systems were forming intercalated complexes. Reinhardt and Krugh (1977) also showed that when complementary mixtures of deoxydinucleotides were used (i.e., pdGpdT + pdCpdA and pdGpdG + pdCpdC) the largest downfield shifted resonance ob-served (~0.75 ppm) was from the internucleotide phosphorus of pdApdC. These data provide several interesting suggestions concerning the asymmetry of the complex at the intercalation site, which need to be explored (e.g., by studying complexes of the actinomycins with complementary mixtures of deoxyoligo-nucleotides).

It seems clear that when actinomycin D intercalates at a (dGpdC)·(dGpdC) sequence, the ^{31}P resonances of the phosphorus-31 nuclei located at the intercalation site move substantially downfield ($\Delta\delta \cong$ -1.7 and -2.5 ppm), but these large changes in the chemical shifts are not characteristic of all intercalating drugs. We also note that the proton NMR spectra of actinomycin D complexes with the block copolymer d($C_{15}A_{15}$)·d($T_{15}G_{15}$) led Kearns and co-workers (Early *et al.*, 1977) to conclude that, while the binding of the drug to the GC portion of the double-stranded oligonucleotide helix may perturb the conformation of A·T base pairs located near the AT-GC junction, there is no effect on the conformation of the majority of the AT base pairs. However, Krugh and co-workers (1978) have observed what appear to be longer range effects in the duponal induced dissociation of actinomycin from actinomycin/poly (dG-dC)·poly (dG-dC) complexes, where the dissociation time constants were found to be a function of the nucleotide/actinomycin ratio (at the start of the experiment). More experimental and theoretical calculations on the correlation of phosphorus-31 chemical shifts with the geometry of the phosphodiester backbone of nucleic acids are required before phosphorus-31 spectroscopy will realize its full potential in the study of the conformation of nucleic acids and drug-nucleic acid complexes.

An anomalous chemical shift versus temperature profile is observed for one of the ^{31}P resonances of d-CpCpGpG as this tetranucleotide goes from the double-stranded form to the single-stranded form (the left panel of Fig. 8). Kastrup *et al.* (1978) have measured the circular dichroism (CD) spectra of the four C and G containing self-complementary deoxytetra-nucleotides and compared these to the corresponding CD spectra calculated from a nearest neighbor analysis. The spectra in Fig. 11 show good agreement between the experimental and calculated CD spectra for all of the tetranucleotide sequences except d-pCpCpGpG, which suggests that this tetranucleotide may exist in an unusual conformation. The unusual temperature profile of the ^{31}P chemical shifts of this tetranucleotide is consistent with this suggestion. Patel (1977a) investigated the proton NMR spectra of d-CpCpGpG and d-GpGpCpC (in the same paper in which the ^{31}P NMR spectra were reported) and observed several discrepancies between calculated (for B form DNA) and experimental chemical shifts as well as the chemical shift changes for the helix to coil transition. The assignment of the phosphorus resonances and a more detailed study of the proton spectra should provide insight into the interesting effect of the sequence of the deoxytetranucleotides on their conformation in the double helical states.

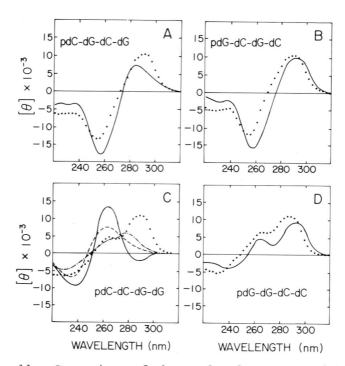

Fig. 11. Comparison of observed and nearest neighbor approximation CD spectra of the four self-complementary C-G tetranucleotides. All spectra were recorded in a 0.1 M NaCl, 0.1 mM
Na EDTA, 5 mM phosphate solution, pH 7.0, 0°C, unless otherwise
indicated. The ... curves represent the nearest neighbor approximation curves for each of the deoxytetranucleotides: (A),
_____44 µM pdC-dG-dC-dG; (B), _____48 µM pdG-dC-dG-dC;
(C) _____56 µM pdC-dC-dG-dG in 5 mM phosphate, pH 7.0, 2°C;
---5.6 µM pdC-dC-dG-dG in·5 mM phosphate, pH 7.0, 0°C; -·-
56 µM pdC-dC-dG-dG in 5 mM phosphate, pH 7.0, 41°C. A spectrum
of 33 µM pdC-dC-dG-dG in a 0.1 M NaCl, 0.1 mM EDTA, 5 mM phosphate solution, pH 7.0, 0°C, had a similar shape, but larger
θ values than the spectrum shown above (———) that was recorded in a solution that did not contain the NaCl. (D),
____45 µM pdG-dG-dC-dC. Reproduced with permission from Kas-
trup et al. (1978).

VII. INDIRECT SPIN-SPIN COUPLING CONSTANT ANALYSIS

In principle, the measurement and analysis of the proton-proton coupling constants of the deoxyribose protons, as well as the proton-phosphorus and the phosphorus-carbon coupling constants will provide a great deal of stereochemical information on the structure of the oligonucleotides and the drug-oligonucleotide complexes. In practice, this is difficult information to extract even from proton spectra recorded at 360 MHz, because of severe overlap of the complicated resonance patterns from the many sugar residues. In addition, there may be substantial linebroadening present in the spectra due to slow rotation of the complexes. For example, an actinomycin D complex with a double-stranded deoxyhexanucleotide (one drug to one double helix) has a molecular weight in the vicinity of 5000, and consequently appreciable dipolar linebroadening is anticipated, especially at low temperatures. Linebroadening may also result from chemical exchange effects. However, Lee and Tinoco (1978) and Patel and Shen (1978) have recently obtained useful information from an analysis of the coupling constants of the ribo and deoxyribo sugar protons of drug-nucleic acid complexes (as will be discussed later). A complete analysis of the proton-proton coupling constants of a drug-oligonucleotide complex may require the synthesis of specifically deuterated oligonucleotides (e.g., see Kondo and Danyluk, 1972; Ezra et al., 1977; and references therein). This approach would overcome some of the difficulties encountered and allow for the determination of the important stereochemical changes that accompany drug-nucleic acid complex formation. The reader is referred to the texts by Ts'o (1974a,b) and the recent articles (and references therein) by Lee and Sarma (1976a,b), Altona and Sundaralingam (1973), Borer et al. (1975), and Ezra et al. (1977) for more specific details on the interpretation of the coupling constants of the ribose and deoxyribose sugar protons in terms of the conformation of the molecule.

VIII. NUCLEAR RELAXATION EXPERIMENTS

The addition of paramagnetic metal ions or the use of spin-labeled drugs or nucleotides can provide valuable information on the structure of macromolecules through the measurement of the paramagnetic induced nuclear relaxation. The interaction between the unpaired electrons and the nuclei are usually dominated by a dipole-dipole interaction, which has an r^{-6} dependence. The increase in the linewidths of the resonances and the simultaneous decrease in the spin lattice relaxation times, T_1, thus provide a sensitive ruler to determine relative distances between the paramagnetic center and various nuclei as was first shown by Shulman and Sternlicht (1965) and Shulman et al. (1965). A number of workers have studied the binding of paramagnetic metal ions to nucleoside mono-, di-, and triphosphates as well as DNA (e.g., see Cohn and Hughes, 1962); Shulman et al., 1965; Reed et al., 1971; Anderson et al., 1971; Reuben and Gabbay, 1975; Chiao and Krugh, 1977; and the many references therein). The reviews by Krugh (1976), Dwek (1973), Swift (1973), and Mildvan and Cohn (1970) provide the necessary background to the theory, control experiments, and illustrative examples of the use of paramagnetic induced nuclear relaxation. Since there have only been limited applications of this approach in the study of drug-nucleic acid complexes, the background material will not be discussed in the present chapter. A recent example of this approach is the work of Chiao and Krugh (1977) in which they exploited the preferential binding of manganese (II) ions to the 5'-terminal phosphate groups of mono- and dinucleotides to measure the preferential binding of adenine and guanine mono- and dinucleotides to the two nucleotide binding sites of actinomycin D. The change in the linewidth (which is proportional to the paramagnetic induced nuclear relaxation) of the 4-CH$_3$, the 6-CH$_3$, and the H(7) and H(8) protons of the phenoxazone ring of actinomycin D (Fig. 1) were measured as Mn(II) was added to various solutions of actinomycin D in which only one of the nucleotides had a 5' (terminal) phosphate group. The preferential broadenings of the actinomycin D resonances were consistent with the interpretation of the chemical shift data for similar systems (Krugh et al., 1977).

We discussed above that ^{13}C chemical shifts of drug-nucleic acid complexes are not likely to be very useful in providing structural information on the nature of the complex. However, the use of a paramagnetic probe bound selectively to the terminal phosphate group of a nucleotide (Chiao and Krugh, 1977) and the measurement of the paramagnetic induced nuclear relaxation of the ^{13}C resonances of the drug may be an ideal way to obtain quantitative structural information. Since the chromophores of the drugs always contain a number of carbon nuclei, it should be possible to measure a number of relative distances and thus determine a detailed molecular structure. It will be important to determine if the selectivity of the binding of Mn(II) ions to the terminal phosphate group (which carries a -2 charge at neutral pH) is maintained in the oligonucleotide systems, or alternatively stated, the location of the Mn(II) binding site must be known. Of course, the same approach can be used with other nuclei (e.g., ^{1}H NMR), but a number of important drugs, such as daunorubicin or adriamycin (Fig. 1) have chromophores with only a limited number of protons. For these drugs the pa-

ramagnetic induced nuclear relaxation experiments may be the
only way of providing detailed structural information. Moehle
et al. (unpublished data) have used fluorescence, optical absorption, and proton NMR spectroscopies to study the complexes
of daunorubicin with deoxy di- and tetranucleotides. The magnitudes of the induced shifts of the H 1, H 2, H 3, and the
methoxy protons on the chromophore were relatively small (<0.3
ppm), which allows one to conclude that these protons are <u>not</u>
located between the G·C or A·T base pairs. This observation
is consistent with several previous model building studies.
Because of the possible complications in the use of ring current analyses, it is generally risky to propose unique structural complexes (without obtaining corroborating data to support the model) for complexes which exhibit induced shifts of
this magnitude.

Reuben *et al.* (1976) have also used Mn(II) induced ^{1}H nuclear relaxation experiments to determine the orientation of
the 3,8-dimethyl-*N*-methyl-phenanthrolinium cation when it is
intercalated into DNA. The addition of DNA to a solution of
the 3,8-dimethyl-*N*-methyl-phenanthrolinium had a negligible effect on the chemical shifts and spin lattice relaxation rates.
The subsequent addition of Mn(II) ion resulted in a marked difference in the paramagnetic induced nuclear relaxation of the
various protons (as much as a 100-fold difference) from which
Reuben *et al.* (1976) concluded that in the intercalation complex the long axis of the 3,8-dimethyl-*N*-methyl-phenanthroli-
nium cation is almost perpendicular to the hydrogen bonds of
the DNA base pairs, in agreement with previous physical and
spectral data for the same system (Gabbay *et al.* 1973b). Although this is an interesting approach, we comment that it will
not be a generally useful approach to the study of drug-DNA
complexes because the intercalation of compounds into DNA results in a severe broadening of the resonances (e.g., see
Gabbay *et al.*, 1973b). Reuben *et al.* (1976) circumvented this
problem by using a solution that contained a large excess of
unbound cation (0.05% bound). The spectral parameters were a
weighted average of the parameters for the bound and free
phenanthrolinium cations, since the mean residence time in the
bound state (<3×10^{-5}sec) is much smaller than the relaxation
rates (and the $\Delta\delta$ values) for the various protons of the DNA
bound molecule. Unfortunately, many important drugs dissociate
relatively slowly from DNA, which vitiates the usefulness of
paramagnetic induced nuclear relaxation experiments with DNA.
Actinomycin D is an extreme example of a slowly dissociating
drug in that the dissociation time constants range from

10 to 1500 sec (Müller and Crothers, 1968). Another complicating problem is that the large excess of unbound cations or
drugs may result in aggregation of the positively charged molecules along the outside of the helix near the negatively charged
phosphate groups (a well-known phenomenon for other positively
charged intercalating chromophores such as the acridines or
ethidium bromide). If the outside bound molecules contributed
to the observed magnitude of the paramagnetic induced nuclear
relaxation, then neglecting this contribution may lead to errors in the interpretation of the data. However, paramagnetic
induced nuclear relaxation experiments can be a valuable tool
in the study of drug-nucleic acid complexes, especially in the
use of oligonucleotide model systems.

IX. KINETIC PARAMETERS

    Nuclear magnetic resonance is an excellent technique for
providing detailed information on the kinetics of certain systems, especially the exchange rate between different chemical
environments. Monitoring the lineshape of the resonances as
a function of the temperature of the solution is a standard
approach for obtaining kinetic and thermodynamic information.
This approach has been particularly useful in studying the
helix-to-coil transition of double-stranded oligonucleotides.
Proton magnetic resonance studies have provided documentary
evidence for the "fraying" that occurs at the ends of the
oligonucleotide duplexes, as well as the "breathing" of the
double helix. The reader is referred to the chapter by
Hilbers, as well as the articles (and references therein) by
Crothers et al. (1973), Hilbers and Patel (1975), Kan et al.
(1975), and Patel (1978) for the details of the use of magnetic resonance experiments to elucidate the structure and dynamics of oligonucleotide duplexes.
    Krugh and Neely (1973b) observed that the H(7) and H(8)
resonances of actinomycin D (Fig. 1) exhibited slow chemical
exchange effects in the titration of actinomycin D with the
self-complementary deoxydinucleotide pdG-dC at 3°C. None of
the other deoxydinucleotide complexes with actinomycin D exhibited slow chemical exchange effects in the proton NMR spectra. Patel (1974b, 1976b) investigated the temperature dependence of the ^{31}P and ^{1}H magnetic resonance spectra of actinomycin D complexes with pdG-dC and calculated the chemical exchange
lifetimes by assuming a model for the exchange process. Davanloo and Crothers (1976) used temperature jump relaxation measurements to study the kinetics of actinomycin and ethidium bromide binding to the deoxydinucleotides. Third-order kinetics

were observed for the binding of actinomycin to pdG-dC, which
is consistent with the highly cooperative binding of this dinucleotide to actinomycin as was previously reported (Krugh,
1972; Schara and Müller, 1972). The nuclear magnetic resonance spectra of actinomycin D bound to tetra- or hexanucleotide duplexes (e.g., Patel, 1974c, 1976a) exhibit slow chemical
exchange, on the NMR time scale, in the temperature range in
which a stable duplex is formed. Although the magnetic resonance experiments have provided kinetic information which has
been useful in understanding the binding of drugs to DNA, the
more direct approaches such as temperature jump relaxation
measurements, stopped flow studies, or dissociation experiments
have provided more direct kinetic information. However, there
are certain areas in which magnetic resonance data may be extremely valuable. One example is to use magnetic resonance
studies on a properly selected series of oligonucleotides to
try and determine the molecular parameters or structural constraints responsible for the extremely slow dissociation of
actinomycin D from DNA, since this slow dissociation has been
associated with the biological activity of the actinomycins.

X. A BRIEF HISTORICAL INTRODUCTION TO THE NMR STUDIES OF
   DRUG-NUCLEIC ACID COMPLEXES

    Before outlining the detailed studies on actinomycin D,
ethidium bromide, and 9-aminoacridine it is appropriate to provide a brief background and selected review of the many papers
that have used nuclear magnetic resonance techniques in the
study of drug-nucleic acid complexes, or more generally, small
molecule-nucleic acid complex formation. In 1966 Chan et al.
published the first in a series of papers in which proton magnetic resonance was used to show that purine formed intercalated
complexes with a series of (single-stranded) dinucleoside monophosphates (e.g., Chan et al., 1966; Chan and Kreishman, 1970).
In 1971, Kreishman et al. published a detailed study in which
they used proton magnetic resonance to show that in aqueous solutions ethidium bromide self-aggregates by the stacking of the
phenanthridinium rings. Ethidium also formed mutually stacked
aggregates with several uracil residues. An analysis of the
chemical shift changes and preferential line broadening of the
H(1) and H(10) proton resonances that were observed upon complex formation with UpU provided evidence that the phenanthridinium ring of ethidium was sandwiched between the adjacent
uracil bases with the H(1) and H(10) protons directed towards
the ribose-phosphate backbone of the dinucleoside monophosphate.
The proton magnetic resonance data of ethidium complexes with

poly U were less conclusive than the ethidium-dinucleotide
data, but the preferential broadening of the H(1) and H(10)
proton resonances again indicated the intercalation of ethidium between adjacent uracil bases of poly U. It is noteworthy that the Kreishman *et al.* (1971) paper is the first report
of the use of nuclear magnetic resonance to study drug-nucleic
acid complexes at the dinucleoside level. Several other groups
had used proton magnetic resonance to study a variety of complexes formed between aromatic chromophores with mononucleo-
sides or mononucleotides, and obtained the general result that
stacked aggregates are formed. However, except for actinomycin D, these studies have not been definitive in providing a
molecular interpretation of the interaction of the drugs with
double-stranded DNA or RNA because the mononucleotides cannot
form a miniature double helical complex. In this context it
is interesting to note that the proposed structure of the ethi-
dium:UPU complex (Kreishman *et al.*, 1971) has the long axis of
ethidium oriented approximately perpendicular to the direction
that the bases would have if they were in a double helix.
Kreishman *et al.* (1971) noted that it should not be surprising
to find different complexes formed with single-stranded nucleic
acids when compared to double-stranded nucleic acids. Krugh
and Reinhardt (1975) studied the proton magnetic resonance,
circular dichroism, and visible absorption spectra of complexes
of ethidium with UpU, ApA, and a mixture of the complementary
dinucleoside monophosphates UpU + ApA. These experiments
showed that in the solution of the complementary dinucleoside
monophosphates the predominant complex formed is one in which
ethidium was intercalated between the two A·U base pairs in the
Apa:UpU:ethidium complex. Thus a comparison of the results of
Kreishman *et al.* (1971) and Krugh and Reinhardt (1975) illustrates the importance of using complementary oligonucleotides
as models for the binding of the drugs to double-stranded DNA
or RNA. Davanloo and Crothers (1976) performed temperature
jump relaxation experiments on a series of actinomycin D and
thidium bromide complexes with dinucleotides and observed that
all reactions of ethidium in which a second dinucleotide is
added to the rapidly formed 1:1 dinucleotide:drug complex to
give a 2:1 dinucleotide:drug complex are limited by a first-
order step at high concentration. The different geometry
proposed by Kreishman *et al.* (1971) for the complex of
ethidium with single-stranded UpU (as compared to the geometry
of ethidium intercalated into a double-stranded complex) was
cited by Davanloo and Crothers (1976) as a possible interpretation to the limiting first-order step. A rearrangement of the
complex in the single-stranded form to an intermediate capable
of accepting a second dinucleotide to complete the miniature

double helix would account for the slow limiting kinetic step.
We cite this series of experiments as another example of the
interaction between various spectroscopic studies. It would
be interesting to determine the geometry of an ethidium complex with several noncomplementary deoxy- and ribotetranucleo-
tides to further clarify the interesting results discussed
above.

We have already mentioned a few studies in which nuclear
magnetic resonance has been used to study the conformation or
the self-aggregation of the drugs and the nucleic acids.
These approaches are still being used, as illustrated, for
example, by the recent paper by Delbarre *et al*. (1976) in
which the aggregation of the intercalating drug ellipticine
was studied by proton magnetic resonance and compared to the
aggregation of actinomycin D, acridine orange, and 9-hydroxy
elliptocine. From the structure of the inverted dimer the
authors concluded that the orientation of the quadrupole axis
of the chromophore is an important aspect in the intermolecular
association of the intercalating drugs. Another recent example
is the study of Cheung *et al*. (1978) on the conformation of
echinomycin in solution. Echinomycin is an antibiotic isolated
from *streptomyces echinatus* (Corbaz *et al*., 1957) and was the
first reported bifunctional intercalating molecule (Waring
and Wakelin, 1974; Wakelin and Waring, 1976). A number of
compounds have recently been synthesized that contain two intercalating chromophores connected by various chemical "chains".
The acridine dimers were the first synthetic bisintercalators
(Barbet *et al*., 1975) and proton magnetic resonance has been
used to study the conformation and interactions of an acridine
dimer with mono- and dinucleotides (Barbet *et al*., 1976).

Krugh and Neely (1973b) reported the nuclear magnetic resonance studies of a series of actinomycin D complexes with both
complementary and noncomplementary deoxydinucleotides. The
following year Patel (1974b,c) used ^{1}H and ^{31}P spectroscopy to
investigate actinomycin D complexes with pdG-dC and the self-
complementary deoxyhexanucleotide d-ApTpGpCpApT. Patel and
co-workers have performed extensive nuclear magnetic resonance
studies on the binding of actinomycin D to deoxyoligonucleotides (see below) as well as the binding of several other drugs
to deoxyoligonucleotides and synthetic polynucleotides (e.g.,
see Patel, 1976a, 1977b; Patel and Canuel, 1976b, 1977a,b,c;
Patel and Shen, 1978). Davidson *et al*. (1977) have recently
reported ^{1}H and ^{13}C NMR and UV-visible spectroscopic studies
on the binding of a methylated quinacrine derivative and propidium diiodide to dinucleoside monophosphates.

Heller *et al*. (1974) studied the interaction of poly A_{20-24}
with poly U_{20-24} to form the poly (A)·poly (U) duplex (20-24
nucleotides in length). The binding of miracil D (1-(2-diethyl-

aminoethylamino)-4-methyl-10-thiaxanthenone) to the poly
(A)·poly (U) (20-24) duplex and the constituent single-stranded
oligomers was also studied by proton magnetic resonance. A
preferential broadening and a 0.3 ppm upfield shift of the
4-methyl resonance on the thiaxanthenone ring was in contrast
to the unshifted resonances for the two terminal methyl resonances of the diethylaminoethylamino chain. These data are
consistent with the intercalation of miracil D into the double-
stranded poly (A)·poly (U) duplex as had been proposed from
other studies (see Heller *et al.*, 1974). A structure for the
complex was proposed in which the miracil D was given a preferred orientation such that the 4-methyl group was "probably
located" between two adenine rings on the poly A side of the
poly (A)·poly (U) double-stranded complex. Although it is possible to construct a complex in which the 0.3 ppm shift would
result from a preferential orientation of the thiaxanthenone
ring, it is equally possible to construct a family of geometries
in which the average chemical shift for the methyl resonance
would be 0.3 ppm. An induced shift of *only* 0.3 ppm for a
methyl resonance attached to a thiaxanthenone ring that is
intercalated between the base pairs might be used to argue
that either the thiaxanthenone ring is not completely intercalated or that there is conformational averaging in which the
4-methyl resonance spends a significant fraction of the time
in the vicinity of the uracil bases with their much smaller
(than adenine) ring currents. Heller *et al.* (1974) are careful to call the structure shown in their Fig. 8 a "probable or
proposed" geometry of the complex. We recommend caution in
the interpretation of limited data in terms of a detailed
geometry of the complex.

     Chien *et al.* (1977) have studied the interaction of
bleomycin A_2 and tripeptide S with DNA using fluorescence spectroscopy and proton magnetic resonance spectroscopy. The bleo-
mycins are a family of glycopeptide antibiotics that exhibit a
wide spectrum of antimicrobial activities. Tripeptide S is a
partial hydrolytic product of bleomycin. Quenching of the
bleomycin fluorescence was used to determine the apparent equilibrium binding constants for DNA binding. The preferential
broadening of the proton resonances from the bithiazole rings
and dimethylsulfonium groups suggested that these parts of the
bleomycin bind most tightly to the DNA polymer.

     Gabbay and co-workers have used proton magnetic resonance
and a variety of other techniques to investigate the interaction of a large number of synthetic oligopeptides with nucleic
acids (e.g., see Gabbay *et al.*, 1973, 1976, and the many references therein). These extensive (and interesting) studies have
provided a great deal of stereochemical and topological information on oligopeptide-nucleic acid interactions, especially
the partial intercalation of aromatic amino acids. Dimicoli

and Helene (e.g., 1974) have also studied the interactions of aromatic residues of proteins with nucleic acids.

XI. 9-AMINOACRIDINE

Reuben *et al.* (1978) have recently studied the proton NMR spectra of 9-aminoacridine and complexes of 9-aminoacridine with dG-dC, dC-dG, and the hexanucleotide d-ApTpGpCpApT. 9-Aminoacridine is a particularly interesting molecule because Sobell and co-workers (Sakore *et al.*, 1977) have reported the x-ray crystallographic structure of a 2:2 complex of 9-aminoacridine with i^5-CpG. This cocrystalline complex contained two different kinds of intercalated 9-aminoacridines. The first of these involved a pseudosymmetric stacking interaction of the 9-aminoacridine between guanine·cytosine base pairs. The second configuration is an asymmetrically intercalated 9-aminoacridine. A schematic illustration of the two types of intercalative structures observed in the crystal structure is shown in Figs. 12A and B. Figures 12C and D represent two additional classes of trial configurations used by Reuben *et al.*

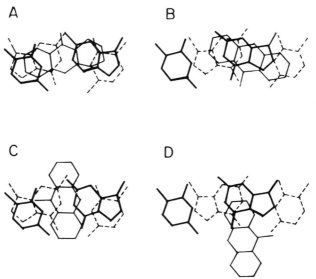

Fig. 12. Scale drawing of four possible complexes of dinucleotide with 9-aminoacridine, based on the crystal structures of Sakore et al. (1977). Reproduced with permission from Reuben et al. (1978).

(1978) to make a comparison between the experimental and calculated values of the induced shifts. It should be noted that
the crystal structure contained the 5-iodinated <u>ribose</u> dinucleoside monophosphate i^5-CpG, while the solution studies were
performed with <u>deoxy</u>nucleotides, and thus a direct comparison
of the results may not be appropriate. However, Sobell has
used the crystal structures obtained with the i^5-UpA
and i^5-CpG molecules as a starting point for building detailed
models for the binding of the drugs to DNA (e.g., see Sobell
et al., 1977), and it seems to be a reasonable starting point
for interpreting the solution data. It is also noteworthy that
the crystal structure of 9-aminoacridine with ApU has been reported (Seeman *et al.*, 1975) and in this structure the 9-amino-
acridines were stacked on A·U base pairs, but the dinucleoside
was not in the normal double-stranded form.

The chemical structure of 9-aminoacridine and a plot of the
chemical shifts of the 9-aminoacridine proton NMR resonances as
a function of the concentration of dG-dC and d-ApTpGpCpApT are
shown in Fig. 13. The shape of the titration curves in Fig.
13B illustrate a strong complex formation with a stoichiometry
of two dG-dC nucleotides binding to each 9-aminoacridine. For
the ^{1}H NMR titration of 9-aminoacridine with d-ApTpGpCpApT the
break in the curves in Fig. 13C occurs at a concentration approximately corresponding to a 3:2 9-aminoacridine:hexanucleo-
tide stoichiometry (Reuben *et al.*, 1978). The continual change
in the chemical shifts of the H(3), H(6), H(2), and H(7) resonances as excess hexanucleotide is added is interesting, since
it suggests that the "average" conformation of the complex is
changing as an excess number of intercalation sites are added
(above the minimum number). Another interesting (and unexpected) aspect of these data is that the induced chemical shifts
for the 9-aminoacridine complexation with dG-dC (2:1 dinucleo-
tide:dye) are temperature dependent; at 3.5°C the values increase by an average of ~60% over the values at 18°C. This
temperature dependence is in contrast to the invariance of the
induced chemical shifts (at 3 and 25°C) for the 2:1 complex of
actinomycin D with pdG-dC (Krugh and Neely, 1973b; Krugh and
Chen, 1975) as well as the 2:1 complex of ethidium with dC-dG
or CpG (Krugh *et al.*, 1975; Reinhardt and Krugh, 1978). The
theoretical induced shifts were calculated on the basis of the
overlap of the 9-aminoacridine with the adjacent C·G or A·T
base pairs as shown in Fig. 12. Reuben *et al.* (1978) concluded
that either configuration 12B or configuration 12D provided good
agreement between the experimental and calculated values of the
induced shifts. The chemical shift data appear to rule out the
pseudosymmetric intercalated structure as the predominant complex formed in solution (Fig. 12A), which contrasts with the
proposal of Sobell (Sakore *et al.*, 1977) that this is probably
the type of complex expected for DNA intercalation. The agree-

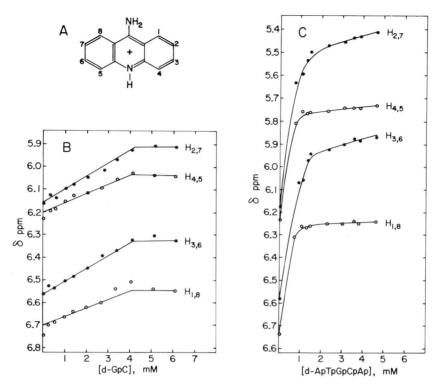

Fig. 13. (A) The chemical structure of 9-aminoacridine. (B) Titration of 2.0 mM 9-aminoacridine with the dinucleotide dG-dC in D₂O, pD 7.5, T = 18°C. (C) Titration of 1.7 mM 9-aminoacridine with the deoxyhexanucleotide d-ApTpGpCpApT in D₂O, pD 7.5, T = 18°C. The chemical shifts are measured relative to tert-butyl alcohol. Reproduced with permission from Reuben et al. (1978).

ment between the experimental and calculated Δδ values for both structures 12B and 12D should serve as a cautionary note to those interpreting chemical shift data in drug-nucleic acid complexes (structure 12D results in the same average chemical shifts as 12B because the rapid chemical exchange produces an averaging of the very large upfield shifts of the 9-aminoacridine resonances that are located between the base pairs and the negligible induced shifts of the 9-aminoacridine resonances on the end of the chromophore away from the base pairs).

Reuben *et al.* (1978) also performed a titration of 9-amino-acridine with dA-dT, under conditions similar to the dG-dC titration (Fig. 13B), and they state that the data are indicative of no strong interaction or intercalation. The miniature inter-

calated complex that would be formed by having two dA-dT molecules wrapped around a 9-aminoacridine molecule would have
four Watson-Crick hydrogen bonds stabilizing the complex as
compared to six hydrogen bonds in the analogous complex with
dG-dC. Thus we would expect that dA-dT would appear to bind
more weakly than dG-dC (e.g., see the ethidium studies of
Krugh and Reinhardt 1975). If, however, the 9-aminoacridine
does have a low affinity for the two dA-dT sites in the hexanucleotide double helix, then the three (presumably) intercalated 9-aminoacridines in the 3:2 dye:hexanucleotide complex
would be located at three adjacent sequences along the double-
stranded hexanucleotide. If this interpretation is supported
by additional experiments, then this would be an example of a
violation of the nearest neighbor exclusion principle, which
states that intercalation of a drug molecule at a given site
prevents the intercalation of a second drug at either of the
two adjacent sites [the two nearest neighbors--see Crothers
(1968) and McGhee and von Hippel (1974) for a general introduction to this area]. In summary, these initial experiments
of Reuben *et al.* (1978) stimulate several interesting questions, many of which may be answered by additional magnetic
resonance studies, while some of them will be more amenable to
optical spectroscopic studies. The optical studies will be especially important in providing the link between the oligonucleotide and the polynucleotide complexes with 9-aminoacridine.

XII. MUTAGEN-OLIGONUCLEOTIDE COMPLEXES WITH A BULGED BASE

A recent communication by Lee and Tinoco (1978) provided
an interesting observation on the structure of drug-nucleic
acid complexes. These authors studied the binding of ethidium
to CpG, CpUpG, GpUpG, and an equimolar mixture of GpUpG+CpC.
The changes in the induced shifts of ethidium were interpreted
as evidence for the formation of complexes whose structures
are shown in Fig. 14. The data by Lee and Tinoco are given
in Table I. A comparison of the $\Delta\delta$ values for the ethidium
resonances in the CpG:ethidium complex with the $\Delta\delta$ values in
the CpUpG:ethidium complex implies that either the average
conformations for the stacking of the two G·C base pairs on
the phenanthridinium ring of ethidium are substantially
different in these two complexes (Fig. 14), or that there is incomplete complex formation under the conditions used to obtain
the data in Table I. We favor the former interpretation because
the T_m of the two complexes were 50 and 35°C, whereas the spectra of both complexes were recorded at 10°C. In addition, the
H(1) and H(10) protons have similar $\Delta\delta$ values in these two com-

TABLE I. *Upfield chemical shifts of ethidium in the presence of oligonucleotides*[a],[b]

	δ(free ethidium)$-\delta$(ethidium in complex) in ppm			
Proton	CpG[c]	CpUpG[d]	GpUpG[e]	GpUpG+CpC[f]
H1	0.58	0.52	0.24	0.45
H2	0.92	0.66	0.24	0.68
H4	1.22	0.75	0.21	0.61
H7	1.05	0.63	0.28	0.62
H9	0.85	0.66	0.22	0.56
H10	0.58	0.49	0.25	0.38

[a]*Shifts are reported at 10°C in D_2O solution containing EDTA*

[b]*Spectra were recorded on a Bruker HXS 360-MHz spectrometer (Stanford Magnetic Resonance Laboratory).*

[c]*5 mM CpG+2.5 mM ethidium.*

[d]*3.2 mM CpUpG+1.6 mM ethidium.*

[e]*2 mM GpUpG+2 mM ethidium.*

[f]*2 mM GpUpG+2 mM CpC+2 mM ethidium. At these concentrations the complex is not fully formed. From Lee, C.-H. and Tinoco, Jr., I. (1978), Nature, 274, 609.*

plexes while the H(2), H(4), and H(7) resonances exhibit substantially different $\Delta\delta$ values. In other words, the structures for the ethidium complex with CpUpG and the ethidium complex with GpUpG + CpC, which are shown in Fig. 14, are only schematic representations. However, the interesting proposal of the formation of bulged based structures (Fig. 14) suggests that ethidium may cause frameshift mutations by intercalation into transient bulges formed during replication, recombination, or repair. It will be interesting to study the binding of ethidium to longer oligonucleotides where the double-stranded form has a preexisting defect due to mismatched base pairing.

XIII. BIPHASIC HELIX-COIL TRANSITIONS OF INTERCALATING DRUG-NUCLEIC ACID COMPLEXES

Patel and Canuel (1976a, 1977a,c) and Patel (1977b) have used proton magnetic resonance spectroscopy and UV-visible spectroscopy to study the helix to coil transition of poly (dA-dT)·poly (dA-dT) and the complexes of ethidium bromide, propidium diiodide, proflavine, and netropsin (a nonintercalating

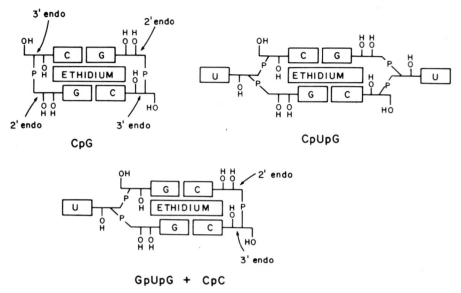

Fig. 14. Proposed structures in aqueous solution for the complexes containing ethidium. In these complexes the 3' linked sugar of each dinucleoside phosphate tends to the 3' endo conformation, whereas the 5' linked sugar tends to the 2' endo conformation. Reproduced with permission from Lee and Tinoco (1978).

oligopeptide) with poly (dA–dT)·poly (dA–dT). The ^{1}H NMR spectra provide evidence for the intercalation of ethidium, propidium, and proflavine and also show that the binding of netropsin alters the average conformation of the helix. Another interesting aspect of three of these papers (Patel and Canuel, 1977a,c; Patel, 1977b) is the observation of biphasic melting curves in the presence of various drug to phosphate ratios. These authors appear to interpret these data in terms of the melting of "drug bound" and "drug free" regions and suggest that ethidium bromide and propidium diiodide bind cooperatively to poly (dA–dT)·poly (dA–dT), with decreased cooperative binding at higher drug to phosphate ratios (Patel and Canuel, 1977a). This is a misleading interpretation of the data because the cooperativity observed in the melting curves is associated with the helix to coil transition (in both the presence and the absence of the drugs). Patel and Canuel (1977a) did not present any evidence which indicates that ethidium binds in a cooperative manner to a stable double helix of poly (dA–dT)· poly (dA–dT). McGhee (1976) had previously investigated the bi-

phasic melting curves observed with netropsin: poly (dA-dT)·poly
(dA-dT) complexes and showed that the biphasic curves were due
to the redistribution of the bound drug along the remaining
double-stranded region of the polymer during the helix to coil
transition. In addition, Aktipis and Martz (1974) had also
studied the thermal denaturation of DNA in the presence of ethidium and used the temperature dependence of the circular dichroism
spectra to provide convincing evidence that ethidium is redis-
tributed along the double-stranded regions during melting. A
somewhat oversimplified picture is that as the polymer is melting the drugs redistribute themselves along the remaining helix
until all available binding sites are occupied, and then there
is a final cooperative melting phase of the complex.

XIV. ACTINOMYCIN D-DEOXYOLIGONUCLEOTIDE COMPLEXES

    The reader is referred to recent review articles on actino-
mycin D (e.g., Meienhofer and Atherton, 1977; Hollstein, 1974;
Lackner, 1975; Sobell, 1973); only selected literature references pertinent to the discussion of the NMR results will be
presented below. The actinomycins bind to double-stranded DNA
with a general, but not absolute, requirement for the presence
of a guanine base (e.g., see Wells, 1971; Krugh and Young,
1977). Müller and Crothers (1968) used hydrodynamic, equilibrium, kinetic, and sedimentation experiments to study the
binding to DNA of several actinomycins and actinomycin derivatives and showed that the actinomycins bind to DNA by intercalation of the phenoxazone ring, in contrast to a previous
model that had the molecule binding on the outside of the helix.
The intercalation model for actinomycin binding received support from the experiments of Waring (1970) and Wang (1971) on
the actinomycin D unwinding of supercoiled DNA. In 1970 Arison
and Hoogsteen published a detailed magnetic resonance study on
actinomycin D and a 2:1 dGMP:actinomycin D complex (see above).
The 2:1 stoichiometry differed from the apparent 1:1 complex
reported by Gellert et al. (1965), but the discrepancy is
easily resolved if one remembers that the optical titrations
are unable to distinguish between 1:1 and 2:1 complex formation if the two binding sites of actinomycin are identical
and noninteracting. Sobell et al. (1971) cocrystallized deoxyguanosine with actinomycin D and determined the x-ray
crystallographic structure of the 2:1 dG:actinomycin D complex.
From this crystal structure Sobell and Jain (1972) presented a
detailed molecular model for the binding of actinomycin D to
DNA. The actinomycin D-deoxyguanosine complex is shown in

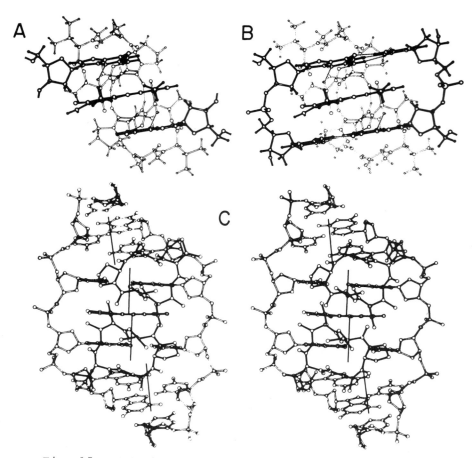

Fig. 15. *(A) The actinomycin D-deoxyguanosine complex used
in the stepwise building of the DNA binding model. (B) The actinomycin D: dGpdC complex proposed by Sobell and Jain (1972).
(C) Stereopairs of the actinomycin D-DNA Model from Sobell et
al., 1977. A stereoviewer is highly recommended for viewing
(C). However, Philip Dean advised me (TRK) that you can
easily view the stereodiagrams without the use of a stereoviewer. Simply look at a spot midway between the two figures
and relax your eyes (i.e., let your eye focus go to infinity).
Make certain that the page is flat and try to have even illumi-
nation on the figure. It may help to adjust the distance from
your eyes to the page (10-21 inches) if the image does not appear to be focused. I first "saw" the stereoview after 1-2
min. of trying. I found that it is well worth the effort if
you do not have a stereoviewer handy. Figs. 15A and B are reprinted from Sobell and Jain (1972). Figure 15C is reprinted
from Sobell et al. (1977).*

Fig. 15A and the model for the actinomycin D complex with two dG-dC molecules is shown in Fig. 15B. Sobell *et al.* (1977) have recently revised the model for the binding of actinomycin D to DNA; a stereo pair of drawings for the actinomycin complex with the hexanucleotide sequence d-ApTpGpCpApT is shown in Fig. 15C. The reader is reminded that while Fig. 15A corresponds reasonably well with the actual crystal structure, Figs. 15B and C are the result of model building studies as discussed in detail by Sobell *et al.* (1977). The X-ray crystal structure and the associated model building (Sobell *et al.*, 1971) provided the stimulus for one of us (TRK) to initiate a program to study the interaction of actinomycin D with deoxyoligonucleotides in aqueous solution to determine if the types of structures proposed in Fig. 15 would be observed in solution. The first experiments (Krugh, 1972) involved the monitoring of the change in the absorbance of actinomycin D at 425 nm as a function of the concentration of the deoxydinucleotides. The guanine containing dinucleotides exhibited three classes of titration behavior, depending on the sequence of the dinucleotides. Using the X-ray structure as a guide, Krugh (1972) interpreted the optical titration data in terms of the four complexes shown in Fig. 16. The highly cooperative binding of pdG-dC was evidence that two dinucleotides were binding to actinomycin D (Fig. 16B). The cooperativity presumably resulted from the formation of the six hydrogen bonds associated with the two G·C Watson-Crick base pairs. An intriguing aspect of the optical titration data is that the self-complementary deoxydinucleotide pdC-dG formed a 2:1 stacked complex (Fig. 16C) instead of a miniature intercalated complex in which tne actinomycin D chromophore would be intercalated at a dC-dG sequence. This suggests that the stacking forces and the hydrogen bonding forces are both important in stabilizing the structures of the dinucleotide complex. By monitoring several of the actinomycin D resonances in the proton NMR spectra of this series of dinucleotide complexes, Krugh and Neely (1973b) were able to verify that in the presence of excess dinucleotides (>2:1 dinucleotide:drug ratios) the predominant complexes formed are the ones shown in Fig. 16. For example, the stacking patterns observed for the N nucleotides in the four pdN-dG deoxydinucleotide complexes with actinomycin D were verified (in part) by monitoring the proton resonances of the two methylvaline N-CH3 resonances that are located on the face of the pentapeptides adjacent to the helix, and are therefore in a location that would be particularly sensitive to the ring current of the nucleotide base one removed from the base that is stacked on the phenoxazone ring. There are two other noteworthy aspects of these spectra.

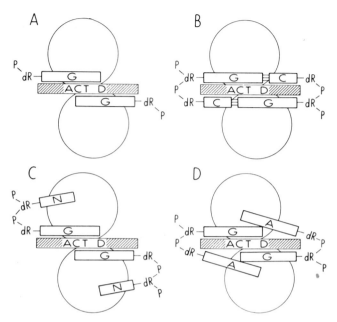

Fig. 16. Schematic illustrations of actinomycin D com-plexes. In these figures the phenoxazone ring projects out of the plane of the paper and is indicated by the dashed lines. The cyclic pentapeptide rings are schematically illustrated as circles. (A) Actinomycin D-deoxyguanosine 5'-monophosphate complex; (B) actinomycin D-pdG-dC complex, illustrating the formation of an intercalated complex; (C) actinomycin D-pdN-dG complex; (D) actinomycin D-pdG-dA complex, illustrating steric interference of the noncomplementary bases. Reproduced with permission from Krugh and Neely (1973b).

First, the induced shifts of the H(7) and H(8) resonances of the phenoxazone ring in the pdN-dG complexes with actinomycin D (Fig. 16C) are consistent with the ring currents calculated from stacking the guanine base in an orientation similar to that observed for deoxyguanosine in the crystal structure (see Figs. 16C and 15A). On the other hand, the induced shifts of the H(7) and H(8) resonances in the pdG-dN complexes (0.63 and 0.16 ppm for pdG-dC) contain, in addition to the ring currents, local diamagnetic anisotropic contributions from the sugar phosphate backbone that result from the proximity of these groups to the H7 and H8 resonances (Figs. 16 and 15B).

Another important point is that all of these spectra are subject to conformational averaging and thus the geometries deduced reflect this averaging. Fortunately, the cyclic pentapeptides greatly reduce the number of possible conformations that are allowed. The stereo view of Sobell's latest model for the actinomycin D-DNA complex nicely demonstrates this point (Fig. 15C). Consequently, for the cooperative binding of pdG-dC to actinomycin D one cannot form the type of complexes illustrated in Figs. 12C and D for 9-aminoacridine. By comparing the chemical shift data for pdG-dC complex formation to the chemical shift data for the actinomycin D complex formation with the other dinucleotides, we may also conclude that the predominant form of the complexes is that illustrated in Fig. 16.

It must also be noted that Schara and Müller (1972) independently studied actinomycin C_3 complex formation with several deoxydinucleotides (and a deoxytetranucleotide) by monitoring the change in the absorbance of actinomycin C_3 as a function of nucleotide concentration. They also determined the average molecular weight of the complexes from sedimentation equilibrium experiments. Schara and Müller (1972) showed that pdG-dC binds cooperatively to actinomycin C3 and proposed a structure for the complex that was identical to the one shown in Fig. 16B. Schara and Müller (1972) observed that the G containing dinucleotides exhibited three classes of binding constants (i.e., different binding for the pdN-dG, pdG-dN, and pdG-dC dinucleotides). The sedimentation experiments were interpreted as evidence that the actinomycin complex with pdC-dG involved the formation of a 2:2 complex in which the two pdC-dG molecules had formed a double helix with the two actinomycin molecules bound on either end of the helix. However, the proton NMR spectra of the actinomycin D complex with pdC-dG (Krugh and Neely, 1973b), the optical titration data (Krugh, 1972; Schara and Müller, 1972), as well as subsequent studies (Davanloo and Crothers, 1976b; Auer *et al.*, 1978) all support the formation of a 2:1 pdC-dG:actinomycin complex (Fig. 16C).

At this point it is informative to consider the proton magnetic resonance data on the complexes of actinomycin D with pdG-dG, pdC-dC, and the binary mixture pdG-dG + pdC-dC. Krugh *et al.* (1977) showed that in a solution of actinomycin D (5 m*M*) with ≥10 m*M* pdG-dG (i.e., ≥2:1 nucleotide:drug ratio) the predominant complex formed is the stacked complex illustrated in Fig. 17A. However, the data also suggested that during the initial portions of the titration a fraction of the pdG-dG dinucleotides are wrapped around the quinoid portion of the phenoxazone ring (as illustrated by the nucleotides in Fig. 17B). Even in the presence of excess pdG-dG (>2:1 dinucleotide:actinomycin D ratios) a fraction of the complexes may be in the form illus-

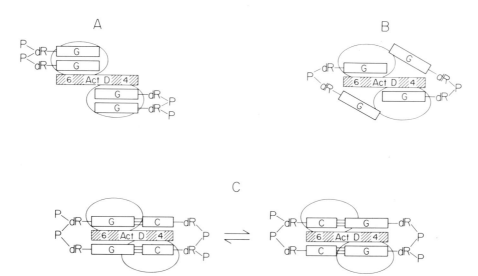

Fig. 17. Schematic illustrations of actinomycin D complexes. (A) A stacked complex for Act D·2(pdG-dG); (B) a
sandwich type complex formation for Act D·2(pdG-dG); (C) two
different conformations for an intercalated type complex of
Act D:pdG-dG:pdC-dC. Reproduced with permission from Krugh
et al. (1977).

trated in Fig. 17B as we cannot eliminate the possibility of
~10% of the complexes having this type of geometry because
the conformational averaging present in these solutions places
an upper limit on the certainty with which one can interpret
the NMR data. In the actinomycin D titration with pdC-dC
there was no evidence for complex formation. However, the
titration of actinomycin D with the binary mixture showed
quite clearly that the predominant complex formed was an intercalated complex as schematically illustrated in Fig. 17C.
It is interesting to note that the formation of the six
Watson-Crick hydrogen bonds between the two G·C base pairs
provides sufficient energy to facilitate the binding of pdC-dC
and to change the manner in which pdG-dG binds. It is also
interesting to note that the two possible complexes that
could be formed (due to the asymmetry of the phenoxazone ring)
are approximately equal in energy since the chemical shift
data are consistent with a 50:50 (±10) distribution of these
two complexes. This observation reflects upon the origin
of guanine recognition when actinomycin D binds to DNA
(e.g., see Krugh et al., 1977) and has proved to be useful

in the interpretation of our recent dissociation kinetics of
the actinomycin complexes with synthetic polynucleotides
(Krugh, T. R. *et al.*, 1978).

Patel (1974c) has extended the actinomycin D binding studies
to the hexanucleotide level by measuring the ^{31}P and ^{1}H NMR
spectra of the 2:1 nucleotide:drug complexes. As already noted,
Patel (1974c) observed large downfield shifted resonances (~1.5
and 2.5 ppm) in the ^{31}P NMR spectra of the complex of actinomycin D with the self-complementary deoxyhexanucleotides
d-(ApTpGpCpApT) (Fig. 7), d-(pGpC)$_3$, and the deoxytetranucleotide d-(CpGpCpG) (Fig. 8). Patel has coupled the similarity in
the ^{31}P chemical shifts and the observation that all three nucleotides contain the dG-dC sequence to conclude that intercalation is occurring at the dG-dC sequence. It is important to
assess the strength of the ^{31}P (and ^{1}H) NMR data concerning the
location of the intercalation site because Allen *et al.* (1976,
1977) have measured the circular dichroism spectrum of 11
double-stranded DNA's in the absence and presence of increasing
amounts of actinomycin D and have concluded that the dC-dG sequence is as favorable a binding site as the dG-dC sequence.
The question is whether or not the evidence to date is definitive with respect to the uniqueness (or sequence preferences)
of the actinomycin D binding to these oligonucleotides as models for DNA. The complex of actinomycin D with d-(ApTpGpCpApT)
might be thought to be the easiest to analyze, since it contains
the dG-dC sequence at the center of the oligonucleotide, and
the presumably unfavorable dA-dT sequences at either end.
Thus the d-(ApTpGpCpApT) double helix has only two likely intercalation sites, the (dT-dG)·(dC-dA) site (which is the same
as a (dC-dA)·(dT-dG) site), and the central (dG-dC) site.
Patel (1974c) states in the abstract of his paper that "the
complexation site is at the (GC)$_{central}$ base pairs since the
two G-N$_1$H resonances exhibit the largest chemical shift difference in the spectrum of 1:2 Act-D:d-(ApTpGpCpApT) in aqueous solution." Although this line of reasoning seems logical
for the intercalation of an asymmetrical chromophore, it includes a number of assumptions concerning the
effect of the helix distortion on the chemical shift differences in N-H-N resonances of adjacent base pairs. The data in
Tables I and II of Patel (1974c) illustrate the situation; for
example, at pH 7, 3°C, in H$_2$O solution the (A·T)$_{internal}$ base
pairs (i.e., the ones between the terminal A·T base pairs and
the central G·C base pairs) resonate at 13.77 ppm in the absence of actinomycin D, and are found at 14.13 and 13.95 ppm
in the 2:1 complex ($\Delta\delta$ = -0.03 and +0.38). We question whether
these chemical shift changes are definitive in terms of locating the intercalation site. Remember that the same asymmetric
phenoxazone ring of actinomycin D (it is actually nonplanar)

used in the "ring current" arguments by Patel (1974c) could
result in asymmetric distortions of the helix in the vicinity
of the intercalation site that could account for the observed
trends in the chemical shifts of the N-H-N resonances if the
actinomycin D molecule intercalated at one of the two (dT-dG)
·(dC-dA) sites. However, Patel does have other data that are
consistent with intercalation at the dG-dC sequence as the
primary intercalation site. For example, in the actinomycin
D complex with intercalation at the dG-dC sequence as the primary intercalation site. For example, in the actinomycin D
complex with d-(ApTpGpCpApT) the temperature dependencies of
the chemical shifts of both of the G-N_1H protons are about
equal, as are the temperature dependencies of the chemical
shifts of the two T-N_3H protons; this suggests that intercalation primarily occurs at the central (i.e., the dG-dC)
intercalation site. In the ^{31}P spectra with the deoxyhexanucleotide d-(GpCpGpCpGpC) there are at least four downfield
shifted resonances (Patel, 1974c), which could result from
either intercalation at the dG-dC sequence or the dC-dG sequences. The deoxytetranucleotide dCpdGpdCpdG shows only
two downfield shifted resonances (which implies a unique intercalation site) whose chemical shifts are the same as the
chemical shifts of the two downfield shifted resonances in
the d-(ApTpGpCpApT) complex (Patel, 1974c, 1976). The slow
exchange 31 p NMR shifts in the actinomycin D complex with
pdG-dC complex (Reinhardt and Krugh, 1977) are also the same
as the shifts cited above and thus it seems likely that actinomycin D does exhibit a preference for dG-dC sequences when
compared to dC-dG sequences. Additional evidence for this interpretation has been obtained from the competitive binding experiments of Reinhardt and Krugh (1978). The optical titrations of actinomycin D with dCpdCpdGpdG, dGpdGpdCpdC, and
dCpdGpdCpdG (Patel and Canuel, 1977b) indicated that actinomycin D bound more strongly to the last two deoxytetranucleotides,
which contain a dG-dC sequence, than it binds to dCpdCpdGpdG,
which does not contain a dG-dC sequence. Patel and Canuel concluded that actinomycin D exhibits a pronounced specificity for
oligonucleotides that contain dG-dC intercalation sites. However, the reader is also referred to the discussion by Kastrup
et al. (1978) for a more detailed analysis of the binding curves
in oligonucleotide titrations where there are several intercalation sites. Nevertheless, the majority of the data at the oligonucleotide level, and the competitive binding experiments with
calf thymus DNA (Lepecq and Paoletti, 1967; Reinhardt and Krugh,
1978) are consistent with actinomycin D exhibiting a range of affinities for the various intercalation sites, with a pronounced
preference for binding to dG-dC sequences when compared to dC-dG

sequences. However, the range of intrinsic binding constants of actinomycin D for the 10 unique intercalation sequences available on DNA, as well as the biological implications of these observations, is still an active area of study.

XV. ETHIDIUM BROMIDE-OLIGONUCLEOTIDE COMPLEXES

The unusual ability of actinomycin D to preferentially bind to various sequence isomers of the guanine containing deoxydinucleotides led us to suspect that other drugs would also exhibit sequence preferences in binding to DNA and RNA. Of the several drugs originally screened, we selected to study ethidium bromide for two reasons:

(1) The binding of ethidium to DNA had been well characterized by fluorescence, visible absorption, and circular dichroism studies (e.g., see LePecq et al., 1964; Waring, 1965; LePecq, 1971; Reinhardt and Krugh, 1978; and the references therein.

(2) Ethidium did not exhibit the same magnitude of problems with adsorption and photochemical degradation associated with some of the other compounds originally tested.

Ethidium was generally believed to intercalate into double-stranded DNA and RNA. The titrations of ethidium bromide with the deoxydinucleotides showed clearly that ethidium binds stronger to pyrimidine(3'-5')purine sequence dinucleotides than to their respective purine(3'-5')pyrimidine sequence isomers (Krugh, 1974; Krugh et al., 1975; Krugh and Reinhardt, 1975), as illustrated by the shape of the titration curves shown in Fig. 18. The visible absorption titrations of ethidium (e.g., Fig. 18) provided the stimulus to Sobell's colleagues to attempt to cocrystallize ethidium with the ribodinucleoside monophosphates. Cocrystalline complexes of ethidium with both UpA and CpG were easily obtained; the crystal structures of 2:2 complexes of ethidium with the iodinated derivatives of UpA and CpG have been reported (e.g., see Tsai et al., 1975 and Sobell et al., 1978).

Circular dichroism (CD) spectroscopy is particularly useful for studying the bonding of ethidium to DNA because ethidium (by itself) is optically inactive and thus it does not exhibit a circular dichroism spectrum. However, when ethidium is bound

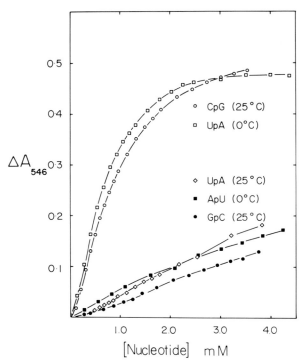

Fig. 18. *Change in absorbance of ethidium bromide at 546 nM as a function of added dinucleoside monophosphate concentration. The ΔA values were scaled to an equivalent ethidium concentration of 2.7×10^{-4} M. Reproduced with permission from Krugh and Reinhardt (1975).*

to double-stranded DNA or RNA, a characteristic circular di-chroism spectrum is observed in the 300-600 nm region (e.g., see the many references in Krugh and Reinhardt, 1975). As a result, circular dichroism spectroscopy provides an important link between the oligonucleotide complexes and the DNA com-plexes. The CD spectra of ethidium solutions with CpG and dCpdG (Fig. 19) exhibit maxima at about 307 nm, 330 nm, and 380 nm; the similarity of these spectra to the CD spectra of ethidium complexes with polynucleotides provided excellent sup-port for the use of these model systems for studying the com-plexes with DNA or RNA. A striking example of the role of com-plementarity and the sequence preference of ethidium in binding the pyr(3'-5')pur sequence dinucleotides as opposed to pur(3'-5')pyr sequences is shown in Fig. 20. Only the solution of ethidium with the <u>complementary</u> dinucleotides of the pyr(3'-5')-pur sequence (UpG + CpA) exhibited the characteristic CD spec-

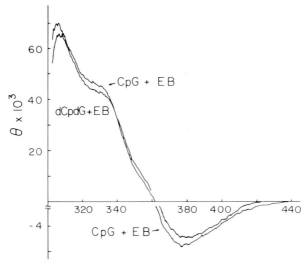

Fig. 19. Induced CD spectra of ethidium bromide upon addition of CpG and dC-dG in a 1-mm thermostatted cell at 3°C.
The samples were 0.75 mM EB and 2.3 mM CpG and 0.75 mM EB and
3.2 mM dC-dG. Both samples were dissolved in a D_2O potassium
phosphate buffer (5 mM) pD = 7.4 (pH meter + 0.4). The θ
values are direct readings from the recorder. Note that the
scale of the ordinate has been expanded below the zero level.
Reproduced with permission from Krugh et al. (1975).

trum observed in polynucleotide complexes. The proton magnetic
resonance spectra of aqueous solutions of ethidium, CpG, as
well as a CpG + ethidium mixture are shown in Fig. 21. In the
ethidium:CpG solution the H7 proton moves upfield more than 0.9
ppm (it disappeared under the nucleotide resonances), while the
H1 and H10 protons are shifted upfield about 0.6 ppm. Lee and
Tinoco (1978) reinvestigated this system using a 360 MHz instrument which has allowed them to follow all of the resonances
during the titration. The large Δδ values observed by Krugh and
Reinhardt (1975) and Lee and Tinoco (1978) for the ethidium
complexes with the dinucleotides (Table I) leads to the conclusion that ethidium bromide forms a miniature intercalated complex with both CpG and dCpdG. The unusual types of complexes
that were considered as possibilities for 9-aminoacridine (i.e.,
Figs. 12C and D) are not consistent with the large ring current

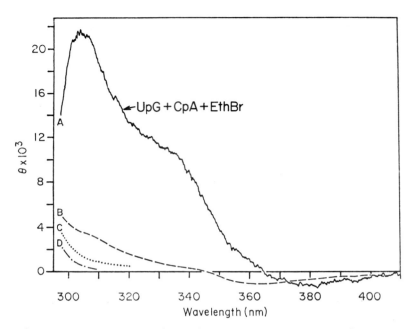

Fig. 20. Circular dichroism spectra for solutions of ethidium bromide with mixtures of complementary (UpG + CpA and GpU + ApC) and noncomplementary (UpG + ApC and GpU + CpA) dinucleoside monophosphates in a 5-mm cell at 1°C. Each solution contained 3.1 × 10⁻⁴ M ethidium bromide and 7.5 × 10⁻⁴ M of each nucleotide. ϵ_L - ϵ_R (305 nm) = 21.4 for the UpG + CpA + ethidium solution. The absorption maxima of these solutions were: curve A, UpG + CpA + ethidium (λ_{max} = 513 nm); curve B, GpU + CpA + ethidium (λ_{max} = 493 nm); curve C, UpG + ApC + ethidium (λ_{max} = 486 nm); curve D, GpU + ApC + ethidium (λ_{max} = 488 nm). Reproduced with permission from Krugh and Reinhardt (1975).

shifts observed for all six phenanthridinium ring protons in the ethidium complexes with CpG (as well as other pyr(3'-5')pur dinucleotides). In addition, the presence of the N-ethyl group and the phenyl ring attached to the phenanthridinium ring restrict the possible geometries for the complex. The Lee and Tinoco paper (1978) appeared just as this chapter was nearing its final form, but an approximate (back of the envelope) calculation of the ring currents expected from the overlap geometry observed in the 2:2 ethidium-iodo[5]-CpG complex suggests to us that the average structure in solution is different from the solid state structure. However, more detailed

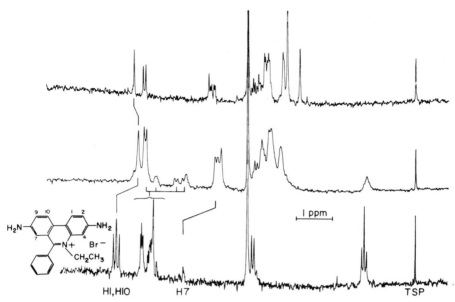

Fig. 21. *100-MHz proton Fourier transform NMR spectra of 0.9 mM CpG (top); 2.3 mM CpG + 0.75 mM EB (middle); and 0.75 mM EB (bottom). The spectra were run at 25°C with a 180-τ-90 (accumulate) pulse sequence to minimize the residual HDO peak. The use of this pulse sequence accounts for the small H7 peak since it has a longer T_1 value than the other EB ring protons. The movement of the peaks has also been followed by the incremental addition of the CpG. The samples were dissolved in a D_2O potassium phosphate buffer (5 mM) and the pD was adjusted to 7.4 (pH meter reading + 0.4). The small amount of dimerization present in the ethidium bromide solution (bottom spectrum) does not significantly affect the appearance of the spectrum. The induced chemical shifts of the various ethidium protons given in the text were calculated on the basis of the infinite dilution chemical shifts. Reproduced with permission from Krugh et al. (1975).*

analysis is required to determine the range of conformations that are consistent with these   values (Table I).

Patel and Canuel (1978a, 1977a) have recently used proton magnetic resonance to study ethidium bromide complexes with dC-dG-dC-dG and with poly (dA-dT)·poly (dA-dT). Both experiments show conclusively that ethidium bromide forms an intercalated complex with these nucleotides.

    Patel and Canuel (1976b) also used visible absorption spec-
troscopy to show that dC-dG-dC-dG, dC-dC-dG-dG, and dG-dG-dC-dC
all form a complex with ethidium. However, determining the
location of the intercalation site (especially for the ethidium
complex with dC-dC-dG-dG) is not as straightforward as suggested
by Fig. 8 of Patel and Canuel (1976b). The reader is referred
to Kastrup et al. (1978) for a discussion of the complex formation of ethidium with the four guanine-cytosine self-complementary deoxytetranucleotides.

    The crystal structures of 2:2 complexes of ethidium bromide
with i^5-CpG and i^5-UpA have been reported and provide detailed
stereochemical information on the structure of the complex in
the solid state (Tsai et al., 1975, 1977; Jain et al., 1977).
All of the ethidium cocrystalline complexes with the ribodinu-
cleosides exhibited a 2:2 stoichiometry, including three separate precipitates analyzed by our group. 9-Aminoacridine also
cocrystallized with both i^5-CpG (Sakore et al., 1977) and ApU
(Seeman et al., 1975). The 2:2 stoichiometry in the crystal
can be rationalized by noting that the two positively charged
drugs provide the necessary electrical balance to the two negatively charged phosphate groups. The question is whether the
2:2 complexes also exhist in solution. Krugh and Reinhardt
(1975) suggested that 2:2 complexes exist when ethidium and
dinucleosides are present at equimolar concentrations, but they
showed that in the presence of excess nucleotides the 2:1 nu-
cleotide:drug complex predominates. More recently, Reinhardt
and Krugh (1978) performed fluorescence lifetimes measurements
on both ethidium bromide-dinucleoside solutions and a cocrystalline complex of ethidium with CpG. The fluorescence decay
of ethidium in a solution containing excess CpG was characterized by a single exponential decay with a lifetime of 23 ns,
which is very close to the value reported for ethidium intercalated into double-stranded RNA (LePecq, 1971). On the other
hand, the fluorescence decay curve of an ethidium cocrystalline
complex with CpG (2:2 stoichiometry) was best fit by two exponential decays with lifetimes of 5 and 11 ns. The occurrence
of two lifetimes in the polycrystalline sample is consistent
with the observation that there are two crystallographically
distinct ethidiums in the X-ray structure of Jain et al. (1977).
Thus the fluorescence experiments help to provide a link between the solution studies and the single crystal X-ray studies.

XVI. ETHIDIUM-tRNA COMPLEXES

     Recent experiments have shown that under appropriate condi-
tions there is a single, unique binding site for ethidium on
yeast tRNA^{phe} and several other tRNAs (e.g., see Tao et al.,
1970; Jones and Kearns, 1975; Wells and Cantor, 1977; Jones et
al., 1978). The presence of a strong ethidium binding site in
the acceptor stem of several tRNA molecules has been demonstrated by fluorescence energy transfer experiments (Wells and
Cantor, 1977) and by monitoring the ethidium induced changes
in the hydrogen bonded resonances (10-15 ppm) of several purified tRNAs (Jones et al., 1978). The nature and location of
the ethidium binding site observed in the single crystals of
tRNA^{phe} (into which ethidium was allowed to diffuse) was different than in solution studies cited above (ethidium was
bound on the outside of the tRNA^{phe} molecule in the crystal,
which may reflect the different conditions under which the
samples were prepared. The readers are referred to Wells and
Cantor (1977) and Jones et al. (1978) for a detailed discussion
of the experiments and a review of the literature.

XVII. SEQUENCE PREFERENCES

     The extensive oligonucleotide binding data cited above
shows that both actinomycin D and ethidium bromide preferentially bind to certain sequences. We have occasionally found that
scientists outside the immediate area misinterpret the X-ray
and solution studies when it comes to extrapolating the results
with the oligonucleotides to the binding of the drugs to
double-stranded DNA or RNA. It is therefore instructive to
outline the limits of our current knowledge. The various intercalation sites are listed in Table II. Note that this list only
considers the base pairs on either side of the intercalated
chromophore, and we have specifically neglected the effects of
the sequence of nucleotides adjacent to the intercalation site.
This assumption is not rigorously correct when one is considering
the intercalation of drugs that interact with the bases adjacent
to the intercalation site (e.g., actinomycin D, daunorubicin,
etc.) or for the binding of bisintercalating molecules. Another
factor which should be kept in mind when considering the preferential binding of drugs to various sequences of DNA is that
the stability of double-stranded DNA and RNA is known to be
sequence dependent (e.g., see Wells et al., 1977; and the references therein). Nevertheless, the following discussion should
serve as a useful orientation. The ethidium-dinucleotide data
show that ethidium prefers to bind to pur(3'-5')pur sequences much
more favorably (at least an order of magnitude) than to the

TABLE II. Intercalation sequences on DNA[a]

pyr(3'-5')pur	pur(3'-5')pyr	(pyr(3'-5')pyr)·(pur(3'-5')pur) = (pur(3'-5')pur)·(pyr(3'-5')pyr)
1 (dC-dG)·(dC-dG)	4 (dG-dC)·(dG-dC)	7 (dC-dC)·(dG-dG) = (dG-dG)·(dC-dC)
2 $\left\{\begin{array}{l}(dT-dG)\cdot(dC-dA)\\(dC-dA)\cdot(dT-dG)\end{array}\right.$	5 $\left\{\begin{array}{l}(dG-dT)\cdot(dA-dC)\\(dA-dC)\cdot(dG-dT)\end{array}\right.$	8 (dT-dT)·(dA-dA) = (dA-dA)·(dT-dT)
		9 (dC-dT)·(dA-dG) = (dA-dG)·(dC-dT)
3 (dT-dA)·(dT-dA)	6 (dA-dT)·(dA-dT)	10 (dT-dC)·(dG-dA) = (dG-dA)·(dT-dC)

[a] Of the 16 total sequences (4^2) there are only 10 unique intercalation sites (due to symmetry). The redundancies are indicated in the Table by brackets or equal signs. If, however, we include the effects of the second neighbors, then the number of intercalation sites is greatly increased. For the present discussion we shall neglect second neighbor effects. The 10 sites are numbered from 1 to 10 for reference.

pur(3'-5')pyr sequence. In terms of the sequence listed in Table II, we predict that, for DNA, the intrinsic binding affinities for intercalation at sites 1,2, and 3 are much larger than the intrinsic binding affinities for intercalation at sites 4,5, and 6. Kastrup *et al.* (1978) have shown that the (dG-dG)·(dC-dC) (site 7 in Table II) is also a favorable binding site for ethidium. The oligonucleotide studies reported to date do not provide information on the remaining intercalation sites. Thus, of the 16 actual sites along a particular DNA, we propose that there are six highly favored sites (1,2,3, and 7), four less favorable sites (4,5, and 6), and six sites of unknown affinity. One important point is that ethidium has a number of favorable binding sites, including sites containing both G·C and A·T base pairs, and thus the oligonucleotide data do not predict that ethidium should exhibit any marked preference for binding to either G·C rich or G·C poor natural DNAs (in agreement with experimental results - see for example, LePecq, 1971; Müller and Crothers, 1975). The proposal that ethidium intercalates into DNA with a variety of intrinsic binding constants that are sequence dependent (Krugh, 1974; Krugh *et al.*, 1975; Krugh and Reinhardt, 1975; Tsai *et al.*, 1975) is not inconsistent with any of the data for the binding of ethidium to DNA. Baguley and Falkenhaug (1978) recently studied the binding of ethidium to six synthetic DNAs (using fluorescence spectroscopy) and observed a range of apparent binding constants; these results are consistent with and supportive of our comments above.

There is no doubt that certain dyes and drugs exhibit preferential binding to base pairs or sequences (e.g., the references above; Müller and Crothers, 1975; Müller *et al.*, 1975; Müller and Gautier, 1975; Wartell *et al.*, 1974). While the presence of sequence recognition and preferential binding creates difficulties in the data analysis, it also creates exciting opportunities that may be exploited in a variety of areas, including the design of bis- or polyintercalating drugs which could, in principle, be as sequence specific as proteins.

XVIII. GEOMETRY OF THE HELIX AT THE INTERCALATION SITE

The X-ray structure of cocrystalline complexes of the ribodinucleoside monophosphates with ethidium bromide, acridine
orange, ellipticine, 9-aminoacridine, and proflavine have been
reported (see Sobell *et al.*, 1978; Jain *et al.*, 1977; Neidle
et al., 1977; Tsai *et al.*, 1975; Seeman *et al.*, 1975). With
the exception of the 9-aminoacridine complex with ApU (Seeman
et al., 1975), all of the structures have had a drug molecule
intercalated between two Watson-Crick base pairs. The 9-amino-
acridine:ApU complex exhibited Hoogsteen type base pairing with
the 9-aminoacridine molecules stacked between two base pairs,
although these base pairs were not connected by a sugar phosphate backbone. Six of the eight crystal structures have been
solved in Sobell's laboratory; in these six cocrystalline complexes there is an equal number of dye (or drug) molecules and
either i^5-CpG or i^5-UpA. This stoichiometry allows the positive charge of the drugs to balance the negative charge of the
phosphates (which also provides electrostatic forces that contribute to the stability of the complex). In five of the six
structures reported by Sobell's group, a mixed sugar puckering
(C3'*endo*(3'-5')C2'*endo*) is observed in each dinucleoside monophosphate. The twist angles between the base pairs range from
8-12° which, when extrapolated to DNA or RNA, represent an unwinding of the helix by about 25°. However, the 2:2 proflavine:
i^5-CpG crystal structure (Sobell *et al.*, 1978) and the 3:2 pro-
flavine:CpG crystal structure (Neidle *et al.*, 1977) are different
from the structures of the other complexes in that all sugar
residues have a C3' *endo* conformation. In these proflavine complexes the twist angle was 33-36° and thus the intercalation of
the proflavine results in a "stretching" (rather than unwinding) of these mini-helices.
    Sobell *et al.* (e.g., Sobell *et al.*, 1978, 1977) have pro-
posed that the mixed sugar puckering is a common feature of intercalation, and that the hydrogen bonds connecting the amino
groups of proflavine and the phosphate oxygen atoms present a
steric constraint in the proflavine complex that is not present
in the other structures. On the other hand, Berman *et al.*
(1978) have examined the range of possible intercalation geometries and conclude that there is a continuum of stereochemi-
cally plausible opened-up dinucleoside phosphate structures;
thus the geometry would depend upon the nature of the intercalating drug. Alden and Arnott (1975, 1977) have used linked-
atom molecular modeling to consider the possible choices available for intercalation into both DNA and RNA. For cohelical
A-DNA their model building studies optimized a structure in
which all the sugar moieties retained the characteristic C3'
endo pucker with virtually no helical unwinding; for B DNA

they predicted that the intercalation site would contain a mixed
sugar puckering. Bond *et al.* (1975) have measured the X-
ray diffraction patterns from DNA fibers that were saturated
with a platinum metallointercalation agent, and concluded that
the platinum-terpyridine complex was intercalated at every other
base pair. However, the Fourier transforms that were calculated
from their DNA intercalation models containing C3' *endo*:C2' *endo*
mixed sugar puckers did not agree with the X ray fiber diffraction diagrams.

It should be clear from the above discussion that there is
no uniformity of thought concerning the stereochemical details
associated with intercalation. As more x-ray structures are
solved, there will probably be a variety of geometries observed.
The use of deoxytetranucleotides will provide a good deal more
confidence in the extrapolation of the results to DNA, but their
use also presents more formidable technical problems. We predict that those drugs that either interact with base pairs adjacent to the intercalation site and/or those drugs that can
both intercalate and bind to the sugar-phosphate backbone (e.g.,
actinomycin D, daunorubicin, or any of the bifunctional intercalating drugs) will exhibit individual geometries at the intercalation site, with the exact stereochemistry depending upon the
sequence of nucleotides in the vicinity of the intercalation
site.

XIX. SUGAR PUCKER IN SOLUTION STUDIES

Lee and Tinoco (1978) and Patel and Shen (1978) have recently utilized 360 MHz NMR spectroscopy to determine the conformation of the sugar rings in complexes of dinucleosides with ethidium bromide and propidium diiodide, respectively. Lee and Tinoco (1978) observed that at 40°C, the ribose sugar of unbound
CpG was 62% *endo* for Cp and 52% *endo* for pG; in the presence of
ethidium, such that about half the dinucleoside phosphate was
complexed, the $J_{1,2}$ values indicate that the 3' sugar (Cp) tends
to increase the proportion in a 3'-*endo* conformation while the
5' sugar (pG) tends to increase the population of the 2'-*endo*
conformer. These data support the model building that incorpo-
rates the mixed sugar puckering observed in the solid state complex of ethidium with iodo-CpG and iodo-UpA (Tsai *et al.*, 1975;
Jain *et al.*, 1977). The values for the percentage of riboses in
each of the two conformations were not given for the ethidium:CpG
complex in this initial communication. It should be noted that Lee
and Tinoco (1978) performed their experiments in the presence of
excess nucleoside, and thus they would have to extrapolate their
results (by a factor of two) to obtain estimates for the sugar

conformation in the complex. Lee and Tinoco (1978) also state
that ethidium complexes with deoxyribodinucleoside monophosphates show similar behavior; that is, the sugars tend toward
a C3' endo(3'-5')C2' endo conformation. Interestingly enough,
they also state that an ethidium complex with hybrid helices
(one strand ribose-one strand deoxyribose) also shows similar
behavior.

    Patel and Shen (1978) monitored the temperature dependence
of the chemical shifts and coupling constants of
propodium diiodide solutions with CpG and with dCpdG. The
chemical shift data show that propidium forms a miniature intercalated complex with the self-complementary dinucleotides
(both ribo and deoxyribo). The temperature dependence of the
$J_{1'2'}$ coupling constants for the CpG:propidium solutions are
shown in the top portion of Fig. 22. The coupling constants
were measurable only at temperatures above ~50°C, at that temperature about 55% of the drug was in the complexed form. Thus
an extrapolation is required to estimate the conformation of
the sugar in the complex. For the propidium experiments with
CpG we note that the largest change is observed for the $J_{1'2'}$
coupling constant of the cytidine sugar. The tendency of
$J_{1'2'}$ to go toward very small values (Fig. 22A) shows that the
Cp residue favors the C3'-endo conformation in the complex. If
we try to quantitate Patel and Shen's data (Fig. 22) by using
the relationship %C3'-endo = 10(10 - $J_{1'2'}$) (e.g., see Lee and
Sarma, 1976a,b) then we estimate that Cp is 55% C3'-endo at
95°C, while at 50°C (when the fraction of the complex, is
≅ 0.55) approximately 75% of the Cp sugars are in the C3'-endo
conformation. Extrapolating these values to 100% complexation
suggests that more than 90% of the Cp residues will be in the
C3'-endo conformation in the propidium:CpG complex. On the
other hand, at 95°C the pG residue of CpG has an approximately
50:50 mixture of the C3'-endo and C2'-endo sugar conformations
which changes to ~43% C3'-endo when the temperature is lowered
to 50°C, where the fraction of the drug in a complex is ~0.55.
An approximate extrapolation to 100% complexation suggests that
35-40% of the pG residues will be in the C3'-endo conformation
(and consequently 60-65% of the pG residues will be in the C2'-
endo conformation). Thus, the data of Patel and Shen (1978)
show that the C3' endo(3'-5')C2' endo conformation is favored
in the propidium diiodide complex with CpG, but the extrapolated data also suggest that a substantial fraction (~30-40%)
of the dinucleosides will have the C3' endo(3'-5')C3' endo conformation in the complex. If this approximate extrapolation is
appropriate, then the solution data indicate that the 5' residue
of CpG has only a relatively small energy difference between the
two different sugar puckers in the intercalated complex with
propidium (0.4-0.7 kcal/mole). We reach the same conclusion
concerning the small energy difference between the two sugar

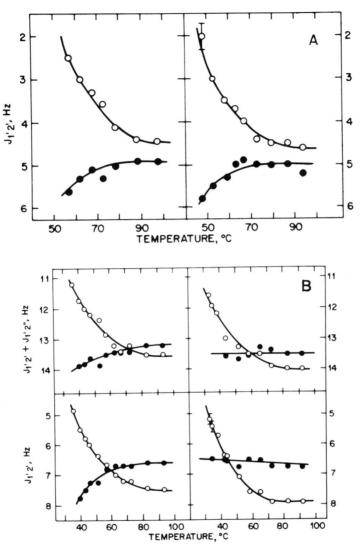

Fig. 22.   (A) The temperature dependence of the vicinal
$J_{1'2'}$ coupling constant in the $PrdI_2 \cdot C-G$ complex (Nuc/D=4) in
the absence (left) and presence (right) of 1 M NaCl in 0.1 M
phosphate/1mM EDTA, pH 6.55, in ^2H_2O. The cytidine and
guanosine coupling constants are designated by o and • , respectively.   (B) Temperature dependence of the vicinal $J_{1'2'}$
+ $J_{1'2''}$ coupling constant sum and the vicinal $J_{1'2'}$ coupling
constant in the $PrdI_2 \cdot dC-dG$ complex in the absence (Nuc/ D =
2.8) (left) and presence (Nuc/ D = 3.2) (right) of 1 M NaCl in
0.1 M phosphate/1mM EDTA,EDTA, pH 6.55, in ^2H_2O. Reproduced
with permission from Patel and Shen (1978).

puckers from an analysis of Patel and Shen's data for the dC-dG
complex with propidium (Fig. 22 bottom). We also note that
Patel and Shen (1978) observed that in the presence of 1 M NaCl,
the pdG residue does not exhibit any change in the coupling constant upon complex formation. While the effect of salt is not
understood at the present time, the observation of the phenomenon is interesting and suggests that caution must be exercised in the interpretation of the data.

In summary, we agree with Patel and Shen (1978) and Lee and
Tinoco (1978) when they state that their data indicate that the
mixed sugar pucker is favored in the miniature intercalated
complexes studied; equally important, though, is our observation that Patel and Shen's data suggest that the mixed sugar
pucker is not a highly favored conformation in the propidium
complexes with the deoxy- and ribodinucleoside monophosphates.
Thus, the role of mixed sugar puckering in solution complexes
may not be as important as has been suggested by Sobell in the
interpretation of the crystal structures (e.g., Sobell et al.,
1977). However, we emphasize that additional experiments on the
solution complexes and crystal structures of several classes of
drugs with longer oligonucleotides (e.g., tetra- or hexanucleotides) are required in order to obtain more definitive answers
to these interesting questions.

XX. CONCLUDING REMARKS

We have tried to describe the application of nuclear magnetic resonance studies in the elucidation of the structure,
kinetic, and thermodynamic properties of drug-nucleic acid complexes. Many of the approaches described here and in the other
chapters of this volume have general utility to wide-ranging
areas of molecular biology, pharmacology, etc. The continual
development of higher frequency NMR spectrometers ($\geq$400 MHz)
has, and will, open exciting new areas to be explored. The use
of convolution difference spectroscopy (e.g., Campbell et al.,
1973) and two dimensional Fourier Transform NMR (e.g., Aue et
al., 1976) as well as other sophisticated NMR techniques, provides a range of techniques that may be used on these difficult problems. The synthesis of selectively deuterated oligonucleotides will also be a tremendous aid toward the interpretation of the conformation of the drug-nucleic acid complexes.
Magnetic resonance techniques can provide information on the
complexes in solution, whereas X-ray crystallography is limited
to the solid state (although the use of synchrotron X-ray
sources is changing this). In the area of drug-nucleic acid
complexes, X-ray crystallography and nuclear magnetic resonance
spectroscopy (in combination with optical and kinetic studies)

have developed an important synergistic relationship that will, it is hoped, grow and provide continuing progress toward the goal of understanding the molecular basis of drug action.

ACKNOWLEDGMENTS

The authors' research discussed in this chapter, and the writing of this review, have been supported by research grants (CA-14103 and CA-17865) and a Career Development Award (CA-00257 to TRK) from the National Cancer Institute, DHEW, and an Alfred P. Sloan Fellowship (to TRK). The authors wish to acknowledge the collaboration and interaction with E. J. Bastian, Jr., Y.-C. Chen-Chiao, J. W. Hook III, R. V. Kastrup, J. W. Laing, K.-R. Lee, S. Lin, W. E. Moehle, E. S. Mooberry, J. W. Neely, C. G. Reinhardt, and F. W. Wittlin. The authors thank R. G. Shulman for many helpful comments on the manuscript.

REFERENCES

Aktipis, S., and Martz, W. W. (1974). *Biochemistry 13,* 112-118.
Alden, C. J., and Arnott, S. (1975). *Nucleic Acids Res. 2,* 1701-1717.
Alden, C. J., and Arnott, S. (1977). *Nucleic Acids Res. 4,* 3855-3861.
Allen, F. S., Moen, R. P., and Hollstein, U. (1976). *J. Am. Chem. Soc. 98,* 864-865.
Allen, F. S., Jones, M. B., and Hollstein, U. (1977). *Biophys. J. 20,* 69-78.
Altona, C., and Sundaralingham, M. (1973). *J. Am. Chem. Soc. 95,* 2333-2344.
Anderson, J. A., Kuntz, G. P. P., Evans, H. H., and Swift, T. J. (1971). *Biochemistry 10,* 4368-4374.
Angerman, N. S., Victor, T. A., Bell, C. L., and Danyluk, S. S. (1972). *Biochemistry 11,* 2402-2411.
Arcamone, F., Cassinelli, G., Franceschi, G., Penco, S., Pol, C., Redaelli, S., and Selva, A. (1972). *In* "International Symposium on Adriamycin: (Carter, S. K., *et al.,* eds.) pp. 9-22. Springer-Verlag, New York.
Arison, B. H., and Hoogsteen, K. (1970). *Biochemistry 9,* 3976-3983.
Aue, W. P., Bartholdi, E., and Ernst, R. R. (1976). *J. Chem. Phys. 64,* 2229-2246.

Auer, H. E., Pawlowski-Konopnicki, B. E., and Krugh, T. R. (1977). *FEBS Lett.* *73,* 167-170.

Auer, H. E., Pawloski-Konopnicki, B. E., Chiao, Y.-C. C., and Krugh, T. R. (1978). *Biopolymers 17,* 1891-1911.

Baguley, B. C., and Falkenhaug, E.-M. (1978). *Nucleic Acids Res. 5,* 161-171.

Barbet, J., Roques, B. P., Le Pecq, J.-B. (1975). *C. R. Acad. Sci. Paris 281,* 851-853.

Barbet, J., Roques, B. P., Combrisson, S., and Le Pecq, J. B. (1976). *Biochemistry 15,* 2642-2650.

Berman, H. M., Neidle, S., and Stodola, R. K. (1978). *Proc. Nat. Acad. Sci. USA 75,* 828-832.

Blears, D. J., and Danyluk, S. S. (1967). *J. Am. Chem. Soc. 89,* 21-26.

Bond, P. J., Langridge, R., Jennette, K. W., and Lippard, S. J. (1975). *Proc. Nat. Acad. Sci. USA 72,* 4825-4829.

Borer, P. N., Kan, L. S., and Ts'o, P. O. P. (1975). *Biochemistry 14,* 4847-4863.

Calascibetta, F. G., Dentini, M., DeSantis, P., and Morosetti, S. (1975). *Biopolymers 14,* 1667-1684.

Campbell, I. D., Dobson, C. M., Williams, R. J. P., and Xavier, A. V. (1973). *J. Mag. Res. 11,* 172-181.

Chan, S. I., Bangerter, B. W., and Peter, H. H. (1966). *Proc. Nat. Acad. Sci. USA 55,* 720-727.

Chan, S. I., and Kreishman, G. P. (1970). *J. Am. Chem. Soc. 92,* 1102-1103.

Cheung, H. T., Feeny, J., Roberts, G. C. K., Williams, D. H., Ughetto, G., and Waring, M. J. (1978). *J. Am. Chem. Soc. 100,* 46-54.

Chiao, Y.-C. C., and Krugh, T. R. (1977). *Biochemistry 16,* 747-755.

Chien, M., Grollman, A. P., and Horwitz, S. B. (1977). *Biochemistry 16,* 3641-3647.

Cohn, M., and Hughes, T. R. Jr. (1962). *J. Biol. Chem. 237,* 176-181.

Corbaz, R., Ettlinger, L., Gaumann, E., Keller-Schierlien, W., Kradolfer, F., Neipp, L., Prelog, V., Reusser, P., and Zahner, H. (1957). *Helv. Chim. Acta 40,* 199-204.

Cozzone, P. J., and Jardetzky, O. (1976a). *Biochemistry 15,* 4853-4859.

Cozzone, P. J., and Jardetzky, O. (1976b). *Biochemistry 15,* 4860-4865.

Crothers, D. M. (1968). *Biopolymers 6,* 575-584.

Crothers, D. M., Sabol, S. L., Ratner, D. I., and Müller, W. (1968). *Biochemistry 7,* 1817-1823.

Crothers, D. M., Hilbers, C. W., and Shulman, R. G. (1973). *Proc. Nat. Acad. Sci. USA 70,* 2899-2901.

Danyluk, S. S., and Victor, T. A. (1970). *Jerusalem Symp. Quantum Chem. Biochem. 2,* 394-410.

Davanloo, P., and Crothers, D. M. (1976). *Biochemistry 15,*
    5299-5305.
Davidson, M. W., Griggs, B. G., Lopp, I. G., and Wilson, W. D.
    (1977). *Biochim. Biophys. Acta 479,* 378-390.
Delbarre, A., Roques, B. P., LePecq, J.-B., Lallemand, J. Y.,
    and Nguygen-Dat-Xuong (1976). *Biophys. Chem. 4,* 275-279.
Dimicoli, J.-L., and Helene, C. (1974). *Biochemistry 13,* 714-
    723.
Dwek, R. A. (1973). *In* "Nuclear Magnetic Resonance in Biochem-
    istry," pp. 48-61; 213-326, Oxford University Press, London.
Dugas, H. (1977). *Acc. Chem. Res. 10,* 47-54.
Early, T. A., Kearns, D. K., Burd, J. F., Larson, J. E., and
    Wells, R. D. (1977). *Biochemistry 16,* 541-551.
Emsley, J. W., Feeny, J., and Sutcliffe, L. H. (1965). *In*
    "High Resolution Nuclear Magnetic Resonance Spectroscopy."
    Pergamon Press, New York.
Ezra, F. S., Lee, C.-H., Kondo, N. S., Danyluk, S. S., and
    Sarma, R. H. (1977). *Biochemistry 16,* 1977-1987.
Gabbay, E. J., Sanford, K., Baxter, C. S., and Kapicak, L.
    (1973a). *Biochemistry 12,* 4021-4029.
Gabbay, E. J., Scofield, R. E., and Baxter, C. S. (1973b). *J.
    Am. Chem. Soc. 95,* 7850-7857.
Gabbay, E. J., Adawadkar, P. D., Kapicak, L., Pearce, S., and
    Wilson, W. D. (1976). *Biochemistry 15,* 152-157.
Gellert, M., Smith, C. E., Neville, D., and Felsenfeld, G. (1965).
    J. Mol. Biol. 11, 455-457.
Giessner-Prettre, C., and Pullman, B. (1976). *Biochem. Biophys.
    Res. Commun. 70,* 578-581.
Giessner-Prettre, C., Pullman, B., Borer, P. N., Kan, L.-S.,
    and Ts'o, P. O. P. (1976). *Biopolymers 15,* 2277-2286.
Gorenstein, D. G. (1975). *J. Am. Chem. Soc. 97,* 898-900.
Gorenstein, D. G., and Kar, D. (1975). *Biochem. Biophys. Res.
    Commun. 65,* 1073-1080.
Gorenstein, D. G., Findlay, J. B., Momii, R. K., Luxon, B. A.,
    and Kar, D. (1976). *Biochemistry 15,* 3796-3803.
Guéron, M. (1971). *FEBS Letters 19,* 264-266.
Guéron, M. and Shulman, R. G. (1975). *Proc. Nat. Acad. Sci. USA
    72,* 3482-3485.
Heller, M. J., Tu, A. T., and Maciel, G. E. (1974). *Biochem-
    istry 13,* 1623-1631.
Henry, D. W. (1976). *In* "Cancer Chemotherapy" (A. C. Sartorelli,
    ed.) pp. 15-57, Am. Chem. Soc., Washington, D.C.
Hilbers, C. W., and Patel, D. J. (1975). *Biochemistry 14,*
    2656-2660.
Hong, S.-J., and Piette, L. H. (1976). *Cancer Res. 36,* 1159-
    1171.
Hollstein, U. (1974). *Chem. Rev. 74,* 625-652.
Jain, S. C., Tsai, C.-C., and Sobell, H. M. (1977). *J. Mol.
    Biol. 114,* 317-331.

James, T. L. (1975). *In* "Nuclear Magnetic Resonance in Biochem-
    istry: Principles and Applications," Academic Press, New
    York.
Jones, C. R., and Kearns, D. R. (1975). *Biochemistry 14,* 2660-
    2665.
Jones, C. R., Bolton, P. H., and Kearns, D. R. (1978). *Bio-
    chemistry 17,* 601-607.
Kallenbach, N. R., and Berman, H. M. (1977). *Quart. Rev. Bio-
    phys. 10,* 137-236.
Kan, L. S., Borer, P. N., and Ts'o, P. O. P. (1975). *Biochemis-
    try 14,* 4864-4869.
Kastrup, R. V., Young, M. A., and Krugh, T. R. (1978). *Biochem-
    istry 17,* 4855-4865.
Kearns, D. R. (1977). *Ann. Rev. Biophys. Bioeng. 6,* 477-523.
Kondo, N. S., and Danyluk, S. S. (1972). *J. Am. Chem. Soc. 94,*
    5121-5122.
Kreishman, G. P., Chan, S. I., and Bauer, W. (1971). *J. Mol.
    Biol. 61,* 45-58.
Krugh, T. R. (1972). *Proc. Nat. Acad. Sci. USA 69,* 1911-1914.
Krugh, T. R. (1974). *In* "Molecular and Quantum Pharmacology"
    (Bergman, E. D., and Pullman, B., eds.) 7, pp. 465-471,
    Reidel Publ. Co., Dordrecht, Holland.
Krugh, T. R., and Chen, Y.-C. (1975). *Biochemistry 14,* 4912-
    4922.
Krugh, T. R., and Neely, J. W. (1973a). *Biochemistry 12,* 1775-
    1782.
Krugh, T. R., and Neely, J. W. (1973b). *Biochemistry 12,* 4418-
    4425.
Krugh, T. R., and Reinhardt, C. G. (1975). *J. Mol. Biol. 97,*
    133-162.
Krugh, T. R., and Young, M. A. (1977). *Nature 269,* 627-628.
Krugh, T. R., Wittlin, F. N., and Cramer, S. P. (1975). *Bio-
    polymers 14,* 197-210.
Krugh, T. R., Mooberry, E. S., and Chiao, Y.-C. C. (1977). *Bio-
    chemistry 16,* 740-747.
Krugh, T. R. (1976). *In* "Spin Labeling: Theory and Applications"
    (Berlinger, L. J., ed.) pp. 339-372, Academic Press, New
    York.
Krugh, T. R., Petersheim, M., Lin, S., and Hook, J. W. (1978).
    Unpublished experiments.
Lackner, H. (1975). *Angew. Chem. Intern. Ed. 14,* 375-386.
LaMar, G. N., Horrocks, Jr., W. Dew, and Holm, R. H. (eds.)
    (1973). NMR of Paramagnetic Molecules: Principles and Ap-
    plications. Academic Press, New York.
Lee, C.-H., and Sarma, R. H. (1976a). *Biochemistry 15,* 697-704.
Lee, C.-H., and Sarma, R. H. (1976b). *J. Am. Chem. Soc. 98,*
    3541-3548.
Lee, C.-H., and Tinoco, I. Jr. (1978). *Nature 274,* 609-610.
LePecq, J.-B. (1971). *In* "Methods of Biochemical Analysis"
    (Glick, D., ed.) *20,* pp. 41-86, John Wiley and Sons, New
    York.

LePecq, J.-B., and Paoletti, C. (1967). *J. Mol. Biol.* 27, 87-
    106.
LePecq, J.-B., Yot, P., and Paoletti, C. (1964). *CRH Acad. Sci.*
    259, 1786-1789.
Lerman, L. S. (1961). *J. Mol. Biol.* 3, 18-31.
Levy, G. C., and Nelson, G. L. (1972). *In* "Carbon-13 Nuclear
    Magnetic Resonance for Organic Chemists," Wiley-Inter-
    science, New York.
McGhee, J. D. (1976). *Biopolymers* 15, 1345-1375.
McGhee, J. D., and von Hippel, P. H. (1974). *J. Mol. Biol.* 86,
    469-489.
Meienhofer, J., and Atherton, E. (1977). *In* "Structure-Activity
    Relationships Among the Semisynthetic Antibiotics" (Perl-
    man, D., ed.), pp. 427-529, Academic Press, New York.
Mildvan, A. S., and Cohn, M. (1970). *Adv. Enzymol.* 33, 1-70.
Müller, W., and Crothers, D. M. (1968). *J. Mol. Biol.* 35, 251-
    290.
Müller, W., and Crothers, D. M. (1975). *Eur. J. Biochem.* 54,
    267-277.
Müller, W., Bunemann, H., and Dattagupta, N. (1975). *Eur. J.*
    Biochem. 54, 279-291.
Müller, W., and Emme, I. (1965). *Z. Naturforsch.* B20, 835-841.
Müller, W., and Gautier, F. (1975). *Eur. J. Biochem.* 54, 385-
    394.
Neidle, S., Archai, A., Taylor, G. L., Berman, H. L., Carrell,
    H. L., Glusker, J. P., and Stallings, W. C. (1977). *Nature*
    269, 304-307.
Olson, W. K. (1975). *Biopolymers* 14, 1797-1810.
Patel, D. J. (1974a). *Biochemistry* 13, 1476-1482.
Patel, D. J. (1974b). *Biochemistry* 13, 2388-2385.
Patel, D. J. (1974c). *Biochemistry* 13, 2396-2402.
Patel, D. J. (1976a). *Biopolymers* 13, 533-558.
Patel, D. J. (1976b). *Biochim. Biophys. Acta* 442, 98-108.
Patel, D. J. (1977a). *Biopolymers* 16, 1635-1656.
Patel, D. J. (1977b). *Biopolymers* 16, 2739-2754.
Patel, D. J. (1978). *Eur. J. Biochem.* 83, 453-464.
Patel, D. J., and Canuel, L. L. (1976a). *Proc. Nat. Acad. Sci.*
    USA 73, 674-678.
Patel, D. J., and Canuel, L. L. (1976b). *Proc. Nat. Acad. Sci.*
    USA 73, 3343-3347.
Patel, D. J., and Canuel, L. L. (1977a). *Biopolymers* 16, 857-
    873.
Patel, D. J., and Canuel, L. L. (1977b). *Proc. Nat. Acad. Sci.*
    USA 74, 2624-2628.
Patel, D. J., and Canuel, L. L. (1977c). *Proc. Nat. Acad. Sci.*
    USA 74, 5207-5211.
Patel, D. J., and Shen, C. (1978). *Proc. Nat. Acad. Sci. USA*
    75, 2553-2557.

Piette, L. H. (1974). *Fed. Proc. 33,* 1371.

Reed, G. H., Leigh, J. S., and Pearson, J. E. (1971). *J. Chem. Phys. 55,* 3311-3316.

Reinhardt, C. G., and Krugh, T. R. (1977). *Biochemistry 16,* 2890-2895.

Reinhardt, C. G., and Krugh, T. R. (1978). *Biochemistry 17,* 4845-4854.

Reuben, J., and Gabbay, E. J. (1975). *Biochemistry 14,* 1230-1235.

Reuben, J., Adawadkar, P., and Gabbay, E. J. (1976). *Biophys. Struct. Mech. 2,* 13-19.

Reuben, J., Baker, B. M., and Kallenbach, N. R. (1978). *Biochemistry 17,* 2915-2919.

Sakore, T. D., Jain, S. C., Tsai, C.-C., and Sobell, H. M. (1977). *Proc. Nat. Acad. Sci. USA 74,* 188-192.

Schara, R., and Müller, W. (1972). *Eur. J. Biochem. 29,* 210-216.

Seeman, N. C., Day, R. O., and Rich, A. (1975). *Nature 253,* 324-326.

Shulman, R. G., and Sternlicht, H. (1965). *J. Mol. Biol. 13,* 952-955.

Shulman, R. G., Sternlicht, H., and Wyluda, B. J. (1965). *J. Chem. Phys. 43,* 3116-3122.

Sinha, B., Cysyk, R. L., Millar, D. B., and Chignell, C. F. (1976). *J. Med. Chem. 19,* 994-998.

Sobell, H. M., Jain, S. C., Sakore, T. D., and Nordman, C. E. (1971). *Nature 231,* 200-205.

Sobell, H. M., and Jain, S. C. (1972). *J. Mol. Biol. 68,* 21-34.

Sobell, H. M. (1973). *Progr. Nucleic Acid Res. Mol. Biol. 13,* 153-190.

Sobell, H. M., Tsai, C.-C., Jain, S. C., and Gilbert, S. G. (1977). *J. Mol. Biol. 114,* 333-365.

Sobell, H. M., Jain, S. C., Sakore, T. D., Reddy, B. S., Bhandary, K. K., and Seshadri, T. P. (1978). *In Proc. Int. Symp. Biomol. Structure, Conformation, Function, Evolution, Madras, India, January 4-7, 1978* (Srinivasan, R., ed.) Academic Press, New York.

Stothers, J. B. (1972). *In* "Carbon-13 NMR Spectroscopy," Academic Press, New York.

Sundaralingam, M. (1973). *In* "Conformation of Biological Molecules and Polymers" (Bergmann, E. D., and Pullman, B., eds.) pp. 417-456. Academic Press, New York.

Swift, T. J. (1973). *In* "NMR of Paramagnetic Molecules" (LaMar, G. N., Horrocks, W. Dew. Jr., and Holm, R. H., eds.) pp. 53-83. Academic Press, New York.

Tao, T., Nelson, J. H., and Cantor, C. R. (1970). *Biochemistry 9,* 3514-3524.

Tewari, R., Nanda, R. K., and Govril, G. (1974). *Biopolymers 13*, 2015-2035.

Tsai, C.-C., Jain, S. C., and Sobell, H. M. (1975). *Proc. Nat. Acad. Sci. USA 72*, 628-632.

Tsai, C.-C., Jain, S. C., and Sobell, H. M. (1977). *J. Mol. Biol. 114*, 301-315.

Ts'o, P. O. P. (1974a). *In* "Basic Principles in Nucleic Acid Chemistry" Vol. I. Academic Press, New York.

Ts'o, P. O. P. (1974b). *In* "Basic Principles in Nucleic Acid Chemistry" Vol. II. Academic Press, New York.

Ugurbil, K., Shulman, R. G., and Brown, T. R. (1979). *In* "Biological Applications of Magnetic Resonance" (Shulman, R. G., ed.). Academic Press, New York.

Wakelin, L. P. G., and Waring, M. J. (1976). *Biochem. J. 157*, 721-740.

Wang, J. C. (1971). *Biochim. Biophys. Acta 232*, 246-251.

Waring, M. J. (1965). *J. Mol. Biol. 13*, 269-282.

Waring, M. (1970). *J. Mol. Biol. 54*, 247-279.

Waring, M. J., and Wakelin, L. P. G. (1974). *Nature* (London) *252*, 653-657.

Wartell, R. M., Larson, J. E., and Wells, R. D. (1974). *J. Biol. Chem. 249*, 6719-6731.

Wells, R. D. (1971). *In* "Progress in Molecular and Subcellular Biology" (Hahn, F. E., ed.) 2, pp. 21-32. Springer Verlag, New York.

Wells, B. D., and Cantor, C. R. (1977). *Nucleic Acid Res. 4*, 1667-1680.

Wells, R. D., Blakesley, R. W., Hardies, S. C., Horn, G. T., Larson, J. E., Selsing, E., Burd, J. F., Chan, H. W., Dodgson, J. B., Jensen, K. F., Nes, I. F., and Wartell, R. M. (1977). *In* "CRC Critical Reviews in Biochemistry" (Fasman, G. D., ed.), pp. 305-340. CRC Press, Cleveland, Ohio.

Yathindra, N., and Sundaralingam, M. (1974). *Proc. Nat. Acad. Sci. USA 71*, 3325-3328.

NUCLEAR OVERHAUSER EFFECTS ON PROTONS,
AND THEIR USE IN THE INVESTIGATION
OF STRUCTURES OF BIOMOLECULES

Aksel A. Bothner-By

Department of Chemistry
Carnegie-Mellon University
Pittsburgh, Pennsylvania

I. INTRODUCTION: DEFINITION OF THE NUCLEAR OVERHAUSER EFFECT

The nuclear Overhauser effect (NOE) is defined as the change in integrated intensity of the resonance signal(s) from one set of nuclei, when one or more of the transitions of a second set are saturated by application of a strong resonant radio-frequency (rf) field. The change in intensity is the result of cross relaxation between the observed and the irradiated nuclei. The irradiated nuclei are in a "hot" non-equilibrium state as a result of the absorption of energy. They may lose this energy by interaction with the set being observed, either by transfer of excitation to them, in which case the observed set will be "heated," the lower and higher energy states will be more nearly equally populated, and the intensity of the signal will be less than normal, or by events in which both nuclei lose energy in a coupled process to the molecular motions of the sample, in which case the observed set will be "cooled" and the intensity of the signal will be greater than normal. The sign and magnitude of the intensity change are thus an indication of the manner and degree to which the nuclei being irradiated contribute to the relaxation of the nuclei being observed. As such, it can give information about the relative spatial proximity of various sets of nuclei in the molecule and about their relative motions.

It should be pointed out that exactly the same information may be obtained by other techniques. The effectiveness of a nucleus in relaxing a neighboring nucleus by magnetic dipole interaction is proportional to the square of its gyromagnetic ratio and to $I(I + 1)$, where I is its spin, so that a deuterium ($I = 1$, $\gamma = 4107$ rad/sec G) will be only 0.063 times as effective as a proton ($I = 1/2$, $\gamma = 26752$ rad/sec G) in relaxing neighboring nuclei (Abragam, 1961). Exploiting this fact, Berry *et al.* (1977), have measured the relaxation rates of the anomeric protons of α- and β-methyl glucosides, each synthesized with normal methyl and trideuteriomethyl groups (Fig. 1). The difference in the relaxation rates yields directly the contribution of the $-OCH_3$ group to the anomeric proton relaxation. This contribution is larger in the β-anomer, suggesting a greater preference for a rotamer with the anomeric proton and methyl protons in close proximity. The same result can be obtained from NOE experiments (Lemieux, 1973).

The theory and application of the NOE have been presented and reviewed in a monograph by Noggle and Schirmer (1971), and in brief reviews by Bachers and Schaefer (1971), and by Bell and Saunders (1973).

In this review, aspects of the basic theory of the NOE will be presented, and the more recent applications of NOEs observed for protons in biomolecules will be surveyed, with a view to demonstrating the methods of approach, and the kinds of information that may be obtained, as well as some of the pitfalls lying in wait for the unwary.

NOEs are also observed for ^{13}C and other nuclei in biomolecules upon irradiation of attached protons, and use has been made of this phenomenon, particularly to study details of motion in complex molecules. Relaxation of ^{13}C nuclei tends to be dominated by the attached protons. The NOE of ^{13}C signals thus tends to be an all or nothing phenomenon, and direct measurements of ^{13}C relaxation times usually provide more useful information. Such applications have been discussed in recent books (Levy, 1976: Stothers, 1972), and these treatments complement this review.

A. *Basic Theory*

The basic theory of the NOE can be presented in several equivalent ways, each of which has some conceptual advantages. A brief presentation of the considerations for a two-spin problem follows. Since nuclei with spin greater than 1/2 generally possess electric quadrupole moments that dominate their relaxation, most useful studies have concentrated on nuclei with spin = 1/2, and only those will be considered here.

A system of two spin-1/2 nuclei, A and B, in a magnetic field may occupy any of four Zeeman levels, which for nuclei of sufficiently different resonant frequency, may be designated as αα, αβ, βα, ββ, using the usual convention (Pople *et al.*, 1959) (Fig. 2). Transitions or relaxation paths are labeled showing

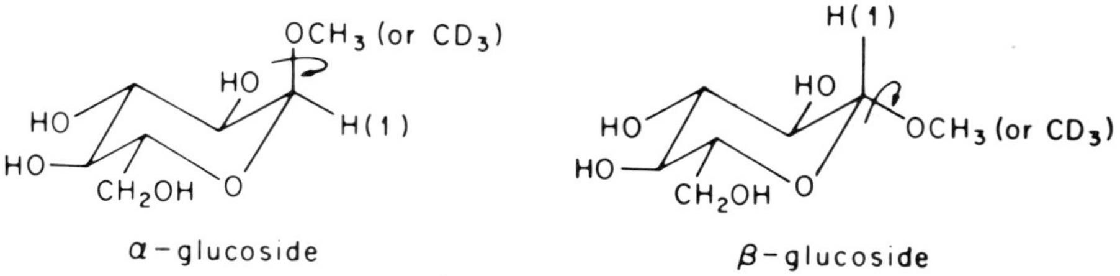

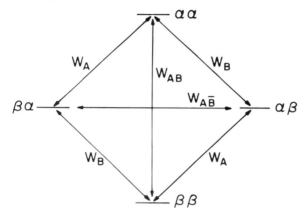

Fig. 1.   α- and β- trideuteriomethyl glucosides.

Fig. 2.   Zeeman levels and relaxation paths for two spin-
1/2 nuclei.

changes of the state of A only, of B only, of the two-quantum
process AB, and the zero quantum process A$\bar{\text{B}}$.

When the spin system is at thermal equilibrium, the four
levels are populated according to the Boltzmann distribution,
so that

$$n_{\alpha\beta}/n_{\beta\beta} = e^{-h\nu_A/kT} \tag{1}$$

where $n_{\alpha\beta}$ is the population of the αβ level, and ν_A is the
resonant frequency for the transition labeled A.   (For these
purposes, the spin-spin coupling J_{AB} is supposed very small
compared to ν_A, ν_B, or $\nu_A - \nu_B$ and may be ignored.)   Similar
equations relate the populations of the other levels.   Since
$h\nu_A \ll kT$, we can expand Eq. (1) and show that the equilibrium
population difference between the levels ββ and αβ is to a
close approximation $(n_{\beta\beta}H_0h/kT)\gamma_A \equiv \delta_A$, where H_0 is the strength

of the applied static field, $\hbar$ Planck's constant/2π, and γ_A the gyromagnetic ratio for the nucleus A.

In an NMR experiment, application of rf excitation at the resonant frequency ν_A will bring about transitions $\beta\beta \to \alpha\beta$ and $\alpha\beta \to \beta\beta$ with equal efficiency. There will be an initial net absorption or emission of energy from the spin system proportional to the difference in the population of the two levels. The rate of absorption or emission corresponds to the intensity of the observed transition, which therefore measure $n_{\beta\beta} - n_{\alpha\beta}$. The intensity of the second A transition $\beta\alpha \to \alpha\alpha$ similarly monitors $n_{\beta\alpha} - n_{\alpha\alpha}$. The two transitions may or may not overlap.

Application of a second strong irradiating field at the resonant frequency of the B nuclei will bring about the transitions $\beta\beta \rightleftarrows \beta\alpha$ and $\alpha\beta \rightleftarrows \alpha\alpha$. If the field is sufficiently strong, the natural relaxation processes of the B nuclei tending to restore the equilibrium population difference δ_B will be swamped. The populations of the two levels will be equalized: $n_{\beta\beta} - n_{\beta\alpha} = 0$ and $n_{\alpha\beta} - n_{\alpha\alpha} = 0$. This situation corresponds to an infinite spin temperature for the nuclei B (Eq. (1)) and we can speak of the B nuclei as "hot." After some time, a new steady state set of populations will be established for the four levels as a result of the operation of the various relaxation processes W_A, W_{AB}, and $W_{A\overline{B}}$ corresponding to the spontaneous rate of transitions between levels, as marked in Fig. 2. After any perturbation, the populations of the levels will tend to return to equilibrium according to the coupled Bloch equations:

$$\frac{dn_{\alpha\alpha}}{dt} = W_A\,(n_{\beta\alpha}-n_{\alpha\alpha}-\delta_A) + W_B\,(n_{\alpha\beta}-n_{\alpha\alpha}-\delta_B)$$

$$+ \; W_{AB}\,(n_{\beta\beta}-n_{\alpha\alpha}-\delta_A-\delta_B) \tag{2}$$

$$\frac{dn_{\alpha\beta}}{dt} = W_A\,(n_{\beta\beta}-n_{\alpha\beta}-\delta_A) + W_B\,(n_{\alpha\alpha}-n_{\alpha\beta}+\delta_B)$$

$$+ \; W_{A\overline{B}}\,(n_{\beta\alpha}-n_{\alpha\beta}+\delta_B-\delta_A) \tag{3}$$

$$\frac{dn_{\beta\alpha}}{dt} = W_A\,(n_{\alpha\alpha}-n_{\beta\alpha}+\delta_A) + W_B\,(n_{\beta\beta}-n_{\beta\alpha}-\delta_B)$$

$$+ \; W_{A\overline{B}}\,(n_{\alpha\beta}-n_{\beta\alpha}-\delta_B+\delta_A) \tag{4}$$

$$\frac{dn_{\alpha\beta}}{dt} = W_A(n_{\alpha\beta}-n_{\beta\beta}+\delta_A) + W_B(n_{\beta\alpha}-n_{\beta\beta}+\delta_B)$$

$$+ W_{AB}(n_{\alpha\alpha}-n_{\beta\beta}+\delta_A+\delta_B) \tag{5}$$

The most usual case involves irradiation of both B transitions at once. To allow for this irradiation, we set $n_{\alpha\alpha} = n_{\alpha\beta} = n_\alpha$ and $n_{\beta\alpha} = n_{\beta\beta} = n_\beta$. We further set all time derivatives equal to zero. Equations (2)-(5) then yield

$$\frac{n_\beta - n_\alpha}{\delta_A} = 1 + \frac{\delta_B}{\delta_A}\frac{W_{AB} - W_{A\overline{B}}}{W_{AB} + 2W_A + W_{A\overline{B}}} \tag{6}$$

$(n_\beta-n_\alpha)/\delta_A$ is just the ratio of the intensity of the $\beta \to \alpha$ transition in the steady state with B saturated to the normal intensity (δ_A). The expression on the right relates this to the equilibrium population differences for the B and A nuclei, and to a function of the several transition probabilities W_A, W_{AB}, and $W_{A\overline{B}}$. The relative magnitudes of these depend on the detailed nature of the relaxation processes that occur in the particular system studied. We shall proceed from simpler to more complex causes of behavior in this review.

The computational technique above allows one to calculate intensities, hence Overhauser effects, for irradiation of any one or more transitions in a two- or multispin system; most often, however, the experiment consists of irradiating all transitions of one set of nuclei (since they are grouped closely together) and observing the changes in the total intensity of the signals from other sets, since transitions within a set may not be resolved. There is then an advantage in combining equations of the type (2)-(5) to give the sum of the population differences for all pairs of levels corresponding to a particular set of nuclei. For the two-spin case above, these are

$$I_A = (n_{\alpha\beta}-n_{\beta\beta}) + (n_{\alpha\alpha}-n_{\beta\alpha}) \tag{7}$$

$$I_B = (n_{\beta\alpha}-n_{\beta\beta}) + (n_{\alpha\alpha}-n_{\alpha\beta}) \tag{8}$$

where I_A represents a net population difference or total magnetization of the A set of nuclei, etc. By forming appropriate sums and differences of (2)-(5), one obtains (Noggle and Schirmer, 1971)

$$\frac{I_A}{I_A^0} = 1 + \frac{I_B^0}{I_A^0}\frac{W_{AB} - W_{A\overline{B}}}{2W_A + W_{AB} + W_{A\overline{B}}} \tag{9}$$

as before.   Here $I_B{}^0$ and $I_A{}^0$ are the thermal equilibrium values
of I_B and I_A.
        In a useful notation introduced by Noggle and Schirmer
(1971), the relative change in intensity of the A nuclei is
symbolized by $f_A(B)$ and thus

$$f_a(B) = \frac{I_B^0}{I_A^0} \frac{W_{AB} - W_{A\bar{B}}}{2W_A + W_{AB} + W_{A\bar{B}}} \tag{10}$$

The cross relaxation $W_{AB} - W_{A\bar{B}}$ is abbreviated as σ_{AB} and the
total relaxation of A, $2W_A + W_{AB} + W_{A\bar{B}}$, is denoted by ρ_A (Solomon), so that a frequently used form of (21) is

$$f_A(B) = \frac{\gamma_B}{\gamma_A} \frac{\sigma_{AB}}{\rho_A} \tag{11}$$

For a three-spin system A, B, X, with X irradiated, this yields

$$f_a(X) = \frac{\gamma_X}{\gamma_A} \frac{\sigma_{AX}\rho_B - \sigma_{AB}\sigma_{BX}}{\rho_A\rho_B - \sigma_{AB}^2} \tag{12}$$

Note that ρ_A is the total relaxation of A by both B and X, as
well as any other sources.

II. LOW-MOLECULAR-WEIGHT COMPOUNDS--THE EXTREME NARROWING
    LIMIT

A. *Intramolecular Effects*

        We consider first a sample consisting of molecules contain-
ing two protons dissolved in a solvent that also contains protons.   The protons in the molecule have an appreciable chemical
shift, so that they may be separately observed and irradiated,
and negligible spin-spin coupling ($|J| \ll |\nu_A - \nu_B|$).   The mole-
cules reorient randomly as a result of molecular collisions,
so that the internuclear vector $\vec{r}_{AB}$ reorients with respect to
the externally applied magnetic field with a single correlation
time τ_c.   As a result of this motion, the magnitude and direction of the magnetic field that is experienced by nucleus A,
arising from the magnetic moment of nucleus B, changes.   The
z-component (parallel to H_0) of this fluctuating field contributes to $1/T_2$ (the transverse relaxation rate) of nucleus A,
while the X and Y components (perpendicular to H_0) contribute
both to $1/T_2$ and $1/T_1$ (the longitudinal relaxation rate).

Nucleus B is similarly relaxed by the fluctuating field produced by nucleus A. The probability of the various relaxation
processes in such a system has been evaluated by Solomon
(1955), who gives

$$W_A = \frac{\gamma^4 \hbar^2}{20 r_{AB}^6} \frac{3\tau_c}{1 + \omega_A^2 \tau_c^2} \tag{13}$$

$$W_{AB} = \frac{\gamma^4 \hbar^2}{20 r_{AB}^6} \frac{12\tau_c}{1 + 4\omega_A^2 \tau_c^2} \tag{14}$$

$$W_{\overline{AB}} = \frac{\gamma^4 \hbar^2}{20 r_{AB}^6} 2\tau_c \tag{15}$$

For small molecules τ_c is of the order of 10^{-10}–10^{-12} sec,
while the forseeable range of field strengths used in NMR spectrometers corresponds to proton-resonant frequencies of
3.6×10^8 to 3.6×10^9 rad/sec. Thus, $\omega^2 \tau_c^2$ in the denomina-
tors of Eqs. (13) and (14) is negligible, and the equations
can be reduced to the simpler form

$$W_A = \frac{\gamma^4 \hbar^2}{20 r_{AB}^6} 3\tau_c \tag{16}$$

$$W_{AB} = \frac{\gamma^4 \hbar^2}{20 r_{AB}^6} 12\tau_c \tag{17}$$

$$W_{\overline{AB}} = \frac{\gamma^4 \hbar^2}{20 r_{AB}^6} 2\tau_c \tag{18}$$

Substituting these values in Eq. (15), we obtain

$$n_\beta - n_\alpha / \delta_A = 1 + 1/2 \tag{19}$$

i.e., the intensity of the signal from the A nuclei is increased
by 50%. The individual relaxation probabilities depend on the
inverse sixth power of the internuclear distance, but the r_{AB}^6
factors have cancelled exactly in Eq. (22), so no distance information is implied.

The observation of the maximum 50% enhancement implies that
all of the relaxation of the observed nucleus is by magnetic
dipole interaction with the irradiated nucleus, i.e., that the
relaxation by other magnetic nuclei, whether attached to the
same molecule or in the solvent is negligible, either because
they are much further away (r^{-6} dependence), because the τ_c for
interaction is much shorter, or because γ^2 is small. Near
maximum effects have been observed in a number of cases. For
example, Fraser and Schuber (1970) report enhancements of 47
and 48% between pairs of protons in 1,11-dimethyl-5,7-dihydro-
dibenz(c,e)thiepin and its dioxide in dilute, degassed solution.

1. Reduction of NOE by Competing Relaxation--The Bell and Saunders Treatment and Application to Peptides

Normally less than the maximum 50% enhancement is observed.
This comes about because some other process is relaxing the A
nuclei. One possible mechanism is that the fluctuating fields
produced by magnetic nuclei in the solvent, as they approach,
dance about and recede from the A nuclei in the solute. Such
a mechanism will contribute to W_A (and to a lesser extent to
W_{AB} and $W_{A\bar{B}}$). If we write this contribution as W_A^* without
evaluating it exactly, we now get

$$\frac{n_\beta - n_\alpha}{\delta_A} = 1 + \frac{W_{AB} - W_{A\bar{B}}}{W_A^* + W_{AB} + 2W_A + W_{A\bar{B}}} \tag{20}$$

or, on substitutions in (16)-(18) and simplifying,

$$f_A(B) = \frac{1}{2 + cr_{AB}^6} \tag{21}$$

where $c = 40W_A^*/\gamma^4\hbar^2\tau_c$. This relation was pointed out by Bell
and Saunders (1970), who have developed a method of measuring
interproton distances based on it. They measured the NOE between pairs of protons, situated in a variety of low-molecular-
weight (<500) organic compounds of rigid structure, dissolved
in deuterochloroform, and obtained reasonable agreement with
Eq. (33) recast in the approximate form $1/f_A(B) = cr^6$, with
$c \sim 1.8 \times 10^{-2}$ Å^{-6}. The graph obtained by plotting log $f_A(B)$
versus log r_{AB} was reasonably linear, and internuclear distances for other proton pairs could presumably be evaluated
from it, after measurement of the NOE between them. It is important to list the assumptions in this model explicitly:

(1) All protons are equally exposed to the solvent and the
relaxation by solvent is equally effective so that W_A^*, hence

$\underline{C}$, is constant. If a proton is in a sterically hindered environment, or protected in any way from normal encounters with solvent molecules, anomalies are therefore expected.

(2) Proton A, which is observed, is not relaxed significantly by any intramolecular protons other than B. Any such contribution would have to be added into W_A^*, and the assumption of its constancy would be invalidated.

(3) No other sources of relaxation for proton A exist. Other possible sources include dissolved magnetic species (oxygen, free radicals), time-varying spin-spin coupling (e.g., with exchangeable protons or with quadrupolar nuclei), spin-rotation interaction, and at sufficiently high fields, chemical shift anisotropy. Each of these can also contribute to $\underline{W}_A^*$.

If these conditions are met, the NOE is a sensitive indicator of distance in the region where $\underline{c}r^6 \sim 2$. Figure 3 illustrates the dependence of NOE on distance. In this plot $\underline{r}$ is

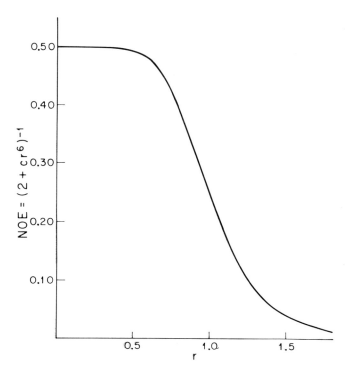

Fig. 3. Plot of $(2+\underline{c}r^6)^{-1}$ versus $\underline{r}$, with $\underline{c} = 2$, giving a +25% NOE at $\underline{r}$ = unit distance. If intensities are measured with a precision of 5%, distances from 0.7 to 1.4 times the unit distance may be estimated.

given in units of $(2/\underline{c})^{1/6}$, which must be established for each
solvent, temperature, etc. employed.

Recently Leach *et al.* (1977) have proposed the application
of this method to the determination of the backbone and side-
chain conformations in oligopeptides. Assuming planarity of
the amide group, the distance between an α-CH and the next following amide NH proton ($\underline{r}_1$, Fig. 4) is a function only of ψ,
the torsional angle about the CH-CONH bond. When values of ψ
corresponding to high energy conformations are excluded, it is
found that in the remaining range, each value of ψ corresponds
to one distance, hence one expected value of the NOE according
to Bell and Saunders' equation (see Fig. 3). They point out
that a different value of $\overset{\cdot}{\underline{C}}$ in the Bell and Saunders equation
may be required, since solvents other than deuterochloroform
may be used for the peptides, leading to different values of
$W_{\underline{A}}^*$. It is also necessary to consider whether other protons
(e.g., the β or γ protons of the side chain or the NH of the
same residue) can approach the α proton as closely or more
closely than the next following NH, violating condition 2 above.
If any of these distances ($\underline{r}_2$, $\underline{r}_3$, $\underline{r}_4$ in Fig. 4) are comparable
to or less than $\underline{r}_1$, the NOE will be sharply reduced.

Rae *et al.* (1977) and Gibbons *et al.* (1975) have measured
NOEs of the α protons of Gramicidin S (Fig. 5) in deuterated
dimethyl sulfoxide with selective saturation of the NH signals.

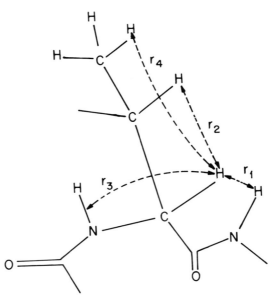

*Fig. 4. Labeling of some interproton distances in a pep-
tide.*

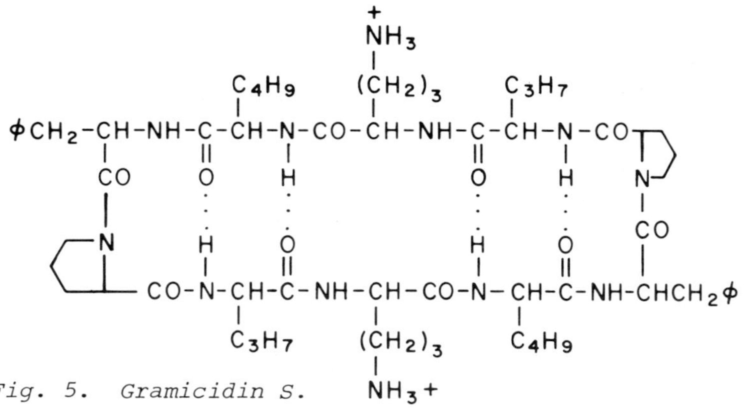

$$\overset{+}{N}H_3$$
$$|$$

$$
\begin{array}{c}
C_4H_9 \qquad (CH_2)_3 \qquad C_3H_7 \\
| \qquad\qquad | \qquad\qquad | \\
\phi CH_2-CH-NH-C-CH-N-CO-CH-NH-C-CH-N-CO \\
| \qquad\quad \| \quad | \qquad\qquad\qquad \| \quad | \quad\; N \\
CO \qquad O \quad H \qquad\qquad\qquad O \quad H \\
| \qquad\qquad \vdots \quad\; \vdots \qquad\qquad\qquad \vdots \quad\; \vdots \qquad CO \\
-N \qquad\quad H \quad O \qquad\qquad\quad H \quad O \quad | \\
| \qquad\qquad | \quad \| \qquad\qquad\qquad | \quad \| \\
CO-N-CH-C-NH-CH-CO-N-CH-C-NH-CHCH_2\phi \\
| \qquad\quad | \qquad\qquad\qquad | \\
C_3H_7 \qquad (CH_2)_3 \qquad C_4H_9 \\
| \\
NH_3{}^+
\end{array}
$$

Fig. 5.   Gramicidin S.

The extreme narrowing condition ($\omega\tau_c \ll 1$) is not met in this
case, which leads to a more complicated situation (see below).
However, Rae *et al.* applied a first-order correction to calculate the effects that would be observed in the extreme narrowing limit, then applied Bell and Saunders' equation. The distances calculated agreed reasonably well with the generally accepted solution structure of Gramicidin S (Dygert *et al.*, 1975).
A difficulty is that the deuterated dimethyl sulfoxide solvent
is of unknown effectiveness in relaxing the solute protons, so
that one might expect a different value of C in Bell and
Saunders' equation.

The same approach has also been employed by Khaled and
Urry (1976), in an investigation of repeat peptides of elastin.
The three peptides observed in DMSO solution were HCO-val-pro-
gly-gly-ome, t-BOC-val-pro-gly-val-gly-ome, and t-BOC-val-ala-
pro-gly-val-gly-ome. Upon irradiation of the pro-α proton in
each case, the NH signal of the adjacent gly residue increased
in intensity by 10 ± 1%; the same change in intensity occurred
if the NH proton was irradiated and the pro-α proton observed.
Application of Bell and Saunders' equation yields an H-H internuclear distance for this pair of 2.87 ± 0.02 Å. It has been
established by other means that the pro-gly sequence in these
peptides are the central residues in a β-turn. If the turn is
of type II (Fig. 6), the protons are separated by 2.4-2.7 Å,
as judged from models; if the β-turn is of type I, the separation would be closer to 3.4 Å. Thus the results are in better
agreement with the presence of a type II β-turn. A type I β-
turn is thought to occur in the valinomycin-K$^+$ complex. (Valinomycin, Fig. 7) In a corresponding experiment Khaled and Urry
observed a 1.9% NOE between the α-CH proton of L-lac and the NH
of L-val, barely larger than experimental error. The limits of
error corresponded to interproton distances of 3.43 and 4.90 Å,
which would agree with the presence of a type I β-turn, but not
with that of type II.

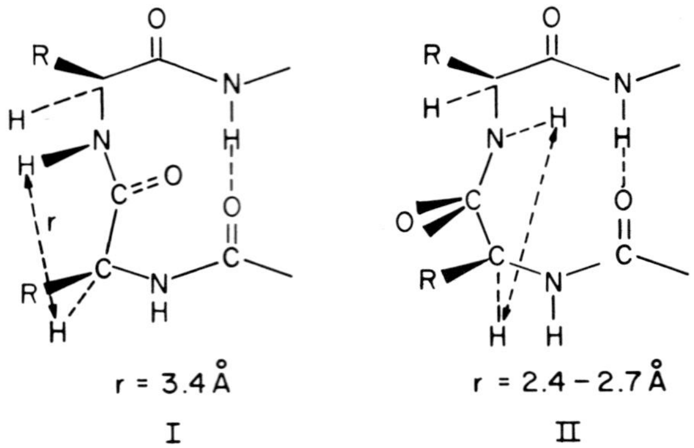

r = 3.4 Å   r = 2.4 – 2.7 Å

I   II

Fig. 6. Types I and II β turns in a peptide. r_1 is in Fig. 4.

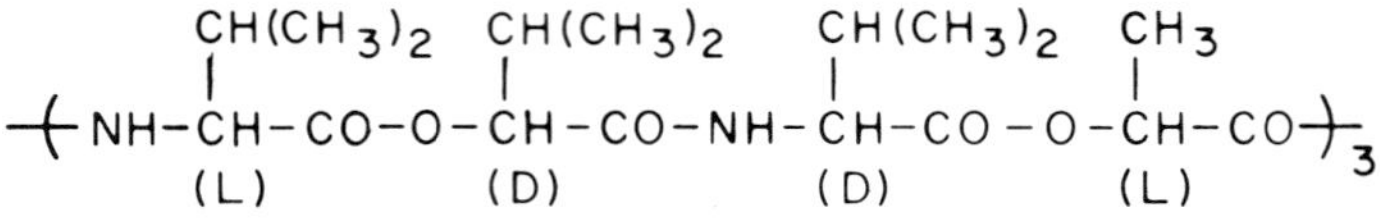

$$\text{(—NH—CH—CO—O—CH—CO—NH—CH—CO—O—CH—CO—)}_3$$

CH(CH$_3$)$_2$ CH(CH$_3$)$_2$ CH(CH$_3$)$_2$ CH$_3$

(L)   (D)   (D)   (L)

Fig. 7. Valinomycin.

2. Application to Nucleosides and Nucleotides

A problem that has seen extensive application of the NOE technique is the determination of conformation and structure in nucleosides and nucleotides. Of particular interest in these molecules are the torsional angles about the bond connecting the sugar to the purine or pyrimidine base (see Fig. 8) and about the C4-C5 bond in the sugar, as well as the anomeric configurations at Cl' (Fig. 9).

A simple qualitative approach to this problem may be taken. It is based on the premise that since magnetic dipole relaxation falls off as the sixth power of the internuclear distance, only the closest proton to the one whose signal is being observed will contribute an important Overhauser effect. Thus comparison of NOEs produced on irradiating either of two protons will establish which one is closer on the average to the observed one. This approach has been adopted in numerous studies. In an early example, Cushley et al. (1972) compared the NOEs produced on H(6) of pyrimidines and H(8) or H(2) of purines when either the H(4') proton or the H(5') protons of the ribose were irradiated. In most cases both anomers of the nucleoside were available, and were used. The results are shown in Table I. A clear pat-

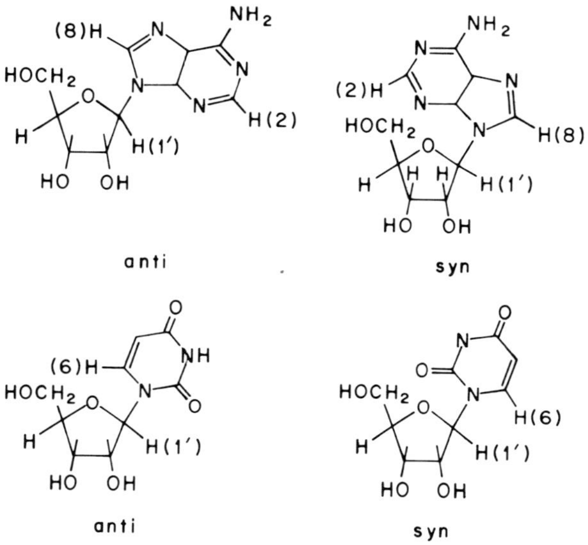

Fig. 8. *Syn* and *anti* conformations of nucleosides.

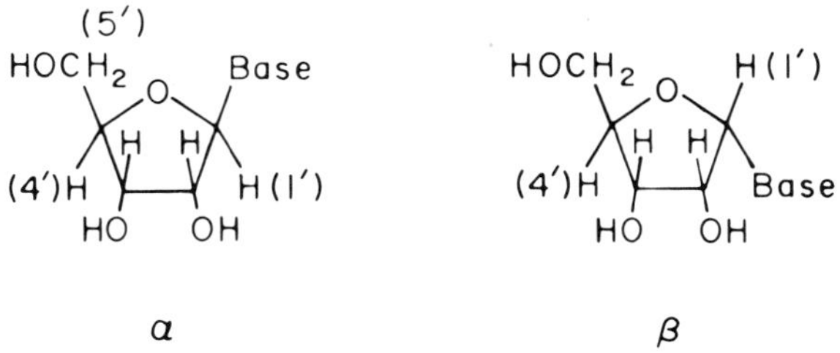

Fig. 9. α and β anomers of glucosides.

tern emerges: for the α anomers, the NOE produced by irradiating the H(5') protons is larger than that produced by irradiating the H(4') proton. For the β anomers, the reverse is
true.

Arnold *et al.* (1974) have applied this method to the determination of the torsional angle in pyridine nucleotides. At
neutral pH and low temperature they find, in general, that the
NOE produced on H(1') of the ribose is greater when H(2) of the
pyridine ring is irradiated than when H(6) is irradiated. If
it is assumed that H(2) in the *anti* conformation occupies the
same position as H(6) in the *syn* conformation, and that effects
from other spins are absent, this indicates a preference for the

TABLE I. NOE enhancements in nucleosides[a]

Compound	Anomer	Solvent	Conc. % wt/vol	Enhancement (%) H6 of pyrimidines H5' irradiated	Enhancement (%) H6 of pyrimidines H4' irradiated
1-D-ribfuranosyluracil	α	D_2O	15	5	--
1-D-ribfuranosyluracil	β	D_2O	15	0	13
1-(D-2'-deoxyribofuranosyl)-5-fluorouracil	α	D_2O	15	6,7	0
1-(D-2'-deoxyribofuranosyl)-5-fluorouracil	β	D_2O	15	-3	9
1-(D-2'-deoxyribofuranosyl)-5-fluorocytidine	α	D_2O	10	5	1
1-(D-2'-deoxyribofuranosyl)-5-fluorocytidine	β	D_2O	20	2	16

Compound	Anomer	Solvent	Conc. % wt/vol	Enhancement (%) H8, H2 of purines H5' irradiated H8	H2	H4' irradiated H8	H2
9-(D-2'-deoxy-2'-fluoroarabinofuranosyl)-adenine	α	d_6-DMSO	8	0	10	0	0
9-(D-2'-deoxy-2'-fluoroarabinofuranosyl)-adenine	β	d_6-DMSO	8	0	0	10	6
9-(D-ribofuranosyl)-hypoxanthine	α	d_6-DMSO	15	14-17	7-9	0-4	3-5

Compound	Anomer	Solvent	Conc. % wt/vol	Enhancement (%) H8 of purine H5' irradiated	H4' irradiated
9-(D-2',3'-isopropylidine ribofuranosyl)-guanine	α	d_6-DMSO	14	10	2

[a]Adapted by permission of the copyright owners from Cushley et al. (1972).

anti conformation of the pyridine ring. At low pH and high
temperatures, some of the *anti* preference is lost. Recently
Plochocka *et al.* (1977) observed with N(6)-dimethyl-2',3'-iso-
propylidineadenosine that the NOE produced on H(8) by irradiation of H(1') is +12%, while no effect is produced on irradiation of H(2') or H(3'), and on a similar basis concluded that
the purine ring is in the *syn* conformation.

   Another method of interpreting the NOE data has been introduced by Son *et al.* (1972). They also assume that there are no
multispin effects, and that the nucleoside may occupy only two
discrete conformations, *syn* and *anti*. Under these circumstances,
because of the $\underline{r}^{-6}$ dependence, H(8) of the purine ring will be
relaxed effectively only by H(1') when the purine is in the *syn*
conformation, and about equally by H(2') and H(3') when in the
anti conformation. If the NOE follows Bell and Saunders' equation, it follows that

$$\underline{P}_{syn} = \frac{r_{8,1'}^{6}\underline{f}_{8}(1')}{\underline{r}_{8,1'}^{6}\underline{f}_{8}(1') + \underline{r}_{8,2'}^{6}\underline{f}_{8}(2') + \underline{r}_{8,3'}^{6}\underline{f}_{8}(3')} \qquad (22)$$

Taking distances from molecular models, this reduces to

$$\underline{P}_{syn} \simeq \frac{3\underline{f}_{8}(1')}{3\underline{f}_{8}(1') + \underline{f}_{8}(2') + \underline{f}_{8}(3')} \qquad (23)$$

Using this relationship they deduced a substantial preference
for the syn form in guanosine-2'-phosphate and guanosine-3'-
phosphate, and about equal amounts of *syn* and *anti* conformers
in guanosine-5'-phosphate.

   The same method has been applied by Karpeisky and Yakovlev
(1975) to cytidine nucleotides, using the NOEs produced on H(6)
by irradiation of H(1'), H(2'), and H(3'). The interproton
distances were taken from x-ray data for two conformations of
the sugar ring, flat and 3'-endo, yielding minimum distances:
   (1) $\underline{r}_{6,1'}$ = 1.91, $\underline{r}_{6,2'}$ = 1.33, $\underline{r}_{6,3'}$ = 2.72.
   (2) $\underline{r}_{6,1'}$ = 1.91, $\underline{r}_{6,2'}$ = 1.33, $\underline{r}_{6,3'}$ = 2.40.
From the measured effects, they deduced that cytidine-2'-
phosphate, cytidine-3'-phosphate, and cytidine-2',3'-cyclic
phosphate exist 95-98% in the *syn* conformation, while, cyti-
dine-5'-phosphate and cytidine itself have only 10-35% *syn* conformation. The enhancements were independent of temperature
from 34 to 88°C, indicating a negligible enthalpy difference
between the conformers.

   Nande *et al.* (1974) used a modification of this approach in
investigating the conformation of β-pseudouridine, concluding

that it occurs approximately equally in *syn* and *anti* conformations.

The most rigorous approach to interpreting NOEs in terms of
conformation in these systems was introduced by Schirmer *et al.*
(1972). They demonstrated first that three different cases may
be envisioned, which require different treatments.

(1) Interconversion between the conformers is slow compared
to the interproton cross-relaxation rates, in which case the
NOEs are the population-weighted averages of the NOEs in the
conformers.

(2) Interconversion between conformers is fast compared to
the interproton cross-relaxation rates, in which case the σ and
ρ parameters themselves must be averaged according to the populations of the conformations and the NOEs calculated from these
average values.

(3) Interconversion is very fast, approaching the Larmor
frequency, in which case the motion itself contributes directly
to the relaxation.

The second case appears most likely for the molecules studied,
while the last case is unlikely except for extremely free motions (e.g., internal rotation of a methyl group). Schirmer *et
al.* tabulated values of σ and ρ for the values of ϕ, the torsional angle about the sugar-base bond. They then assumed that
two conformers were present, for each of which the fraction
with angle ϕ was given by a gaussian error curve, and for a
particular choice of the center and width of the gaussians, calculated (ρ_{ij}) and (σ_{ij}) for all proton pairs. Using the complete multispin theory, they calculated the expected Overhauser
effects. The position, width, and relative areas of the two
gaussians were varied to minimize the error in fit between the
observed and calculated NOEs. The assumption was made that
other sources of relaxation for each proton were equal; the
magnitude of this relaxation was also varied. The best fit indicated that it was small. For 2',3'-isopropylidine inosine,
the best fit was obtained with 0.8 population of the *syn* conformer ($\phi = 115°$) and 0.2 of the *anti* conformer ($\phi = 74°$).
Single gaussian fits were obtained for 2',3'-isopropylidineuri-
dine, for which more limited data were available, and indicated
a preferred *syn* conformation with $\phi = 141°$.

Davis and Hart (1972) have applied this exact method of
analysis to NOE data obtained on adenosine, inosine, guanosine,
and their 2',3'-isopropylidine derivatives, as well as xanthosine in DMSO solution. The results are displayed in Fig. 10,
which shows the probability distribution for the conformers of
the several compounds, as they were deduced.

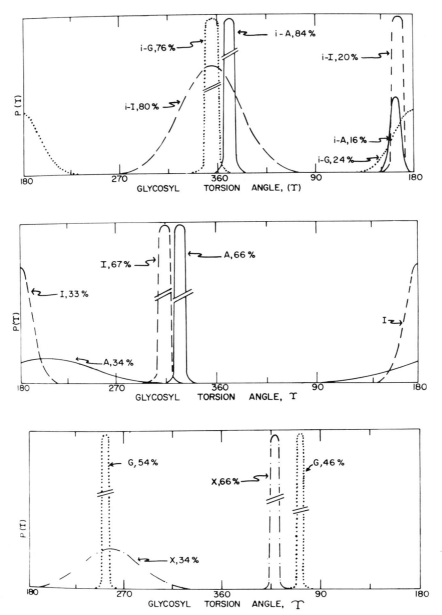

Fig. 10. Computed fits of glycosyl torsional angle to experimentally measured NOEs, assuming fast conformer interchange.
G, guanosine; A, adenosine; X, xanthosine; i, the 2',3'-0-iso-
propylidene derivative. (Reprinted by permission of the copyright owners from Davis and Hart, 1972).

3. Applications to Miscellaneous Small Biomolecules

a. Pyridoxal phosphate oxime. Tumanyan *et al.* (1974) have observed the NOEs between the oxime protons of the -CH=N-O-D group, the carbinol protons of the -CH$_2$OD and the aromatic proton in position 6 of the pyridine ring. Especially large effects were produced on the oxime proton (+22%) and the aromatic proton (+28%) on irradiating the carbinol protons. Such a result suggests that the oxime group occupies a conformation in which it is coplanar with the ring, with its proton close to the carbinol group, since in the coplanar conformation with H *anti* to the carbinol group, the interproton distance would be at least 4 Å, too far to produce a large effect. The quantitative interpretation is rendered difficult in this case by multispin effects and possible interactions with solvent deuterium oxide hydrogen bonded to one or more sites in the molecule.

b. N-acetylproline trideuteromethyl ester. Roques *et al.* (1976) have investigated this compound at sufficiently low temperature (-20 to +30°C) that distinct signals could be observed for the two conformers differing by the orientation of the N-acetyl group (Fig. 11). They irradiated each of the acetyl group proton signals and observed the NOEs produced on each of the α proton signals of the proline. At the higher temperatures, interconversion between the conformers was faster than methyl group relaxation, so that the methyl groups of one conformer, saturated by irradiation, entered the pool of methyl groups of the second conformer (saturation transfer), and NOEs were produced on the proline α-Hs of both conformers. At lower temperatures the slow conformer interconversion rendered this saturation transfer ineffective; in the minor conformer the NOE on the α-H was then +14%, in the major conformer 7%. Thus the minor conformer is the *cis* conformer, the major conformer is the *trans*. The 7% effect in the *trans* case probably arises from unavoidable excitation of the pro-β-proton. Combrisson and Roques (1976) have applied the same method to acetylpyrroles, formylindoline, and formyl and acetyltetrahydroquinoline, and were able in each case to establish the assignments of the signals from the frozen conformers and to demonstrate the onset of saturation transfer at higher temperatures.

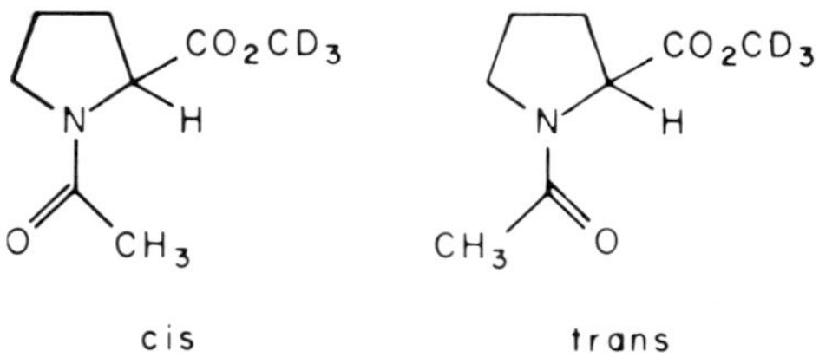

cis   trans

Fig. 11. Cis and trans isomers of N-acetylproline trideu-teriomethyl ester.

c. Dicoumarols. Laruelle and Godfroid (1976) have observed the spectra of dicoumarols substituted on the 1-carbon bridge (Fig. 12) dissolved in deuterochloroform. In this solvent, the exchange of the hydroxyl protons is slow, and the asymmetry in the molecule results in two distinct observable -OH•••O= signals. On irradiation of the bridge methyl group, an NOE of +25% is observed on the less shielded -OH•••O= group, while on irradiation of the bridge proton, a near maximum enhancement of +49% is observed for the more shielded proton. An unambiguous assignment of the two signals is thus possible.

B. Solvent Saturation Studies

1. Peptides

In the model of Bell and Saunders (1970), a competing relaxation by protons in the solvent reduces the observed NOE between a pair of protons in the solute below its maximum of 50%. In principle, one should be able to reverse this experiment: irradiation of the solvent should produce an NOE in a nucleus of the solute. It will be reduced below its maximum value of +50% by competing relaxation produced by neighboring protons in the solute. Thus, the magnitude of the effect will measure the relative "accessibility" of the solute proton to the solvent,

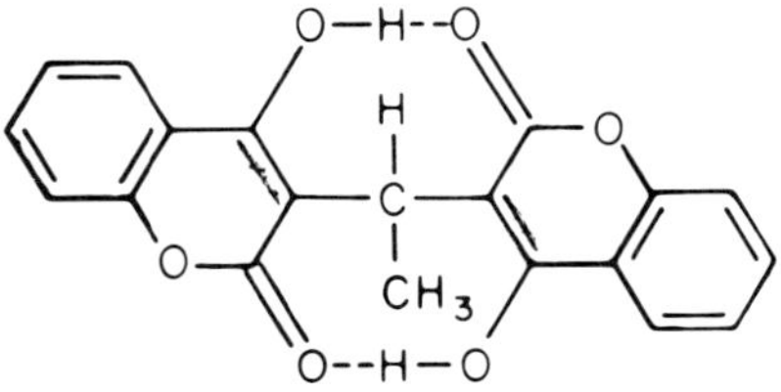

Fig. 12.   Bridge-substituted dicoumarol.

and "isolation" of the observed proton from other protons of
the solute. The idea forms the basis of a series of papers by
Pitner et al. (1974, 1975a,b). In the first of these, the
changes in the appearance of the spectrum of Angiotensin II'
(Asn-Arg-Val-Tyr-Val-His-Pro-Phe) in water solution on irradiation of the water resonance (Fig. 13) are discussed. Both
positive and negative effects are observed. The negative effects for arg^2NH, asn^1-trans-NH, and arg guanidine protons
arise by exchange of water protons into the amide or guanidine
groups of the peptide, rather than by magnetic dipolar interaction. If the exchange is fairly rapid, so that "hot" protons
from the water do not relax appreciably during their residence
in the amide grouping, the signal from the amide proton will
be decreased in intensity or vanish. The positive peaks, on
the other hand, must arise from a normal NOE, coming either
from protons of the solvent, or from other protons of the solute that are "hot." That the enhanced peaks are those of the
protons on the histidine, phenylalanine, and tyrosine rings (in
the latter case the protons ortho to the -OH only) speaks for
solvent water as the origin of the cross relaxation, and suggests that these protons are quite accessible to water. It is
possible that the histidine ring protons are enhanced by cross
relaxation with rapidly exchanging "hot" protons on the imidazole nitrogens (since the experiment was done at pH 3.0). The
lack of effect on the protons meta to the -OH in tyrosine might
arise from inaccessibility to water, or because competing relaxation with other protons of the solute (e.g., by β-tyr protons) is dominant.

Gramicidin S (Fig. 5) dissolved in methanol showed similar
behavior (Pitner et al., 1974a,b). Saturation of the methanol
OH resonance resulted in a decrease in the signal assigned to
phe-NH (Stern et al., 1968) as a result of proton exchange, and
no change in the phe ring proton signal. When the CH_3 resonance of the methanol was saturated, however, the phe ring proton signal intensity increased; qualitatively it is concluded
that the benzene ring protons are, on the average, closer to
the CH_3 than to the OH of methanol. A small decrease in the
phe NH signal also occurs, because the -OH of the methanol is
slightly "warmed," as a result of exchange modulation of its
spin coupling with the saturated CH_3 protons (Noggle and Schirmer, 1971). When the solvent is dimethyl sulfoxide, solvent
saturation leads only to enhancement of the phe ring proton signal, since no exchangeable protons are present. The solvent
saturation technique has been applied to oxytocin (Fig. 14) in
aqueous solution by Glickson et al. (1976a). When solvent
water was irradiated at 250 MHz, several changes were observed:

(1) The tyr NH signal decreased in intensity by 57%. This
is very probably a result of rapid exchange of this NH with
solvent water; if the average residence time of the protons in

$Asn^1-Arg^2-Val^3-Tyr^4-Val^5-His^6-Pro^7-Phe^8$

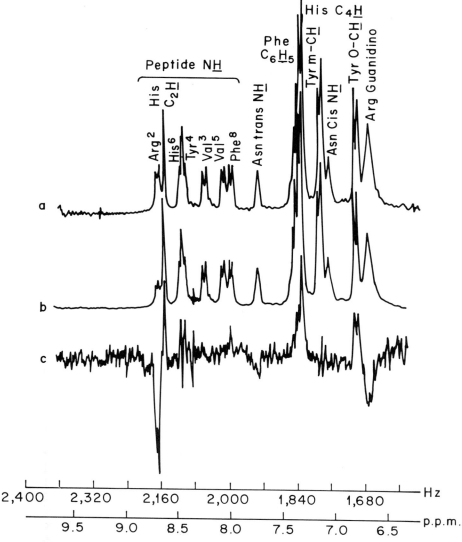

Fig. 13. (a) Normal 250 MHz spectrum of Angiotensin II in H_2O, showing the NH and aromatic proton signals. (b) The same, but with irradiation of the solvent H_2O. (c) Difference spectrum, (b - a). Reprinted by permission of the copyright owners from Pitner et al. (1974).

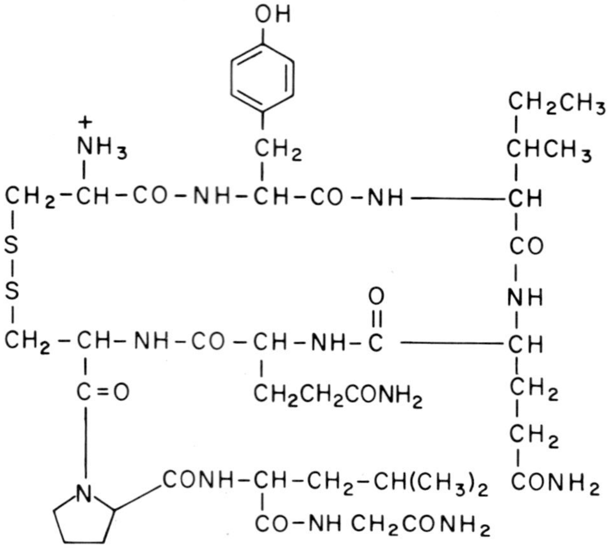

Fig. 14. Oxytocin.

the tyr NH site is short compared to their relaxation times,
the imported "hot" protons from the water will not relax appreciably, and the signal will diminish in intensity.

(2) Patterns corresponding to decoupling from α-CH protons
appear for other CONH protons (cys[6] and asn[5]). This is because the corresponding α-CH proton signals are very close to
the water signal, and it is not possible to irradiate the water
signal without applying rf power to them as well.

(3) The signals from the tyrosine ring protons *meta* and
ortho to the OH group increase in intensity by 9 and 14%, respectively. This is apparently a direct NOE between the water
and the ring protons, and suggests that the tyr ring protons
are relatively accessible to water and not in close contact
with other protons of the oxytocin skeleton. About the same
effect is observed when deamino-oxytocin is used in place of
oxytocin, demonstrating that it is not the exchangeable protons
on the terminal amino group that produce the NOE, but rather
the water itself.

2. Nucleosides

In investigations on nucleosides, the probability of aggregation has mainly been neglected. However, Luedemann and Von
Goldammer (1973) have demonstrated that aggregation can affect

the results. They reported that the NOEs observed for deoxyadenosine and for inosine were significantly different at 0.1
and at 0.025 $\underline{M}$ concentrations, with larger effects at the
higher concentrations. They ascribed the changes to base-
stacking. In a very interesting experiment, they observed an
intermolecular NOE: when benzene was introduced into an aqueous solution of deoxyadenosine and irradiated, intensity changes
in the H(8), H(2), and H(1') signals were observed. This suggests strongly that the stacking interactions occurring in solutions of nucleosides would be accompanied by the occurrence
of intermolecular NOEs, which would complicate the interpretation.

III. BIOPOLYMERS AND INTERMEDIATE TO LONG CORRELATION TIMES

As molecular weight of solute increases, or as the viscosity of solvent increases, τ_c becomes longer. When a higher static magnetic field strength is used in an NMR spectrometer, the
Larmor frequency ω of the observed nucleus is larger. At some
point, the product $\omega\tau_c$ becomes appreciable compared to unity.
After this point, Eqs. (13)-(15) cannot be approximated by (16)-
(18). Substituting the exact expressions into Eq. (6), we now
obtain for the change in intensity in a two-spin system

$$\frac{n_\beta - n_\alpha}{\delta_A} = 1 + \frac{\delta_B}{\delta_A} \frac{\dfrac{12\tau_c}{1+4\omega^2\tau_c^2} - 2\tau_c}{\dfrac{12\tau_c}{1 + 4\omega^2\tau_c^2} \dfrac{6\tau_c}{1 + \omega^2\tau_c^2} + 2\tau_c} \tag{24}$$

whence

$$f_A(B) = \frac{5 + \omega^2\tau_c^2 - 4\omega^4\tau_c^4}{10 + 23\omega^2\tau_c^2 + 4\omega^4\tau_c^4} \tag{25}$$

Figure 15 shows a plot of this function versus log $\omega\tau_c$. One
sees that as long as $\omega\tau_c < 0.1$, the terms in $\omega^2\tau_c^2$ contribute
very little and the NOE is close to +50%. When $\omega\tau_c = 10$ or
more, the numerator and denominator are dominated by the last
terms, so that $f_A(B) \approx -100\%$, which corresponds to disappearance
of the A signal. The change from positive to negative effect occurs at $\omega\tau_c = 1.118$. Pitner *et al.* (1976) and Glickson *et al.*
(1976b) report a study of valinomycin (Fig. 7) illustrating this

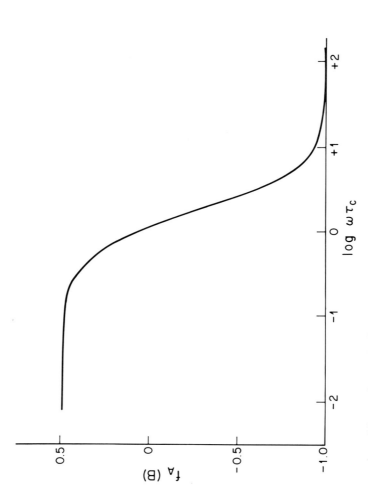

Fig. 15. Plot of the homonuclear NOE in a two-spin system, $f_A(B)$, versus log $\omega\tau_C$.

transition. The 90 and 250 MHz spectra of valinomycin in dimethyl sulfoxide solution with and without irradiation of the
D-val-NH resonance, and the difference spectra showing the
NOEs are displayed in Fig. 16. The effect on the hydroxyvaline
α proton is positive at 90 MHz, negative at 250 MHz. Amide-H
exchange rates are expected to be slow in dimethyl sulfoxide,
and spin-spin coupling between the protons in question negligible, so exchange-modulated spin-spin coupling is ruled out
as a source of the positive effect. No NOEs on other proton
signals are observed when irradiating the D-val-NH, so multispin effects are likewise ruled out as a source of the positive
effect. From Eq. (25), one deduces that τ_c for this proton
pair lies between 0.7 and 2.0 × 10^{-9} sec.

A. *Theory of Proton Magnetic Relaxation and Spin Diffusion in
Proteins*

Kalk and Berendson (1976) have presented a comprehensive
analysis of the expected relaxation behavior in higher molecular weight proteins, including a discussion of the expected
behavior of the interproton NOEs. As a model for a protein,
they took a rigid framework to which the protons were attached,
and assumed that the framework tumbles isotropically in solution with a single correlation time τ_c. Special allowance was
made for the possible rapid rotation of methyl groups about
their three-fold axes, with correlation time τ_i. Cross-correlation effects were assumed small. (Cross-correlation effects
must be taken into account when the relaxations of one nucleus
A by two other nuclei X and Y are not independent, because the
motion of the internuclear vectors $\vec{r}_{AX}$ and $\vec{r}_{AY}$ are not independent. The effects are most important if X and Y are identi-
cal. This can be seen intuitively from a consideration of the
motion, which simply interchanges the position of X and Y. If
X and Y are, for example, in the symmetrical αα state, the
field at A will be unchanged by this motion, and no relaxation
of A would be brought about. The assumption of independence
would, however, predict just twice the relaxation produced by
the presence of X or Y alone.)
Kalk and Berendson recast the master equations (2)-(5) in
theform

$$\frac{dI_A}{dt} = \sum R_{AB}(I_A-1) - \sum C_{AB}(I_A-I_B) \tag{26}$$

In this equation Kalk and Berendson's notation has been modified slightly to conform to the notation in this review. The
equilibrium magnetization for all protons, A, B, C has been set

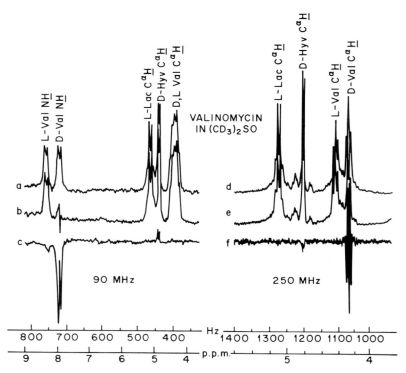

*Fig. 16. Demonstration of positive NOE at 90 MHz and nega-
tive NOE at 250 MHz for the D-hydroxyvaline α-protons in valinomycin in deuterodimethylsulfoxide solution. (a, d) Normal
spectra, (b, e) spectra with irradiation of the D-val NH resonance, (c, f) difference spectra. Reprinted by permission of
the copyright owners from Pitner et al. (1976).*

equal to 1, which can be done without loss of generality.   In
this equation,

$$R_{AB} = \frac{\gamma^4 \hbar^2}{10 r_{AB}^6} \left(\frac{3\tau_c}{1 + \omega^2 \tau_c^2} + \frac{12\tau_c}{1 + 4\omega^2 \tau_c^2} \right) \tag{27}$$

and

$$C_{AB} = \frac{\gamma^4 \hbar^2}{10 r_{AB}^6} \left(\tau_c - \frac{6\tau_c}{1 + 4\omega^2 \tau_c^2} \right) \tag{28}$$

Returning to the two-spin case, we can determine the NOE on A when irradiating B, by setting $\underline{I}_B = 0$ and $d\underline{I}_A/dt = 0$. From Eq. (26) we then get

$$-\underline{R}_{AB}(\underline{I}_A - 1) - \underline{C}_{AB}(\underline{I}_A) = 0 \qquad (29)$$

so

$$\underline{I}_A = \underline{R}_{AB}/(\underline{R}_{AB} + \underline{C}_{AB}) = 1/(1+\rho) \qquad (30)$$

and

$$\underline{f}_A(B) = -\rho/(1+\rho) \qquad (31)$$

where $\rho = \underline{C}_{AB}/\underline{R}_{AB}$. The values of ρ, obtained by substituting (27) and (28) in (30), vary from $-1/3$ when $\omega\tau_c \ll 1$, through zero, when $\omega\tau_c = 1.118$, to large positive values at long τ_c. According to Eq. (31), the NOE correspondingly changes from $+50\%$ through zero to -100%.

Another way of looking at Eq. (31) is thus. At long τ_c, $\underline{C}_{AB}$ becomes large and the second term in Eq. (26) dominates the relaxation behavior. This dominant relaxation process tends to equalize the magnetization $\underline{I}_A$, $\underline{I}_B$, $\underline{I}_C$, so that all protons relax together in a coupled way. Many intermediate weight proteins represent borderline cases where this phenomenon is important, but not dominant.

Kalk and Berendson demonstrate that methyl groups (and perhaps some other rapidly rotating groups) play a special role in the proton relaxation in proteins. From observation on analogous compounds, it is expected that the methyl groups will be relatively freely rotating about their internal three-fold axes, with a correlation time of the order of 10^{-10} sec. The problem of relaxation in such a situation has been considered by Woessner (1962), Werbelow and Marshall (1973), and Matson (1976). The situation is complicated by the necessity to take cross-correlation effects into account, but the results are clear. Even in the limit of slow overall isotropic tumbling of the protein framework, the rapid internal rotation of the methyl group results in a short $\underline{T}_1$ for the methyl protons. Thus, excitation applied to other protons of the protein will be transferred to methyl groups by spin diffusion, then lost by the relatively rapid internal relaxation in the methyl group.

The correctness of this picture was confirmed by Sykes *et al.* (1978), who measured the overall longitudinal relaxation rates for 11 proteins varying in molecular weight from 6500 (Bovine pancreatic trypsin inhibitor) to 150,000 (yeast alcohol dehydrogenase). The relaxation rate ranges for protons in these molecules followed closely the theoretical curve for relaxation of

the protons of a rapidly rotating methyl group attached to a
large rigid molecule of the corresponding molecular weight
(Woessner, 1962).

1. Proteins and Protein-Small Molecule Complexes

Proteins, even of low molecular weight (10,000-30,000),
will have correlation times sufficiently long that $\omega\tau_c > 1$, and
negative effects will in general be observed. Redfield and
Gupta (1971) first observed such effects in horse heart cytochrome C, between methyl groups and meso protons of the heme
ring. Recently, Keller and Wüthrich (1978) have made a thorough study of these effects, and have exploited them to obtain
complete assignments for the methyl, thioethyl, and meso proton
resonances of the porphyrin ring.

Negative effects were also observed in complexes of neurophysin-II with tripeptide analogs of oxytocin and vasopressin
(Balaram et al. 1972a,b, 1973). For example, it was observed
that irradiation at $\delta = 1.9$ ppm caused no change in the intensity of the protons ortho to the hydroxyl of the tyrosine in
ala-tyr-phe-NH$_2$ alone in solution, but a 36% decrease in intensity in the presence of 0.07 equivalents of neurophysin-II
(Fig. 17). The observed NOE is thus associated with the formation of the peptide-neurophysin complex, and it is distributed
into the pool of free peptide by exchange between the bound and
free states (see below). Maxima in the NOEs were observed for
several peptides that bind to neurophysin-II while irradiating
at $\delta = 1.90$, 3.10, and 6.86 ppm, and these presumably correspond to resonances of protons on the protein, which approach
closely the aromatic ring of the bound peptide.

The long correlation time required for the observation of
such negative effects is generally connected with the presence
of a more or less rigid structure in the protein. If the protein is in a random coil state, rapid segmental motion will
cause the effective τ_c to decrease, so that negative effects
are no longer observed. This was demonstrated by Bothner-By
and Gassend (1973) with a synthetic poly benzylglutamate,
(MW ~ 84,000), which gave a 60% decrease in the intensity of
the phenyl ring proton signal on irradiation of the benzylic
-CH$_2$-protons when the polymer was in trichloroethylene solution, but which gave no negative effect when the α-helix was
disrupted by the addition of trifluoroacetic acid.

The structure of the complex of N-formyltryptophan and α-
chymotrypsin has been determined by x-ray diffraction (Steitz
et al., 1969); in this structure, the 5 and 6 protons of the
indole ring approach closely the methyl groups of valine 213.
The complex of D-tryptophan and α-chymotrypsin was studied as
a test case by Bothner-By and Gassend (1973); maxima in the

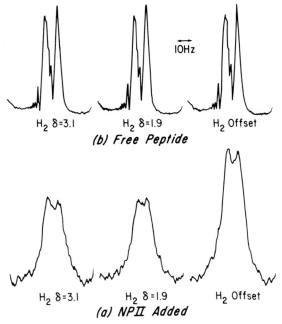

(b) *Free Peptide*

H$_2$ δ=3.1   H$_2$ δ=1.9   H$_2$ Offset

10Hz

H$_2$ δ=3.1   H$_2$ δ=1.9   H$_2$ Offset

(a) *NPII Added*

Fig. 17. Ala-tyr-phe amide tyr ortho protons. Comparison of signal intensities of the protons ortho to the tyrosine -OH in ala-tyr-phe-NH$_2$ under various conditions of irradiation. Above, peptide in the absence of neurophysin II. Below, in the presence of 0.066 equivalents of neurophysin II. H$_2$ offset denotes irradiation at δ ≈ 14 ppm, remote from any neurophysin resonances. Reprinted by permission of the copyright owners, from Balaram et al., 1973.

NOEs for protons 5 and 6 of the D-tryptophan were indeed observed on irradiation at δ = 1.0 ppm, where valine methyl signals are expected. Lesser effects were observed on the other tryptophan resonances. Again much less than the stoichiometric amount of α-chymotrypsin was required to elicit the full effect. They derived an equation relating fractions bound and free ($\underline{b}$ and $\underline{f}$), the lifetimes in the bound and free states (τ_b and τ_f), and the relaxation times for cross relaxation in the complex ($\underline{T}_{\overline{AB}}$) and total relaxation in the unbound state ($\underline{T}_{1f}$) to the observed signal intensity:

$$\underline{I}_t = \frac{\underline{fT}_{\overline{AB}}\underline{I}_0 + \underline{bT}_{1f}\underline{I}_{max} + \underline{f}\tau_b(\underline{fI}_0 + \underline{bI}_{max})}{\underline{fT}_{\overline{AB}} + \underline{bT}_{1f} + \underline{f}\tau_b} \tag{32}$$

where $\underline{I}_0$ is the normal intensity and $\underline{I}_{max}$ the intensity if all inhibitors were bound. Note that if exchange is fast the last terms in the numerator and denominator may be dropped. Under

these circumstances, half of the maximum effect will be ob-
served with $fT_{A\overline{B}} = bT_{1f}$, and since T_{1f} is normally long compared to $T_{A\overline{B}}$, only a small fraction of the inhibitor need be
bound.

If it can be safely assumed that the relaxation of a proton
in a protein is dominated by interaction with other well-iden-
tified nuclei, then the magnitude of the NOE may provide a
means for determining the internuclear τ_c. Campbell *et al.*
(1974) have reported the observation of a 45 ± 10% decrease in
the intensity of a signal at $\delta = 6.28$ arising from a benzenoid
ring proton of tryptophan on irradiation of the two flanking
protons. Solution of Eq. (25) yielded a τ_c of 1.5×10^{-9} sec,
which corresponded well with the times calculated from measurements of relaxation rates at two frequencies (1.1×10^{-9} sec)
or from hydrodynamic calculations (1.2×10^{-9} sec). The experiments were performed on lysozyme in aqueous solution at
68°C.

The possibility of using this method for identifying groups
at a binding site has been nicely exploited by James and Cohn
(1974) and by James (1976) in two studies on the catalytic site
of creatine kinase. In the first of these, the complexes of
creatine kinase with creatine, Mg^{2+}, ADP, and formate was examined. A large negative NOE on the intensity of the formate
proton was recorded when irradiation was applied at $\delta = 2.6$
ppm. The magnitude of the NOE was diminished by addition of ni-
trate, which displaces formate from the complex, and also by
omission of either Mg^{2+}, ADP, or creatine from the system. The
NOE was eliminated by dansylation of an essential lysine. From
the coincidence of the chemical shift of the irradiated group
with that of the ε-CH_2 of lysine, and the above evidence it was
concluded that the lysine terminal-NH_3^+ was involved in the binding of formate in the complex. Formate appears to occupy the
site of the transferable phosphate in the ATP-creatine complex,
so that the lysine is identified as the binding site for the
phosphate.

In the second study, attention was centered on the nucleotide binding site. On irradiation at 0.9 or 1.7 ppm to low
field of DSS, negative NOEs were observed for the H(2) of AMP,
ADP, ATP, and IDP of approximately equal magnitude, and the
magnitude was not affected by the presence or absence of creatine, Mg^{2+}, phosphate, formate, etc., nor by inactivation of
the enzyme with iodoacetemide. The NOE was eliminated by modification of an arginine. It was therefore concluded that the
effect arose from the β and γ methylene group of an arginine
that binds ADP via the guanidinyl group and the phosphate linkages.

Karpeisky and Yakovlev (1976) have applied this method to
the investigation of the complexes of cytidine -3'-phosphate
and cytidine-2'-phosphate to ribonuclease. They observed a 25%

decrease in intensity of the H(1') of cytidine-2-phosphate on
irradiation of the H(6) of the pyridine ring, and a 13% decrease when irradiating the H(2) of histidine-12. They conclude that the nucleotide is not in the *syn* conformation, that
the histidines are relatively immobile in the complex, and that
the distance between the ribose H(1') and the H(2) of his-12 is
~3.3 Å. Smaller effects were observed in the complex of ribonuclease with cytidine-3'-phsophate, and yielded a torsional
angle $\phi = -30°$ and a distance between the H(1') and H(2) of
his-12 of 3.6 Å.

   An extensive application of these effects to the assignment
of the 20 methyl resonances in the proton spectra of bovine pancreatic trypsin inhibitor has been reported by Wüthrich *et al.*
(1978). Three types of experiments were performed, permitting
all assignments to be made. In the first, NOE effects produced
by irradiating the methyl groups were observed under conditions
where the spin-coupling with the methyl groups was eliminated
(see Section IV). Under these conditions alanines should yield
singlets of reduced intensity corresponding to the α-protons,
while threonines should yield doublets of reduced intensity cor-
responding to the β-protons (unless $J_{\alpha\beta}$ or $\delta_{\alpha\beta}$ happens to be
small). In this way two of the three threonines were unambigu-
ously identified. In the second experiment, observations were
made at a temperature where the protein was half in the native
and half in a random-coil conformation. Irradiation of the
threonine peaks in the random-coil form resulted in transfer of
magnetization and consequent reduction of the intensity of the
threonine peaks in the native form, identifying the remaining
threonine resonances and confirming the first two. Finally, in
the third experiment, NOE effects on the methyls produced by irradiation of the previously assigned tyrosine ring resonances
were observed. From a model based on x-ray diffraction measurements the proximity of the methyl groups to the four rings could
be estimated, and the assignments completed on this basis. Additional confirmatory evidence was obtained by the use of lanthanide shift reagents and chemical modification of the protein.

B. *Higher-Molecular-Weight Proteins and Long Correlation Times*

   For proteins of higher molecular weight, correlation times
are of the order of tens of nanoseconds, so that $\omega\tau_c \gg 1$, and
the individual transition probabilities in the two-spin system
tend toward the limits

$$\underline{W}_A \to 0 \tag{33}$$

$$\underline{W}_{AB} \to 0 \tag{34}$$

$$\underline{W}_{A\overline{B}} \rightarrow \frac{\gamma^4 \hbar^2}{10 r_{\underline{A}B}^6} \tag{35}$$

Thus, $\underline{W}_{A\overline{B}}$, the adiabatic "flip-flop" transition, dominates. Substituting these values of $\underline{W}$ in the master equation for the homonuclear two-spin system (15) gives $\underline{f}_A(B) = -100\%$. A simple intuitive way to look at the situation is this: Nuclei B are turned antiparallel to $\underline{H}_0$ by the irradiating field, gaining energy and becoming hot. They lose the energy by the $\underline{W}_{A\overline{B}}$ process, heating A nuclei, which thereby become saturated. Loss of energy from A or B nuclei to the lattice is slow, since $\underline{W}_A$, $\underline{W}_B$, and $W_{A\overline{B}}$ are all very small.

In the multispin system of the protein, this effect may be propagated through the whole solute molecule: B heats A, which in turn heats C, which heats D, etc., a process called spin diffusion. Thus multispin effects can become very important.

The comparative success of the treatment of NOEs in low-molecular-weight systems ignoring multispin effects rests on the relatively small changes in magnetization produced in each spin. The effects are normally reduced by relaxation by the solvent in any event, so even in two-spin systems they seldom exceed 25%. If spin A on saturation produced a 20% change in the magnetization of spin B, then spin B will produce only 20% of its normal NOE on spin C (say $0.20 \times 0.20 = 0.04$), modifying the effect of spin A on spin C. Thus indirect effects will be of the order of $\underline{f}_d(\underline{s})^2$, usually close to experimental error. This will not be the case if the effect is close to -1, and careful attention to multispin effects will be necessary.

This consideration has been beautifully demonstrated by Hull and Sykes in a series of papers (1974, 1975a,b). They observed the fluorine resonances of a fluorinated alkaline phosphatase (MW $\simeq$ 84,000) obtained from bacteria grown on a culture containing 3-fluorotyrosine. The 11 tyrosines of the enzyme were replaced to a large extent by the fluorotyrosine, and 11 distinct fluorine signals could be observed. On irradiation at any point within the absorption envelope of the proton spectrum, most of the fluorine signals disappeared (Fig. 18). While this is not strictly a homonuclear case, the Larmor frequencies of ^1H and ^{19}F are sufficiently close that $(\omega_H - \omega_F)^2 \tau_c^2 \ll 1$, so that the relations (33), (34), (35) still apply to a good approximation.

From measurement of $\underline{T}_1$ on the various fluorine signals, Hull and Sykes were able to show that the fluorinated tyrosine rings were held essentially rigidly in the molecular framework, and that the effective correlation time was that for tumbling of the whole molecule, with a value of about 70 nsec. With a proton frequency of 100 MHz, this implies $\omega^2 \tau_c^2 \simeq 1700$, so that

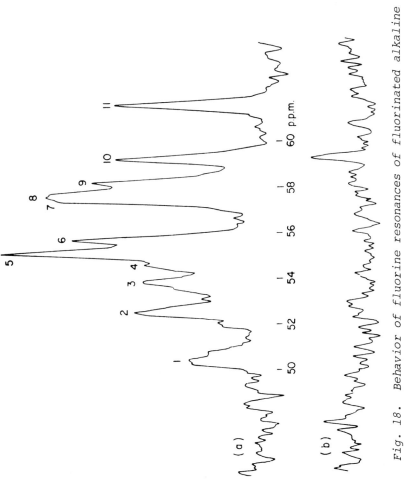

Fig. 18. Behavior of fluorine resonances of fluorinated alkaline phosphatase on irradiation of the protons. (a) Normal spectrum, (b) with irradiation of protons. Reprinted by permission of the copyright owners from Hull and Sykes (1975b).

$W_{\underline{AB}} \gg W_A$, $W_{\underline{AB}}$ and spin diffusion in the protons is very effective. Excitation applied at any point in the molecule diffuses without appreciable loss to the immediate vicinity of the fluorine atoms, and H-F cross relaxation then results in loss of the fluorine signal. In order to obtain a qualitative picture of the degree of selectivity of the NOE that might be observed in a typical system, Gerig (1977) has performed calculations on an arbitrarily selected array of spins, representing a p-fluorophenyl group in the neighborhood of a chain of three methylene groups (Fig. 19).

It was assumed that the entire array was rigidly held in the molecule, and that a single isotropic rotational correlation time τ_c characterizes the motion. Calculations were made with normal distances between nuclei within the fluorophenyl and trimethylene groups and for several intergroup distances $\underline{r}$. No relaxation from the solvent was included, but the effect of relaxation of the F by chemical shift anisotropy was included and becomes important at sufficiently high spectrometer frequencies. At a proton frequency of 100 MHz, the selectivity observed depends both on τ_c and $\underline{r}$.

For $\underline{r} = 2$ Å, and $\tau_c = 3 \times 10^{-9}$ sec, a selective effect is observed: irradiation of methylene group (1) produces a negative effect of about 30%, while irradiation of methylene group (2) or (3) produce negligible effects. For the same distance and $\tau = 10^{-8}$ sec, the effect produced on irradiation of (1) is -80%, while irradiation of group (2) gives an effect of -30% and of group (3) gives -8%. With sufficiently long correlation time ($\tau_c = 10^{-7}$ sec) irradiation of any of the methylene groups gives essentially complete loss of F signal (-100%).

As $\underline{r}$ increases, the effectiveness of cross relaxation within thetrimethylene group becomes much greater than the relaxation of F by H, so selectivity is lost. With $\underline{r} = 5.0$ Å, little effect is seen until τ_c exceeds 10^{-8} sec, and essentially the same diminution in F signal is observed regardless of which methylene group is irradiated.

Spin diffusion and transfer of magnetization may also take place at the interface between protein molecules and solvent water, as shown lv experiments reported by Stoesz et al. (1978).

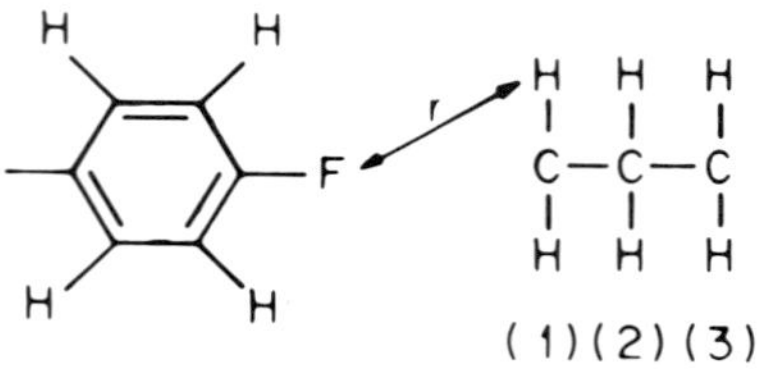

Fig. 19. Model for selectivity calculations performed by Gerig (1977).

If H_2O rather than D_2O is used as a solvent, the investigator often arranges for the H_2O to be effectively saturated, so that the very large peak from H_2O (110 $\underline{M}$ protons) will not interfere with the signals from the proteins (at millimolar concentrations). However, water may be hydrogen bonded to groups on the surface of the protein, and the effective τ_c then becomes long enough so spin diffusion from the water to the protein occurs, distorting and weakening its spectrum. Alternatively, the water protons may exchange into exchangeable sites on the surface, and spin diffusion within the protein will then cause intensity decreases in the protein spectrum.

Stoesz et al. observed such effects on the six observable hydrogen-bonded protons of the histidine side chains in superoxide dismutase. By studying deuterium exchange kinetics and the Overhauser effects accompanying irradiation of the protein side chain resonances in the region 0.0-3.5 ppm from DSS, they could establish that two of the resonances become saturated as a result of rapid exchange with solvent water, while four became saturated as a result of spin diffusion from the water, via the outer layers of the protein.

A test of the behavior in systems with long correlation times is possible by studying simpler compounds with well-separated resonances in viscous solvents, where the higher frictional coefficient leads to longer τ_c. Bothner-By and Johner (1977) adopted this approach in a study of Gramicidin S (Fig. 5) in deuterated ethylene glycol. From the correlation times in methanol and dimethylsulfoxide solutions earlier determined from ^{13}C relaxation times (Allerhand and Komoroski, 1973; Komoroski et al., 1975) and the Stokes formula for the frictional coefficient, τ_c at 25°C is calculated to be 14.3 nsec. With a proton frequency of 250 MHz, this yields $\omega \tau_c = 22$: thus spin diffusion may be expected to be important.

In a typical set of experiments, they observed that irradiation of individual resonances of the proline or phenylalanine protons produced an NOE of -15 to -20% in the intensity of the resonance from the protons of the phenylalanine ring, while irradiation of protons in the val and leu residues produced a smaller effect of -2 to -9%, and irradiation of protons of the orn residue near zero effects.

The behavior observed is understandable in the light of the behavior calculated for several idealized limiting systems. The systems considered are the "chain," "island," and "two islands in weak contact."

In the chain, it is supposed that a single chain of equally spaced nuclei can be traced through the molecule. Because of the $1/r^6$ dependence of the relaxation rates, interaction between nonnearest neighbors is ignored. Using the formalism of Kalk and Berendson, one obtains

$$2R(I_m-1) + 2CI_m - C(I_\ell+I_n) = 0 \tag{36}$$

where ℓ, m, n refer to successive protons in the chain. It follows that

$$I_m = \frac{2 + (I_\ell+I_n)}{2 + 2\rho} \tag{37}$$

where $\rho = C/R$. This recursion formula gives

$$I_k = 1 - f^k \tag{38}$$

where k is the position along the chain, with the saturated nucleus assigned $k = o$, and

$$f = (1 + \rho - \sqrt{1+2\rho})/\rho \tag{39}$$

Figure 20 is a plot illustrating how the NOE falls off along the chain of nuclei for various values of $\omega\tau_c$.

In the "island" model, several nuclei are supposed closely grouped and isolated from all others. The island contains $n + 1$ nuclei, the last of which is saturated. All nuclei are mutually relaxed by magnetic dipole interaction with the same parameters, R and C. The formalism of Kalk and Berendson gives

$$nR(I_a-1) + CI_a = 0 \tag{40}$$

since by symmetry $I_a - I_k = 0$ for all nonirradiated nuclei. This yields

$$I_a = n/(n+\rho) \tag{41}$$

Table II gives some typical values.

With two islands in weak contact, it is supposed that there are two identical islands, each with $(n+1)$ nuclei, each uniformly relaxed internally with parameters R and C; in addition, each nucleus in group one (a, b, c) is weakly relaxed by each nucleus in group two (m, n, o) with parameters R' and C'. Since the entire rigid assembly is governed by one correlation time τ_c, it also follows that $R/R' = C/C' = r'^6/r^6 = \gamma$. Solution of the appropriate equations for the case where one nucleus in group two is irradiated, gives

$$I_a = \frac{(n+1)\gamma+n^2 + (2n^2+3n+1)\gamma^2+(2n^2+2n+1)\gamma+n\rho}{(n+1)\gamma+n^2 + (n+1)\gamma+n\ 2(n+1)\ +1\ \rho+(n+1)(\gamma+1)\gamma\rho^2} \tag{42}$$

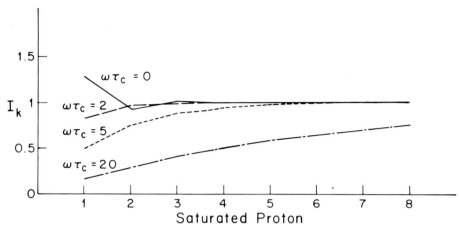

Fig. 20.   *Signal intensity expected for a proton in a uni-*
form linear chain of protons, when the proton n *positions away*
in the chain is saturated; the several curves show the behavior
expected for selected values of $\omega\tau_c$.

TABLE II.   *NOES expected in "island" model*

$\omega\tau_c$	n=2	4	6	8
0	+0.200	+0.091	+0.059	+0.043
1.118	0	0	0	0
2	-0.198	-0.110	-0.076	-0.058
5	-0.667	-0.501	-0.401	-0.334
10	-0.920	-0.809	-0.738	-0.679
20	-0.971	-0.943	-0.917	-0.893
50	-0.995	-0.990	-0.985	-0.980

and

$$\frac{I_m}{} = \frac{(n+1)\gamma + n^2 + 2\ (n+1)^2\gamma^2 + n(n+1)\gamma\ \rho}{(n+1)\gamma + n^2 + (n+1)\gamma + n\ 2(n+1)\gamma + 1\ \rho + n+1)\ (\gamma+1)\gamma\rho^2} \quad (43)$$

These equations reduce to the appropriate form of equation (41)
if $\gamma = 0$ or 1, corresponding to complete separation or merging
of the islands.   Figure 21 displays a plot of data for a selected case of weakly coupled islands with islands of three pro-

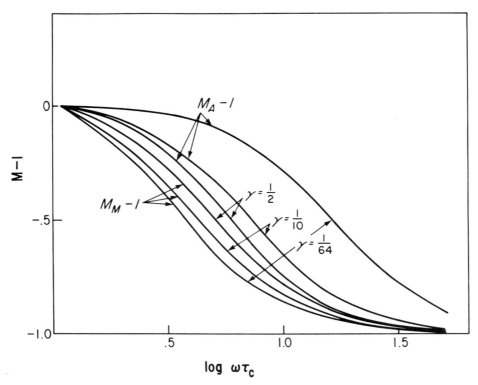

Fig. 21. NOEs expected in two weakly coupled "islands" of three protons each, as a function of γ, the ratio of interisland to intraisland relaxation rates.

tons each, for values of γ = 0.5, 0.1, and 0.0156 corresponding to $\underline{r}'/\underline{r}$ = 1.12, 1.47, and 2.00.

These three cases described limiting behavior from which a picture of the behavior in proteins may be obtained. Firstly, it should be noted that in all cases, if τ becomes long enough so that ρ is very large, specificity is lost, and all effects become -100%. In the range $\omega\tau_c$ = 2 to 30, specificity may occur. Because of the mutual tight coupling in islands, the effects tend to be large and to extend to more nuclei than in chains. The behavior of real groups of protons in proteins should be intermediate between these two extreme cases.

For Gramicidin S in deuterated ethylene glycol with $\omega\tau_c$ = 22 and ρ = 80, we expect little selectivity within islands, which we may equate the side chains or if two side chains are in intimate contact, all protons of both. Figure 22 shows a view of a conformation of Gramicidin S that fits well with the observed NOEs. The α proton of the phenylalanine residue is in close

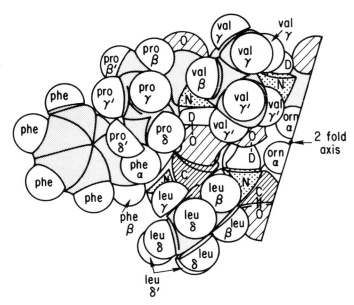

*Fig. 22. View of one-half of the symmetrical molecule Gra-
micidin S. The conformation shown is the generally accepted
one, except that the val side chain has been turned to demonstrate the gap between val-β and pro protons at closest approach.*

contact with a δ proton of the proline residue, coupling these
islands together, and this explains the large effects observed
on the intensity of the phenylalanine ring protons on irradiation ofproline protons. There are gaps, separating the val and
leu side chain protons from the phe-pro island, so lesser effects are observed on irradiating the leu or val protons. Finally, the orn protons are well removed from all others, and no
NOE is observed arising from them. It should be pointed out
that the peptide, as a result of solution in the deuterated solvent has exchanged all -NH- protons for deuterons, further isolating the islands from each other.

NOTE ON TECHNIQUES OF MEASUREMENT

The measurement of NOEs involves the measurement of integrated steady-state intensities of resonance signals in molecules in which a second set of nuclei are saturated. Measurement of NMR signal intensities is a difficult business at best,
and accuracy rarely exceeds ±5%. The situation is rendered
more difficult when there is spin-spin coupling between the observed and irradiated nuclei, since partial or complete decoupling may occur, and the shape as well as the area of the observed signal will change, causing area comparisons to be less
easily made. Difference spectra are also less easily interpretable under these circumstances.

A technique for circumventing these difficulties was suggested by Freeman and Hill (1971) and Freeman *et al.* (1972).
It was originally applied to separate observation of the NOEs
and decoupling in ^{13}C-H systems, but has since been applied to
homonuclear systems as well (for example, by Campbell *et al.*,
1974). The technique is applicable with the Fourier transform mode of observation and consists of either: (1) Gating
the irradiating power on during a time long compared to the
various $\underline{T_1}$ in the system, so that all levels become populated
at the required steady state, then gating the irradiating power off while applying a nonselective 90° pulse and recording
the free induction decay. (In this way, a normal (not decoupled) spectrum is obtained with the full Overhauser effect.);
(2) allowing the system to acquire normal thermal equilibrium,
then simultaneously gating the irradiating field on and applying the nonselective 90° pulse. Fourier transformation of the
free induction decay now yields a spectrum with decoupling, but
no Overhauser effect. In either case, appropriate difference
spectra may be used to estimate the total NOE.

REFERENCES

Abragam, A. (1961), "The Principles of Nuclear Magnetism."
    Clarendon Press, Oxford.
Allerhand, A., and Komoroski, R. A. (1973). *J. Am. Chem. Soc.*
    95, 8228-8232.
Arnold, L. J., Jr., Lee, C., and Van Sant, E. (1974). *Fed. Proc.*
    33, 1546.
Bachers, G. E., and Schaefer, T. (1971). *Chem. Rev. 71,* 617-26.
Balaram, P., Bothner-By, A. A., and Breslow, E. (1972a). *J. Am.
    Chem. Soc. 94,* 4017-4018.
Balaram, P., Bothner-By, A. A., and Breslow, E. (1973). *Bio-
    chemistry 12,* 4695-4704.

Balaram, P., Bothner-By, A. A., and Dadok, J. (1972b). *J. Am. Chem. Soc. 94*, 4015-4017.

Bell, R. A., and Saunders, J. K. (1970). *Can. J. Chem. 48*, 1114-1122.

Bell, R. A., and Saunders, J. K. (1973). *Topics Stereochem. 7*, 1-92.

Berry, J. M., Hall, L. D., Wilder, D. W., and Wong, K. F. (1977). *Carbohyd. Res. 54*, C22-C24.

Bothner-By, A. A., and Gassend, K. (1973). *Ann. N.Y. Acad. Sci. 222*, 668-676.

Bothner-By, A. A., and Johner, P. E. (1977). *Proc. XX CSI 7 ICAS (Prague)*, 355-372.

Campbell, I. D., Dobson, C. M., and Williams, R. J. P. (1974). *J. Chem. Soc. Comm.*, 888-889.

Combrisson, S., and Roques, B. P. (1976). *Tetrahedron 32*, 1507-1516.

Combrisson, S., Roques, B. P., and Oberlin, R. (1976). *Tetrahedron Lett. 38*, 3455-3458.

Cushley, R. J., Blitzer, B. L., and Lipsky, S. R. (1972). *Biochem. Biophys. Res. Commun. 48*, 1482-1488.

Davis, J. P., and Hart, P. A. (1972). *Tetrahedron 28*, 2883-2891.

Dickerson, R. E. (1971). *J. Biol. Chem. 246*, 1511.

Dygert, M., Gō,N., Scheraga, H. A. (1975). *Macromolecules 8*, 750-761.

Fraser, R. R., and Schuber, F. J. (1970). *Can. J. Chem. 48*, 633-640.

Freeman, R., and Hill, H. D. W. (1971). *J. Mag. Res. 5*, 278-280.

Freeman, R., Hill, H. D. W., and Kaptein, R. (1972). *J. Mag. Res. 7*, 327-329.

Gerig, J. T. (1977). *J. Am. Chem. Soc. 99*, 1721-1725.

Gibbons, W. A., Crepaux, D., Delayre, J., Dunand, J. J., Hajdukovich, G., and Wyssbrod, H. R. (1975). In "Peptides, Chemistry, Structure and Biology" (Walter, R., and Meienhofer, J., eds.), pp. 123-137. Ann Arbor Science Publ., Ann Arbor, Michigan.

Glickson, J. D., Gordon, S. L., Pitner, P. T., Agresti, D. G., and Walter, R. (1976a). *Biochemistry 15*, 5721-5729.

Glickson, J. D., Rowan, R., Pitner, P. T., Dadok, J., Bothner-By, A. A., and Walter, R. (1976b). *Biochemistry 15*, 1111-1119.

Hart, P. A. (1976). *J. Am. Chem. Soc. 98*, 3735-3737.

Hull, W. A., and Sykes, B. D. (1974). *Biochemistry 13*, 3431-3437.

Hull, W. E., and Sykes, B. D. (1975a). *J. Chem. Phys. 63*, 867-880.

Hull, W. E., and Sykes, B. D. (1975b). *J. Mol. Biol. 98*, 121-153.

James, T. L. (1976). *Biochemistry 15*, 4724-4730.

James, T. L., and Cohn, M. (1974). *J. Biol. Chem. 249*, 2599-2604.

Aksel A. Bothner-By

Kalk, A., and Berendson, H. J. C. (1976). *J. Mag. Res. 24,* 343-366.

Karpeiski, M. Y., and Yakovlev, G. I. (1975). *Bioorg. Khim. 1,* 749-757.

Karpeiski, M. Y., and Yakovlev, G. I. (1976). *Bioorg. Khim. 2,* 1221-1230.

Keller, R. M., and Wüthrich, K. (1978). *Biochim. Biophys. Acta 533,* 195-208.

Khaled, M. A., and Urry, D. W. (1976). *Biochem. Biophys. Res. Commun. 70,* 485-491.

Komoroski, R. A., Peat, I. R., and Levy, G. C. (1975). *Biochem. Biophys. Res. Commun. 65,* 272-278.

Laruelle, C., and Godfroid, J. J. (1976). *Can. J. Chem. 54,* 813-816.

Leach, S. J., Némethy, G., and Scheraga, H. A. (1977). *Biochem. Biophys. Res. Commun. 75,* 207-215.

Lemieux, R. V. (1973). *Ann. N.Y. Acad. Sci. 222,* 915-934.

Levy, G. C. (1976). "Topics in Carbon-13 NMR Spectroscopy. Wiley, New York.

Luedemann, H. D., and Von Goldammer, E. (1973). *Z. Naturforsch. Pt. C. 28,* 361-369.

Matson, G. B. (1976). *J. Chem. Phys. 65,* 4148-4154.

Nande, R. K., Tewari, R., Govil, G., and Smith, I. C. P. (1974). *Can. J. Chem. 52,* 371-375.

Noggle, J. H., and Schirmer, R. E. (1971). "The Nuclear Overhauser Effect--Chemical Applications," pp. 103-164. Academic Press, New York.

Pitner, T. P., Glickson, J. D., Dadok, J., and Marshall, G. R. (1974). *Nature 250,* 582-584.

Pitner, T. P., Glickson, J. D., Rowan, R., Dadok, J., and Bothner-By, A. A. (1975a). *J. Am. Chem. Soc. 97,* 5917-18.

Pitner, T. P., Glickson, J. D., Rowan, R., Dadok, J., and Bothner-By, A. A. (1975b). *In* "Peptides: Chemistry, Structure and Biology" (R. Walter and J. Meienhofer, eds.) pp. 159-164. Ann Arbor Science Publ., Ann Arbor, Michigan.

Pitner, T. P., Walter, R., and Glickson, J. D. (1976). *Biochem. Biophys. Res. Commun. 70,* 746-751.

Plochocka, D., Rabczenko, A., and Davis, D. B. (1977). *Biochim. Biophys. Acta 476,* 1-15.

Pople, J. A., Schneider, W. G., and Bernstein, H. J. (1959). "High Resolution Nuclear Magnetic Resonance." McGraw Hill, New York.

Rae, I. D., Stimson, E. R., and Scheraga, H. A. (1977). *Biochem. Biophys. Res. Commun. 77,* 225-229.

Redfield, A. G., and Gupta, R. K. (1971). *Cold Spring Harbor Symp. Quant. Biol. 36,* 405-411.

Roques, B. P., Combrisson, S., and Wasylishen, R. (1976). *Tetrahedron 32,* 1517-1521.

Schirmer, R. E., Davis, J. P., Noggle, J. H., and Hart, P. A. (1972). *J. Am. Chem. Soc. 94,* 2561-2572.

Solomon, I. (1955). *Phys. Rev. 99,* 559-563.

Son, T.-D., Guschlbauer, W., and Gueron, M. (1972). *J. Am. Chem. Soc. 94,* 7903-7911.

Steitz, R., Henderson, R., and Blow, D. M. (1969). *J. Mol. Biol. 46,* 337-347.

Stern, A., Gibbons, W. A., and Craig, L. C. (1968). *Proc. Nat. Acad. Sci. 61,* 734-741.

Stoesz, J. D., Redfield, A. G., and Malinowski, D. (1978). *FEBS Lett.* (in press).

Stothers, J. B. (1972). "Carbon-13 NMR Spectroscopy." Academic Press, New York.

Sykes, B. D., Hull, W. E., and Snyder, G. H. (1978). *Biophys. J. 21,* 137-146.

Tumanyan, V. G., Mameave, O. K., Bocharov, A. O., Ivanov, V. I., Karpeiski, M. Y., and Yakovlev, G. I. (1974). *Eur. J. Biochem. 50,* 119-127.

Werbelow, L. G., and Marshall, A. G. (1973). *J. Mag. Res. 11,* 299-309.

Boessner, D. E. (1962). *J. Chem. Phys. 36,* 1-6.

Wüthrich, K., Wagner, G., Richards, R., and Perkins, S. J. (178). *Biochemistry 17,* 2253-2263.

ACKNOWLEDGMENT

I am much indebted to my colleagues and collaborators whose work in this area has formed and given shape to the ideas outlined here: P. Balaram, K. Bose, J. Dadok, R. Gassand, J. D. Glickson, P. E. Johner, B. M. Harina, T. P. Pitner, and R. Rowan. Support from National Institutes of Health grant AM-16532 is gratefully acknowledged.

PULSED EPR STUDIES OF METALLOPROTEINS

W. B. Mims
Bell Laboratories
Murray Hill, New Jersey

J. Peisach
Departments of Molecular Pharmacology and Molecular Biology
Albert Einstein College of Medicine
Yeshiva University
Bronx, New York
and
Bell Laboratories
Murray Hill, New Jersey

In this chapter we describe some specialized EPR techniques that have recently been applied to biological problems such as the investigation of active site structures in paramagnetic metalloproteins. These methods are unique in that they make use of the relatively long coherence times of spin precession in dilute paramagnetic materials at low temperatures. In hydrogen-containing materials such as proteins, these coherence times are typically 2 or 3 μsec and correspond to spectral widths (in magnetic field units referred to $g = 2$) of $\simeq 0.1$ G.

Narrow lines such as these are not generally observed in EPR measurements of metalloprotein spectra. However, this is because the observed lines are really bands consisting of a large number of close-lying spectral components. In the accepted terminology, the lines are said to be "inhomogeneously broadened" and to be made up of a distribution of "spin packets." The intrinsic narrowness of the spin packets can be demonstrated by means of electron spin echo techniques (Geschwind, 1972; Salikhov *et al.*, 1976). Spin echo techniques can also be used, under the appropriate conditions, to detect perturbations of the resonance frequency that are small compared with the overall linewidth such as, for instance, perturbations originating from laboratory applied stresses or from the magnetic fields of weakly coupled nearby nuclei.

In order to give a more concrete picture of how these techniques are applied, and to illustrate the advantages and limitations of the method, we begin with a brief description of the spin echo mechanism, noting in particular those conditions which apply to materials of interest in biology. Following this, we describe two types of electron spin echo experiment that have proved useful for the study of paramagnetic centers in proteins.

I. THE ELECTRON SPIN ECHO MECHANISM

The interaction of spins with an oscillating electromagnetic field is most easily described by adopting a system of Cartesian coordinates that rotates at the resonance frequency. If the microwave magnetic field amplitude is resolved into two circularly polarized components, one of them can then be made to appear stationary. (The other is without any resonance effect). The situation is as shown in Fig. 1, where the z axis is taken along the dc magnetic field H_0 (the Zeeman field) and the x axis along the microwave field vector H_1. It can be seen that the dc field in the rotating coordinate system is reduced to a small residual component ΔH_0 along the z axis. This residual field represents any discrepancy that may exist between the microwave frequency ω_0 and the free precession frequency ω_S of the group of electron spins under consideration (both frequencies in radian units). On exact resonance, ΔH_0 disappears. Otherwise, it is given by

$$\Delta H_0 = (\omega_S - \omega_0)/\gamma \tag{1}$$

where $\gamma = g\mu_B/\hbar$ is the gyromagnetic ratio. In the absence of H_1, spin precession occurs about the z axis with an angular frequency $\gamma \Delta H_0$, this frequency together with the angular frequency ω_0 of the coordinate system making up the total spin precession frequency ω_S referred to the fixed or "laboratory" coordinate frame.

The effect of introducing H_1 is to change the axis of precession so that it lies along the resultant H_{eff} and H_1 and ΔH_0. Precession about H_{eff} reorients the spins in such a way that they no longer possess a constant component of magnetic moment along the z axis. Clearly the change will only be significant if H_{eff} makes an appreciable angle with the z axis (that is, if $\Delta H_0 < H_1$). Substantial reorientations only occur for those spins which lie in a band $\cong 2H_1$ wide, centered on the exact resonance setting. Elsewhere the effect of the microwave

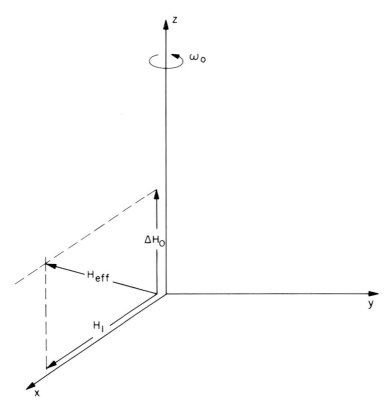

Fig. 1.   *Diagram showing magnetic fields in a coordinate
system that rotates with an angular velocity* $\omega_0 = 2\pi f_0$, *where
f_0 is the frequency of the resonance spectrometer.   In this
rotating system of coordinates, the microwave magnetic field
vector $\underline{H}_1$ is stationary.   The field $\Delta \underline{H}_0$ represents the discrepancy between the Larmor frequency of the spins and the
frequency of the spectrometer [Eq. (1)].   For spins that are
exactly on resonance H_0 vanishes and precession takes place
about $\underline{H}_1$.   Otherwise spin precession takes place about $\underline{H}_{eff}$,
the resultant of $\underline{H}_1$ and $\Delta \underline{H}_0$.*

field is a minor one.   It is negligible if $\Delta \underline{H}_0 \gg \underline{H}_1$, that is
to say, for spins with frequencies ω_S displaced so far from ω_0
as to be effectively off resonance.   It may be noted that in a
standard EPR experiment utilizing microwave field intensities
of a few milligauss, only a very narrow band of spins is at any
instant being driven by $\underline{H}_1$.   The band is much wider, however,
in electron spin echo experiments where power levels corres-
ponding to amplitudes $\underline{H}_1 \stackrel{\sim}{=} 10$ G are commonly employed.

Electron spin echoes are generated by applying two microwave pulses in succession to a sample placed in a suitable cavity or microwave circuit. If these two pulses are separated
by a time interval τ, then a third pulse, the spin echo, will
be emitted by the sample as an interval τ after the second microwave pulse as shown in Fig. 2. In the general case where
the resonance line is broad and the pulse durations arbitrary,
the effect of applying these pulses can be somewhat complicated
and difficult to analyze. In order to keep the description
simple, we shall therefore assume the following restrictive
conditions:

(a) $\Delta H_0 \ll H_1$ (i.e., the resonance line is narrow in relation to H_1). This will enable us to ignore ΔH_0 during the
relatively short times for which microwave power is applied,
and hence to assume that during the pulses the spins nutate
about H_1 (the $\underline{x}$ axis).
(b) Pulse I is timed to last long enough for a 90° nutation about H_1.
(c) Pulse II is twice as long as pulse I and therefore
causes a 180° nutation about H_1.

Echoes can, in fact, be generated in broad resonance lines by
pulses that do not have the 90-180° property. (For practical
reasons, it is often easiest to use equal-length pulses as in
Fig. 2.) It should be noted, however, that the pulses must be
intense enough and last long enough to produce substantial reorientations of the spins if the spin echo is to be observable.

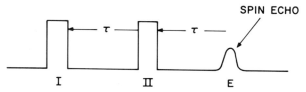

*Fig. 2. Sequence of pulses in a spin echo experiment. I
and II are pulses of microwave power generated by a transmitter
such as a magnetron or traveling-wave tube and applied to the
sample in a resonant cavity. The third pulse, the spin echo,
is emitted by the sample itself. This pulse is relatively weak
and is detected by a sensitive receiver. In the simple description of the echo-generating mechanism given in the text, pulse
II is twice as long as pulse I. This is not essential, however, and in practice it is usually more convenient to make the
two pulses equal in length.*

For the same reason spins must lie within $\simeq \pm H_1$ of exact resonance to make any substantial contribution to the echo signal.

The sequence of events initiated by applying microwave
pulses under the conditions noted above is illustrated in Fig.
3. Figure 3a shows the situation at the beginning of the spin
echo cycle. The net magnetization $\underline{M}$ due to the whole assembly
of precessing spins lies along the $\underline{z}$ axis. (Components along
other axes cancel on account of the randomness of the phases of
precession of individual spins.) The effect of pulse I is to
cause $\underline{M}$ to nutate about $\underline{H}_1$, the pulse ending when $\underline{M}$ lies along
the $+\underline{y}$ axis. A relatively long period follows in which there
is no microwave field and the only fields are the small residual fields $\Delta \underline{H}_0$. Since these fields are spread over a band
$\simeq 2\underline{H}_1$ wide and correspond to spins for which ω_S can be either
above or below ω_0, they cause varying degrees of precession.
The situation some time after pulse I is illustrated in Fig.
3c. The vector $\underline{M}$ has broken up into a number of components of
which $\underline{M}_1$ and $\underline{M}_2$ represent spins precessing faster than the coordinate system (direction F) and $\underline{M}_4$ and $\underline{M}_5$ represent spins
precessing more slowly (direction S).

In practical cases the breakup shown in Fig. 3c generally
occurs quite soon after pulse I, and the magnetization is spread
out evenly in the $\underline{xy}$ plane long before pulse II is applied.
For convenience in the illustration, let us suppose, however,
that pulse II begins when the magnetization vectors are distributed as in Fig. 3c. Pulse II will then cause each of the vectors to nutate by 180° about $\underline{H}_1$, resulting in the new distribution shown in Fig. 3d. The vectors are, of course, still in
the $\underline{xy}$ plane but, as can be seen by comparing Figs. 3c and 3d,
the effect of the pulse has been to handicap the more rapidly
precessing spins (vectors $\underline{M}_1$, $\underline{M}_2$) by setting them behind the
average (i.e., behind vector $\underline{M}_3$), and to confer an advantage on
the slower spins (vectors $\underline{M}_4$, $\underline{M}_5$) by setting them ahead. This
artificial arrangement will, of course, at once begin to break
down since the spin precession frequencies remain the same as
they were before the pulse; $\underline{M}_1$ and $\underline{M}_2$ will catch up and $\underline{M}_4$ and
$\underline{M}_5$ will fall back until, as in Fig. 3e, all the spins are in
phase. The resultant magnetization vector appears along the
$-\underline{y}$ axis in Fig. 3e. In the laboratory coordinate frame it is
rotating with an angular frequency of ω_0 and generates a signal
in the microwave circuit by the familiar dynamo mechanism. The
signal will be a short one, building up to a maximum as the
vectors come into phase and decaying as they fan out once more
in the $\underline{xy}$ plane.

It may be noted that, insofar as the mechanism described
above is concerned, echoes might be generated for any value of
τ, however long. This is because it has been tacitly assumed
that each spin has its own characteristic unvarying precession
frequency and does not interact with any fields other than $\underline{H}_1$

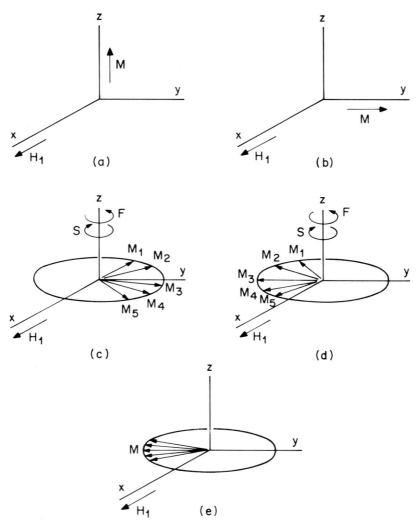

Fig. 3. *Magnetic moments representing the spins that take part in the microwave resonance interaction at various stages in the spin echo cycle of events. The coordinate axes rotate at ω_0 as in Fig. 1. (a,b) Before and after the $90°$ pulse. (c,d) Before and after the $180°$ pulse. F denotes the direction of rotation of spins whose Larmor frequency is slightly higher than the spectrometer frequency. S denotes the direction of rotation of spins whose Larmor frequency is lower. (e) Phase convergence of all spins at the time of the echo signal. (Reprinted with permission from the Physical Review.)*

and $\underline{H}_0$. Lifetime broadening (associated with spin lattice relaxation) and the random fluctuations in the local magnetic
field that occur in all paramagnetic materials have been ignored. In actual fact the spin echo signals will become weaker as τ is lengthened, the whole echo experiment being limited
by a characteristic "phase memory time" that describes the
duration of the coherent precession induced by the first microwave pulse. At helium temperatures and in protein samples, the
limiting factor is usually the random local field fluctuation
due to spin flip-flops between protons. Typically this results
in a phase memory time ≈ 3 µsec measured from pulse I. Shorter
times may be encountered in materials with large $\underline{g}$ values, the
spins being more susceptible to local magnetic field disturb-
ances in such cases, and very short times will occur if the
sample is concentrated enough for electron spin-spin interaction to be a significant factor. Frozen glassy unoriented
samples of up to 10 $\underline{mM}$ concentration have been found to be
usable for materials containing low spin Fe^{3+}, Cu^{2+}, or rare
earth ions, but attention must be paid to the possibility of
aggregation, which could significantly increase the spin-spin
interaction. Interactions that would reduce the phase memory
time in an electron spin echo experiment correspond to broadening effects ≈ 0.1 G and are, as pointed out earlier, orders of
magnitude smaller than those which would show up in a typical
EPR study.

The brevity of the phase memory time in biological materials
establishes a time scale that dictates the duration of the mi-
crowave pulses $\underline{t}_p$, the microwave power levels, the cavity $\underline{Q}$,
and the performance standards of the associated circuitry. The
values $\underline{t}_p = 20$ nsec, $\underline{H}_1 \sim 10$ G, $\underline{Q} = 100$ have proved satisfactory. Large values of $\underline{H}_1$ taken in conjunction with the low $\underline{Q}$
(essential in order to maintain a short cavity ring-down time
after each pulse) dictate the use of microwave sources such as
magnetrons or traveling-wave tubes to provide the necessary
several hundred watts of power. Heating of the sample is usually negligible, however, since the duty cycle of the system
(limited by the spin lattice time $\underline{T}_1$) is less than 1 in 10^4 for
a liquid helium experiment.

The signal to noise ratio is generally more than adequate
in single-crystal electron spin echo experiments but it can become a source of difficulty in unoriented metalloprotein samples
if some care is not taken with cavity design. The problem
arises because only a small portion of the overall resonance
line (e.g., ~10 G out of 1000 G) is sampled at any one field
setting. An idea of the signal to noise ratio that is attainable with a broad line material can be obtained from the
oscilloscope tracing of the spin echo signal for a cytochrome c sample shown in Fig. 4 (Mims, 1974). The cavity (or
more correctly the microwave resonator) used to obtain this

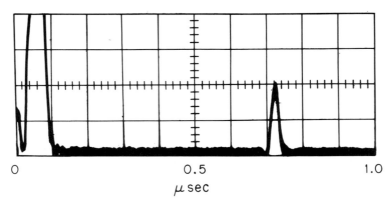

O   0.5   1.0

μ sec

Fig. 4. Electron spin echo signal obtained with a 5 mM sample of cytochrome c at 4.2°K. The resonance frequency f_0 = 9.4 GHz and the field setting H_0 = 3100 G. The oscilloscope was triggered in synchronism with pulse II. The signal at the left-hand end of the trace is caused by overloading of the receiver system. The duration of the overload signal depends on the cavity Q. It lasts until the power level in the cavity due to pulse II has fallen from its initial value $\approx$500 W to $\approx 10^{-9}$ W. (Reprinted with permission from the Review of Scientific Instruments.)

tracing was designed to minimize the volume of the microwave circuit (Mims, 1974). Confinement of the microwave field in a small volume increases the efficacy of the dynamo effect in generating echo signals, and thus improves the signal to noise ratio. It also means that less power is needed in order to generate high microwave field amplitudes in the sample during pulses I and II. High Q values such as are desirable in standard EPR experiments might indeed not be attainable with a cavity of such a design, but as we have pointed out above, high Q values are not needed in an electron spin echo experiment.

In concluding this section, we note briefly that there is an alternative and more general way of explaining the echo phenomenon and of calculating the echo amplitudes. This is based on the density matrix method (Slichter, 1963), a standard quantum mechanical technique that is, in this context (Geschwind, 1972, pp. 269-273) a bookkeeping device to keep account of the populations of the quantum states and the phase relations between them. Detailed calculations based on this method can be found in the literature (Rowan *et al.*, 1965; Mims, 1972a,b) and are too lengthy to be reproduced here. The notions of phase of quantum states and of coherence are, however, useful in understanding what happens when more than two

levels are simultaneously involved, and we shall therefore attempt to give a simple qualitative outline of the argument underlying the density matrix analysis.

Let us suppose that the simple "bar magnet" diagram in
Fig. 3 represents an ensemble of two-level systems such as one
might find if Kramer's doublet paramagnetic ions were doped into a material containing no nuclear magnetic moments. The situation shown in Fig. 3a corresponds to there being no coherence
between the pairs of states forming the doublets. The application of pulse I induces a coherent relationship between the two
eigenstates belonging to each ion. The corresponding eigen-
functions contain time-dependent phase factors of the form
$\exp[-E_i t/\hbar]$, which evolve according to the energies E_i of the
eigenstates concerned. Immediately before pulse II the total
phase difference between any pair of upper and lower states
with energies $E_\uparrow$ and $E_\downarrow$ is $(E_\uparrow - E_\downarrow)\tau/\hbar$. By the quantum relation $\hbar\omega_S = E_\uparrow - E_\downarrow$ we see that this is merely the classical
phase $\omega_S\tau$ of a spin precessing at ω_S. [In the actual density
matrix calculation, it is also possible to eliminate ω_0 leaving
only $(\omega_S-\omega_0)\tau$ just as was done in the rotating coordinate picture.] Pulse II brings about an interchange of off-diagonal
elements in the density matrix, which is equivalent to reversing
the signs of the phase differences $(\omega_S - \omega_0)$. Further evolution of the phase factors $\exp -E_i t/\hbar$ then reduces these phase
differences to zero at a time τ after pulse II. At this time
the phase factors for all the two-level systems are the same,
making it possible for the spins to interact coherently with
the electromagnetic field in the microwave circuit. At all
other times the phase factors are distributed over a range of
values depending on the distribution of the energies $E_\uparrow - E_\downarrow$
of the spins belonging to the inhomogeneous resonance line.

We return to this picture in Section III.

II. THE LINEAR ELECTRIC FIELD EFFECT IN EPR

A. *The Linear Electric Field Effect and Site Symmetry*

EPR studies performed by standard continuous wave microwave methods have proved widely useful for identifying paramagnetic centers in metalloproteins and for indicating the local
symmetry of the site (Swartz .*et al.*, 1972). The information
obtainable in this way is of a somewhat restricted nature, however. The symmetry can only be classified as isotropic, axial,
or rhombic, and no inference can be drawn regarding to odd component of the crystal field (i.e., regarding any possible lack

of balance between oppositely located ligands). This is a serious shortcoming in biological studies, where most of the
complexes are noncentrosymmetric and where any imbalance between ligands is likely to be a factor of importance in the
functioning of the system.

To obtain a more concrete idea of what is involved, let us
consider the schematic drawing of a heme center shown in Fig.
5. The four nitrogen atoms at the corners of the square belong
to the pyrrole groups of the porphyrin, and the two transverse
ligands L_1 and L_2 stand for charged ions or for protein side
chains such as the imidazole group of histidine or the -CH$_3$S
group of methionine. In most cases, L_1 and L_2 are different
from one another. But this difference is not revealed by a
standard EPR study of the center, which will merely show the
presence of an axial field corresponding to the sum of the two
contributions from L_1 and L_2. A given measurement might be interpreted by supposing that L_1 and L_2 are the same and contribute equally, or by supposing that L_1 is the dominant ligand
determining the value of the g or D parameter and that L_2 has
only a negligible effect.

This limitation can be overcome by performing a supple-
mentary experiment in which an electrostatic field is applied

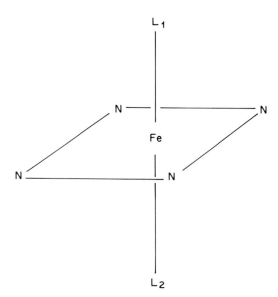

*Fig. 5. Schematic drawing of a heme complex. Standard
EPR experiments detect the axial ligand field arising from the
summed contributions of ligands L_1 and L_2 but do not reveal
differences between the two ligands. These differences can be
detected by performing EPR experiments in the presence of an
applied electric field.*

in conjunction with the EPR measurement (Roitsin, 1971; Mims, 1976). The way in which the electrostatic field distinguishes between the two cases of equivalent and inequivalent transverse ligands can be understood by considering a complex of the form shown in Fig. 5. In this case, let us suppose that L_1 and L_2 are charged ions at two different distances R_1 and R_2. Also, to illustrate the argument let us adopt the point charge model of the crystal field (Abragam and Bleaney, 1970). According to this model, the lowest-order crystal field component capable of mixing states in the 3d manifold of Fe^{3+} is the second-degree potential

$$\underline{V} = \underline{V}_1 + \underline{V}_2 = \left(\frac{4\pi}{5}\right)^{1/2}\left[\frac{q_1}{R_1^3} + \frac{q_2}{R_2^3}\right] r^2 \underline{Y}_2^0 (\theta,\phi) \qquad (2)$$

where $\underline{V}_1$ and $\underline{V}_2$ are the contributions from the two ligands, q_1 and q_2 the charges, r a distance measured from the center of the Fe^{3+} ion, and $\underline{Y}_2^0(\theta,\phi)$ a spherical harmonic. If we now apply an electrostatic field in a direction along the line joining L_2 and L_2, thus displacing the Fe^{3+} ion by a small amount δR, the new value of the second-degree potential is $\underline{V} + \delta\underline{V}$, where

$$\delta\underline{V} = \delta\underline{V}_1 + \delta\underline{V}_2 = -3\left(\frac{4\pi}{5}\right)^{1/2}\left[\frac{q_1}{R_1^4} - \frac{q_2}{R_2^4}\right] r^3 \underline{Y}_2^0 (\theta,\phi) \ \delta\underline{R} \qquad (3)$$

Reversal of the electrostatic field will cause reversal of δR and reversal of the sign of $\delta\underline{V}$. Provided that δR is small compared with $\underline{R}$, we have therefore a change in axial potential that is linear in the applied electrostatic field. This will in turn lead to linear electric field induced shifts in the $\underline{g}$ tensor or in the fine structure terms in the spin Hamiltonian. For brevity, these shifts are commonly denoted as the "linear electric field effect" (LEFE) in EPR. It will be seen from Eq. (3) if the two ligands are the same and are situated at the same distance from the Fe^{3+} ion, that the change in potential is zero to the first order in δR. The contribution $\delta\underline{V}_1$ due to the ligand L_1 is exactly balanced by the contribution $\delta\underline{V}_2$ due to L_2. Electric field effects can still in principle occur since the next term in the Taylor expansion of $\delta\underline{V}$ i.e., the term in $(\delta R)^2$ is independent of the sign of δR and would not cancel. But the resulting quadratic electric field effects are exceedingly small for all practical values of the applied field and they have only been observed in one or two exceptional cases.

The above example is a very rudimentary one and ignores a number of important factors that are likely to influence the magnitude of the LEFE in actual cases. The applied electrostatic field may, for example, polarize the ligand wave function, a significant consideration when L_1 and L_2 are molecules or molecular ions such as imidazole or CN^-, or it may polarize the Fe^{3+} ion, mixing the 3d wave functions with wave functions belonging to the unfilled 4p manifold. Various additional complications can arise when the bonding is partially covalent and when the bulk polarizability of the material surrounding the paramagnetic complex is taken into account. The fundamental principle remains the same, however. Linear effects are observed only for noncentrosymmetric sites and the magnitude of the effect affords a measure of the deviation from centrosymmetry. Electrostatic fields applied in selected directions can reveal imbalances in the coordination pattern and contribute information about the odd part of the crystal field potential, thus complementing standard EPR studies, which explore the even part.

The electric-field-induced changes in the $\underline{g}$ values can be represented formally in the spin Hamiltonian by a set of g-shift coefficients

$$\underline{T}_{ijk} = \partial(\underline{g}_{jk})/\partial\underline{E}_i \tag{4}$$

where the $\underline{g}_{jk}$ are elements of the $\underline{g}$ tensor and the $\underline{E}_i$ the three Cartesian components of the applied field. In the most general case where the paramagnetic complex has the lowest possible ($\underline{C}_1$) point symmetry, there are 18 coefficients $\underline{T}_{ijk}$ and, for spins $\underline{S} > \frac{1}{2}$, there are 15 more coefficients denoting changes in the fine structure terms {such as $\underline{D}[S_z^2 - \frac{1}{3}\underline{S}(S+1)]$ and $\underline{E}(S_x^2 - S_y^2)$}. Such complexity suggests that LEFE experiments might enable one to characterize a paramagnetic center in a far more detailed and specific manner than can be done by standard EPR methods. This is indeed true for single-crystal samples. But for the majority of biological materials, which are only available as powders or as frozen solutions, the amount of information obtainable by means of an LEFE experiment is much more limited and the result tends to be dominated by one or two of the largest terms. This will be apparent from the illustrations given later on.

B. Experimental Methods

The linear electric field effect in EPR was first observed by Ludwig and Woodbury (1961) in a single crystal of silicon doped with iron (see Fig. 6). In this material the resonance

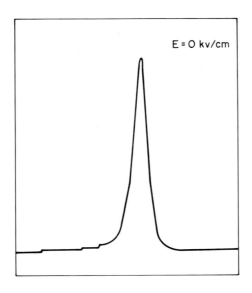

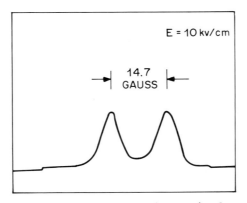

Fig. 6. The LEFE as observed in a single-crystal sample (Ludwig and Woodbury, 1961). The paramagnetic center is Fe in silicon. The line is split because there are two sites that are indistinguishable in EPR but that undergo equal and opposite shifts in an applied electric field. (Reprinted with permission from the Physical Review.)

line is narrow and is easily seen to split into two components when the EPR experiment is performed with an electric field applied to the sample. The two components correspond to two sites related to one another by the inversion operation. (The geometry of one site can be inferred from the geometry of the other by reversing the x, y, z coordinates of all the ligands).

Such "inversion image" sites give a single resonance line and cannot be distinguished from one another in the usual kind of EPR experiment but give equal and opposite line shifts when an electric field is applied.

Subsequent LEFE experiments performed on a variety of crystalline materials showed that the effect is not always easy to detect by straightforward EPR methods and that fields in excess of 10^5 V cm^{-1} are generally needed to achieve a spectral displacement by as much as a linewidth. The reason for this is not hard to understand. The field due to a neighboring charged ion is of the order of 10^8 V cm^{-1} and it is therefore to be expected that the shifts induced by laboratory fields will be quite small. Estimates for the shifts due to the displacement of ions in an electrically stressed ionic crystal also indicate that the effect will be a small one. Only in exceptional cases, when the waveform of the paramagnetic ion is very easily polarized or when the EPR parameters are critically dependent on the geometry of the crystal field environment will larger shifts occur. Even in cases such as these, the shift may be hard to detect since sensitivity to applied electric fields is often found in conjunction with a considerable degree of line broadening.

Notwithstanding these difficulties, numerous measurements have been made on single inorganic crystals by the straightforward direct method described above. The direct method is, however, much less likely to succeed with biological materials, since the lines themselves tend to be broader by an order of magnitude than the lines in inorganic crystals. Fortunately, the problem can be overcome by adopting a different technique based on electron spin echoes, which enables one to detect shifts comparable with the spin packet width ($\simeq 0.1$ G) rather than the width of the overall inhomogeneously broadened line.

The experimental procedure and the mechanism underlying the spin echo method of measuring electric field induced shifts, as they apply to single crystal samples, are illustrated in Fig. 7. The electric field is applied as a step at the halfway point in the spin echo cycle of events. If we assume that this field causes a uniform shift of δf in the spin precession frequency, for example, the shift

$$\delta \underline{f} = \delta \underline{g} \mu_B \underline{H}_0 \tag{5}$$

associated with a change $\delta \underline{g}$ in the $\underline{g}$ factor (μ_B is the Bohr magneton), then the magnetization responsible for generating the spin echo will be shifted in phase by an amount

$$\Delta \phi = 2 \pi \tau \, \delta \underline{f} \tag{6}$$

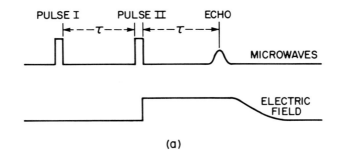

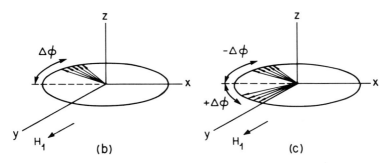

Fig. 7.   Procedure for making an LEFE measurement by the
spin echo method.   The electric field is applied at the same
time as transmitter pulse II and is held on until after the
appearance of the echo signal.   (b) Phase convergence of spins
showing the phase change that occurs when the precession frequencies are shifted as a result of applying the field (compare with Fig. 3e).   (c) If there are two sites that undergo
equal and opposite frequency shifts, the net magnetization responsible for generating the echo signal is the resultant of
two vectors shifted with phases ±Δϕ.   Partial or complete cancellation of the echo can occur.   (Reprinted with permission
from the Physical Review.)

This phase shift is illustrated in Fig. 7b, which may be compared with Fig. 3e. Actually, it is unusual to find a material
in which there is a single type of noncentrosymmetric site
uniquely oriented with respect to $\underline{H}_0$. More common is the situation encountered in the Ludwig and Woodbury experiment in which
there are two inversion image sites. The magnetization responsible for the generation of the echo is then split into two vectors shifted by ±Δϕ as shown in Fig. 7c, and the net effect is
a reduction in amplitude of the echo signal rather than a shift

in phase. Exact cancellation of the echo signal will occur if the time τ and the electrostatic field amplitude are adjusted to give phase shifts $\Delta\phi = \pm 90$, $\pm 270^\circ$, etc., thus affording a particularly easy and straightforward means of measuring the $\underline{g}$ shifts.

In powders and frozen solutions, the essential mechanism is the same but the physical situation is usually more complicated. With $\underline{H}_0$ set to a given point in the EPR spectrum, a distribution of complexes with various orientations is selected for observation (Poole and Farach, 1972). All have a common resonance frequency and all contribute to the spin echo signal, but the differences in orientation usually correspond to shifts of different magnitudes. The effect on the spin echo signal is determined by

$$\underline{R}(\tau) = \int \underline{S}(\Delta\omega)\, \underline{e}^{-i\Delta\omega\tau} \underline{d}(\Delta\omega) \qquad (7)$$

In Eq. (7), $\Delta\omega = 2\pi\, \Delta\underline{f}$ is the shift in radian frequency units, $\underline{S}(\Delta\omega)$ the distribution of shifts at a given $\underline{H}_0$ setting, and $\underline{R}(\tau)$ the ratio of echo signal amplitudes observed with and without the application of the electric field step in Fig. 7a. $R(\tau)$ is the Fourier transform of $S(\Delta\omega)$. Equation (7) can also be written in a slightly different form, which is more useful when experiments are performed by varying the electrostatic field rather than by varying τ. If we write $\overline{\Delta\omega} = \Delta\omega/E$ for the shift per unit applied field $\underline{E}$, then

$$\underline{R}(\underline{E}\tau) = \int \underline{S}(\overline{\Delta\omega})\, \underline{e}^{-i\overline{\Delta\omega}E\tau} \underline{d}(\overline{\Delta\omega}) \qquad (8)$$

where

$$\int \underline{S}(\overline{\Delta\omega})\, \underline{d}(\overline{\Delta\omega}) = 1$$

It can be shown that for observations made on a powder or frozen solution, the distribution function $\underline{S}(\overline{\Delta\omega})$ is symmetric about $\overline{\Delta\omega} = 0$ (i.e., that for each group of spins undergoing a shift $\Delta\omega$ there is, at the same point in the EPR spectrum, an equally large group undergoing a shift $-\Delta\omega$) (Mims and Peisach, 1976a). This is a convenient property since it means that $\underline{R}(\underline{E}\tau)$ is real and contains no net phase shift. All that is needed, therefore, is to measure the ratio of echo signal amplitudes obtained with and without the applied electric field.

An apparatus for making LEFE measurements is described in detail elsewhere (Mims, 1974), and we need therefore only comment briefly on the experimental conditions. The samples are introduced in liquid form into two cells of total volume ≈ 0.1 ml in a specially designed microwave resonator containing an insulated high voltage electrode. The samples are frozen and examined at liquid helium temperatures. At these temperatures, fields of up to 10^5 V cm^{-1}, obtained by applying a 15 Kv step to the electrode, can be applied to protein samples without causing voltage breakdown. Signal to noise ratios are in many cases large enough to enable one to measure $R(E\tau)$ by observing the spin echo signal directly on an oscilloscope. A 1 mM metalloprotein sample typically yields a signal that is >20 dB above noise (see, e.g., Fig. 4). It is more accurate and more convenient, however, to make the measurements with a simple data collecting and averaging system of the "boxcar" type. In the system described in the above reference, the voltage step is applied during alternate spin echo cycles and the echo signals are sorted into two channels where their amplitudes are averaged over times ≈ 1 sec. The resulting two voltage levels are then divided into one another electronically and the ratio $R(E\tau)$ is read directly on a meter.

The complete distribution function $S(\overline{\Delta\omega})$ for the shifts could be determined by plotting the ratio $R(E\tau)$ and taking its Fourier transform. Experiments with protein samples have shown, however, that there is little useful information contained in the actual shape of $R(E\tau)$. In most instances the function decays monotonically to zero with a form that is intermediate between that of a Gaussian and a Lorentzian curve. The corresponding distribution $[S(\overline{\Delta\omega})]$ is broad and might perhaps be described by a half-width parameter representing the mean shift. More convenient, however, since it does not involve the labor of recording and Fourier transforming $R(E\tau)$ for each experimental point, is to specify the shift in terms of some property of $R(E\tau)$ itself. The half-fall product, i.e., the value of the electrostatic field multiplied by the time τ that halves the echo amplitude is an easily measured quantity and is, in some simple cases, related directly to the fractional change in $\underline{g}$. Thus, in the case of a single-crystal sample of the inversion image type doped with Kramer's doublet ions, the half-fall product $(E\tau)_{1/2}$ is related to the fractional $\underline{g}$ shift per unit field by the expression

$$\left|\frac{\delta \underline{g}}{\underline{g}}\right| \div E = \left|\frac{\delta \underline{f}}{\underline{f}}\right| \div = E = [6\underline{f}(E\tau)_{1/2}]^{-1} \qquad (9)$$

In Eq. (9), $\underline{f}$ is the microwave frequency and $\delta \underline{f}$ the frequency shift as in Eqs. (5) and (6). The same expression holds good when measurements are made with $\underline{E}$ parallel to $\underline{H}_0$ and $\underline{H}_0$ set at either end of the EPR spectrum for Kramer's ions (such as low-spin Fe^{3+}) in a glassy material.

More generally we can use Eq. (9) to define a shift parameter σ by means of the equation

$$\sigma = \frac{d}{6(\underline{V\tau})_{1/2}\underline{f}}$$
(10)

where $\underline{d}$ is the thickness of the sample and $(\underline{V\tau})_{1/2}$ the product of applied voltage $\underline{V}$ and time τ that causes the echo signal to be halved. The mean fractional $\underline{g}$ shift $|\delta\underline{g}/\underline{g}|_{Av}$ will be related to σ by

$$\sigma = \underline{K}\left|\frac{\delta\underline{g}}{\underline{g}}\right|_{Av} \div \underline{E}$$
(11)

where $\underline{K}$ is a numerical constant. (In the special cases noted above $\underline{K} = 1$.)

C. LEFE Measurements in Biochemistry

To show how LEFE measurements are used to elucidate problems of structure and bonding in biochemistry, we give three examples. The first is taken from a study of electric field effects in ferric heme compounds (Mims and Peisach, 1976a). Since the heme system is relatively well understood, this will serve to illustrate how LEFE data can be interpreted and how they depend on structure and bonding in a paramagnetic complex. The second involves the iron sulfur proteins (Peisach et al., 1977) and shows how LEFE data may be used to provide evidence for or against a particular model. The last involves the blue copper proteins (Mims and Peisach, 1976b). It will be seen that, in addition to identifying the symmetry of a complex, LEFE measurements can provide clues as to the nature and polarizability of the coordinating bonds.

1. Measurements on Heme Complexes

LEFE measurements have only been made on the low-spin ($\underline{S} = 1/2$) group, the effect in this case being manifested as a shift in $\underline{g}$. As an example, let us consider the result for myoglobin azide shown in Fig. 8. The integral EPR spectrum that appears as a broken line in the figure was computed from the known $\underline{g}$ values (Helcké et al., 1968). The curve might also, in principle, have been obtained by plotting the amplitude of the electron spin echo signal as a function of the Zeeman field setting $\underline{H}_0$, but spectra obtained in this way are liable to be distorted by the dependence of the phase memory time on $\underline{H}_0$ and by the

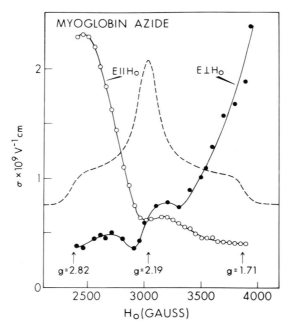

Fig. 8. Shift parameter σ as a function of Zeeman field measured in a frozen solution of sperm whale myoglobin azide. Shifts were determined as in Eq. (10) with the electric field aligned parallel (E ‖ H_0) and perpendicular (E ⊥ H_0) to the Zeeman field. The g values are indicated by the vertical arrows. The dotted line is a computer simulation of an EPR absorption spectrum with the same g values as myoglobin azide. (Reprinted with permission from the Journal of Chemical Physics.)

phenomenon of the nuclear modulation effect (see Section III). Two sets of LEFE measurements are shown. In one set (E ‖ H_0) the electric field E is aligned parallel to the magnetic field axis and in the other set it is aligned in a perpendicular direction (E ⊥ H_0). (In practice the magnetic field is rotated relative to the electric field, which remains fixed in relation to the experimental cavity.)

These results can be understood by referring to the diagram in Fig. 9, where the square, shown in perspective, indicates the positions of the four nitrogen ligands belonging to the porphyrin. It is known from EPR studies of single-crystal samples (Helcké et al., 1968) that the principal axis associated with the largest g value (g_{max}) is approximately perpendicular to the plane of the porphyrin; the middle g value (g_{mid}) and the

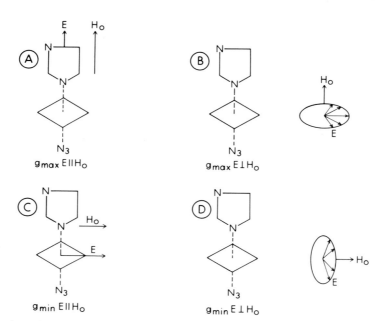

Fig. 9. *Diagram showing the relative orientation of magnetic field H_0, electric field E, the plane of the porphyrin,
and the approximate orientations of the axial ligands of myoglobin azide. In A, the electric field E is aligned parallel
to the magnetic field H_0, at the g_{max} setting and the electric
field is therefore aligned approximately in the direction of
the axial ligands to the heme. When the electric field is
aligned perpendicular to the magnetic field, at the g_{max} setting, as in B, the electric field is aligned approximately in
the plane of the porphyrin and perpendicular to the axial
ligands. In C, the electric field is aligned parallel to the
magnetic field at the g_{min} setting and both fields are therefore approximately in the plane of the porphyrin and in the
plane of the proximal imidazole ligand. In D, E and H_0 are
perpendicular to one another at the g_{min} setting. The electric
field therefore lies in a plane perpendicular to the planes of
the porphyrin and of the proximal imidazole. The assumptions
regarding the orientations of H_0 at the g_{max} and g_{min} settings
in relation to the planes of the porphyrin and the proximal
imidazole are derived from single-crystal studies by Helcké et
al. (1968). (Reprinted with permission from Biochemistry.)*

lowest g value (g_{min}) lie more or less in the plane. Assuming
exact perpendicularity, the geometrical relationships are
therefore as indicated in the figure when H_0 is set at either
end (g_{max} and g_{min}) of the EPR spectrum. It will be seen that

these H_0 settings do not specify unique orientations of the
heme complex, but distributions of orientations having a com-
mon g_{max} or g_{min} axis. This is unimportant if $E \parallel H_0$, but it
results in there being a range of different shifts when $E \perp H_0$.
The experiment then gives a weighted average shift.

For myoglobin azide, the largest shift in the $E \parallel H_0$ curve
is observed when the electric and magnetic fields are both
aligned along the g_{max} axis. A large shift is also observed at
the g_{min} setting for $E \perp H_0$. In both these cases it may be
noted that large shifts are obtained when E either lies along
the line joining the imidazole and N_3^- ligands or has a component in this direction. The effect is small when E is applied
in the plane of the porphyrin as for $E \perp H_0$ at the g_{min} setting
and $E \parallel H_0$ at the g_{max} setting.

This result can be interpreted in the following way. The
LEFE is due to an imbalance between the ligand fields due to
the imidazole and the azide. When an electric field is applied
in the direction of the line joining these ligands, charge is
withdrawn from one ligand toward the Fe^{3+} ion and is simultaneously withdrawn from the Fe^{3+} ion toward the opposite ligand.
Since the two ligands are different, the two effects do not
cancel and there is a net change in the constitution of the 3d
system of the Fe^{3+} ion, which manifests itself as a shift in g.
One effect of the electric field is thus to modify the covalency of the Fe^{3+} ion in the myoglobin azide complex.

Other mechanisms can also contribute to the LEFE. The
electric field can cause a small relative displacement of the
charged Fe^{3+} and N_3^- ions, thus once again modifying the covalency of the bonds and changing the g value. The electric
field can also polarize the Fe^{3+} ion itself and change the mixture of 3d and 4p orbitals in its wave function. The distinction between these mechanisms is to a certain extent an artificial one and is hard to demonstrate experimentally. It
should be noted, however, that polarizations of the porphyrin
and movements of the Fe^{3+} ion caused by electric fields oriented
in the plane of the porphyrin will not lead to a linear shift in
g. Withdrawal of charge from one side of the porphyrin toward
the Fe^{3+} ion is counterbalanced by an equivalent withdrawal of
charge from the Fe^{3+} ion to the opposite side of the porphyrin.
There is also no first-order change in covalency or in crystal
field strength resulting from physical displacements of the
Fe^{3+} ion in the plane of the porphyrin.

In the above discussion we have ignored the possibility that
the axis of the complex might be tilted as a result of the electric field exerting a sideways pull on the N_3^- ion. A tilt of
this kind would rotate the principal axes of the g tensor. Although this would not produce any observable effects at the ends
of the EPR spectrum, it would result in g shifts at intermediate
settings. The small irregularity in the shape of the LEFE
curves between the g_{max} and g_{min} limits may arise in this way.

2. *Measurements on the Iron Sulfur Centers in Ferredoxin*

The two-iron center in spinach ferredoxin and the four-iron center in the bacterial ferredoxins exist in forms in which there is one unpaired electron observable by EPR methods. (For a review of the various iron-sulfur proteins see Orme-Johnson and Sands, 1973.) The resonance properties are describable by an $\underline{S}$ = 1/2 spin Hamiltonian, and any electric field effects will therefore take the form of $\underline{g}$ shifts as in the case of low-spin heme compounds. The geometry of the complexes is shown in idealized form in Fig. 10. If the magnetic electron were equally shared between Fe atoms, its wave function would possess D_{2h} point symmetry for the two iron center and T_d point symmetry for the four iron center.

The results of LEFE experiments on typical iron sulfur containing proteins are shown in Fig. 11. It will be seen that for the two-iron center in spinach ferredoxin the shift parameters somewhat smaller than those for the low-spin heme complex myoglobin azide (Fig. 8). This is in itself not surprising since the $\underline{g}$ values are fairly close to the free electron value of 2, suggesting that the magnetic properties are not especially susceptible to changes in the ligand field. More noteworthy is the fact that there is indeed an LEFE for the two-iron center. Since the $\underline{D}_{2h}$ point group is centrosymmetric, this could only arise as a result of geometrical distortions or as a result of partial localization of the electron on one or the other of the Fe atoms. Evidence from ENDOR (Fritz *et al.*, 1967) and from Mössbauer studies (Dunham *et al.*, 1967) suggests that charge localization does in fact occur.

Less can be concluded from the actual form of the LEFE curve than in the myoglobin azide case since there is no single-crystal EPR result that would establish the orientations of the principal $\underline{g}$ axes in relation to the axes of the complex. It would appear that polarization of the two-iron center occurs most easily in a direction that lies approximately along the g_{min} ($\underline{g}$ = 1.89) principal axis and perpendicular to the g_{max} ($\underline{g}$ = 2.05) principal axis. The largest effects are seen when the electric field is applied in this direction, i.e., for $\underline{E} \parallel \underline{H}_0$ at g_{min} and $\underline{E} \perp \underline{H}_0$ at g_{max}. If, as seems likely, the LEFE is associated with an electric field induced displacement of the partially localized electron away from one Fe atom and toward the other, this result implies that the g_{min} principal axis lies along the line joining the two Fe atoms.

In the case of the four-iron center (as found in the ferredoxin from *B. polymyxa*), one must postulate some lowering of the symmetry to account for the anisotropy of $\underline{g}$ since the T_d group corresponding to the geometry in Fig. 10b is cubic. T_d symmetry would indeed permit the observation of an LEFE as would also the $\underline{D}_{2d}$ symmetry ("flattened tetrahedral"), which

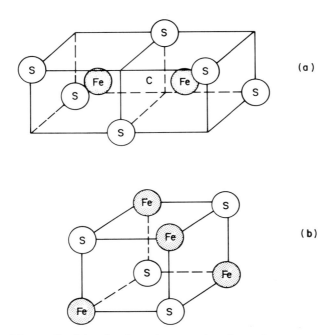

Fig. 10.   *Idealized structures showing the symmetry of the
iron-sulfur complexes in (a) two-iron ferredoxins, (b) four-iron
ferredoxins.   In (a) the Fe atoms are individually located at
noncentrosymmetric sites but the system as a whole has a center
of inversion at the point C.   The system may nevertheless give
an LEFE if the charge is unequally divided between the two Fe
atoms as a result of distortions or extraneous forces not shown
in the figure.   In (b) the system as a whole has tetrahedral
symmetry but the symmetry may be lowered further if charge is
unequally distributed.*

is obtained by shortening one dimension of the cube in Fig. 10b.
Both of these point groups are noncentrosymmetric. However,
the LEFE curves do not have, even approximately, the form required for D_{2d} symmetry (Peisach and Mims, 1976).   No symmetry
assignment has been made on the basis of the LEFE curves but it
seems unlikely that they can be explained as resulting from
small distortions of the cubane model (Holm and Ibers, 1976).
A more plausible interpretation can be arrived at by supposing
that the charge is partially localized, as in the two-iron case.
The applied electric field would then act by causing a small redistribution of the unpaired electron wave functions over the
complex.   A charge transfer mechanism of this kind would also
account for the magnitude of the LEFE, which is surprisingly

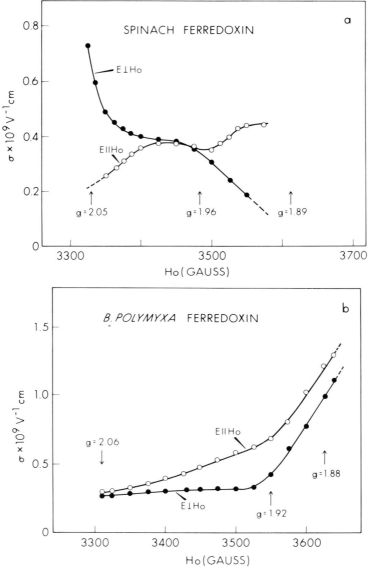

Fig. 11. Linear electric field effect as a function of
Zeeman field for (a) the two-iron ferredoxin from spinach and
(b) the four-iron ferredoxin from B. polymyxa. The shift parameter σ, defined in Eq. (10), was measured with the electric
field aligned parallel to ($E \parallel H_0$) or perpendicular to ($E \perp H_0$)
the Zeeman field. The g values are indicated by the vertical
arrows. (Reprinted with permission from the Journal of Biological Chemistry.)

large for a complex having $\underline{g}$ values so close to the free-elec-
tron value of 2. The hypothesis of partial localization of the
unpaired electron in four-iron ferredoxin conflicts with earlier
Mossbauer evidence (Thompson *et al.*, 1974) but is supported by
more recent studies (Dickson *et al.*, 1976).

3. Measurements on Copper Complexes and on the Paramagnetic Center in Blue Copper Proteins

In the two preceding cases, it has been possible to relate
LEFE data to currently accepted structural models of the complexes concerned. LEFE experiments can, however, provide useful clues even in cases where relatively little is known. They
may also contradict some commonly accepted structural assumptions.

The curves in Fig. 12 show LEFE data for a Cu:aquo complex,
obtained by freezing a solution of $Cu(ClO_4)_2 \cdot 6H_2O$ in water
mixed with glycerol, and for the blue copper protein stellacyanin (Peisach *et al.*, 1976). The Cu:aquo result, although of
lesser biochemical interest, will be discussed first. According to the common viewpoint, Cu^{2+} is octahedrally coordinated
with the nearer ligands arranged on the corners of a square coplanar with the Cu^{2+} ion and two more remote ligands along a
transverse axis (Cotton and Wilkinson, 1972). (A picture may
be obtained by placing four water molecules at the corners of
the square in Fig. 5 and replacing the imidazole and N_3^- ligands
by two equidistant water molecules lying somewhat further away
from the central ion.) This arrangement is characterized by
the centrosymmetric point group $\underline{D}_{4h}$ and cannot in its ideal
form give rise to an LEFE. However, since an effect is observed, we must look for possible modifications of the geometry
that would account for the result.

Fortunately, it is easy to show how LEFE data depend on
symmetry by comparing a number of copper complexes in frozen
solution. Curves similar in form to those shown in Fig. 12b
have been obtained for a wide variety of complexes (Table I),
the shift being largest for complexes in which the square planar

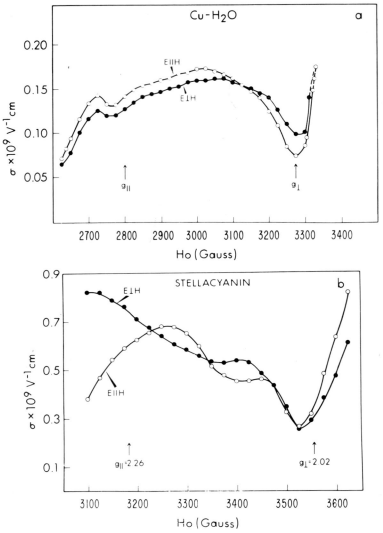

Fig. 12. Linear electric field effect as a function of
Zeeman field for (a) the hydrated Cu^{2+} complex and (b) the blue
copper protein, stellacyanin. The shift parameter σ, defined
as in Eq. (10), was measured with the electric field aligned
parallel to ($\underline{E} \parallel \underline{H}_0$) or perpendicular to ($\underline{E} \perp \underline{H}_0$) the Zeeman
field. The presence of an electric-field-induced $\underline{g}$ shift
shows that both complexes are noncentrosymmetric. The rise
of σ at the low-field end of the EPR spectrum at the $\underline{E} \perp \underline{H}_0$
setting for stellacyanin is characteristic of all blue copper
sites and is indicative of a $RS^- - Cu^{2+}$ charge transfer. (Reprinted with permission from Chemical Physics Letters and from
the European Journal of Biochemistry.)

geometry is distorted into a nearly tetrahedral configuration (i.e., the geometry arrived at by raising and lowering alternate corners of the square in Fig. 5 and moving the two axial ligands away to greater distances). The form of the LEFE curves is also consistent with this geometry, as can be shown by computer simulations (Peisach and Mims, 1976). In only one of the complexes examined was it possible to obtain a nearly vanishing LEFE, this being in the monomeric copper porphyrin, turacin (Blumberg and Peisach, 1965). The maximum shift here was <0.04 × 10^{-9} V^{-1}cm, which is four times smaller than for the Cu aquo complex, and 16 times smaller than for the nearly tetrahedral complex Cu-o-phenanthroline dichloride (Kokoszka et al., 1967). These LEFE results indicate therefore that Cu complexes in frozen solution tend to adopt a weakly tetrahedral coordination rather than the square planar coordination that is frequently assumed. A rigid environment such as is provided by the porphyrin in turacin or by the lattice in some crystals can, however, force the system to assume a square planar configuration.

The LEFE for the complex found in blue copper proteins is larger than for simple Cu^{2+} complexes and is characterized by curves of a unique shape (Fig. 12b). The largest observed shift i.e., $\sigma = 1.25 \times 10^{-9}$ V^{-1}cm, for the blue copper site in laccase (Mondoví et al., 1977) is twice as large as for Cu-o-phenanthroline dichloride and larger by almost an order of magnitude than the shifts observed for "nonblue" copper centers in proteins (unpublished observations). Comparison with optical data for a series of blue copper sites shows, in addition, that there is a rough correlation between the LEFE and the extinction coefficient.

Although there is as yet no generally accepted model of the blue copper site, it has been suggested that the complex includes a mercaptide sulfur ligand, and that charge transfer between this ligand and the Cu^{2+} ion is responsible for the intense blue coloration (McMillin et al., 1974). In addition to this, it can be shown by an analysis of EPR data that the site

TABLE I. Magnitude of linear electric field effect in EPR
for various Cu^{2+} complexes (after Peisach and Mims, 1978)[a]

Type	Copper complex	Ligand atoms	$\sigma \times 10^9$
Essentially centrosymmetric	turacin	4N	0.04
	bis dimethylglyoxime	4N	0.07
Weakly tetra-hedral	H_2O	4O	0.17
	bis acetylacetonate	4O	0.20
	bis (glycine)	2N,2O	0.14
	bis ethylenediamine	4N	0.18
	tetrakis (imidazole)	4N	0.18
	bis (o-phenanthroline)	4N	0.20
Strongly tetra-hedral	bis (N-t-butylsalicyl-diamine)	2N,2O	0.29
	o-phenanthroline dichloride	2N,2Cl	0.68

[a]The shift parameter σ is taken as the maximum in the
$\underline{E} \parallel \underline{H}_0$ curve in the linear electric field effect (see Fig.
12a, for example).

symmetry is almost tetrahedral (Peisach and Blumberg, 1974).
The LEFE is readily interpreted in terms of this model, the
correlation with the blue color suggesting in particular that
both properties of the complex are related to the same feature
(i.e., charge transfer from RS⁻) of the ligand field. Charge
transfer would also help to explain the surprisingly large magnitude of the shifts.
    Polarization of the unpaired electron would presumably be
greatest when the electric field was applied in the direction
of the sulfur ligand. Since the largest shifts occur for
$\underline{E} \perp \underline{H}_0$ at the g_{max} end of the EPR spectrum, this direction
should be perpendicular or nearly perpendicular to the g_{max}
principal axis. The rapidly increasing shifts at the high
field end of the EPR spectrum are observed in all copper complexes and are a consequence of electron nuclear coupling in
the Cu^{2+} ion. Diagonalization of the spin Hamiltonian for the
case of $\underline{A}_\perp \ll \underline{A}_\parallel$ shows that the extreme high-field portion of
the spectrum is contributed by complexes oriented a few degrees
away from the $\underline{g}_\perp$ principal axis. LEFE behavior here is therefore similar to that which is observed at $\underline{H}_0$ settings on the
low side of $\underline{g}_\perp$.

III. ELECTRON NUCLEAR COUPLING IN SPIN ECHO EXPERIMENTS

A. *The Nuclear Modulation Effect*

In Section I it was stated that spin echo signals become weaker as time τ is lengthened, the overall decay being describable by a coherence time or phase memory time characteristic of the material concerned. This is an oversimplification. In most cases, the decay of echo amplitude with τ proceeds in a more irregular fashion as illustrated in Figs. 13a and 13b. These figures consist of time exposure photographs of a large number ($\approx 10^4$) of electron spin echo signals, in which the oscilloscope is triggered in synchronism with pulse II and the time τ is gradually increased to give a composite picture outlining the form of the echo decay envelope. These photographs were obtained with two different Cu^{2+}-containing materials as noted in the legend. The manner in which these photographs were obtained is indicated in Fig. 14.

It will be seen that the echo amplitude does not decay monotonically with τ but is "modulated" in a periodic fashion. Analysis of the pattern shows that the periods visible in the echo envelopes are related to the precession frequencies of nuclei situated in the immediate vicinity of the electron spin. Thus, in Fig. 13a the fundamental period is 84 nsec, which corresponds to the precession period of H^1 nuclei in the experimental field of 2800 G. The same period can also be seen in Fig. 13b, but in this case the pattern contains additional low-frequency components associated with N^{14} nuclei (Mims and Peisach, 1976b). In general, it can be stated that the periods seen in the echo envelope correspond to the superhyperfine intervals in the level scheme that describes the coupling between the electron spin and its neighboring nuclei. The echo envelope thus contains information of a similar kind to that which would be derived in an ENDOR experiment.

The mechanism responsible for the appearance of nuclear precession frequencies in the echo envelope can be understood from the diagram in Fig. 15, which shows the magnetic fields at the site of an electron and of an adjacent nucleus in a paramagnetic complex. In Fig. 15a the electron spin is aligned along the Zeeman field $\underline{H}_0$, which makes an angle θ_H with a reference axis in the complex. The nucleus is not aligned along $\underline{H}_0$, however, since it also sees a strong local field $\underline{H}_{loc}$ due to the magnetic moment of the electron. Alignment is along the resultant effective field $\underline{H}_{eff}$. Let us suppose now that the electron spin is instantaneously reversed by applying a 180° microwave pulse. This will reorient $\underline{H}_{eff}$ and will leave the nucleus no longer aligned along the static magnetic field (Fig. 15b). The nucleus will therefore begin to precess about the new

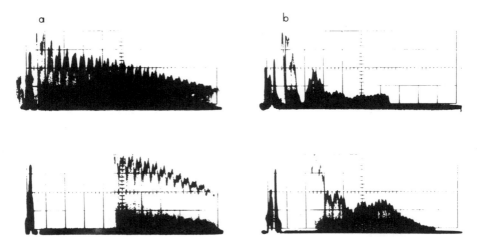

Fig. 13.   *Time exposure photographs of an oscilloscope display comprising altogether several thousand electron spin echo signals and showing the echo decay envelope. The oscilloscope is triggered at the end of transmitter pulse II. The interval τ is gradually increased during the time exposure so that each spin echo occurs at a slightly later time than its predecessor. (a) Echo signals from the Cu:glycine complex in a frozen glass. The longer period (84 nsec) in the echo envelope is the precession period of H^1 nuclei coupled to the Cu^{2+} ion. The shorter period (42 nsec) corresponds to twice the H^1 precession frequency. The N^{14} nuclei in the ligating molecules do not contribute to the pattern.   (b) Echo signals from the blue copper protein stellacyanin. The two short periods are due to H^1 nuclei as in (a). The longer and more irregular period is due to coupling between Cu^{2+} and an N^{14} nucleus at the N-1 position in an imidazole ligand.*

$\underline{H}_{eff}$ and a periodically varying local magnetic field will be seen by the electron. Although the local field component due to the precessing nucleus is quite small, it can, over many cycles of the microwave frequency, cause changes in phase and modify the amplitude of the echo signal in much the same way as the electric field effects discussed in Section II. The overall result will be an alternation in echo amplitude occurring at the nuclear precession frequency.

Although Fig. 15 is helpful in illustrating the physical mechanism underlying the nuclear modulation effect, it does not afford a satisfactory basis for making a calculation. Such a

Fig. 14. *Diagram showing how the time exposure photographs in Fig. 13 were obtained. Pulse II corresponds to the
extreme left-hand side of the oscilloscope scale. The echo
signal moves to the right as τ is lengthened. Pulse I does
not appear in the oscilloscope photographs. (Reprinted with
permission from Biochemistry.)*

calculation can only be made by analyzing the echo-generating
mechanism in terms of the coherence between wave functions belonging to the states involved in the microwave resonance.
This type of analysis was described briefly at the end of Section I for the case where there are only two levels. Here it
must be extended to apply to a more complex level system such
as the four-level system belonging to a proton ($I = \frac{1}{2}$) coupled
with a Kramers electron spin ($\underline{S} = \frac{1}{2}$) (Fig. 16). In a typical
case of an experiment performed at X band ($\simeq$9.4 GHz) the super-
hyperfine splitting between states $|a>$ and $|b>$ and between
states $|c>$ and $|d>$ is primarily due to the Zeeman field ($\simeq$3 kG)
and is approximately 12.8 MHz. In Fig. 16, the transitions
$|a> \rightarrow |c>$ and $|b> \rightarrow |d>$ are both allowed. But the transitions
$|a> \rightarrow |d>$ and $|b> \rightarrow |c>$ are not wholly forbidden (unless $\theta = 0$
or $90°$ in Fig. 15) and they will also be excited by the micro-
wave pulses, thus inducing coherent relationships between all
four states.

The time evolution of the four wave functions can be calculated by the density matrix method, the sequence of events being, in essence, as follows. Let us consider the top two
states $|a>$ and $|b>$. During the interval τ after pulse $\underline{I}$, these
states evolve with time factors $\exp(iE_a t/\cancel{h})$, $\exp(iE_b t/\cancel{h})$.
These factors attain the values $\exp(iE_a \tau/\cancel{h})$, $\exp(iE_b \tau/\cancel{h})$ at the
end of the interval. Pulse II brings about an exchange of popu-
lations and transfers the time factors from states $|a>$ and $|b>$
to the lower pair of states $|c>$ and $|d>$ with a reversal of
phase. In these new states time evolution is governed by the
complex exponentials $\exp(iE_c t/\cancel{h})$, $\exp(iE_d t/\cancel{h})$, resulting in the

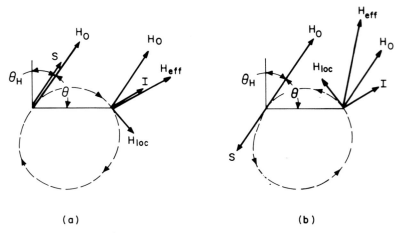

(a)   (b)

Fig. 15.   Mechanism underlying the nuclear modulation effect.   The figure shows the magnetic fields at the site of a
nuclear spin $\underline{I}$ coupled to an electron spin $\underline{S}$. θ_H is the angle
between the Zeeman field $\underline{H_0}$ and an axis in the paramagnetic
complex. θ is the angle between the Zeeman field and the line
joining electron and nucleus.   (a) The electron spin is aligned
along $\underline{H_0}$.   The nuclear spin is aligned along the resultant
$\underline{H_{eff}}$ of $\underline{H_0}$ and the local field $\underline{H_{loc}}$ due to the electron.   (b)
The electron spin is reoriented by $180°$ in a time that is short
compared with the nuclear precession period.   The nucleus,
which remains in its former orientation, now no longer lies
along $\underline{H_{eff}}$.   Nuclear precession about $\underline{H_{eff}}$ gives rise to a
small, periodically varying magnetic field, which is seen by
the electron and which modifies the electron spin echo amplitude.   (Reprinted with permission from the Journal of Chemical
Physics.)

attachment of additional phase factors $\exp(\underline{i}E_c\tau/\cancel{h})$ and
$\exp(\underline{i}E_d\tau/\cancel{h})$ by the time that the echo signal is due to appear.
Altogether, when all transitions between the levels in Fig. 16
are taken into account, it is found that the system contains
phase factors

$$\exp[\pm\underline{i}(E_a - E_b)\tau/\cancel{h}], \quad \exp[\pm\underline{i}(E_c - E_d)\tau/\cancel{h}]$$

$$\exp\{\pm\underline{i}[(\underline{E}_a - \underline{E}_b) + (\underline{E}_c - \underline{E}_d)]\tau/\cancel{h}, \quad \exp\{\pm\underline{i}[(\underline{E}_a - \underline{E}_b)$$
$$- (\underline{E}_c - \underline{E}_d)]\tau/\cancel{h}$$

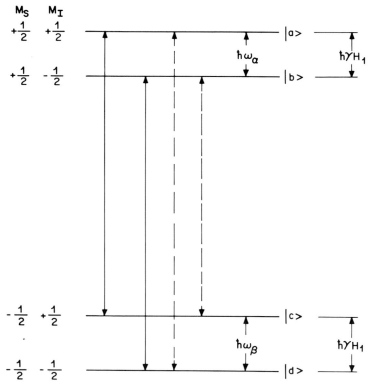

Fig. 16. Four energy level scheme for a nucleus $\underline{I}$ = 1/2
weakly coupled to an electron spin $\underline{S}$ = 1/2. Transitions
$|a\rangle \rightleftarrows |c\rangle$ and $|b\rangle \rightleftarrows |d\rangle$ are allowed and are excited by the microwave transmitter pulses, thus inducing coherent relation-
ships between the wave functions belonging to these pairs of
levels. For all geometries except the end-on and broadside-on
(θ = 0 or 90° in Fig. 15), the transitions $|a\rangle \rightleftarrows |d\rangle$ and
$|b\rangle \rightleftarrows |c\rangle$ are also partially allowed. In this case the microwave pulses produce coherent relationships between all four
wave functions. Time evolution of these wave functions results
in interference effects that give rise to the nuclear modula-
tion phenomenon. Unless transitions $|a\rangle \rightleftarrows |d\rangle$, $|b\rangle \rightleftarrows |c\rangle$ are
partially allowed, there is no way in which the microwave pulses
can make all four states coherent with respect to one another,
and there will be no nuclear modulation effect. It is also essential that the microwave field $\underline{H}_1$ should be large enough so
that $\gamma\underline{H}_1 > \omega_\alpha, \omega_\beta$. Otherwise, the allowed and semiforbidden
transitions cannot be simultaneously excited in a given para-
magnetic complex.

(as well as the phase factors $\exp[\pm i(E_a - E_c)\tau/\hbar]$, etc., which correspond to precession at the microwave frequency itself). Interference between these phase factors accounts for the modulation phenomenon.

It is important to note that branching transitions such as $|a> \rightarrow |c>$, $|a> \rightarrow |d>$ must actually take place if the nuclear modulation effect is to be observed. This implies that two or more transitions must originate from a given level, and also that the microwave field H_1 must be large enough to span the difference between the two transition intervals (e.g., $\gamma H_1 \geq (E_c - E_d)/\hbar$). If branching does not occur, there will be no means of achieving a coherent relationship between the pairs of states $|a>$, $|d>$ or $|b>$, $|c>$ and hence between the states $|a>$, $|b>$ or $|c>$, $|d>$. The coherences that are induced in this case (i.e., those between the pairs of states $|a>$, $|c>$ and $|b>$, $|d>$) are merely those which would be induced in a simple two-level system and, while they contribute to the generation of the echo signal, as outlined in Section I, they do not yield interference patterns containing the superhyperfine frequencies $(E_a - E_b)/\hbar$ and $(E_c - E_d)/\hbar$. The phase evolution picture, and the density matrix calculations that underlie it, are readily extended to cases where $I > 1/2$ and to cases where the upper and lower spin states are split into more than two superhyperfine levels (Mims, 1972a,b). It is also applicable to cases in which contact interactions or nuclear quadrupole interactions enter into the problem. As might be expected, however, the density matrix calculation becomes considerably more complicated in such instances.

For $I = 1/2$, $S = 1/2$ (as in Fig. 16), we have the result

$$V_{mod} = (1 - \frac{1}{2}k) + \frac{1}{2}k [\cos \omega_\alpha \tau + \cos \omega_\beta \tau]$$

$$- \frac{1}{2} \cos(\omega_\alpha + \omega_\beta)\tau - \frac{1}{2}\cos(\omega_\alpha - \omega_\beta) \tag{12}$$

where $\omega_\alpha = (E_a - E_b)/\hbar$, $\omega_\beta = (E_c - E_d)\hbar$ are the frequencies in the upper and lower superhyperfine intervals in radian units. The constant k is a measure of the degree of forbiddenness in the $|a> \rightarrow |d>$ and $|b> \rightarrow |c>$ intervals. If the electron nuclear coupling is purely dipolar, it is given by

$$k = (\omega_I B/\omega_\alpha \omega_\beta)^2 \tag{13}$$

where

$$\omega_I = (1/\hbar)(g_n \beta_n H_0) \tag{14}$$

and

$$\underline{B} = \frac{1}{\hbar} \left(\frac{g g_n \beta \beta_n}{r^3} \right) (3 \cos \theta \sin \theta) \qquad (15)$$

The quantities $\underline{g}$, β, $\underline{g}_n$, β_n are the $\underline{g}$ factors and the values of the magneton for the electron and nucleus, $\underline{H}_0$ the Zeeman field, $\underline{r}$ the distance between the electron and nucleus, and θ the angle between the line joining the electron and nucleus and the Zeeman field. In Eq. (14), ω_I is the precession frequency of the nucleus in the absence of any electron nuclear coupling. $\underline{B}$ is the electron nuclear coupling term responsible for mixing the superhyperfine states, which causes the transitions $|a> \rightarrow |d>$ and $|b> \rightarrow |c>$ to be partially allowed. (Both ω_I and $\underline{B}$ are in radian frequency units.)

A general formula such as Eq. (12) cannot be given for the case where $\underline{I} > 1/2$ since the calculation involves the diagonalization of a matrix containing nuclear quadrupole terms, but an approximate result is sometimes good enough. For nuclei such as H^2 in which the quadrupole energy is small, Eq. (12) can be used with $\underline{k}$ modified according to the equation

$$\underline{k}(\underline{I}) = \frac{4}{3} \underline{I}(\underline{I} + 1)\underline{k} \left(\frac{1}{2} \right) \qquad (16)$$

where $\underline{k}(1/2)$ is the value of $\underline{k}$ for an $\underline{I} = 1/2$ nucleus as in Eq. (13) (Mims et al., 1977). For the case in which a number of nuclei are coupled to the same electron, the overall modulation function is given by the product of the modulation functions for each nucleus considered separately. Thus

$$\underline{V}_{mod,I_1,I_2,I_3} \cdots = \underline{V}_{mod,I_1} \times \underline{V}_{mod,I_2} \times \underline{V}_{mod,I_3} \cdots \qquad (17)$$

Since the theory of the modulation effect enables one to compute $\underline{V}_{mod}$ for any number of nuclei and any set of coupling parameters, it is possible to check a proposed model of a paramagnetic complex by simulating the echo envelope curve and comparing it with experiment. A decay factor to take account of the slow monotonic decay of echo amplitude due to factors other than the nuclear modulation effect can be introduced, if desired, to facilitate the comparison. However, this approach tends to become unwieldy and expensive when the coordination geometry is not already known in advance since there are many variations in nuclear distances and angular coordinates to be examined.

A major reduction of effort can be achieved by constructing a model with spherical symmetry. This approximation is most

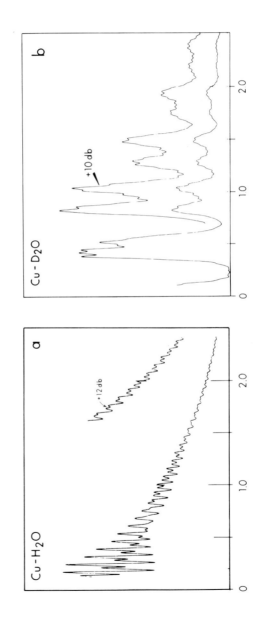

256

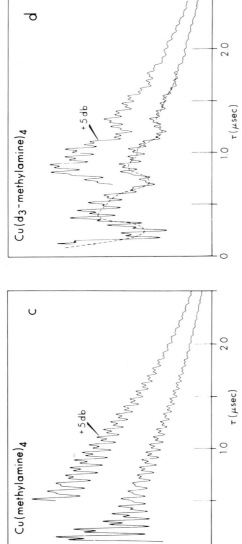

Fig. 17. X-Y recordings of the electron spin echo decay envelopes for (a) hydrated Cu^{2+} complex, $\underline{H_0} = 3220$ G, $\underline{f} = 9.355$ GHz, (b) hydrated Cu^{2+} complex in D_2O, $\underline{H_0} = 3220$ G, $\underline{f} = 9.355$ GHz, (c) Cu^{2+}-methylamine complex, $\underline{H_0} = 3210$ G, $\underline{f} = 9.292$ GHz, and (d) Cu^{2+}-methylamine complex where the methyl protons have been replaced with deuterons, $\underline{H_0} = 3200$ G, $\underline{f} = 9.112$ GHz. Solutions were prepared in 1:1 glycerol:H_2O except (b), which was prepared in 1:1 glycerol:D_2O. In (a), the characteristic pattern for protons can be seen. This pattern is largely replaced in (b) by one characteristic of deuterons. Here, the residual proton pattern is ascribed to the protons of glycerol. In (c), the proton pattern is not very different from that in (a), while in (d), the effects of dipolar coupling to protons on the amine nitrogen and to the methyl deuterons can be seen. (Reprinted with permission from the Journal of Chemical Physics.)

successful when $\underline{I}$ = 1/2 and when there are a number of nuclei
situated at similar distances from the electron and coupled by
the magnetic dipolar interaction. According to the spherical
model the modulation depth parameter is inversely proportional
to the sixth power of the radius $\underline{r}$ of the first coordination
sphere, and directly proportional to the number $\underline{n}$ of nuclei in
the sphere. The quantity $\underline{n}/\underline{r}^6$ can be determined from experimental data with an accuracy of several percent.

A similar procedure can be used where electrons are
coupled with H^2 nuclei although in this case the accuracy is
reduced e.g., to $\simeq (\pm 10\%)$ owing to the presence of the quadrupolar coupling. More serious difficulties are encountered with
N^{14} since the quadrupolar term is large and the coupling is
likely to include contact and pseudodipolar terms. Useful biochemical information can, however, be obtained in this particular case by making comparisons with a series of carefully chosen
model compounds. This approach is illustrated in Section
III,B,b.

B. Experimental Applications of the Nuclear Modulation Effect

It is difficult to perform any kind of experiment involving electron spin echoes without observing a nuclear pattern
of some kind. In favorable cases the electron spin echo envelope does indeed provide a kind of instant ENDOR measurement,
limited in scope but capable of answering some questions rapidly and unambiguously. A certain amount of care is needed, however, if these effects are to be interpreted correctly and
made to yield useful biochemical information. We give here
several illustrations.

1. Protons and Deuterons

Figure 17 shows the echo decay envelopes for four copper
complexes (Mims et al., 1977). The result for the Cu:aquo complex (Fig. 17a) is dominated by the proton precession frequency,
which at the experimental setting H_0 = 3220 G has a period of
73 nsec. Coupling to the Cu^{2+} is weak, and the electron-nuclear
dipolar interaction serves merely to broaden the spectrum of the
protons, causing a rapid dying away of the proton pattern in the
resulting curve. The shorter period of $\simeq 36.5$ nsec corresponds
to twice the proton precession frequency and represents the sum
frequency $\omega_\alpha + \omega_\beta$ in Eq. (12). This component is spectrally
narrower since the dipolar perturbation cancels out to the first
order in the sum $\omega_\alpha + \omega_\beta$, and it can therefore be seen for longer times τ in the echo envelope. A similar result is obtained
for the Cu:D_2O complex (Fig. 17b) except that in this case the

periods (475 and 238 nsec) correspond to deuteron precession in the field H_0 = 3220 G. Fewer complete cycles are observed since the duration of the pattern is limited by the phase memory time of the system. It should also be noted that the modulation is deeper than for protons, as is predicted by Eq. (16).

Figure 17c,d shows how measurements of the echo decay envelope can be used to detect coordination with molecules specifically labeled with deuterons. For the Cu methylamine:H_2O complex only the proton pattern is observed (Fig. 17c) since the directly coordinated N does not produce a modulation effect (Mims and Peisach, 1976b). When deuterons are substituted for the methyl protons the characteristic deuteron pattern is seen (Fig. 17d).

Another experiment of this kind (Peisach et al., 1977) is illustrated in Fig. 18. The tracing in Fig. 18a was obtained with a sample of a reduced four-iron ferredoxin from B. polymyxa, which was prepared in H_2O. In Fig. 18b the sample has been exchanged against D_2O before reduction. The curve shows that deuterium, belonging either to a labile bond or to a water molecule, has taken up a position adjacent to the iron sulfur centers. The same phenomenon is observed for two-iron ferredoxins. In the case of a high potential iron protein (HIPIP) sample from C. vinosum, however, exchange against D_2O produced virtually no alteration in the echo envelope (Fig. 18c,d). It is interesting to note that even though four-iron ferredoxins and HIPIPs are believed to contain essentially the same iron sulfur centers (Carter et al., 1971; Adman et al., 1973) proton exchange can be demonstrated for the former but not for the latter.

2. Nitrogen 14

It is sometimes possible to observe a fine structure due to coupling between N^{14} nuclei and paramagnetic ions in the EPR spectrum (Kivelson and Neiman, 1961). In the case of Cu^{2+} directly coordinated with ^{14}N, the splittings are $\simeq$10 G and the coupling is represented by a term $\underline{A}\underline{I}.\underline{S}$ in the spin Hamiltonian, with $\underline{A} \simeq$ 30 MHz. In other cases where coupling is weaker or where the ^{14}N atom lies further away, $\underline{A}$ may be too small for the splitting to be seen in the EPR spectrum. Pronounced effects can, however, often be seen in the echo decay envelope.

The tracing in Fig. 19a was obtained with the blue copper protein stellacyanin (Mims and Peisach, 1976b) and is similar to that obtained in a number of other natural and artificial copper containing proteins, including galactose oxidase (Kosman et al., submitted) and transferrin (Zweier et al., in press). It is also similar to the tracing obtained for various models in which the Cu ligands are known. These include copper di-

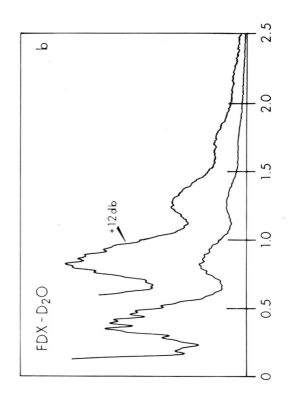

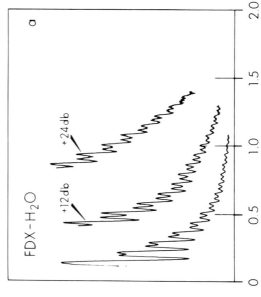

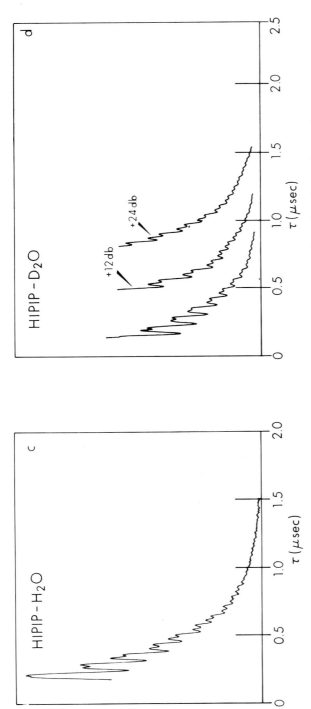

Fig. 18. X-Y recording of the spin echo decay envelopes for the four-iron ferredoxin from B. *polymyxa* (a) in H_2O and (b) in D_2O. These are to be compared with the envelopes for the high potential iron protein from C. vinosum (c) in H_2O and (d) in D_2O. It can be seen that a modulation pattern ascribable to close-lying deuterons can be observed for the ferredoxin but not the high potential iron protein. (Reprinted with permission from the Journal of Biological Chemistry.)

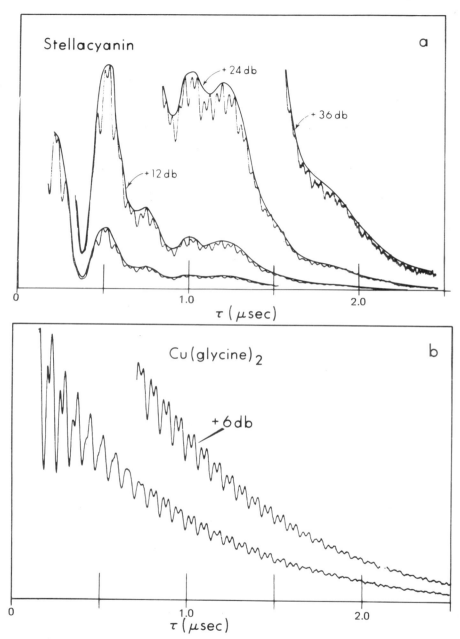

Fig. 19. *X-Y recording of the echo decay envelope for (a) stellacyanin and (b) Cu²⁺ bis(glycine). In both traces, the high-frequency pattern is due to protons. In (a), the low-frequency pattern is due to imidazole ^{14}N. Directly coupled ^{14}N as in (b) does not contribute. (Reprinted with permission from Biochemistry.)*

ethylene triamine:imidazole, which has three -NH$_2$ ligands and
one imidazole ligand (Mondoví *et al.*, 1977), the Cu^{2+}-bovine
serum albumen complex, in which the metal ion is coordinated
by one amino nitrogen, two peptide nitrogens, and one imidazole group (Peters and Blumenstock, 1967), Cu^{2+}-(imid)$_2$ oxa-
late, and Cu^{2+}-(imid)$_4$. Comparison with additional models in
which copper is ligated by amino or peptidic nitrogen shows
that the deep modulation pattern is not due to nitrogen nuclei
that are directly coordinated with Cu^{2+}. Thus the tracing for
the Cu:glycine complex in Fig. 19b contains the short proton
periods seen in Fig. 19a but not the longer period. This longer period is only obtained for nitrogen nuclei that are more
weakly coupled, and it serves in this case as an indicator of
the imidazole group. In the imidazole group, the nitrogen responsible for the nuclear modulation pattern is situated at
the N-1 position. The nitrogen at the N-3 position which is
directly coupled to the Cu^{2+}, has a large contact interaction
A$\underline{I}$.$\underline{S}$ and can be seen in the EPR spectrum (Peisach, unpublished
observation) but not in the nuclear modulation pattern. The
reason for the absence of any contribution to the modulation
pattern is that transitions between the two electron spin states
$\underline{M}_S$ = ±1/2 are either allowed or strictly forbidden when $\underline{A}$ is
larger than the dipolar or pseudodipolar coupling coefficients
in the spin Hamiltonian.

    An exact analysis of the result in terms of spin Hamilto-
nian parameters is hard to make in such cases since there are
many unknown quantities to be fitted to the data. The problem
is an especially difficult one since the ^{14}N quadrupole splittings are large ($\approx$2 MHz) and comparable with the Zeeman splitting (0.92 MHz at 3000 G) in many molecules of biological interest. The separations between the superhyperfine levels
therefore depend in a complex manner on the relative orientations of the Zeeman field, the quadrupolar axes, and the line
joining the electron to the nucleus. Fortunately, it is possible to obtain a certain amount of useful biological information by making comparisons with a series of carefully chosen
models without proceeding to a full solution of the resonance
problem. Thus, in the case discussed above, it was possible
to identify imidazole as one of the ligands in the copper protein stellacyanin.

3. *Phosphorus-31*

    Coupling with ^{31}P, and with ^{13}C and ^{15}N in isotopically en-
riched samples, can also be observed by means of the nuclear
modulation effect. The case of ^{31}P is an especially interesting one since coupling with ^{31}P occurs in a number of systems
that are models for the complexes formed between ATP, metal

ions, and the kinases (Mildvan and Cohn, 1970) and the effect
is easy to see. Some preliminary experiments of this kind have
been reported by Shimizu et al. (1977).

The general form of the echo envelope in a case where ^1H
and ^{31}P are both simultaneously coupled to a paramagnetic ion
is illustrated in Fig. 20, which shows the echo decay envelope
for a complex of Ce^{3+} and ATP formed by freezing a solution of
$CeCl_3$ and ATP in a mixture of water and glycerol. Since ^{31}P
has a spin $I = 1/2$, the analysis of the pattern is made in the
same way as for ^1H, the fundamental period at a field of 3200
G being 181 nsec. The beat pattern observed in the figure in-
volves the sum frequency $\omega_\alpha + \omega_\beta$ for ^{31}P see Eq. (12) and the
fundamental frequencies ω_α, ω_β for the protons. At 3200 G,
$\omega_\alpha + \omega_\beta \simeq 11.0$ MHz for ^{31}P whereas ω_α, $\omega_\beta \simeq 13.6$ MHz for ^1H.
Since the dipolar interaction is smaller than the Zeeman split-
ting for both nuclei, the frequencies are distributed in a
band centered on the free nuclear precession frequencies. This
band is broader for the ^{31}P nuclei than for the ^1H nuclei since
they are more strongly coupled to the paramagnetic ion, and the
fundamental frequencies ω_α, ω_β for ^{31}P are therefore more rapidly attenuated than the corresponding frequencies for ^1H in the
modulation tracing. The fundamental frequencies for ^{31}P are
not apparent in Fig. 20. The sum frequency $\omega_\alpha + \omega_\beta$ for ^{31}P
(period $\simeq$ 91 nsec) remains in evidence for longer, however
(Section III,B,1) and is able to beat with the frequencies ω_α,
ω_β associated with the protons. A curve with the same form as

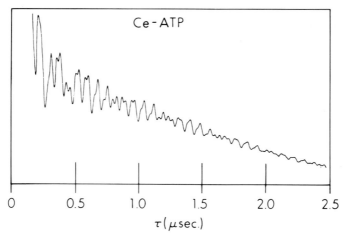

Fig. 20. *X-Y recording of the echo decay envelope for Ce-
ATP complex. H_0 = 3200 G. 1 mM Ce^{3+} was reacted with 7 mM ATP
in 1:1 glycerol water, pH 7.0. The beat pattern is due to the
interference of the sum frequency $\omega_\alpha + \omega_\beta \simeq 11.0$ MHz for ^{31}P
and the fundamental frequency ω_α, $\omega_\beta \simeq 13.6$ MHz for ^1H.*

that in Fig. 20 can be obtained by making a computer simulation for the case in which ^{1}H and ^{31}P nuclei are simultaneously coupled to an electron spin $\underline{S} = 1/2$. As pointed out earlier the computed curve must be multiplied by a negative exponential or by some other suitable monotonic decay function to allow for phase memory decay due to mechanisms other than the nuclear modulation effect.

C. Fourier Analysis of the Echo Envelope

We comment here on what would appear to be one of the more obvious ways of analysing nuclear modulation data. It would seem that contributions to the modulation pattern due to different nuclei might be easily resolved by taking a Fourier transform and presenting the results as a frequency spectrum. Attempts to use this method (Blumberg et al., 1973) have shown, however, that it is liable to introduce a number of artifacts, most of them connected with the fact that the experimental observation begins some time after the actual commencement of the echo envelope at $\tau = 0$. The missing portion of the echo envelope constitutes an appreciable fraction of the function that is to be transformed (see, e.g., Fig. 18) and its omission may result in line distortion and the appearance of spurious peaks in the frequency spectrum. These unwanted features can be smoothed out or eliminated by suitable computational procedures, but an artificial element is thereby introduced and the transformed data must be treated with some reservations.

Further difficulties arise when there is more than one nucleus coupled with the electron. The theoretical modulation function is in this case given by the product formula Eq. (17), which, in the frequency domain, corresponds to a multiple convolution of the spectra associated with each nucleus. This convolution can introduce certain extra features (e.g., harmonics and beat frequencies), which, though not artifacts, may be difficult to recognize and assign correctly. Obviously, the best and most reliable results are obtained in cases where frequencies are already easy to identify in the time waveform. In such cases it may be possible to obtain a cleaner spectrum by making a trial completion of the echo envelope and by taking the Fourier cosine transform of the resulting curve.

IV. THE DESIGN OF ELECTRON SPIN ECHO EXPERIMENTS

Although EPR techniques are widely familiar, the use of electron spin echo techniques in biology is relatively new. To make it clear what experimental conditions are encountered and what kind of apparatus is required, we have therefore prepared the following summary. This is based on experience gained in working in the X band range ($\sim$9.4 MHz). Some of the conditions may be different in other ranges.

Phase memory times: These are limited by the presence of hydrogen nuclei and are of the order 1-2 μsec.

Magnetic dilution: Samples must be magnetically dilute to attain the above limiting phase memory times. The dilution factor in metalloproteins is already sufficient in most cases, but some care may be needed when preparing model compounds. Samples $\simeq$5 mM strength are usually satisfactory unless the resonance line is very narrow.

T_1: The spin lattice relaxation time T_1 should at least exceed 10 μsec (i.e., it should exceed the limiting phase memory time by a comfortable margin). Even longer T_1 may be needed in fairly concentrated materials. Experiments generally have to be performed at helium temperatures.

Microwave transmitter: A high-power microwave source delivering short pulses with a low-duty cycle is needed in order to generate the two transmitter pulses. Cavity Q values must be small in order to handle these short microwave pulses and avoid long ringing times. Power levels of 1 kW, pulse widths of 20 nsec, duty cycles $<10^{-4}$, and Q values $\sim$100 have been found satisfactory in practice.

Cavity filling factor: A cavity with a small volume and a high filling factor is needed to obtain good echo signals with the protein materials available. The conventional TE 101 and TE 201 cavities are not satisfactory (in the X band range) and special designs are needed. Cavities with a small volume tend to have lower Q values but this constitutes no problem in short-pulse experiments.

Receiver system: The receiver must recover as soon as possible after being overloaded by the transmitter pulse. A low-noise traveling-wave tube provides a convenient first stage in the receiver system. Recovery from overload is virtually immediate.

Pulse timing system: A pulse timing system capable of making adjustments within $\sim$1 nsec is needed in order to trigger the transmitters, to gate the echo pulses, and to set up a suitable oscilloscope display. Most of the required units are available commercially. Others are easily made using TTL integrated circuits.

LEFE experiment: In addition to the spin echo apparatus,
a high-voltage step generator is required. This can be constructed with a high voltage (~15 kV) power supply and a hydrogen thyratron. Power requirements are minimal since current
flows only during the charging of circuit capacities.

Nuclear modulation effect: In order to record the echo delay envelope it is necessary to sample the receiver output and
gate the echo signal at a precisely determined time τ after the
second transmitter pulse. This time, which is the same as the
time between the two transmitter pulses, should be continuously
variable. A simple circuit performing the necessary functions
can be constructed using integrated circuits and other small
electronic components (Blumberg *et al.*, 1973). Programmable
delay units are also available commercially.

REFERENCES

Abragam, A., and Bleaney, B. (1970). "Electron Paramagnetic
    Resonance of Transition Ions." Clarendon Press, Oxford.
Adman, E. T., Sieker, L. C., and Jensen, L. H. (1973). *J. Mol.*
    Biol. 248, 3987-3996.
Blumberg, W. E., and Peisach, J. (1965). *J. Biol. Chem. 240,*
    870-876.
Blumberg, W. E., Mims, W. B., and Zuckerman, D. (1973). *Rev.*
    Sci. Instrum. 44, 546-555.
Carter, C. W., Jr., Freer, S. T., Xuong, N. G., Alden, R. A.,
    and Kraut, J. (1971). *Cold Spring Harbor Symp. Quant. Biol.*
    36, 381-385.
Cotton, F. A., and Wilkinson, G. (1972). "Advanced Inorganic
    Chemistry," 3rd Ed. Wiley (Interscience), New York.
Dickson, D. P. E., Johnson, C. E., Middleton, P., Rush, J. D.,
    Cammack, R., Hall, D. O., Mollinger, R. N., and Rao, K. K.
    (1976). *J. Phys. 37, Coll. C6,* 171-175.
Dunham, W. R., Bearden, A. J., Salmeen, I. T., Palmer, G.,
    Sands, R. H., Orme-Johnson, W. H., and Beinert, H. (1967).
    Biochim. Biophys. Acta 253, 134-152.
Fritz, J., Anderson, R., Fee, J., Palmer, G., Sands, R. H.,
    Tsibris, J. C. M., Gunsalus, I. C., Orme-Johnson, W. H.,
    and Beinert, H. (1967). *Biochim. Biophys. Acta 253,* 110-
    133.
Geschwind, S. (1972). "Electron Paramagnetic Resonance," Chap.
    4. Plenum, New York.
Helcké, G. A., Ingram, D. J. E., and Slade, E. F. (1968). *Proc.*
    Roy. Soc. (London) B169, 275-279.
Holm, R. H., and Ibers, J. A. (1976). *In* "Iron Sulfur Proteins
    III" (W. Lovenberg, ed.), pp. 205-281. Academic Press, New
    York.

Kivelson, D., and Neiman, R. (1961). *J. Chem. Phys.* *35,* 149-155.

Kokoszka, G. F., Reiman, C. W., and Allen, H. C., Jr. (1967). *J. Phys. Chem.* *71,* 121-126.

Kosman, D., Mims, W. B., and Peisach, J., submitted to *Biochemistry.*

Ludwig, G. W., and Woodbury, H. H. (1961). *Phys. Rev. Lett.* *7,* 240-241.

McMillin, D. R., Rosenberg, R. C., and Gray, H. B. (1974). *Proc. Nat. Acad. Sci. USA 71,* 4760-4762.

Mildvan, A. S., and Cohn, M. (1970). *Advan. Enzymol. 33,* 1-70. Wiley (Interscience), New York.

Mims, W. B. (1972a). *Phys. Rev. B5,* 2409-2419.

Mims, W. B. (1972b). *Phys. Rev. B5,* 3543-3545.

Mims, W. B. (1974). *Rev. Sci. Instrum.* 45, 1583-1591.

Mims, W. B. (1976). "The Linear Electric Field Effect in Paramagnetic Resonance." Clarendon Press, Oxford.

Mims, W. B., and Peisach, J. (1974). *Biochemistry 13,* 3346-3349.

Mims, W. B., and Peisach, J. (1976a). *J. Chem. Phys. 64,* 1074-1091.

Mims, W. B., and Peisach, J. (1976b). *Biochemistry 15,* 3863-3869.

Mims, W. B., Peisach, J., and Davis, J. L. (1977). *J. Chem. Phys. 66,* 5536-5550.

Mondoví, B., Graziani, M. T., Mims, W. B., Oltzik, R., and Peisach, J. (1977). *Biochemistry 16,* 4198-4202.

Orme-Johnson, W. H., and Sands, R. H. (1973). *In* "Iron Sulfur Protein," Vol. II (W. Lovenberg, ed.), pp. 195-235. Academic Press, New York.

Peisach, J., and Blumberg, W. E. (1974). *Arch. Biochim. Biophys. 165,* 691-708.

Peisach, J., and Mims, W. B. (1976). *Chem. Phys. Lett. 37,* 307-310.

Peisach, J., and Mims, W. B. (1978). *Eur. J. Biochem. 84,* 207-214.

Peisach, J., Levine, W. G., and Blumberg, W. E. (1967). *J. Biol. Chem. 242,* 2847-2858.

Peisach, J., Orme-Johnson, N. R., Mims, W. B., and Orme-Johnson, W. H. (1977). *J. Biol. Chem. 252,* 5643-5650.

Peters, T., and Blumenstock, F. A. (1967). *J. Biol. Chem. 242,* 1574-1578.

Poole, C. P., and Farach, H. A. (1972). "The Theory of Magnetic Resonance," p. 95. Wiley, New York.

Roitsin, A. B. (1971). *Usp. Fiz. Nauk 105,* 677-705 (transl. *Sov. Phys. Usp. 14,* 766-782).

Rowan, L. G., Hahn, E. L., and Mims, W. B. (1965). *Phys. Rev. 137,* A61-A71.

Salikhov, K. M., Semenov, A. G., and Tsvetkov, Yu. D. (1976).
    "Electron Spin Echoes and their Application." Science,
    Novosibirsk.
Shimizu, T., Mims, W. B., Peisach, J., and Orme-Johnson, W. H.
    (1977). *Fed. Proc. 36*, 720.
Slichter, C. P. (1963). "Principles of Magnetic Resonance."
    pp. 127–134. Harper & Row, New York.
Swartz, H. M., Bolton, J. R., and Borg, D. C. (1972). "Biologi-
    cal Applications of Electron Spin Resonance." Wiley, New
    York.
Thompson, C. L., Johnson, C. E., Dickson, D. P. E., Cammack,
    R., Hall, D. O., Weser, U., and Rao, K. K. (1974). *Bio-
    chem. J. 139*, 97–103.
Zweier, J. A., Aisen, P., Mims, W. B., and Peisach, J., *J.
    Biol. Chem.*, in press.

ACKNOWLEDGMENTS

    The portion of this investigation carried out at the Albert
Einstein College of Medicine was supported in part by U.S. Pub-
lic Health Service Research Grant HL 13399 from the Heart and
Lung Institute and by National Cancer Institute Contract N01-
CP55606 to J. Peisach and as such is Communication No. 387 from
the Joan and Lester Avnet Institute of Molecular Biology.

AN ANTIBODY BINDING SITE:
A COMBINED MAGNETIC RESONANCE
AND CRYSTALLOGRAPHIC APPROACH

Steven K. Dower
Raymond A. Dwek

Department of Biochemistry
Oxford University
Oxford, England

I. INTRODUCTION

A. *The Role of Antibodies in the Immune Response*

The immune response is the major defence system of higher animals. It can specifically recognize and destroy any foreign substance (antigen) present in the body, for example, a pathogenic microorganism that gains access to the bloodstream. The recognition is accomplished by antibodies or immunoglobulins. An antigen stimulates small lymphocytes in the bloodstream (B-cells), to differentiate and replicate into an expanded population of plasma cells, which produce antibodies that specifically recognize the stimulating antigen. Antibodies bound to antigen activation effector systems such as complement (a series of proteins that is triggered to cause lysis of cells to which antibody is bound) or macrophages (cells that phagocytose antigen/antibody complexes). Antibodies thus label antigens as targets for destruction by the effector systems of the immune response (Porter, 1973). Hence the biological role of antibodies unites two functions: the first is to combine specifically with a potentially unlimited range of antigenic structures; the second is to activate the effector systems.

An important aspect of the immune response is that many different B-cells can recognize the same antigenic determinant using a range of different sites, thus the population of antibodies produced against a single antigen is highly heterogenous.

It has been estimated (Kreth and Williamson, 1973) that the total number of different antibodies produced against a hapten is
3000-30,000. (A hapten is a small antigen that is immunogenic
only when coupled to a macromolecule.) This great heterogeneity severely limits the information that can be obtained from
studying induced antibody populations, since it has not yet
proved possible to resolve such a preparation into pure components.

Pure antibodies, however, do result from a pathological
condition termed myelomatosis. This is a carcinoma of the
small lymphocytes in which a single cell proliferates in an
uncontrolled fashion to produce a very large clone of identical
cells. These cells resemble the plasma cells of the normal immune response in that they produce immunoglobulin, but most importantly, since they are all genetically identical they all
produce the same immunoglobulin. Myelomatosis was first observed in human patients at the beginning of this century (Geschichter and Copeland, 1928) and human myeloma proteins have
been extensively studied (Nisonoff et al., 1975). Myeloma proteins have been isolated from many species, and most of the data discussed below have been obtained from a study of such proteins.

B. General Antibody Structure

The structural features of the antibody molecule are summarized in Fig. 1; the molecule is composed of two heavy
chains (molecular weight 50,000) and two light chains (molecula:
weight 25,000). The heavy chains of IgG (immunoglobulin G) contain ~450 amino acid residues, and the light chain contains
~220. All immunoglobulin sequences show a repeated structure
of 110 amino acid residue sections termed domains, which show a
significant degree of homology with one another. These fold relatively independently and associate in pairs (Fig. 1). Thus,
the immunoglobulin molecule is composed of six globular regions
linked together by short sections of relatively less structured
polypeptide chain.

The greatest sequence variability between immunoglobulins
is confined to the N-terminal domain of both chains (heavy and
light), and these are thus termed variable (V) domains. Comparison of a large number of V domain sequences (Wu and Kabat,
1970) has shown that most of the sequence differences are confined to short sections of the chains: residues ~32-35(H_1),
~50-55(H_2), ~86-91$(H)_3$), and ~95-110(H_4) in the heavy chain;
and residues ~27-35(L_1), ~52-57(L_2), and ~91-98(L_3) in the ligh
chain. Affinity-labeling studies have confirmed that some of
these are combining site residues (Strausbach et al., 1971), an
thus that the V domains contain the antibody-combining site.

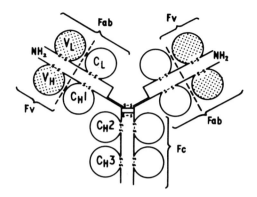

*Fig. 1.   Immunoglobulin structure from chemical studies.
Schematic representation of an immunoglobulin G molecule.   The
molecule is composed of two identical (light) (L) chains each
composed of two homology domains ($\underline{V}$, $\underline{C}_L$), and two identical
heavy chains each composed of four homology domains ($\underline{V}_H$, $\underline{C}_H1$,
$\underline{C}_H2$, $\underline{C}_H3$).   The domains are indicated by circles with the
intrachain disulfide bridges shown.   The domains are associated
pairwise to form globular regions.   Thus the $\underline{V}_L$ and $\underline{V}_H$ domains
associate to form the variable region, which in protein 315 can
be separated from the rest of the molecule by proteolysis to
yield the Fv (fragment variable) fragment.*

C. The Immunoglobulin Fold

This picture of the antibody molecule has recently been
confirmed and extended by the results of several X-ray diffraction studies (Davies et al., 1975; Poljak et al., 1974; Schiffer et al., 1973; Epp et al., 1974; Huber et al., 1976).   Most
importantly these studies reveal that all antibody domains for
which high-resolution X-ray data have been obtained have a similar tertiary structure.   This structure is called the immunoglobulin fold, and consists of two layers of β-sheet connected
by a disulfide bond.   There are 12 such units in an IgG molecule, each containing 110 amino acid residues and corresponding
to the homology domain described earlier.   The hypervariable
loops of the variable domains (Fig. 1) are attached to the immunoglobulin fold, which is a rigid framework.   The heavy- and
light-chain $\underline{V}$ domains are paired in the variable region, so that
the hypervariable loops all lie close together in space at the
front of the molecule, where they form the antibody-combining
site.

II. MODEL BUILDING AND MAGNETIC RESONANCE STUDIES: AN APPROACH TO ANTIBODY SPECIFICITY

Since all antibody domains share the immunoglobulin fold as a common structural feature, this is used as a basis for predicting the structure of the V domains and thus of the antibody combining site. The predicted site is refined with data obtained from physical studies, in particular NMR. This combination of a range of physical methods, including ^1H NMR, ESR, ^{31}P NMR, chemical modification studies, fluorescence and absorption spectroscopy, with model building, provides a flexible multidisciplinary approach to the problem of the structural basis of antibody specificity. The procedure can be summarized as follows:

(1) The sequences of the V_L and V_H domains of the antibody are aligned with those of an antibody for which the crystal structure is available. Homologies are maximized, allowing for insertions and deletions.

(2) The structure of the immunoglobulin fold framework is built using the α carbon coordinates of an antibody of known crystal structure.

(3) The hypervariable loops are built onto the framework using any sequence homologies with antibodies of known structure, and where none are available, general rules of protein folding. In this way, it is possible to generate a predicted structure of the combining site.

(4) Data from physical studies are used to refine the predicted model of the site, by using the structural information obtained as constraints on the orientations of amino acid residues.

In this chapter the use of the approach outlined above, to determine the structure of the combining site of the Dnp-binding mouse myeloma IgA protein 315, is reviewed (Dwek *et al.*, 1977). The model of the Fv fragment combining site (Padlan *et al.*, 1976), based on the α carbon coordinates of the phosphoryl choline binding mouse myeloma IgA protein McPC 603 (Segal *et al.*, 1974), is illustrated in Fig. 2.

The use of model building to provide a structure that is expected to be a close approximation to the true structure of the combining site of protein 315 is an important aid to proton NMR studies of this protein. The Fv fragment of protein 315 has a molecular weight of 25,000. Thus line widths of the proton resonances are so broad that they obscure the fine structure due to spin-spin coupling. Patterns of spin coupling are of vital importance in assigning resonances to a particular type of amino acid residue (Dwek, 1973; Wüthrich, 1976).

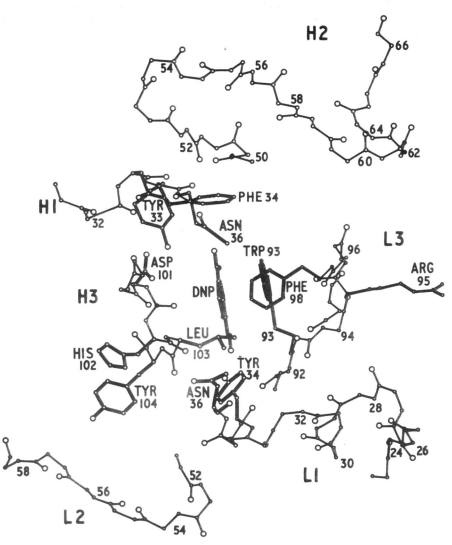

Fig. 2. The predicted combining site of protein 315. This has been constructed from the coordinates of Padlan et al. (1976), which have been very kindly made available to us prior to publication. The hypervariable loops and amino acid residues predicted to be important in the combining sites are indicated.

Hence, in the case of protein 315 it is necessary, if assignments are to be made at all, e.g., on the basis of hapten binding or chemical modification studies, to use the model in order to assign a particular resonance directly to a particular residue.

III. GENERAL FEATURES OF Fv FRAGMENT-HAPTEN INTERACTIONS

A. *The Histidine Residues of the Fv Fragment*

The Fv fragment of protein 315 contains three histidine residues, at positions 44_L, 97_L, and 102_H, respectively. In the structure of Padlan *et al.*, these are arranged such that one is far from the combining site (44_L), one is close to the site but would not be in contact with the bound hapten (97_L), and one is in the site and would be expected to form contacts with hapten side chains.

The three histidines in the Fv fragment can be detected by pH difference spectroscopy (Fig. 3). It has been found that three histidines have pK_as of 5.9 (3), 6.9 (2), and 8.2 (1), in the absence of hapten. Studies using paramagnetic broadening show that histidines 3 and 2 are closer to the combining site than 1 (Dwek *et al.*, 1975) (and see later). Studies in which the pK_as of the histidines were measured in the presence of a variety of haptens, showed that only histidine 2 has its pK_a perturbed (Wain-Hobson *et al.*, 1977), by bound hapten.

Thus the structure of Padlan *et al.* correctly predicts the properties of the three histidines, and a set of assignments consistent with experiment and prediction are that His 1 is 44_L, His 2 is 102_H, and His 3 is 97_L.

B. *The Nature of the Combining Site*

The 270-MHz proton NMR difference spectra of the Fv fragment with several different haptens are shown in Fig. 4. In al these spectra less than 20 protons are perturbed. This suggest that only those residues close to the bound hapten are perturbed by it, and therefore no major change in conformation occurs on hapten binding. Further, since all the difference spec tra are very similar, the mode of binding for all the haptens must be similar. Finally, most of the perturbed resonances arise from aromatic residues, and thus the combining site of protein 315 is highly aromatic, in agreement with the structure of the model (Padlan *et al.*, 1976).

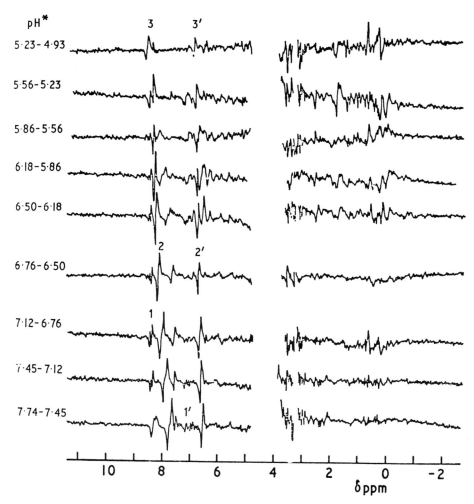

Fig. 3. *270-MHz proton NMR difference spectra of the Fv*
fragment of protein 315 in the presence of Dnp-aminomethyl
phosphonate. *Conditions are [Fv] = 1.7 mM, [Hapten] = 1.7 mM,*
temperature = 303°K, [NaCl] = 150 mM, solutions in ²H₂O.
Spectra obtained by subtracting as indicated. *Resonances as-*
signed as in text. *Chemical shift is measured from the methyl*
residue of 3-(trimethylsily-pentane) sulfonic acid.

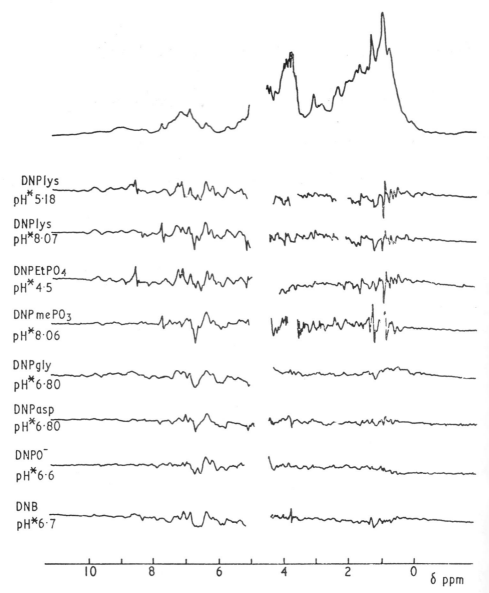

Fig. 4.   270-MHz proton NMR.   Difference spectra of the Fv
fragment of protein 315, with several different haptens.   Conditions as described in Fig. 3.   Hapten and pH are indicated; all
samples contained hapten and Fv fragment at equal concentrations
of ≃1.2 mM.   All spectra were generated by subtracting from the

IV. DIMENSIONS AND POLARITY OF THE COMBINING SITE

A. *Electron Spin Resonance Studies*

The estimation of the dimensions of the combining site using spin labels involves the comparison of the observed hyperfine splitting in the ESR spectrum with splittings calculated for different postulated motions of the spin label. In this analysis the Dnp group of the hapten is assumed to be rigidly held in the site, and thus to act as an anchor group for the spin label group, which does not interact specifically with the combining site.

The depth of the site was estimated to be 1.1-1.2 nm, by the use of a homologous series of six-membered spin-labeled haptens, by observing at what length of hapten the ESR spectrum became characteristic of almost complete isotropic motion. The ESR spectra of haptens contained within the site showed that flexing of the nitroxide-containing ring must be permitted, and it was concluded that the minimum dimensions of the entrance to the site are 0.6 × 0.9 nm (Dwek *et al.*, 1977).

This analysis was extended by the use of five-membered spin labels with chiral centers. It was observed that the ESR spectra of the two enantiomers showed different hyperfine splittings when the haptens were bound to the Fv fragment of protein 315 (Sutton *et al.*, 1977). This is illustrated in Fig. 5 showing the two enantiomers in the combining site. The differential immobilization of the two enantiomers shows that the combining site is asymmetric. The dimensions of the site from ESR mapping studies are summarized in Fig. 6.

The constraints obtained from the ESR mapping studies provide the first refinement of the predicted model. The constraints are summarized as follows:

(1) The mobility of short six-membered and five-membered spin labels can only be explained if they are contained completely within the combining site. Thus the tryptophan 93_L must be positioned parallel to the side of the site, allowing the Dnp ring to enter 1.2 nm into the site.

(2) Since the hyperfine splitting of the more immobilized enantiomer is greater than can be explained on the basis even of complete immobilization, the nitroxide group of this hapten must be in Van der Waals contact with a positively charged

spectrum of ~1.2 mM Fv alone, a spectrum of that sample of Fv alone, a spectrum of that sample of Fv with an equal concentration of hapten added. The difference spectra are twice the scale of the ordinary spectrum.

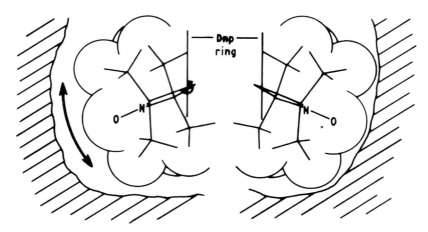

*Fig. 5.   Two enantiomers of the five-membered spin label
bound to protein 315 (see Fig. 6).   The arrows indicate the
bonds about which free rotation is possible, due to the greater
width of the left side of the site, and the extent of the allowed motions.*

residue, which would increase the hyperfine splitting (Griffith
et al., 1974).   Inspection of the model shows that this residue
must be Arg 95_L, which can be positioned with respect to the
Dnp ring.   This assignment is further confirmed by the observation that glyoxalation of all the arginine residues of the Fv
fragment abolishes hapten binding, and that one arginine residue is protected by bound hapten (Klostergaard *et al.*, 1977).

B. *Phosphorus NMR Studies*

The positive subsite assigned to Arg 95_L can be detected by
the use of phosphate and phosphonate haptens (Dwek *et al.*,
1977).   The pK_a of such a group is determined by observing the
change in the chemical shift of the ^{31}P resonance as the pH is
changed.   The pK_as of these haptens both free in solution and
bound to the Fv fragment are shown in Table I.   The maximum
change in pK_a is seen with Tnp propyl phosphonate, and assuming
that the bound hapten has a fully extended conformation, this
places the positive charge 0.9 nm from the Tnp ring, in agreement with the position of Arg ^{95}L as predicted by the model
(Fig. 2).

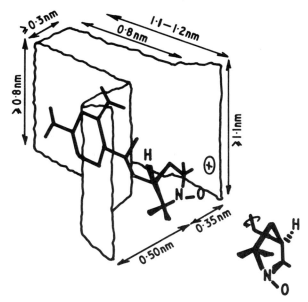

Fig. 6. The overall dimensions of the combining site of protein 315 as determined by spin-label mapping. The two enantiomers of one of the five-membered spin labels are indicated. Enantiomer (a) is completely immobilized but (b) is free to undergo a restricted rotation as shown.

V. THE ENVIRONMENT OF THE HAPTEN AROMATIC RING

A. The Aromatic Box

A number of studies have shown that tryptophan and nitroaromatic compounds stack in solution, and that an interaction of this type occurs in the combining site of many anti-Dnp and anti-Tnp antibodies (Little and Eisen, 1967; Eisen et al., 1968; Johnston et al., 1974). Optical difference spectra characteristic of such interactions show a red shift and hyperchromicity. The optical difference spectra shown in Fig. 7 thus show that Dnp-aspartate stacks with a tryptophan residue in the combining site of protein 315, in a complex similar to that formed in solution.

The structure of the Dnp/Trp complex was determined by observing the mutual upfield shifts produced in the 270-MHz NMR

TABLE I. *pKₐ values of phosphonates and phosphate groups of phosphohaptens free and bound to the Fv fragment from protein 315[a]*

Phosphohapten	pK_a free	pK_a bound to Fv	Changes in chemical shift (ppm)	
			Acid extreme	Base extreme
$Tnp-NH-CH_2-CH_2-CH_2-PO_3H^-$	7.7	6.9 ± 0.2	0.7	0.7
$Tnp-NH-CH_2-CH_2-PO_3H$	6.9	7.1 ± 0.2	0.5	0.2
$Tnp-NH-CH_2-PO_3-H$	5.7	5.9 ± 0.2	0	0
$Dnp-HN-CH_2-PO_3-H$	6.1	6.1 ± 0.2	0	0
$Dnp-NH-CH_2-CH_2-OPO_3H$	6.3	6.6 ± 0.1	0.2	1.3

[a] pK_a values were determined from ^{31}P NMR titrations at 129 MHz. All measurements were made at 20°C in the presence of 0.15 $\underline{M}$ NaCl. The errors are those obtained from analysis of the titration curves. The chemical shifts are the differences between the extremes of the titration curves.

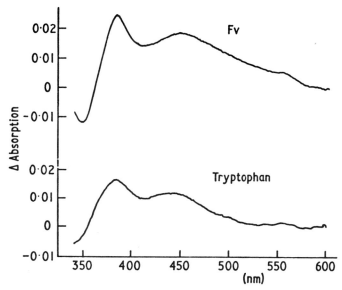

Fig. 7. A comparison of the difference absorption spectra of Dnp-aspartate on interaction with either the Fv fragment or tryptophan. Conditions were Dnp-aspartate = 70 μM, Fv = 40 μM, Trp = 20 mM, pH = 6.8, PIPES buffer = 40 mM, NaCl = 0.15 M, T = 298 K.

spectra of both molecules on complex formation (Fig. 8). From these data the structure shown in Fig. 9 was determined with a search program using ring current theory based on a modified form of the Johnson-Bovey equation, which allows for the different properties of different aromatic amino acids (Johnson and Bovey, 1958; Dwek *et al.*, 1977).

The interaction of Dnp-aspartate with the Fv fragment was studied by observing the changes in protein and hapten 270-MHz proton NMR spectra on hapten binding (Dwek *et al.*, 1977). The spectral changes are illustrated in Fig. 10 by a set of cumulative difference spectra, and are summarized in Table II. The most striking feature of these data is the large shifts observed of the hapten aromatic resonances, compared with the small shifts of the protein resonances.

The structure of the Dnp-aspartate/tryptophan complex shown in Fig. 9 was taken as a starting point for determining the geometry of the Dnp contact residues. It was assumed that, since the shifts on the hapten, in particular, were so large, these arose almost entirely from ring current effects (Dower *et al.*, 1977). Thus further aromatic rings were placed around

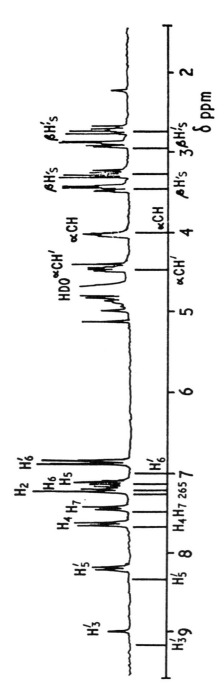

Fig. 8. Proton spectrum at 270 MHz of a mixture of Dnp-aspartate and tryptophan. The shifts on complex formation are all upfield for the aromatic protons. The shifts of the protons in the free compounds are indicated by the stick diagram below. Conditions were 10 mM-Trp, 10 mM-Dnp-aspartate, 150 mM-NaCl, pH 6.0, and T = 298 K.

(a)

(b)

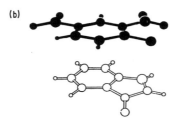

Fig. 9. Structure of the Dnp-aspartate and tryptophan
complex. The two views are obtained from a computer search of
the possible structures, with a separation between the two aromatic rings of 0.33 nm, which give calculated ring current
shifts in agreement with those obtained experimentally. The
shifts are calculated using the Johnson-Bovey (1958) equation,
modified to take into account the aromatic nature of the indole
ring. The two views represent (a) a vertical view down the z
axis and (b) that vertical view rotated 30 and 60° about the x
axis. The ellipse and the arrows indicate the extent of the family of solutions, of which the structure shown is a member.

the Dnp/Trp complex so as to generate a set of ring current
shifts (based on the Johnson-Bovey equation) that matched those
observed both on the protein (small) and on the hapten (very
large). The structure thus obtained, is shown in Figs. 11 and
12, which illustrate that the bound hapten is located in an
aromatic box of amino acid residues. Comparison of the structure with that of the predicted site (Fig. 2) shows that these
residues that form the box are Trp 93_L, Tyr 34_L, and Phe 34_H.
It was also observed that a nuclear overhauser effect occurs
between the Dnp H_3 proton and a proton on the protein, giving
rise to a resonance at 6.2 ppm that is downfield shifted on
hapten binding. Thus at least one proton on the aromatic ring
producing the shift on H_3 must be 0.3 nm from it. This further
limits the orientation of this ring.

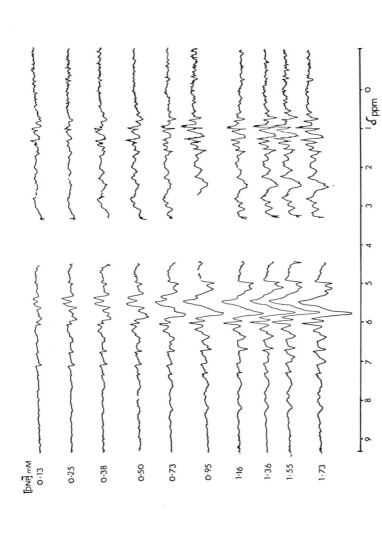

Fig. 10. Cumulative proton difference spectra at 270 MHz of the Fv fragment of protein MOPC 315 resulting from addition of Dnp-aspartate at pH 6.9 and $\underline{T}$ = 303 K in 0.15 M NaCl. The differences are obtained by subtracting the spectrum recorded in the presence of the concentration of hapten indicated, from the spectrum of the Fv alone.

TABLE II.   Chemical shift and change in chemical shift of
resonances perturbed on binding of Dnp-aspartate on the Fv
fragment from protein 315 at pH 6.9[a]

	δ_0 (ppm)	δ_0 (ppm)[b]	No. of protons
Hapten resonances			
H_3	9.18	1.68	1
H_3	8.36	2.30	1
H_6	7.04	1.31	1
β-CH_2-	2.95	0.23	1
	2.76	0.1	1
α-CH-	not observable		
Protein resonances			
His 97$_L$ H2	7.82	0.06	1
	7	broadens out	~2
His 97$_L$ H4	6.95	0.07	1
	6.9	0.19	1
	6.62	-0.08	1
	6.37	-0.30	2
Trp 93$_L$	6.33	0.18	~2
H_4, H_5, H_6, H_7	6.16	0.26	2
decreases	7.20	-	~2
increases	7.42	-	2

[a]Measurements were made at 270 MHz and 303°K, in the presence of 0.15 M NaCl. Shifts were measured from the sodium salt
of 3-(trimethyl-silyl propane) sulfonic acid, as in external
standard.
[b]A negative sign indicates that the shift is downfield;
otherwise the shift is upfield.

The structure of the aromatic box (Fig. 11) was built into
the predicted site, using the relative positions of the Dnp and
the three rings as constraints on the conformation of the Dnp
contact residues and obtaining the best compromise with possible
orientations of combining the residues. The final structure
gives calculated ring current shifts that are in agreement with
the observed shifts, and shows a slightly altered tryptophan/Dnp
stacking geometry. A further constraint is obtained from the
large shifts observed of the α-CH_2 resonances of Dnp-glycine
(Table III). This requires a fourth ring to be placed close to
these protons of the bound hapten. This can be identified with
Tyr [33]H. The magnitude of shift from this residue allows its
position to be fixed. The site, refined to include these
features, is shown in Fig. 12.

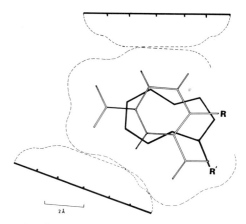

Fig. 11. Relative geometry of aromatic rings around the Dnp ring in the combining site of protein MOPC 315. The geometry is calculated on the basis of ring-current interactions on the hapten protons, assuming that the Dnp and tryptophan interaction is as shown in Fig. 9, but with one of the rings rotated through 180°. The Van der Waals surfaces of the Dnp ring and of the two perpendicular aromatic rings are indicated. R and R^1 represent the side chain of the Dnp and Trp rings. In the text it is argued that the Trp residue points back into the site.

Table III shows that the chemical shift positions of several of the resonances in the 6–6.5 ppm region perturbed on hapten binding are considerably upfield of the random coil positions for aromatic amino acid ring proton resonances (Fig. 13). This would be explained if there were a second layer of aromatic amino acid residues surrounding the aromatic box, as predicted by the model (Figs. 2 and 12), which had shifted the resonances of the inner layer of aromatics upfield. Comparison of the refined structure (Fig. 12) with the predicted structure (Fig. 2), shows that the major difference is the conformation of H3 (third hypervariable sequence of the heavy chain) loop.

B. Hydrogen Bonding Interactions

An important feature of the refined model of the combining site (Fig. 12) is that the positions and orientation of the side chains of Asp 36_L and Tyr 34_L are such that they can hydrogen bond to the Dnp ring 4-nitro and 2-nitro groups, respectively.

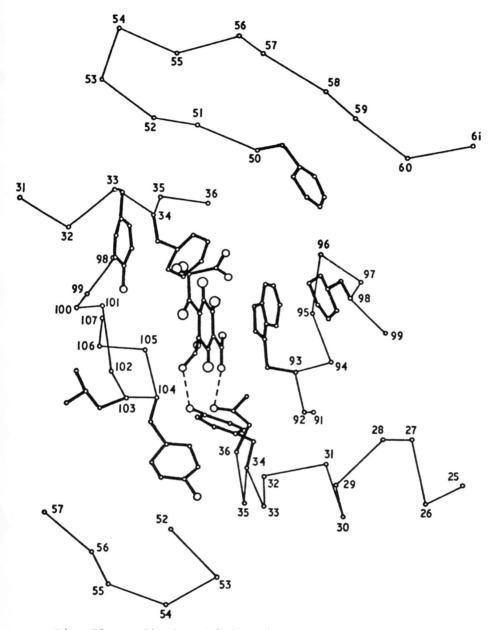

Fig. 12. Refined combining site of protein MOPC 315. This
is based on the NMR data presented here. The positioning of
Phe-34$_H$, Tyr-34$_L$, Trp-93$_L$, and Tyr-33$_H$ are obtained from the
ring-current calculations. Tyr-104$_H$ can interact with Tyr-34$_L$.
Some of the second layer of aromatic residues is shown. Thus
the orientation of the hapten is such that the two -NO$_2$ groups
can form H-bonds to Asn-36$_L$ and Tyr-34$_L$.

TABLE III.   Change in chemical shift of Dnp hapten resonances perturbed on binding of haptens to the Fv fragment from
protein 315[a]

Hapten	Proton resonance	Shift change (ppm)	pH*	K_D (μm)
Dnp-L-aspartate	H_3	1.68	6.9	24
	H_5	2.30		
	H_6	1.30		
	$\beta-CH_2-$	0.16		
Dnp-glycine	H_3	1.21	6.8	12
	H_5	2.20		
	H_6	1.77		
	$\alpha-CH_2$			
Dnp-O$^-$	H_3	1.25	6.7	100
	H_5	2.00		
	H_6	1.15		

[a]Measurements were at 270 MHz, $T = 303^\circ K$, in the presence
of 0.15 M NaCl.   All shifts are upfield.

Data obtained in this laboratory suggest that the nitro
groups of the bound Dnp ring contribute directly to hapten affinity for the Fv fragment.   This was shown by comparing the
binding of two series of dichloronitrobenzene derivatives to the
Fv fragment (Table IV).   Replacement of a nitro group by a chlorine leaves the ground state electronic structure of the aromatic ring virtually unchanged, while eliminating a potential hydrogen-bonding interaction.   The binding data show that 1.5 kcal
m^{-1} of the binding energy are lost when either of the two -NO$_2$
groups is replaced.   This is unlikely to be due to steric hin-
drance as chlorine is smaller than nitro.   Furthermore, proton
NMR data (Gettins et al., 1978), indicate that all of the haptens interact with the Fv fragment similarly and have similar
binding constants to tryptophan.   From these studies it is concluded that at least two H-bonds may be formed between hapten
and protein.   Comparison between the interaction of Dnp and Tnp
derivatives with the Fv fragments indicates that there may be
two rather than three H-bonds formed (see below).   These conclusions are supported by ^{19}F NMR studies on fluorinated Dnp
analogs binding to the Fv fragment (Kooistra and Richards, 1978).
Several laboratories have investigated the effect of chemical modification of Tyr 34$_L$ on the interaction of haptens with

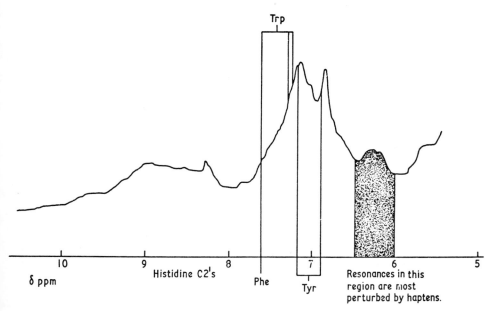

*Fig. 13. 270-MHz proton NMR spectrum of the aromatic resi-
dues of the Fv fragment of proton 315, showing the upfield
shifted position of the resonances most sensitive to hapten
binding. The sticks indicate the random coil shift positions
of the aromatic amino acids.*

the Fv fragment. It has been shown that nitration of Tyr 34_L
has little effect on the binding of Dnp derivatives to the Fv
fragment, and that the pK_a of this nitrotyrosine is only slightly perturbed by the presence of bound hapten; the formation of
an H-bond from this residue should result in a marked raising
of the pK_a when hapten binds. Furthermore, ionization of the
NO_2-tyr 34_L should result in the loss of the H-bond to the
bound hapten and thus a lowering in affinity at pHs above the
pK_a of the nitrotyrosine; this effect was not observed.

In contrast to this, Klostergaard *et al.* (1977) have shown
that iodination of this residue results in a marked decrease
in the binding constant of Dnp-lysine to protein 315 and that
bound hapten protects Tyr 34_L from iodination by iodine chloride.

This residue has also been affinity labeled by m-nitroben-
zenediazonium fluoroborate (Goetzl and Metzger, 1970) and by
N-bromoacetyl-N'-dinitrophenyl ethylene diamine. In both cases
the modified protein has a lowered affinity for haptens. Furthermore, the presence of bound hapten protects the tyrosine
34_L from modification.

TABLE IV. Binding of chloronitrobenzene derivatives to the Fv fragment from protein 315

DNP aniline analogs	K_d (μm)	$\Delta G'^{\circ}$ k_{cal}	DNP glycine analogs	K_d (μm)	$\Delta G'^{\circ}$ (k_{cal})
NO_2 / O_2N benzene–NH_2	4	7.4	NO_2 / O_2N benzene–NH–CH_2–$CO\bar{O}$	12	6.8
Cl / O_2N benzene–NH_2	90	5.6	Cl / O_2N benzene–NH–CH_2–$CO\bar{O}$	100	5.5
NO_2 / Cl benzene–NH_2	70	5.7	NO_2 / Cl benzene–NH–CH_2–$CO\bar{O}$	100	5.5
Cl / Cl benzene–NH_2	~600	4.4	Cl / Cl benzene–NH–CH_2–$CO\bar{O}$	~300	4.8

Most of the data are thus in agreement with the proposed structure of the combining site. Furthermore, the contradictory nature of the chemical modification data implies that some caution should be exercised in interpreting the results of such studies.

C. Placing Residues by Paramagnetic Mapping

The picture of the binding site in Fig. 12 results from the use of ring current perturbations. Another independent method of obtaining structural data is from the perturbations of paramagnetic ions. The Fv fragment contains a specific lanthanide binding site (Dwek et al., 1975, 1976) but not a Mn(II) site. However, the binding of lanthanide results in conformational changes in the protein (Dwek et al., 1976). Our approach to circumventing this difficulty has been to synthesize haptens containing polyphosphate side chains (Dower, 1979) that bind Mn(II) (see Table V). This creates a specific metal binding site close to the combining site.

Some difference spectra obtained by adding increasing concentrations of Mn(II) to the Fv Dnp-aminoethyl-diphosphate complex are shown in Fig. 14. The difference spectra are serial ones. Resonances close to the metal ion are initially broadened and appear in difference spectra with increasing intensity (e.g., His 102$_H$ C2$\underline{H}$ proton). As the metal concentration increases, the broadening of the resonances increases, and in the limit the resonances become undetectable. The net result in the serial difference spectra is that resonances arising from protons that are close to the bound metal ion initially increase in intensity, then decrease, and finally disappear. In this way the resonances from residues at different radii from the metal ion can be distinguished.

Comparison of the serial difference spectra in Fig. 14 with that obtained simply from addition of hapten shows that the resonances perturbed by Mn(II) are also probably those perturbed on hapten binding. Using a correlation time of 7 nsec (Dwek et al., 1976) the distances of the protons giving rise to these resonances can be calculated (Table VI). Most of these resonances are sufficiently close to the metal to indicate that they come from combining site residues. We note from Fig. 14 that several resonances are broadened on addition of hapten in the absence of metal, presumably from an exchange effect. No information on these can be obtained from this paramagnetic mapping.

The predicted model of the combining site places the two histidine residues (His 97$_L$ and His 102$_H$) about 20 Å apart. The Mn(II) broadening data show that this would be consistent

TABLE V.   Binding constants of manganous ion and polyphosphate haptens to the Fv fragment of protein 315

Hapten	K_M (μM)[a]	K_H (μM)[b]
NO_2-phenyl(NO_2)-$NHCH_2CH_2$-O-$P(=O)(OH)$-OH	5000-10,000	0.88
NO_2-phenyl(NO_2)-$NHCH_2CH_2$-O-$P(=O)(OH)$-O-$P(=O)(OH)$-OH	230	0.62
NO_2-phenyl(NO_2)-$NHCH_2CH_2$-O-$P(=O)(OH)$-O-$P(=O)(OH)$-O-$P(=O)$-OH	35	0.50

[a] Binding constant of Mn(II) to the Fv/Hapten complex, determined by ESR measurements. Measured at 30°C and pH 7.0, in the presence of 0.15 M NaCl.
[b] Binding constant of Hapten to the Fv fragment, determined by fluorescence measurements.

with the experimental observation if the metal ion were on a line joining the centers of the two histidine residues.   Some of the resonances broadened by the metal ion correspond in shift position to resonances perturbed by the bound hapten (compare Tables II and VI), which were assigned to residues in contact with the bound hapten (see above).   Since the hapten Dnp ring is approximately 9-11 Å from the metal ion binding site in the hapten, the distances to these protons are constant with the assignment to contact residues made earlier, on the basis of shift data.   Of particular note, too, is that the resonances upfield of the main aromatic envelope at 6.21 ppm (A) are close to the metal ion, thereby providing additional support that these resonances arise from contact residues.

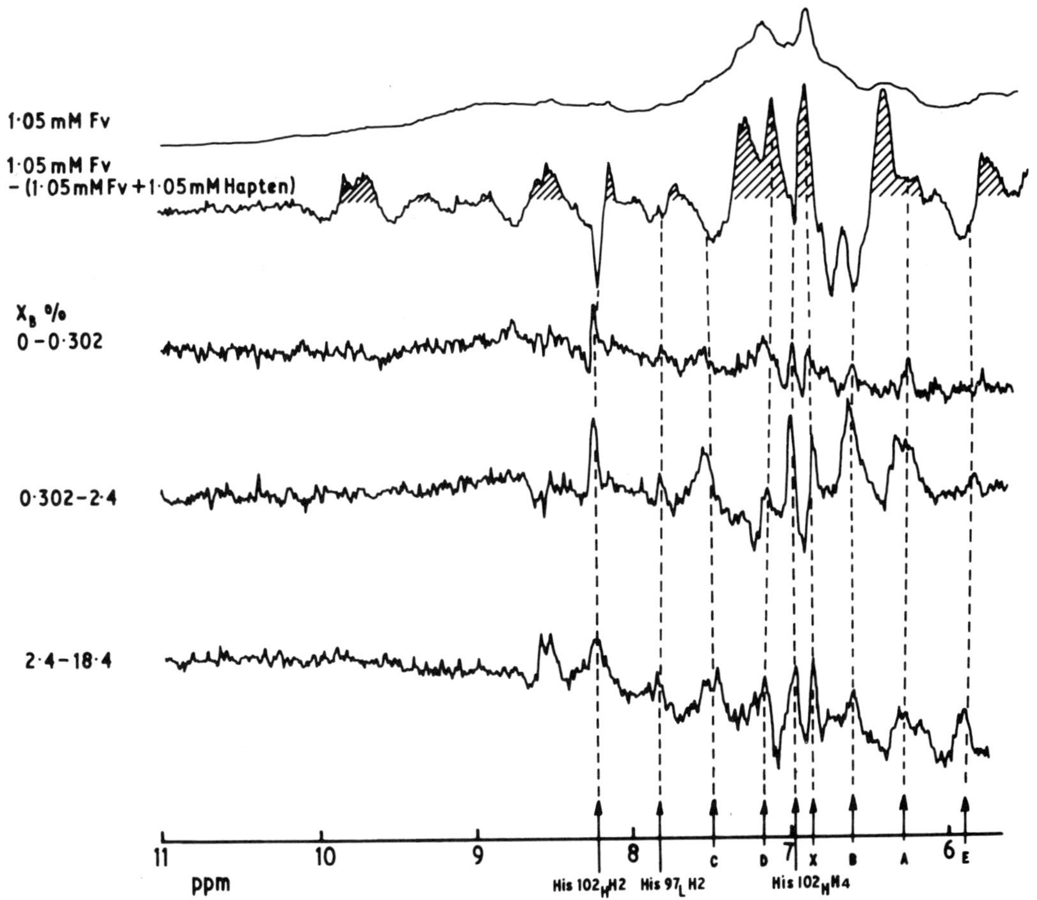

Fig. 14. The effect of the binding of manganese (II) to the Fv/Dnp-aminoethyldiphosphate complex on the aromatic region of the 270 MHz H^1 NMR spectrum of the complex. All spectra are 2K transients, samples at pH 7, 0.15 $\underline{M}$ NaCl, $\underline{T}$ = 303°K; X_B% is the fractional saturation of protein-hapten complex with metal ion. The Fv spectrum is ×1, the hapten difference spectrum is ×8, the metal ion serial difference spectra are ×16. The hapten difference spectrum is produced by subtracting the spectrum of an equimolar protein-hapten solution from the spectrum of a 1-mm Fv fragment solution. The hatching indicates intensity above the baseline. Minima below the baseline are finishing positions of resonances shifted on hapten binding.

TABLE VI. *Chemical shifts and calculated distances from the metal ion of resonances broadened when Mn(II) binds to the Fv fragment/Dnp-aminoethyldiphosphate complex.*

Resonance		δ (ppm)	T_{2M}^{O-1} (sec^{-1})	r (Å)[a]
His 102_H	H2	8.23	1.76×10^4	7.9
His 102_H	H4	6.96	0.69×10^4	9.2
His 97_L	H2	7.79	1.9×10^3	11.4
His 97_L	H4	6.72	3.5×10^3	10.3
A (2)[b]		6.21	7.1×10^3	9.15
B (2)		6.59	4.4×10^3	9.9
C (2)		7.54	4.4×10^3	9.9
D (2)		7.14	3.5×10^3	10.3

[a]*These distances are calculated assuming fast exchange and a correlation time of 7 nsec.*
[b]*The number in brackets indicates the estimate of the number of protons associated with the resonance.*

VI. THE BINDING OF Tnp HAPTENS TO THE Fv FRAGMENT: A STUDY OF CROSS-REACTION BY NMR

A set of criteria that have been used for characterizing the specificity of antibodies follows the approach of Land-steiner (1945) and is based on measuring the affinity of an antiserum or myeloma protein for a series of haptens in which the structure of the immunodominant group is systematically varied. The specificity of protein 315 has been investigated by studying the binding of Tnp and Dnp derivatives, and comparing the binding constants and limiting fluorescence quenching with those for induced anti-Dnp and anti-Tnp antisera (Eisen *et al.*, 1970). The detailed structural information yielded by the NMR/model building approach can be used to investigate the structural basis for this pattern of specificity.

The binding of the Tnp derivatives, Tnp-L-aspartate, Tnp-aminomethylphosphonate, Tnp-glycine and ε-N-Tnp-α-N-acetyl-L-lysine, was studied by a similar approach to that described for Dnp derivatives in the previous section (see also Dower *et al.*, 1977). The changes in the 270-MHz proton NMR spectra of the Fv fragment and Tnp-aspartate on hapten binding are summarized in Table VII. Ring current calculations (see below) were then used to position the Tnp ring in the aromatic box of the combining site and the resulting structure is shown in Fig. 15.

*TABLE VII.   Chemical shifts and change in chemical shift
of resonances perturbed on binding of TNP-aspartate to the Fv
fragment from protein 315 at pH 7.1[a]*

	δ_0 ppm	δ_0 ppm[b]	No. of protons
Hapten resonances			
$H_3 + H_5$	9.07	2.38	2
CH	3.83	1.24	1
CH_2	2.50	~0.88	2
Protein resonances			
6.9 His C_2	8.07	-0.07	1
	7.91	0.10	1
5.9 His C2	7.76	0.06	1
	7.01	broadens out	2
5.9 His C4	6.93	<0.05	1
	6.85	0.20	1
	6.30	-0.22	2
	6.25	0.05	~1
	6.16	0.07	2
	5.65	0.13	1
decreases	7.20	-	2
increases	7.40	-	2

[a]*Measurements were made at 270 MHz and $303°K$, in the presence of 0.15 M NaCl. Shifts were measured from the sodium salt
of 3-(trimethyl-silyl) propane sulfonic acid, as an external
standard.*
[b]*A negative sign indicates that the shift is downfield;
otherwise the shift is upfield.*

The fully bound shifts of ε-N-Tnp-α-N-acetyl-L-lysine, which
binds tightly to the Fv fragment (K_D = 0.2 μM) have been determined by locating the positions of the bound resonances by
a transfer of saturation experiment. These shifts (Table VIII)
indicate that the mode of binding of this hapten is similar to
that of Tnp-L-aspartate, despite there being a 100-fold difference in the affinities of these two haptens.

Binding of two structurally different haptens to an antibody is termed cross reaction, and these data indicate that
protein 315 is highly cross-reactive for Dnp and Tnp haptens,
the structural basis of this being that both types of haptens
can stack with Trp 93_L in the aromatic box and possibly also
form two hydrogen bonds to Tyr 34_L and Asn 36_L. Furthermore,
from dissociation constants and the shifts on the hapten pro-

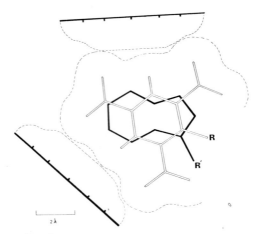

Fig. 15.   Relative geometry of the aromatic rings around the Tnp ring in the combining site of protein 315.   The geometry was calculated on the basis of ring current effects on the hapten protons by fitting the Tnp ring into the site as given in Fig. 12.

tons presented in Tables IV and VIII, it follows that, despite a wide range in affinities of the Fv fragment for the haptens, the same interactions are maintained between the nitro-aromatic ring of the hapten and the aromatic box of protein 315.

VII. SUMMARY AND CONCLUSIONS

This chapter has described the use of physical methods to refine a model of the combing site of the Fv fragment of protein 315, built on the basis of framework invariance in antibodies. These studies have shown that the trial model built on this basis is a good approximation to the structure of the combining site and that the major change between this (Fig. 2) and the refining model (Fig. 12) involves the repositioning of the residues in the third hypervariable loop of the heavy chain (residues 99-104). The most important pieces of evidence used in determining the orientation of particular combining site residues are summarized in Table IX. A wide range of other evidence has also been used; for example, it was found that Tyr 33_H is not involved in hydrogen bonding, even in the case of Tnp haptens, by showing that nitration of this residue does not affect hapten binding.

TABLE VIII. *Change in chemical shift of Tnp hapten resonances on binding to the Fv fragment of protein 315*[a]

	K_D (μm)	Proton resonance	Shift change (ppm)	No. of protons
Tnp-L-aspartate		$H_3 + H_5$	2.38	2
(pH* 7.1)	19	α-CH-	1.24	1
		β-CH$_2$-	~0.88	2
Tnp-glycine	8	$H_3 + H_5$	~2.6	2
(pH* 6.7)		α-CH$_2$-	~1.4	2
-N-Tnp- -N-acetyl-	0.2	H_3	~1.4	2
L-lysine		H_5	~2.2	
(pH* 6.95)				

[a]*All shifts are upfield. All spectra recorded at 270 MHz
and 303°K, in ^2H_2O in the presence of 0.15 M NaCl.*

The structural basis of the specificity of protein 315 has
been shown to be the stacking of the nitro-aromatic haptens
with a tryptophan residue (93$_L$) in a hydrophobic environment
of Phe 34$_H$ and Tyr 34$_L$ surrounded by a high concentration of
aromatic residues in the combining site. There are also two
hydrogen bonds to the protein, between the hapten 2 and 4 nitro groups and Tyr 34$_L$ and Asn 36$_L$, respectively.

The stacking interaction with a tryptophan residue seems
to be a general feature of anti-Dnp and anti-Tnp antibodies,
as demonstrated by optical difference spectroscopy (Little and
Eisen, 1967). It must be stressed that this may not be the only type of high affinity nitro-aromatic binding site that occurs in proteins, but it is, so far, the only kind that has
been detected in antibodies.

Finally, the evidence discussed shows that all the haptens
studied bind to protein 315 in essentially the same way, despite a range of three orders of magnitude in dissociation constants. This has been shown quantitatively for Dnp-aspartate,
Dnp-glycine, DnpO⁻, Tnp-aspartate, Tnp-glycine and ε-N-Tnp-α-
N-acetyl-L-lysine, for which shift data are available (Tables
III and VIII). Also the protein NMR difference spectra for
other haptens, given in Fig. 4, are all very similar. Such
spectra, which are a useful fingerprint of the structural features of the antibody-hapten interaction, indicate that all
these haptens bind very similarly to the Fv fragment.

TABLE IV. A summary: The physical data used to define the positions of particular amino acid residues in the combining site of the Fv fragment of protein 315

Residue	Data used
Tyr 34_L	Proton NMR, Overhauser effect, affinity labeling (Strausbach et al., 1971), chemical modification
Asn 36_L	Modification of haptens, proton NMR
Trp 93_L	Optical difference spectroscopy, proton NMR, ESR
Arg 95_L	^{31}P NMR pK_a mapping, ESR, chemical modification (glyoxalation (Klostergaard et al., 1977)
His $^{97}_L$	Proton NMR, paramagnetic mapping
Tyr 33_H	Proton NMR, chemical modification (nitration)
Phe 34_H	Proton NMR
Lys 52_H	Affinity labeling (Strausbach et al., 1971), chemical modification (maleylation) (Klostergaard et al., 1977)
His 102_H	Proton NMR, paramagnetic mapping
Trp 48_H, Tyr 104_H, Phe 105_H, Phe 98_L	Inferred as the "second layer" of aromatic residues from NMR
Asp 99_H	Necessarily repositioned when model is modified to accommodate above data, Lanthanide binding (Sutton et al., 1977)
Leu 103_H	Proton NMR (no evidence of extensive methyl perturbation) (Dower et al., 1977)

VIII. THE FUTURE

If indeed the basic structure unit of each antibody domain, the immunoglobulin fold, is conserved, then on a common three-dimensional background a vast number of specificities can be generated. This diversity in recognition is the essence of the immune system and a comparative analysis of a large number of combining sites is required to provide the details of molecular recognition by antibodies. The method of model building will be of immense help in this instance provided the resulting combining site and the mode of binding of ligand to it can be tested by various physical methods. The main limitation to using NMR is the assignment problem. The Fv fragment (MW ~ 25,000) can only be made for a very few antibodies and most studies will have to be done on Fab fragments (MW ~ 50,000). Some of the assignment problems may be eased by growing the myelomas in mice whose diet has one or more of the essential amino acids replaced by the deuterated analog (Gettins, 1979). The hapten binding can also be studied by a variety of NMR techniques such as ^{13}C and ^{31}P (Goetze and Richards, 1978, and references therein).

The study of protein 315 has shown that by suitable tricks it is possible to obtain a great deal of information. Crystals of the Fv fragment have been obtained by Aschaffenburg, Sutton, and Phillips (private communication) and if good X-ray data can be obtained this should provide a good test of the structure in Fig. 12, and of the assumptions of the NMR theories.

The recognition of antigen by antibody is only the first stage in the immune response. The mechanism of the subsequent triggering of the complement cascade of proteins that control many secondary immunological functions is also another central biochemical problem. The activation of these secondary functions occurs via the Fc region of the immunoglobulin, some 100 Å away from the antigen combining site. The mechanism of this information transfer is also a field in which NMR and other physical methods have a contribution to make (Easterbrook-Smith et al., 1978).

ACKNOWLEDGMENTS

We thank Professor R. R. Porter for his help, advice, and many useful suggestions about the work. We also thank Professors R. J. P. Williams and D. C. Phillips, and Dr. I. D. Campbell for stimulating discussions, and Miss E. M. Press, Mrs. C.

Wright, Mr. K. Willan, Mr. C. Sunderland, and Drs. D. Marsh,
R. Jones, A. C. McLaughlin, N. C. Price, and A. I. White for
their help with the early stages of this work.

   We thank B. Sutton, P. Gettins, R. Jackson, S. Wain-Hobson,
and S. J. Perkins, whose results we have drawn on extensively
in writing this review, and Professor D. Givol in collaboration
with whom this work has been done. We thank the M.R.C. and the
S.R.C. for financial support. R.A.D. is a member of the Oxford
Enzyme Group.

REFERENCES

Davies, D. R., Padlan, E. A., and Segal, D. M. (1975). *In* "Con-
   temporary Topics in Molecular Immunology,: Vol. 4 (Inman,
   F. P., and Mandy, W. J., eds.), p. 127. Plenum Press, New
   York.
Dower, S. K. (1978). Ph.D. Thesis, Oxford.
Dower, S. K., Wain-Hobson, S., Gettins, P., Givol, D., Jackson,
   R., Perkins, S. J., Sunderland, C., Sutton, B., Wright, C.,
   and Dwek, R. A. (1977). *Biochem. J. 165,* 207.
Dwek, R. A., Jones, R., Marsh, D., McLaughlin, A. C., Press,
   E. M., Price, N. C., and White, A. I. (1975). *Phil. Trans.
   R. Soc. (London) 2 Ser. B 272,* 53.
Dwek, R. A., Givol, D., Jones, R., McLaughlin, A. C., Wain-
   Hobson, S., White, A. I., and Wright, C. (1976). *Biochem.
   J. 155,* 37.
Dwek, R. A., Wain-Hobson, S., Dower, S., Gettins, P., Sutton,
   B., Perkins, S. J., and Givol, D. (1977). *Nature 266,* 31.
Easterbrook-Smith, S. B., Zavodsky, P., Willan, K. J., Gettins,
   P., and Dwek, R. A. (1978). *Biochem. J.,* in press.
Eisen, H. N., Simms, E. S., and Potter, M. (1968). *Biochemistry
   7,* 4126.
Eisen, H. N., Michaelides, M. C., Underdown, B. J., Schulen-
   berg, E. P., and Simms, E. S. (1970). *Fed. Proc. 29,* 78.
Epp, O., Colman, P., Fehlhammer, H., Bode, W., Huber, R., and
   Palm, W. (1974). *Eur. J. Biochem. 45,* 513.
Geschichter, R., and Copeland, S. (1928). *Arch. Surg. 16,* 807.
Gettins, P. (1979). Ph.D. Thesis, Oxford.
Gettins, P., Givol, D., and Dwek, R. A. (1978). *Biochem. J.*
   in press.
Goetze, A. M., and Richards, J. M. (1978). *Biochemistry 17,* 1733.
Goetzl, E. J., and Metzger, H. (1970). *Biochemistry 9,* 3862.
Griffith, O. H., Dehlinger, P. J., and Van, S. P. (1974). *J.
   Membrane Biol. 15,* 159.
Huber, R., Deisenhofer, J., Colman, P. M., Matsushima, M., and
   Palm, W. (1976). *Nature 264,* 415.

Johnson, C. E., and Bovey, F. A. (1958). *J. Chem. Phys. 24,* 1012.

Johnston, M. F. M., Barlsas, B. G., and Sturtevant, J. M. (1974). *Biochemistry 13,* 390.

Klostergaard, J., Krausz, L. M., Grossberg, A. L., and Pressman, D. (1977). *Immunochemistry 14,* 107.

Koostra, D. A., and Richards, J. H. (1978). *Biochemistry 17,* 345.

Kreth, H. W., and Williamson, A. R. (1973). *Eur. J. Immunol. 3,* 141.

Landsteiner, K. (1945). "The Specificity of Serological Reactions,: 2nd ed. Harvard Univ. Press, Cambridge, Massachusetts.

Little, J. R., and Eisen, H. N. (1967). *Biochemistry 6,* 3119.

Metzger, H. (1974). *Advan. Immunol. 18,* 169.

Nisonoff, A., Hopper, E. J., and Spring, S. B., eds. (1975). "The Antibody Molecule." Academic Press, London.

Padlan, E. A., Davies, D. R., Pecht, I., Givol, D., and Wright, C. (1976). *Cold Spring Harbor Symp. Quant. Biol. 41,* 627.

Poljak, R. J., Amzel, L. M., Chen, B. L., Phizackerly, R. P., and Saul, F. (1974). *Proc. Nat. Acad. Sci. USA 71,* 3400.

Porter, R. R., ed. (1973). "Defence and Recognition," MTP Int. Rev. Sci. Biochem. Ser. *1,* Vol. 10.

Schiffer, M., Girling, R. L., Ely, K. R., and Edmundson, A. B. (1973). *Biochemistry 12,* 4620.

Segal, D. M., Padlan, E. A., Cohen, G. H., Rudikoff, S., Potter, M., and Davies, D. R. (1974). *Proc. Nat. Acad. Sci. USA 71,* 4298.

Strausbach, P. H., Weinstein, Y., Wilchek, M., Shaltiel, S., and Givol, D. (1971). *Biochemistry 10,* 4342.

Sutton, B., Dwek, R. A., Gettins, P., March, D., Wain-Hobson, S., Willan, K. J., and Givol, D. (1977). *Biochem. J. 165,* 177.

Wain-Hobson, S., Dower, S. K., Gettins, P., Givol, D., McLaughlin, A. C., Pecht, I., Sunderland, C. A., and Dwek, R. A. (1977). *Biochem. J. 165,* 277.

Wu, T. T., and Kabat, E. A. (1970). *J. Exp. Med. 132,* 211.

Wüthrich, K. (1976). "NMR in Biological Research: Peptides and Proteins." Elsevier, New York.

MODEL COMPOUNDS AS AIDS IN INTERPRETING
NMR SPECTRA OF HEMOPROTEINS

Gerd N. La Mar
Department of Chemistry
University of California
Davis, California

I. INTRODUCTION

Heme proteins unquestionably comprise the most studied and best understood class of biological macromolecules. This relative success is due largely to the ready availability of these proteins, their ubiquitous occurrence, and the fact that they contain a well-defined prosthetic group, the iron porphyrin, which can often be reversibly removed and studied independently (Dolphin, 1978; Smith, 1975). The study of this prosthetic group, or a near facsimile, called a model compound, has provided a wealth of physical-chemical and spectroscopic information that may be utilized directly in interpreting structure-function relationships in the intact proteins.

Next to crystallography (Perutz, 1976) NMR, particularly proton NMR (Wüthrich, 1970; Phillips, 1973; Morrow and Gurd, 1975) has had the most important impact on our understanding of heme proteins. This is due largely to the ability to clearly detect very small changes in environment of a given nucleus, as controlled by protein conformation and iron oxidation/spin state. The three hemoproteins studied most extensively are myoglobin, hemoglobin (Antonini and Brunori, 1971; Shulman et al., 1975), and cytochrome c (Lemberg and Barrett, 1973) although data have also been reported on peroxidases (Williams et al., 1976) and other cytochromes (Keller et al., 1973, 1976; Smith and Kamen, 1974; McDonald et al., 1973). The biological functions and physical properties of these proteins have been adequately reviewed elsewhere (Antonini and Brunori, 1971; Shulman et al., 1975; Lemberg and Barrett, 1973).

The property that must be considered first in discussing NMR spectra of a hemoprotein is the spin state of the iron.

TABLE I. Oxidation and Spin States for Hemoproteins

Oxidation State	Spin State	Electron Configuration	Typical T_{1e}(sec)	Proteins
Fe(II)	$S = 0$	$(d_{xy})^2(d_{xz},d_{yz})^4$	----	$Mb^{II}_L, Hb^{II}_L\ (L=O_2,CO)$ [a] cyto $\underline{c}^{II}$ [b]
	$S = 1$	$(d_{xy})^2(d_{xz},d_{yz})^3 d_{z^2}$	5×10^{-13} [c]	Mb^{II} at low pH (?) [d]
	$S = 2$	$(d_{xy})^2(d_{xz},d_{yz})^2(d_{z^2})-(d_{x^2-y^2})$	$5\times10^{-12}-5\times10^{-13}$ [e]	Mb^{II}, Hb^{II}, [a]
Fe(III)	$S = 1/2$	$(d_{xy})^2(d_{xz},d_{yz})^3$	$2-8\times10^{-12}$ [f]	$Mb^{III}_{CN}, Hb^{III}_{CN}$, [a] cyto $\underline{c}^{III}$, cyto $\underline{c}^{III}_{CN}$ [b]
	$S = 3/2$	$(d_{xy})^2(d_{xz},d_{yz})^2(d_{z^2})$	?	cyto $\underline{c}_2^{III}$ (?)
	$S = 5/2$	$(d_{xy})(d_{xz},d_{yz})^2(d_{z^2})-(d_{x^2-y^2})$	$3-9\times10^{-11}$ [g]	$Mb^{III}_{H_2O}, Mb^{III}_{H_2O}$, [a] HRP^{III} [h]
Fe(IV)	$S = 2$	$(d_{xy})(d_{xz},d_{yz})^2(d_{z^2})$	$<10^{-11}$ [i]	peroxidase compound I [h]

[a] Antonini and Brunori, 1971.   [b] Iemberg and Barrett, 1973.
[c] Goff et al., 1977; Mispelter et al., 1977.   [d] Giacometti et al., 1977.
[e] Johnson et al., 1976; Goff and La Mar, 1977.   [f] Wüthrich et al., 1968c; Kurland et al., 1977; La Mar and Walker, 1973a; La Mar et al., 1977a.   [g] Kurland et al., 1968; La Mar et al., 1973a; La Mar and Walker, 1973; Walker and La Mar, 1973; Perutz et al., 1974.
[h] Yamazaki, 1974.   [i] Felton et al., 1973.

Table I lists the known oxidation/spin states of iron and their known or suspected occurrence in various hemoproteins. Since only one oxidation/spin state is diamagnetic, i.e., low-spin, LS, Fe(II) ($Mb^{II}O_2$, $Hb^{II}O_2$, Cytochrome $\underline{c}^{II}$), the NMR spectra of the majority of the states of hemoproteins involve the analysis of the effects of paramagnetism (La Mar *et al.*, 1973b). The origin and interpretability of shifts in the diamagnetic protein are relatively straightforward, since the peaks that are most readily resolved are upfield ring-current shifted residues of amino acids above and below the heme plane. Interpretation of the shifts is readily performed based on a wide literature on nonprotein NMR (James, 1975). The effect of the porphyrin on the protein is not readily modeled outside the protein, and most of the porphyrin resonances themselves are relatively insensitive to the effect of the neighboring protein structure. An exception is in the case where an aromatic amino acid side chain is approximately parallel to and in contact with the porphyrin (Redfield and Gupta, 1971). Thus studies on model compounds, except for the original identification of resonances, have not significantly contributed to the understanding of the NMR of diamagnetic hemoproteins in terms of their functions.

In the case of paramagnetic proteins, however, the unpaired spin of the iron perturbs the proton resonance (for review, see Jesson, 1973) primarily of the prosthetic group, i.e., the coordinated ligands, permitting the resolution of many of these peaks outside the generally poorly resolved diamagnetic polypeptide envelope in the region 0 to -10 ppm from DSS. These paramagnetic or hyperfine shifted resonances for the prosthetic group are inordinately sensitive to the iron spin/oxidation state and the contact between the heme and the apo-protein. It is precisely this sensitivity that has produced such concentrated research efforts on the various paramagnetic forms of Mb, Hb, and cytochromes. However, the very nature of these extraordinarily structure-sensitive hyperfine shifts provides the major obstacle to extracting the wealth of structural information they contain, in that there exists no satisfactory basis on which the paramagnetic shifts can be interpreted. As we shall demonstrate here, studies on model compounds provide us with some guidelines for such interpretation.

Over the past decade a considerable wealth of structural information on NMR of paramagnetic hemoproteins has appeared. In most cases the changes in hyperfine shifts have been used empirically to mark various states of the protein. The empirical use of these shifts and their value toward understanding important properties of various hemoproteins has been reviewed elsewhere (Wüthrich, 1970; Morrow and Gurd, 1975) and we shall therefore not attempt to interpret the mass of protein NMR data. Neither shall we attempt a comprehensive review of NMR

studies of model heme complexes (La Mar and Walker, 1978). Instead, we focus here on the utility of model compounds for ab-
stracting the maximum information from the NMR spectra of paramagnetic hemoproteins. We shall demonstrate that such studies
greatly enhance the value of protein studies and that, in many
cases, a study should ideally involve both the model compounds
as well as the protein. The value of the model NMR studies, as
well as with other spectroscopic techniques, is that response
of shifts and linewidth to systematic, controlled environmental
perturbations can be more readily interpreted.

Model iron porphyrin complexes have been extensively studied by NMR because of their utility in facilitating the location and interpretation of the hyperfine shifted resonances in
the hemoproteins, and because NMR permits the convenient characterization of a number of interesting porphyrin properties that
are relevant to their biological function but do not necessarily
manifest themselves in the NMR spectra of the intact protein.
This latter category is much larger and diverse, having covered
such properties as π interactions with aromatic acceptors and
donors (Fulton and La Mar, 1976), self-aggregation (La Mar and
Viscio, 1974), thermodynamics and kinetics of axial ligation
(Satterlee *et al.*, 1977), chemical properties of heme substituents (Evans *et al.*, 1977), antiferromagnetic coupling in porphyrin dimers (La Mar *et al.*, 1973a), and electron spin relaxation mechanisms (La Mar and Walker, 1973b). In this discussion
we are interested primarily in work in the first category,
where the model studies directly lead to understanding of the
protein NMR spectra.

Although the utility of model compounds in analyzing the
NMR spectra of hemoproteins was recognized a decade ago with
the characterization of some LS ferric complexes (Wüthrich *et
al.*, 1968c), only recent advances in direct (Cavaleiro *et al.*,
1974; Mayer *et al.*, 1974; Viscio, 1977; G. N. La Mar *et al.*,
1978c, D. L. Budd *et al.*, to be published) and indirect (Brassington *et al.*, 1975) methods for peak assignments have provided a basis for interpreting the effects of a protein environment on the porphyrin chemical shifts. These developments
emphasized the meticulous synthesis of porphyrins with deu-
terium-labeled functional groups, which permit the assignment
of specific resonances in both the model compounds and the
reconstituted proteins. It is precisely this combination of
model compound and labeled prosthetic groups that is expected
to provide us with quantitative interpretation of the NMR
spectra of hemoproteins in the near future.

II. PRINCIPLES

The theory behind NMR spectroscopy in paramagnetic systems has been treated in detail elsewhere (Jesson, 1973). Only an abbreviated qualitative description is given here, which will suffice to appreciate the data presented.

The resonance position of a nucleus in a paramagnetic system is shifted from its diamagnetic position by an amount defined as the hyperfine shift $(\Delta\nu/\nu_0)_{hf}$, in parts per million (ppm). This hyperfine shift may arise through two distinct physical interactions. The Fermi contact shift is due to delocalization of unpaired metal spin(s) to the ligand nucleus $(\Delta\nu/\nu_0)_{con}$. In the ideal case of a single populated level for the iron, which approximates a large number of cases, we have

$$(\Delta\nu/\nu_0)_{con} = -\underline{A}\, \frac{g\beta\underline{S}(\underline{S}+1)}{3(\gamma/2\pi)\underline{kT}} \tag{1}$$

where $\underline{A}$ is the hyperfine coupling constant, $\underline{S}$ the spin quantum number, $\underline{g}$ the spectroscopic splitting factor, β the Bohr magneton, $\underline{k}$ the Boltzmann constant, and $\underline{T}$ the absolute temperature. Determination of $\underline{A}$ yields the transferred unpaired spin density, and mapping $\underline{A}$ over a ligand indicates which ligand molecular orbital(s) are involved in the iron-porphyrin bonding (La Mar, 1973).

In cases where the iron has an anisotropic magnetic susceptibility, a dipolar shift $(\Delta\nu/\nu_0)_{dip}$ results, given by (Jesson, 1973)

$$(\Delta\nu/\nu_0)_{dip} = \frac{1}{3\underline{N}}\left\{ \chi_{\underline{z}} - \frac{1}{2}(\chi_{\underline{x}} + \chi_{\underline{y}})\,[(3\cos^2\theta - 1)\underline{r}^{-3}] \right.$$

$$\left. + \frac{3}{2}(\chi_{\underline{x}} - \chi_{\underline{y}})\sin^2\theta\,\cos 2\Omega\underline{r}^{-3} \right\} \tag{2}$$

where $\chi_{\underline{z}}$, $\chi_{\underline{x}}$, $\chi_{\underline{y}}$ are the principal components of the magnetic susceptibility tensor, θ the angle between the iron-nuclear vector and the $\underline{z}$ axis, $\underline{r}$ the length of this vector, and Ω the angle between the projection of $\underline{r}$ on the $\underline{xy}$ plane and the $\underline{x}$ axis. The first term is due to axial anisotropy while the second term is due to the rhombic or in-plane anisotropy. The shifts for nonequivalent nuclei in a given compound are proportional therefore to the relative values of the geometric factors $(3\cos^2\theta - 1)\underline{r}^{-3}$ and/or $(\sin^2\theta\,\cos 2\Omega)\underline{r}^{-3}$, and their magnitude can yield the magnetic anisotropy if $\underline{r}$, θ, Ω are known, or the position of a nucleus if $\chi_{\underline{x}}$, $\chi_{\underline{y}}$, $\chi_{\underline{z}}$ are known. Because the observed hyperfine shift is the vector sum, i.e.,

$$(\Delta\nu/\nu_0)_{hf} = (\Delta\nu/\nu_0)_{con} + (\Delta\nu/\nu_0)_{dip} \tag{3}$$

the interpretation of shift changes in proteins invariably requires the knowledge of the physical origin of the shifts.

Since spin transfer occurs only to ligands directly coordinated to the iron (La Mar, 1973) contact shifts should arise
only for the porphyrin, proximal histidine, and possible coordinated ligand or substrate. The dipolar shift varies as r^{-3}
and hence falls off rapidly with distance from the iron. Thus
both interactions predict that hyperfine shifts are observed
primarily for nuclei close to the iron, i.e., the prosthetic
group. However, the dipolar term is through space, and if
large enough, can affect amino acid side chains which line the
heme pocket above and below the heme.

Paramagnetism also involves enhanced nuclear relaxation
(Swift, 1973), T_{1N}^{-1}, T_{2N}^{-1}, which also operate via dipolar and
contact terms. In the simplest case, the nuclear relaxation
rates are related to the electron spin-lattice relaxation time
T_{1e}, i.e.,

$$T_{1N}^{-1} = T_{2N}^{-1} = 4\gamma^2 g^2 \beta^2 S(S + 1)/3r^6 \, T_{1e}$$

$$+ \quad 2S(S + 1)A^2/3\hbar^2 \, T_{1e} \tag{4}$$

In general, the dipolar interaction, which is the first term,
dominates so that $T_{1N}^{-1} \propto r^{-6}$, and the relative relaxation rates
therefore yields relative values for r^{-6} for nonequivalent nuclei in the same complex. It is clear from Eq. (4) that in order to have narrow proton lines (long T_{1N} we need very short
T_{1e}. This case is met for all known iron porphyrin systems.
Knowledge of r and T_{1N} for a nucleus permits determination of
T_{1e}; typical values for model hemes and hemoproteins are included in Table I. In some cases the T_{1N} for a given heme proton has been found to depend on protein tertiary structure (La
Mar *et al.*, to be published), but insufficient work has been
done even on model compounds to permit any interpretation at
this time. The potentially high information content of paramagnetic relaxation in heme proteins remains yet to be tapped.

III. MODEL COMPOUNDS

Since the prominent resonances in paramagnetic hemoproteins
arise from the coordinated ligands, we seek to characterize the
origin of structure-sensitivity of the planar porphyrin substi-

tuents, the ubiquitous proximal histidine, and the coordinated
ligand L (which may be an extraneous ligand as the O_2, CO in
HbO_2, HbCO, or another amino acid side chain as the methionine
in cytochrome $\underline{c}$). In Fig. 1 we illustrate two of the states of
the heme pocket generally considered in describing the functions
of hemoproteins. In the ligated state, where the sixth position is occupied by either substrate or another amino acid side
chain, the iron tends to be low-spin (LS) and in-plane in both
oxidation states (an exception is the met-aquo form of Hb and
Mb, where the sixth position is occupied by a water; the iron
is slightly out-of-plane but is primarily in the high-spin, HS,
form). The unligated state has the iron invariably in the HS
form and significantly displaced from the heme plane in the direction of the proximal histidine.

The porphyrin in the heme pocket is modeled by a variety of
derivatives of either synthetic or natural porphyrins, which
contain substituents at the various positions. The natural porphyrin derivatives can all be considered as derived from deuteroporphyrin IX, which is illustrated in Fig. 2, with a variety of
substituents at the 2,4 positions. The variety of substituents,
as well as the resultant trivial names and abbreviations for the
porphyrins are listed in Table II. The axial ligands used to
generate the model should resemble the protein ligands as
closely as possible when the emphasis is on the porphyrin resonances.

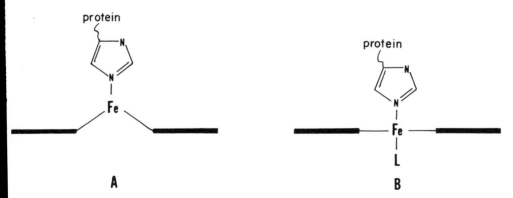

Fig. 1. *Structural features of the heme pocket in (A) unligated, five-coordinate, high-spin form, where the iron is out
of the heme plane, and (B) ligated, six-coordinate, generally
low-spin form, where the iron is either close to or in the heme
plane.*

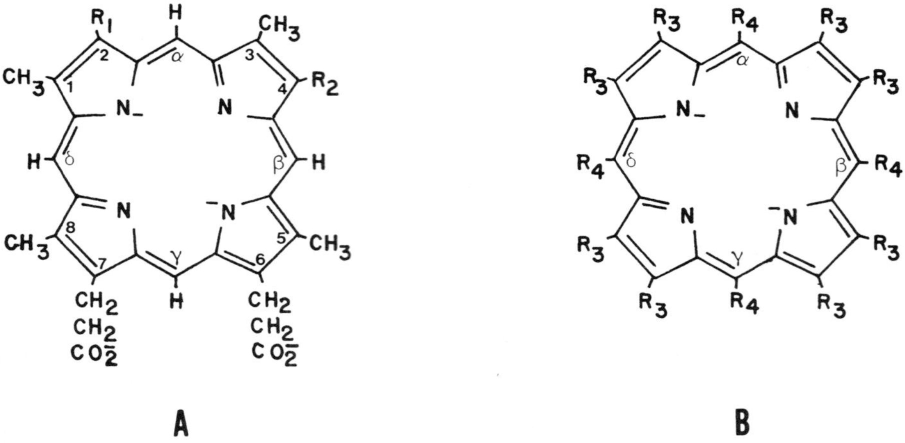

A   **B**

Fig. 2. Structural formulas for (A) natural porphyrin derivatives, and (B) the fourfold symmetric synthetic porphyrin
derivatives. See Table II for variety of R_1, R_2, R_3, R_4 and the
resultant names and symbols.

TABLE II. Names and abbreviations for natural and synthetic
porphyrins[a]

R_1 R_2	Name	Abbreviation
vinyl, vinyl	protoporphyrin	PP^b
proton, proton	deuteroporphyrin	DP
ethyl, ethyl	mesoporphyrin	MP
bromine, bromine	dibromodeuteroporphyrin	Br_2DP
acetyl, acetyl	diacetyldeuteroporphyrin	Ac_2DP
acetyl, proton	2-acetyldeuteroporphyrin	2-AcDP
proton, acetyl	4-acetyldeuteroporphyrin	4-AcDP
R_3 R_4		
proton, phenyl	tetraphenyl porphyrin	TPP
ethyl, proton	octaethyl porphyrin	OEP

[a]See Fig. 1 for structure of porphyrins.
[b]Dimethyl esters of the propionic acid side chains are
indicated by the appended letters DME.

In the situation where information on the axial ligand is
desired, it is often convenient to use synthetic porphyrins,
whose general structure is depicted in Fig. 2B, and whose resulting names and abbreviations are also included in Table II.

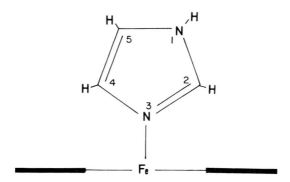

Fig. 3.   Structure and numbering system for an axial imidazole.

The natural and synthetic porphyrins produce very similar in-plane fields, but the latter have the advantage that the much greater symmetry yields many fewer resonances, improving the chances of resolving and assigning the axial ligand peaks (Satterlee and La Mar, 1976; La Mar *et al.*, 1976). The axial ligand, of course, is dependent on whether we are interested in modeling the ubiquitous histidyl imidazole or an extraneous ligand such as cyanide. Table III lists the various possible axial ligands used to model paramagnetic hemoproteins. The axial ligand of greatest interest is a coordinated histidine that is adequately modeled by imidazoles or 5-alkyl imidazoles; the numbering system depicted in Fig. 3 will be used in this text.

IV. LOCATION AND ASSIGNMENT OF RESONANCES

A. Porphyrin Resonances

Since the hyperfine shifts in the paramagnetic system are unique to the particular electronic/molecular structure of the iron, and the range of paramagnetic shifts is exceedingly large (300 ppm for protons, 2600 ppm for ^{13}C) compared to analogous diamagnetic systems, the models comprise the primary source of information as to likely protein assignments. To date, in every case studied, the general direction (upfield

or downfield) and approximate magnitude of the porphyrin shifts
are qualitatively the same in model compounds and proteins, although there are important differences (see below). The simi-
larities of the general shift pattern in ferric LS species
$PPFe(CN)_2^-$ (Viscio, 1977), $Mb^{III}CN$ (Sheard et al., 1970), and
cytochrome $\underline{c}^{III}$ (Wüthrich, 1969) are illustrated in Fig. 4; the
HS ferric model (Budd, 1978) $PPFe(L_2)^+$ (L, dimethylsulfoxide)
and $Mb^{III}H_2O$ (Budd, 1978) are given in Fig. 5, and the porphyrin resonances of the HS Fe(II) model (La Mar et al., 1977c),
$PPDMEFe(2-CH_3Im)$, are compared to those of deoxy Mb^{II} (La Mar et
al., 1978d) and deoxy Hb^{II} (Shulman et al., 1969a) in Fig. 6.

Easiest to assign in the protein spectra are the heme methyls, since all other resonances from heme peripheral substituents or the axial ligand (except for the methionine in cytochrome $\underline{c}^{III}$) yielded only single-proton resonances. The vinyl-
H yield characteristic deviations from the Curie law in LS
Fe(III) models (Wüthrich et al., 1968c) and similar behavior of
single-proton resonances in $Mb^{III}CN$ (Wüthrich et al., 1968a)
have led to their assignment to H_α. Unambiguous assignments of
vinyl protons and specific heme methyls have relied on specific
deuterium labeling of the heme (La Mar et al., to be published)
or resolution of the characteristic spin multiplet structure for
vinyl protons (La Mar et al., 1978a).

The most obvious difference between the models and proteins
in each case is that the spread of the resonances (particularly
for the four heme methyls, one on each pyrrole) is always larger
in the protein, indicating a more asymmetric in-plane environment in the protein. The Hb spectra analogous to those in Figs.
4 and 5 are again similar but with generally larger shift
spreads for the heme methyls (Wüthrich et al., 1968a; Perutz
et al., 1974); the Hb spectra are also more complicated because
of the inequivalence of the α,β subunits. The assignment of
heme resonances to individual subunits in Hb derivatives has en-
joyed some measure of success via the use of valency hybrids
(Ogawa and Shulman, 1971; Ogawa et al., 1972b) and mutant Hbs
(Davis et al., 1971), where the oxidation/spin state of one
chain can be controlled.

The similarity of the shift pattern in models and proteins
is particularly emphasized in some reconstituted hemoproteins.
Thus upon replacing protoporphyrin with deuteroporphyrin, we

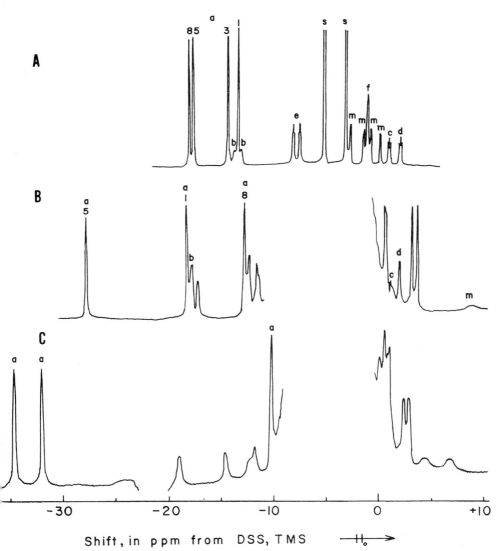

Shift, in ppm from DSS, TMS

Fig. 4. Proton NMR traces of low-spin ferric model and hemo-proteins: (A) PPFe(CN)$_2^-$ in $C^2H_3O^2H$ at 25°C (Viscio, 1977); (B) sperm whale Mb^{II}CN in ^2H_2O, p^2H 7.0, 29°C, (reprinted with permission from Sheard et al., 1970); and (C) cytochrome c^{III} in ^2H_2O, p^2H = 6.0, 25°C; only the portion below +10 ppm is shown (reprinted with permission from Wüthrich, 1969). In all cases, the heme substituents are labeled a, methyl; b, vinyl H_α; c, vinyl H_β(cis); d, vinyl H_β(trans); e, propionic acid α-CH$_2$; f, propionic acid β-CH$_2$; m, meso-H; s, solvent; x, impurity. In the cases where the heme methyls have been assigned, the appropriate position is designated by the number in Fig. 2A.

TABLE III. Potential Axial Ligands for Modeling Pragmatic Hemoproteins

Oxidation state	Spin state S	Model ligands		Proteins
		L	L'	
Fe(II)	1	none	none	Mb^{II} at low pH[a] (?)
	2	Im	none	Mb^{II}, Hb^{II}[b]
Fe(III)	1/2	Im	CN^-	$Mb^{III}CN$, $Hb^{III}CN$,[b] cytochrome c^{III}-CN,[c] $HRP^{III}CN$[d]
	1/2	Im	Im	$Mb^{III}Im$,[b] cytochromes b^{III}[c] c_3^{III}[c]
	1/2	Im	R-S-CH$_3$	cytochrome c^{III}[c]
	1/2, 5/2	Im	N_3^-, OH^-	$Mb^{III}L$, $Hb^{III}L$ L = N_3^-, OH^-[b]
	5/2	Im	H_2O,F^-,SCN$^-$,OCN$^-$	$Mb^{III}L$, $Hb^{III}L$, L = H_2O, F^-, SCN$^-$, OCN$^-$[b]
	5/2	Im	phenoxide	Hb^{III}-Iwate,[e]

[a]Giacometti et al. (1977).
[b]Antonini and Brunori (1971).
[c]Lemberg and Barrett (1973).
[d]Yamazaki (1974).
[e]Mayer et al. (1973).

have two pyrrole protons that experience larger coupling to the
unpaired electron of the heme and hence yield larger hyperfine
shifts than the heme methyls. In LS Fe(III), the 2,4-H resonances 15-25 ppm upfield from DSS are found in both models and
proteins (Shulman et al., 1969b). The proton spectrum of the
HS Fe(II) model, DPDMEFe(2-CH$_3$Im) (Goff and La Mar, 1977) has
the 2,4-H protons ~50 ppm downfield, outside the region of all
other heme substituents, as illustrated in Fig. 7A. The analogous trace for HS deuteroheme-Mb^{II} (Budd, 1978), shown in
Fig. 7B, exhibits 2,4-H resonances in the same region except
that they are slightly further apart. Hence the assignment is
unambiguous. The appearance of two sets of 2,4-H peaks in this
region in the case of the protein has been interpreted as indicating that there are two forms of the reconstituted protein
(Budd, 1978). This pair of interconvertible resonances yields
valuable information on disorder in the heme pocket (see below).

The model complexes for the unligated Fe(II) state, which
has intermediate spin $\underline{S} = 1$, have also been reported and analyzed (Goff et al., 1977; Mispelter et al., 1977). Though
their NMR spectra were investigated in order to establish the
electronic/magnetic properties of the iron in this unusual spin
state, more recent kinetic data on ligation of Mb at low pH
have been interpreted (Giacometti et al., 1977) in terms of
such a state, where the proximal histidine-iron bond is rup-
tured. The proton spectrum of PPDMEFe^{II} ($\underline{S} = 1$) is illustrated
in Fig. 8. The well-resolved resonances are indicative of
large dipolar and contact contributions (Goff et al., 1977).
Proteins in the $\underline{S} = 1$ state should exhibit many nonheme hyper-
fine-shifted resonances, with the amino acid side chains above
and below the plane being shifted upfield, rather than downfield
as for the LS ferric systems, since the signs of the axial anisotropy differ. Thus the shifts in the model compounds will
provide a guide as to where to look for the resonances characteristic of this ground state in proteins.

Some proton NMR spectra on synthetic porphyrin complexes of
iron(IV) have been reported (Felton et al., 1973). The resonances appeared sufficiently well resolved to suggest that similar studies on the oxidized forms of peroxidases should be
useful.

Relatively little ^{13}C work has been done on proteins in paramagnetic forms, and their analyses have not emphasized heme
resonances (Oldfield and Allerhand, 1973; Wilbur and Allerhand,
1976). In the case of paramagnetic models, spectra of LS

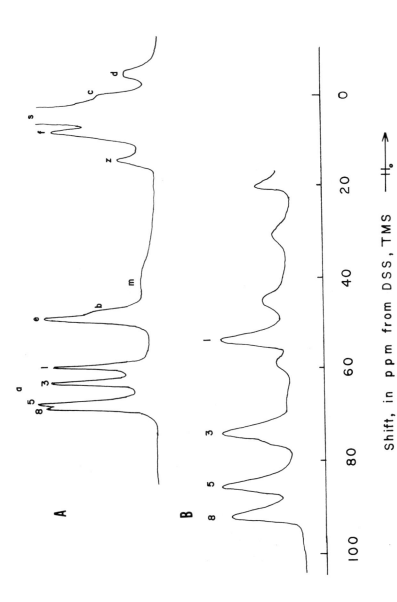

Fig. 5. Proton NMR traces of high-spin ferric model and hemoproteins. (A) $PPFe(L)_2'$ (L, dimethyl sulfoxide), in $(C^2H_3)_2SO$ at 25°C (Budd, 1978); (B) sperm whale $Mb^{III}H_2O$, $p^2H = 5.0$, 25°C (Budd, 1978). The heme resonances are designated by letters as described in Fig. 4.

Shift, in ppm from DSS, TMS $\longrightarrow H_o$

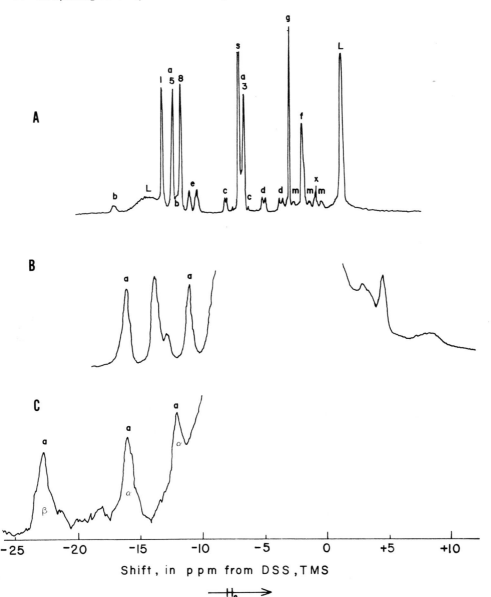

Fig. 6. Proton NMR traces of high-spin ferrous model and hemoproteins. (A) PPDMEFe^{II}(2-CH$_3$Im) in C$_6^2$H$_6$, 25°C (reprinted with permission from Goff and La Mar, 1977); (B) sperm whale deoxy Mb^{II} in ^{2}H$_2$O, p^{2}H = 8.0, 25°C (Budd, 1978); (C) human deoxy Hb^{II} in ^{2}H$_2$O, p^{2}H = 7, 25°C (reprinted with permission from Davis et al., 1971). The heme resonances are designated by letters as described in Fig. 4.

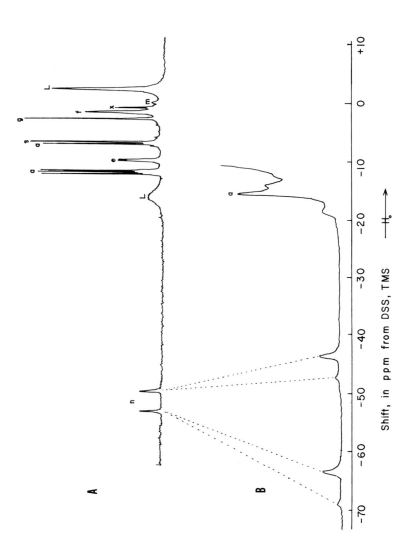

Fig. 7. Proton NMR traces of (A) DPDMEFe^{II}(2-CH$_3$Im) in C$_6^2$H$_6$, 25°C (reprinted with permission from Goff and La Mar, 1977); and (B) deuteroporphyrin reconstituted sperm whale deoxy Mb^{II} in ^{2}H$_2$O, p^{2}H = 7.6, 25°C (Budd, 1978). The observation of two sets of pyrrole-H resonances (designated n) in the protein is indicative of "disorder" of the heme in the heme pocket. The heme resonances are designated by letters as described in Fig. 4; OCH$_3$, g.

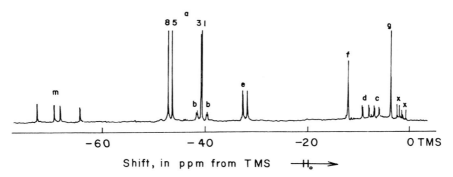

Fig. 8. *Proton NMR trace of the S = 1 ferrous model com-*
pound, PPDMEFe^{II} *in C*$^2_6$*H*$_6$*, 25°C (reprinted with permission from*
Goff et al., 1977). The heme resonances are designated by
letters as described in Fig. 4; OCH$_3$ *= g.*

Fe(III) models have been reported (Wüthrich and Baumann, 1973,
1974) and some data on HS ferric systems have been obtained
(H. Goff, private communication). At least the heme methyls
should be observable in LS proteins. Some ^{13}C heme resonances
have been located in cytochrome c^{III}, but the data were not interpreted (Oldfield and Allerhand, 1973). More work can be expected in this area in the near future.

B. Axial Ligand Resonances

The location and probable assignments of resonances from
the axial ligand, particularly the ubiquitous proximal histidine, pose much more of a problem because this ligand is expected to yield only single-proton resonances, as do all porphyrin substituents except the methyls. Furthermore, the proximal histidine is not as readily amenable to deuterium labeling
as the porphyrin. Location of the resonances in model compounds was facilitated using the synthetic porphyrin, TPP,
whose simplified porphyrin spectrum enabled the ready solution
of all imidazole peaks in both LS ferric and HS ferrous complexes (Satterlee and La Mar, 1976; La Mar et al., 1976; Goff
and La Mar, 1977). Comparable data in HS ferric systems have
not been reported, probably due to the lack of any well-characterized model compound.
    The proton spectra for two LS Fe(III) models, PPDMEFe(Im)$_2^+$
and TPPFe(Im)$_2^+$, are compared (Satterlee and La Mar, 1976; La
Mar et al., 1976) in Fig. 9 with the analogous region of sperm
whale Mb^{III}CN in H$_2$O (Sheard et al., 1970). Since the regions

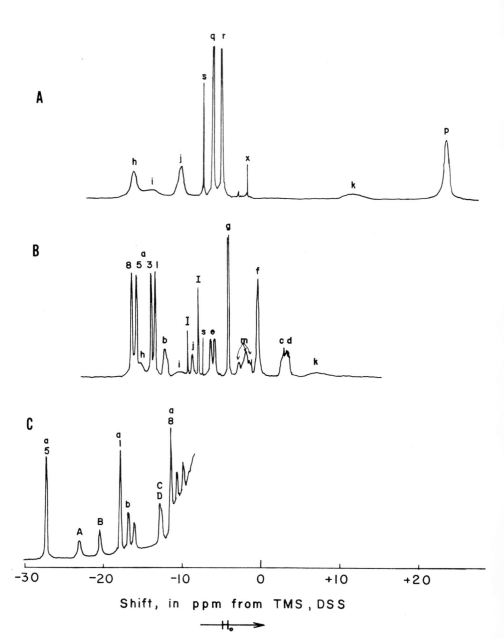

Fig. 9. Proton NMR traces of low-spin ferric model compounds and hemoproteins emphasizing the axial imidazole resonances. (A) TPPFe^{II}(Im)$_2^+$ in C^{2}HCl$_3$, 25°C (reprinted with permission from Satterlee and La Mar, 1976); (B) PPDMEFe^{III}(Im)$_2^+$

of the imidazole and porphyrin resonances overlap completely, unambiguous assignments are not possible. The strongest evidence for an assignment to the protein exists for the exchangeable imidazole 1-H since this resonance can be seen only in H_2O. The comparison of the two model compounds, however, verifies that the nature of the porphyrin is not important in modeling the axial position. Work with complexes containing one imidazole and one cyanide has suggested (La Mar et al., 1977b) that the coordinated imidazole shifts are not significantly altered by a trans-ligand and therefore may be expected in similar regions for $Mb^{III}CN$, cytochrome c^{III}, cytochrome b^{III}. The most useful probes of the proximal histidines are likely to be the exchangeable 1-H and the 5-CH_2 peaks. The former may provide information on bonding (La Mar et al., 1977c), while the latter pair of resonances (α-CH_2 is diastereotopic) can be expected to yield information on the changes in orientation of the imidazole plane relative to this CH_2 group (La Mar et al., 1976). The assignment of these α-CH_2 resonances, however, will have to depend on eliminating all other porphyrin resonances via systematic deuteration, and the assignment of other nonheme nonhistidyl resonances from the heme pocket. Such work may be possible in the near future.

The clearest evidence for the advantage of looking at models occurs in the case of deoxy proteins. Although much work had been done on deoxy Hb^{II} (Morrow and Gurd, 1975), until the results on models were available, the range of resonances was thought to occur exclusively ±25 ppm from DSS, as illustrated in Fig. 6. Experimental conditions permitted (Goff and La Mar, 1977; La Mar et al., 1977c) the detection and unambiguous assignment of all imidazole resonances for the model TPPFe(R-Im), (where R is a methyl group at the 2 or 4 position that inhibits formation of the diamagnetic bis complex), as illustrated in Fig. 10. Most importantly, this work showed that the three histidyl imidazole protons, 2-H, 4-H, and the exchangeable 1-H (Fig. 3) should all resonate well outside the range where the heme resonances are known to appear (Fig. 5). Thus the assign-

in C^2HCl_3, 25°C (reprinted with permission from La Mar et al., 1976); (C) sperm whale $Mb^{III}CN$ in H_2O, pH = 8.2, 29°C (reprinted with permission from Sheard et al., 1970). The natural heme resonances are designated by letters as described in Fig. 4; for TPP, the letters $\underline{p}$ = pyrrole-H, $\underline{q}$ = $\underline{o}$-, $\underline{m}$-H, $\underline{r}$ = $\underline{p}$-H pertain. The axial imidazole peaks in the models are designated $\underline{h}$ = 1-H, $\underline{i}$ = 4-H, $\underline{j}$ = 5-H, $\underline{k}$ = 2-H. In $Mb^{III}CN$, four exchangeable peaks, A, B, C, D, have been identified with peak (B) assigned to 1-H of proximal histidyl imidazole, peak (A) to the 1-H of the distal imidazole, and (C) to the peptide proton of the proximal histidine.

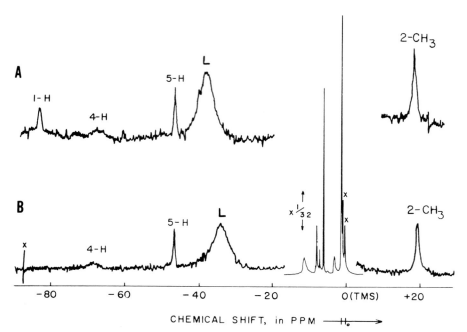

Fig. 10. *Proton NMR traces of high-spin ferrous model compounds emphasizing axial imidazole resonances (reprinted with permission from La Mar et al., 1977c).* (A) *TPPFe^{II}(2-CH$_3$Im) in C$_6^2$H$_6$, 25°C, and* (B) *TPPFe^{II}(1-^{2}H,2-CH$_3$Im) in C$_6^2$H$_6$, 25°C. Deuteration of 1-H in B clearly assigns this exchangeable peak. Peaks marked L are due to the porphyrin resonances, which, under the experimental conditions, are exchange broadened.*

ment of such peaks, when located in proteins, is considerably more certain than if the ranges had overlapped. The most prominent imidazole peak, the exchangeable 1-H, which shifts ~80 ppm downfield from DSS, can always be assigned since it will appear only in H$_2$O solutions.

In Fig. 11, we illustrate the downfield portion of sperm whale Mb^{II} and Hb^{II}A in H$_2$O and D$_2$O (La Mar et al., 1977c). The previously characterized regions (Morrow and Gurd, 1975) containing the heme methyls appears in the right-hand margin. The single exchangeable proton peak for Mb in H$_2$O and the pair of peaks in the H$_2$O solution of Hb in the region -70 to -80 ppm can be unambiguously assigned to the proximal histidyl 1-H (the peaks for the α and β subunits in Hb^{II} are thus resolved). The broad two-proton resonance at -40 ppm (found resolved in other

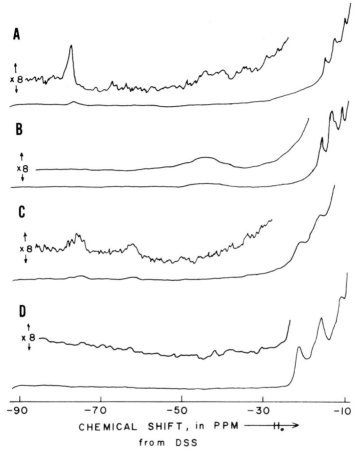

Fig. 11.   Low-field proton traces of deoxy myoglobin and
hemoglobin (reprinted with permission from La Mar et al.,
1977c).   (A) Sperm whale Mb^{II} in H_2O, pH = 6.8; (B) sperm whale
Mb^{II} in ^2H_2O, p^2H = 6.9; (C) human $Hb^{II}A$ in H_2O, pH = 6.0; (D)
human $Hb^{II}A$ in ^2H_2O, p^2H = 6.3.   Temperature, 25°C in all
traces.

single chain Hbs) is consistent with the 2,4-H imidazole peaks,
both by position and by the unusual temperature dependence that
was also found in the models for these positions (Goff and La
Mar, 1977).   Based on this model work, the characteristic 1-H
peak has been found in a number of Hbs, Mbs, and has even been
used to demonstrate that the axial ligand in *aplesia* Mb is a
histidyl imidazole (Budd, 1978).   The use of mutant Hb and va-

lency hybrids has provided evidence for the assignment of the
two exchangeable 1-H peaks to individual subunits in $Hb^{II}A$.

In certain favorable cases, the presence of a potential
axial ligand in a protein can be established on the observation
of highly characteristic hyperfine shifted resonances for the
axial ligand in a model compound. Several mutant hemoglobins
exist in their native form with one of their subunits perma-
nently oxidized to the ferric state (Lehmann and Kynoch, 1976).
In the case where the proximal histidine F8 is replaced by a
tyrosine, it is assumed that the ferric state is strongly
stabilized by coordination to the tyrosine. The proton NMR
spectrum of a model HS ferric compound containing the phenoxide
ion (Fig. 12), as axial ligand has been reported (Caughey and
Johnson, 1969), with the 2,6-H and 3,5-H peaks located at +95
and -83 ppm, respectively. Since the protoporphyrin models do
not have resonances in this region, observation of resonances
in this region can be taken as evidence for the coordinated tyrosine. Unfortunately, the NMR spectra of $Hb(\alpha_2^{III}\beta_2^{II})$-Iwate
(Mayer et al., 1973) reported to date have not included the
range where these resonances may be expected.

The ^{13}C and ^{15}N shifts for coordinated labeled cyanide in
LS ferric models have been reported (Goff, 1977; Morishima and
Inubushi, 1977; Morishima et al., 1977), with large hyperfine
shifts of +2600 and +700-1000 ppm, respectively. These shifts
are potentially very powerful probes of the met-cyano proteins.
However, the ^{13}C resonances in PPFe(CN)(Im) have not yet been
located, either due to exchange line broadening or to excessive
natural linewidth. Similarly, studies with $Mb^{III}CN$ and cytochrome c^{III}-CN have also failed to resolve the coordinated cyanide signal (Goff, 1977). Based on the model compound work
(Morishima and Inubushi, 1977), the ^{15}N peaks in both $Mb^{III}CN$
and $Hb^{III}CN$ enriched in ^{15}N have been located (Morishima et al.,

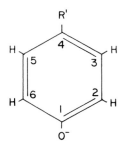

Fig. 12. Structure and numbering of phenoxide or tyrosinate anion. For phenoxide R' = H_1, while for tyrosinate R' =
$-CH_2-CH(CO_2H)NH_2$.

1977). It is likely that these shifts will provide much useful future information on environmental effects on the coordinated cyanide (see below).

Although model compounds have satisfactorily accounted for the types of resonances and their range of shifts in most cases, some interesting aspects of the heme protein proton spectra have so far eluded assignment and interpretation. Thus deoxy Hb yields several resonances ~20 ppm upfield from DSS (C. Ho, private communication). To date no porphyrin resonances in a model has been found anywhere in this region. Thus more work on model compounds is indicated.

V. HYPERFINE SHIFT ORIGINS

The ability to introduce probes into the heme periphery without significantly altering the electronic/magnetic properties of model compounds provides a direct method for determining the physical origin of the shifts, as discussed in greater detail elsewhere (La Mar and Walker, 1973; Goff et al., 1977; Goff and La Mar, 1977; La Mar et al., 1977a,b). The similarity of the shifts for given porphyrin functional groups in model compounds and hemoproteins indicates that the conclusions reached for the model compounds are relevant to the proteins.

The origin of the hyperfine shift, contact or dipolar, must be known before shift changes in the protein can be interpreted in terms of structural changes. Thus it is necessary to know if the metal iron is magnetically anisotropic. Since contact shifts affect only coordinated ligands, the absence of significant dipolar shifts dictates that all resonances with hyperfine shifts arise either from the porphyrin or axial ligand. Conversely, large magnetic anisotropy will induce sizable dipolar shifts in a number of amino acid residues that line the heme pocket (Sheard et al., 1970). Even for pure contact shifts, it is useful to know whether the transferred spin resides in σ or π orbitals (La Mar, 1973), and if possible, which orbital. This type of analysis requires the use of model compounds.

In LS Fe(III), earlier ESR data (La Mar and Walker, 1973a; Peisach et al., 1973) on models as well as proteins indicated that anisotropy, and hence dipolar shifts are significant (Shulman et al., 1971). The use of model compounds of tetraphenyl porphyrins, where the dipolar shifts could be obtained empirically, quantitatively confirmed the importance of these dipolar shifts. The resulting contact shifts were found to be indicative of spin transfer into a π molecular orbital. The known orbitally nondegenerate ground state of HS Fe(III) predicts small dipolar shifts, as was also confirmed by variable temperature

studies (Walker and La Mar, 1973) of such models with synthetic
porphyrins. The downfield pyrrole proton and methyl contact
shifts were interpreted as reflecting σ delocalization (Kurland
et al., 1971).

For the HS Fe(II) deoxy models, however, where much work
had been done on proteins, recent NMR studies using synthetic
porphyrins were needed to provide a basis for characterizing
the shift origin (Goff and La Mar, 1977). Detailed analysis
of these models showed that the magnetic anisotropy was very
small, contributing less than 10% of the shifts in most cases.
Therefore the hyperfine shifts in deoxy Mb^{II} and Hb^{II} arise
predominantly from contact interaction, and hence reflect covalency effects in the iron-porphyrin bond.

Once the dipolar shifts for the porphyrin are known, the
dipolar contribution to the shifts for the axial ligand in the
same complex is readily established by the relative values of
the geometric factor in Eq. (2). For a coordinated imidazole,
the dipolar and contact shift contributions are comparable in
the LS ferric system (Satterlee and La Mar, 1976), with the
contact shift reflecting π spin transfer. In the case of the
HS ferrous models, dipolar shifts were found to be negligible
and the contact shifts were accounted for by σ spin transfer
(Goff and La Mar, 1977).

The relative importance of dipolar and contact contributions
for the heme methyls and imidazole 1-H, as well as the dominant
porphyrin and axial imidazole spin transfer mechanisms responsible for the contact shifts for various model compounds are
summarized in Table IV. The similarity of the heme hyperfine
shift patterns in models and proteins suggest that the data in
Table IV are qualitatively applicable to proteins. Although no
protein in the $\underline{S}$ = 1 spin state has been clearly characterized,
recent reports indicate that the proximal histidine bond is rup-
tured at very low pH in Mb, which could give rise to the intermediate spin state (Giacometti *et al.*, 1977). Future NMR work
may in fact provide evidence for or against this hypothesis.

VI. HYPERFINE SHIFT CHANGES WITH MOLECULAR/ELECTRONIC STRUCTURE

The fact that the proton NMR spectra of the porphyrin and
axial imidazole in models are very similar, although not identical, to those of the hemoproteins, reflects the influence of
the globin on the electronic/molecular structure of the prosthetic group. An understanding of the nature of this globin-
heme interaction can be obtained from studies of the effect of
controlled perturbations on the NMR spectra of model compounds.
Several cases where detailed studies of such models proved useful in understanding the protein spectra follow.

TABLE IV. *Hyperfine shift origin and spin transfer mechanism for porphyrin and axial imidazole in model complexes[a]*

Model complex	Spin state S	Shift Origin (ppm)				Porphyrin spin transfer	Imidazole spin transfer
		Heme Methyl		Imidazole 1-H			
		$(\Delta\nu/\nu_0)con$	$(\Delta\nu/\nu_0)dip$	$(\Delta\nu/\nu_0)con$	$(\Delta\nu/\nu_0)dip$		
PPFe^{II} b	1	-27	-13	--	--	π (highest) filled)	--
PPFe^{II}(2-CH$_3$Im)[c]	2	-9	+1	-67	-6	σ	σ
PPFe^{III}(Im)$_2^+$ [d]	1/2	-20	+4	+10	-12	π (highest filled	π
PPFe^{III}Cl[e]	5/2	-70	-7	--	--	σ	--

[a] All PP complexes are in the form of dimethylesters.
[b] Goff et al. (1977).   [c] Goff and La Mar (1977).
[d] La Mar and Walker (1973a).   [e] La Mar et al. (1973).

A. *Asymmetry in Heme-Protein Linkages*

    Comparison of the heme methyl hyperfine shifts in protopor-
phyrin models and in proteins in every oxidation/spin state reveals that the spread of the four heme methyls is always much
larger in the protein environment (see Figs. 5, 6, 7). Since
there is a methyl group on each pyrrole, the methyl shifts provide a measure of the nonequivalence of the four pyrrole environments (La Mar *et al.*, 1978c). Differences in the axial dipolar
shifts can be completely eliminated as the source of this methyl
shift spread since it cannot change the spread of the resonances. Changes in rhombic (in-plane) anisotropy can similarly be
discounted since the patterns of the methyl peak spread, as determined by deuteration studies for each oxidation/spin state,
fail to obey the symmetry imposed by the second term in Eq. (2).
Hence the greater shift asymmetry in the protein must reflect
the contact interaction in terms of a more assymetric distribution of the unpaired spin among the four pyrroles. This spread
of the methyl shifts must reflect the asymmetric nature of the
heme-protein linkage. Since the increase in asymmetry on going
to the protein environment occurs in all paramagnetic states,
the analysis of the origin of this asymmetry carried out for one
system is likely to be relevant to the other oxidation/spin
states.
    The analysis of the protein-induced asymmetry can be per-
formed most readily on the LS ferric systems, which have been
studied most thoroughly, where the contact shifts are consistent
with porphyrin $\rightarrow$ iron π charge transfer. Studies on dimerization (La Mar and Viscio, 1974; Viscio and La Mar, 1978a,b) of LS
ferric bis cyano complexes has also indicated that the porphyrin
asymmetry arises not only in the highest bonding orbital, but
probably occurs in all π orbitals. One of the initial interpretations (Shulman *et al.*, 1971) of the asymmetry of the methyl
contact shifts in both models and proteins was based on the
raising of the orbital doublet ground state by low-symmetry perturbations. In the protein, the perturbation was proposed to be
the π bonding to the histidyl imidazole. Although the model compounds fitted this theory, assignment of individual methyl groups
by deuteration in sperm whale $Mb^{III}CN$ showed (Mayer *et al.*, 1974)
this interpretation to be inappropriate. Subsequent deuteration
studies of HS ferric models revealed (Budd, 1978) an asymmetry
similar to that of the LS models, demonstrating that the methyl
shift spread reflects asymmetry in the molecular orbital containing the unpaired spin(s).
    The asymmetry of the heme in the protein can be induced by
the protein either by the axial imidazole or via the peripheral
van der Waals contacts with amino acid side chains. The orientation of the axial imidazole as a controlling factor in the

low-symmetry perturbation can be largely discounted since
fixing the orientation of the axial imidazole in models (via
linking to porphyrin side chains) does not significantly increase the methyl shift spread (K. Migita and G. N. La Mar,
to be published). Furthermore, forming the LS Fe(III) protein
by addition of an extraneous imidazole (which is also rota-
tionally constrained), i.e., $Mb^{III}Im$, does not alter the methyl
shift pattern from that of $Mb^{III}CN$ (Budd, 1978). It can therefore be concluded that the methyl shift pattern must be induced
primarily by asymmetric peripheral van der Waals contacts with
the protein.

The best evidence that heme-protein peripheral contacts can
control the asymmetry in LS ferric systems is obtained from studies of the effect of porphyrin 2,4-substituents on the methyl
shift patterns (La Mar et al., 1978c). The data on the effect
of variable R (Fig. 2) in order of increasing electron with-
drawing power, ethyl < proton $\tilde{<}$ vinyl < bromine < acetyl, are
presented schematically for both the LS and HS ferric models
in Fig. 13; also included are the methyl shifts in sperm whale
Mb in the analogous spin states. In each case, increasing
asymmetry in the model, as controlled by 2,4-R, increased the
spread of the four methyl signals, with R = acetyl models
closely resembling the spread of the methyl peaks in the protein. In fact, the LS Fe(III) models of $4\text{-AcDPFe}^{III}(CN)_2^-$
mimics the order of the Mb^{III}-CN shifts (Fig. 13). In the protein, of course, there are many heme-protein contacts with the
systems, but the net effect of the protein can be simulated by
electron withdrawal at pyrrole II.

Based on this analysis, we can interpret differences in
heme methyl contact shift patterns in various LS ferric hemoproteins in terms of differences in asymmetry in the electronic
system that must reflect the characteristic differences in the
net protein-heme van der Waals contacts. Thus individual
methyl deuteration has revealed that $Mb^{III}CNs$ tend to exhibit
highly asymmetric π spin distributions (Mayer et al., 1974;
Budd, 1978), while monomeric $Hb^{III}CN$ chains appear to have an
in-plane asymmetry with pairwise similar methyls (La Mar et
al., 1978a, La Mar et al., manuscript in preparation). More
quantitative interpretations of the data must await further
theoretical developments. Some qualitative uses of this π
asymmetry are exemplified by the ability to characterize heme
"disorder" in the monomeric Hbs (La Mar et al., 1978a) and reconstituted Mbs (La Mar et al., 1978e) using the heme cavity
asymmetry as a marker (see below).

Another important ramification of the interpretation of the
heme methyl shift asymmetry in terms of protein-induced asymmetric porphyrin π electron distribution deals with the ferric
cytochromes. Since the unpaired spin in cytochrome c^{III} is in
the highest filled π orbital (the same orbital that must re-

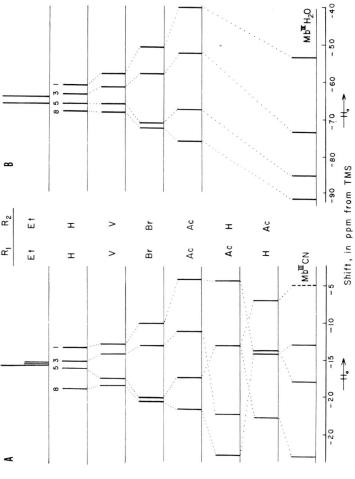

Fig. 13. Schematic proton NMR trace of the heme methyl region for model compounds with variable substituents, R_1, R_2, in the 2,4 position of deuteroporphyrins. (See Fig. 2.) (A) Low-spin, ferric bis-cyno complexes in $C^2H_3O^2H$, 25°C (Viscio, 1977); (B) high-spin, ferric bis-dimethyl-sulfoxide complexes in $(C^2H_3)_2SO$, 25°C (Budd, 1978). The sperm whale myoglobin complexes in the analogous oxidation/spin states are included for comparison. The individual heme methyl peaks have been assigned by specific deuteration in both the models and the proteins.

ceive the added electron upon reduction) it is reasonable that the protein-heme contacts distort the heme so as to point this orbital to the exposed heme edge. Assignment of the methyl peaks in cytochromes c^{III} and c_2^{III}, using saturation transfer techniques (Redfield and Gupta, 1971) and extrinsic dipolar relaxation reagents (Smith and Kamen, 1974), have shown that it is precisely the pyrrole on the exposed heme edge that exhibits the largest π methyl contact shifts. Further interpretation of the NMR spectra of cytochromes requires specific assignment of the individual methyls as done in sperm whale myoglobins.

B. Heme Disorder in Hemoproteins

The proton NMR spectra of the native monomeric hemoglobins for the insect *Chironomus thummi thummi* were found to exhibit two sets of protein resonances (La Mar *et al.*, 1978a), as illustrated in Fig. 14, indicating the presence of two conformations that can be interconverted at pH $\geq$ 10 and 40°C. Results on the model compounds 2-Ac-DPFe^{III}(CN)$_2^-$ and 4-AcDPFe^{III}(CN)$_2^-$ have shown (La Mar *et al.*, 1978e) that moving the low-symmetry perturbating acetyl group between pyrroles I and II (Fig. 13), pairwise interchanges the relative values of the contact shifts for 1-CH$_3$ and 3-CH$_3$, as well as 5-CH$_3$ and 8-CH$_3$. The models therefore demonstrate that the order of the shifts is indicative of the asymmetric peripheral substituent perturbation. In the case of the insect Hbs, specific deuterium labeling of the methyls revealed (La Mar *et al.*, to be published) that the methyl peaks with similar contact shifts in the major (X) and minor (Y) components (Fig. 14) always involved the exchange within the 1-CH$_3$/3-CH$_3$ and 5-CH$_3$/8-CH$_3$ pairs, again indicating a movement of the asymmetric perturbation ~180° with respect to the porphyrin α-γ meso axis. Since the protein tertiary structure is essentially identical for the two components, this requires that the heme orientation for the two components differ by 180° relative to the α-γ meso axis. The observation of such heme disorder in native proteins is unprecedented. Since the discovery of this disorder in native insect Hbs (La Mar *et al.*, 1978a,b), it has been shown that some reconstituted Hbs and Mbs also exhibit this disorder (La Mar *et al.*, 1978e), in spite of the fact that x-ray studies of the identical material have been interpreted in terms of a unique heme orientation (Seybert and Moffatt, 1976).

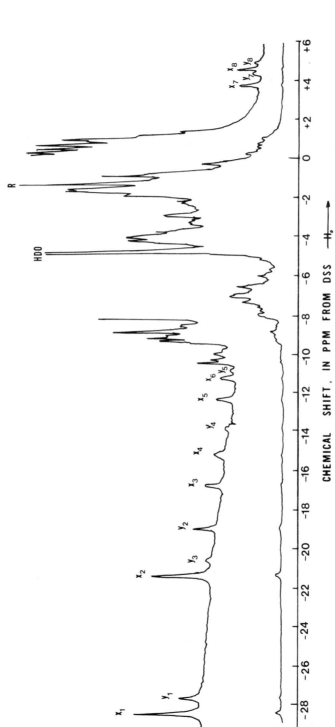

Fig. 14. Proton NMR trace of met-cyano form of monomeric hemoglobin IV from the insect Chironomus thummi thummi in ^2H_2O, $p^2H = 7.4$, $25°C$ (reprinted with permission from La Mar et al., 1978a).
There are two components (interconvertible), whose peaks are labeled x for the major and y for the
minor component. Peak assignments (La Mar et al., 1978a; La Mar et al., to be published) are
$x_1 = 8$-CH_3; $x_2 = 3$-CH_3; $y_1 = 5$-CH_3; $y_2 = 1$-CH_3. $\underline{x_3y_3} = $ vinyl-H_α, $\underline{x_8y_8} = $ vinyl-H_β(trans), $\underline{x_8y_7} = $ vi-
nyl-H_β(cis); R, t-butanol.

C. Vinyl Group Rotation in Proteins

As was indicated in Section I, the theory for both contact and dipolar shifts indicates that the shifts vary approximately as the reciprocal absolute temperature T^{-1}. The plot of the hyperfine shifts vs. T^{-1} for the model $PPFe^{III}(CN)_2^-$, illustrated in Fig. 15, shows that, while most functional groups indeed follow an approximate T^{-1} rule, the α-CH_2 and vinyl-H_α deviate considerably from this expected behavior (La Mar et al., 1978b). This unusual behavior for α-CH_2 and vinyl-H_α was accounted for by the fact that these two substituents can adopt a number of rotational positions with respect to the heme plane for which the degree of coupling (contact shift) with the porphyrin π system is critically dependent on the rotational angle. It was possible to show that the magnitude of the vinyl-H_α shift can indicate the average rotational position of the vinyl group, while the degree of the deviation of the shift from a T^{-1} behavior provided information on the oscillatory mobility of the vinyl group.

Two of the monomeric Hbs from the insect *Chironomus thummi thummi* exhibit marked Bohr effects (Gersonde et al., 1972). The heme hyperfine shifts for the met-CN form were found to be pH-sensitive, indicative of a one-proton transition

$$Hb^{III}CN(\underline{t}) \; \overset{\rightarrow}{\leftarrow} \; H^+ + Hb^{III}CN(\underline{r})$$

where, as in the tetrameric Hbs, $\underline{t}$ and $\underline{r}$ indicate the low-affinity (tense) and high-affinity (relaxed) states (La Mar et al., 1978a). The largest pH sensitivity was found for a heme vinyl, which suggested that the protein structural changes altered the relative orientation of the vinyl and heme planes. Detailed analysis (La Mar et al., 1978b) of the vinyl shift changes in the protein, based on the above model system, revealed that the vinyl group was more in-plane and had less oscillatory mobility in the acidic ($\underline{t}$) form than in the alkaline ($\underline{r}$) form of the protein. Since molecular models indicate clearly that steric interactions with other porphyrin substituents favor an out-of-plane, oscillatory mobile vinyl group, the NMR data provide direct support for the appropriateness of a "tense" discription for the low-affinity state. The steric interaction of a heme vinyl with amino acids side chain has also been proposed (Gelin and Karplus, 1977) in the mechanism of cooperativity of vertebrate Hbs. It appears worthwhile to determine if the rotational position of the vinyl group differs characteristically for the T and R states of tetrameric Hbs.

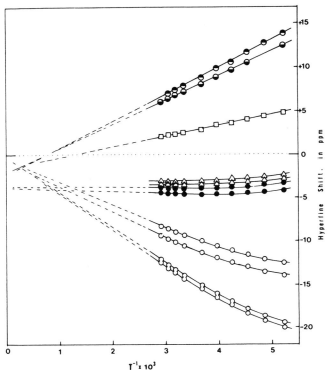

Fig. 15. Plot of observed $(\Delta\nu/\nu_0)_{hf}$ versus reciprocal temperature for $PPFe^{III}(CN)_2^-$ in $C^2H_3O^2H$ (reprinted with permission from La Mar et al., 1978b). Heme methyls; ●, vinyl-H_α; ◐, vinyl-H_β(*cis*); ◑, vinyl-H_β(*trans*); △, propionic acid α-CH_2.

D. Axial Ligand Tension

The ability to assign all peaks in coordinated imidazoles in deoxy model compounds and to show that they arise directly from covalency with the iron d_{z^2} orbital (Goff and La Mar, 1977) leads to the potential use of these peaks as probes for tension in the proximal histidine-iron bond in hemoglobins. The introduction of tension into this bond in the T state of tetrameric hemoglobins has been proposed (Perutz and Ten Eyck, 1971) to account for their low O_2 affinity. Since the proximal histidyl imidazole 1-H peak has been resolved in a variety of Mbs and Hbs (Fig. 11), the change in the shift of this peak can be expected to be a measure of changes in the iron-imidazole bond tension. The observation (La Mar *et al.*, 1977c) of very similar 1-H shift in sperm whale Mb^{II}, the deoxy model compound, and one of the chains in Hb^{II} argues against significant tension in the <u>deoxy</u> state of tetrameric Hbs. In Table V, we list the values for the proximal histidyl imidazole 1-H hyperfine shifts (Budd, 1978), as well as the oxygen pressure for half-saturation, $\underline{p}1/2(O_2)$ and the on-rate for oxygenation, $\underline{k}_{on}(O_2)$, for a number of monomeric Mbs and Hbs (Hoffmann and Mangum, 1970; Antonini and Brunori, 1971; Amiconi *et al.*, 1972).

Although the 1-H shifts varies over 15 ppm, suggesting a more compressed bond in the insect Hbs, there is no simple correlation with either $\underline{p}_{1/2}(O_2)$ or $\underline{k}_{on}(O_2)$. Furthermore, in the case of the monomeric insect Hbs that exhibit significant Bohr effects, the 1-H shift is invariant during the $\underline{t} \longrightarrow \underline{r}$ transition, although the heme methyl peaks change significantly (La Mar et al., to be published). These results again suggest that either the imidazole-iron bond is not strained, or changes in affinity are not due to changes in strain in the deoxy form. Further work on model compounds is needed in order to assess the quantitative interpretability of the imidazole shifts.

E. Solvent Effects

Proton (Frye and La Mar, 1975; La Mar et al., 1977b), as well as ^{13}C (Goff, 1977) and ^{15}N (Morishima and Inubushi, 1977; Morishima et al., 1977) studies on LS ferric model compounds of both natural and synthetic porphyrins revealed that the hyperfine shifts are highly solvent sensitive. The effect of going from aprotic ($CHCl_3$) to protic (CH_3OH, H_2O) solvents was to cause a downfield bias for all heme resonances, which was interpreted in terms of a much larger axial anisotropy and greater iron-porphyrin π bonding in aprotic solvents. The solvent effect was suggested to arise from hydrogen-bonding to the coordinated cyanide. The coordinated cyanide ^{13}C and ^{15}N shifts in PPFe(CN)$_2^-$ also exhibit marked solvent effects. Both experience upfield shifts on going from H_2O to less protic solvents. Since the geometric factor, $(3 \cos^2\theta-1)\underline{r}^{-3}$ in Eq. (2) for the axial ligand has the opposite sign from that of the heme nuclei, the dominant shift changes for the cyanide must reflect changes in the iron-ligand bonding and/or in the molecular orbitals in the cyanide ligand due to the proton donation to the nitrogen. Further work is needed to clearly establish the origin of the ^{13}C and ^{15}N shift changes.

This demonstrated sensitivity of the heme and cyanide shifts to solvent suggests that the coordinated cyanide ^{13}C and ^{15}N peaks, the latter of which have been observed in both Mb^{III}CN and Hb^{III}CN (Morishima et al., 1977), may serve as sensitive indicators of the role of the distal imidazole in stabilizing the coordinated ligand via hydrogen-bonding. The reported pH-sensitivity of the ^{15}N resonances may also provide some definitive information on the pK for the distal imidazole. More work has to be done on the ligand resonances in model compounds before the full scope of the utility in proteins can be assessed, but preliminary results are very encouraging.

TABLE V. *Proximal histidyl imidazole 1-H hyperfine shifts, oxygen affinity, and oxygen on-rate in various hemoproteins*

Protein	Proximal histidine imidazole 1-H shift[a] from DSS (ppm)	$P_{1/2}(O_2)$ (mm)	$k_{on}(O_2) \times 10^7$ sec^{-1}
Hb^{II} (CTT-I)[b]	-95	1.8[d]	--
Hb^{II} (CTT-III)	-92.5	0.68-1.6[d]	30[e]
Hb^{II} (CTT-IV)	-91.7	0.4 -2.8[d]	--
Hb^{II} ($glycera$)	-93.3	1.45[f]	--
Hb^{II} ($aplesia$)	-85.5	2.7[e]	1.5[g]
Mb^{II} ($horse$)	-79.8	0.7[e]	1.4[e]
Hb^{II} ($sperm whale$)	-77.8	0.21[e]	1.9[e]
deuteroheme-Mb^{II} ($sperm whale$)	-79	0.5[e]	--
α_{PMB}^{II} $_C$	-77.8	0.65[e]	5[e]
β_{PMB}^{II} $_C$	-77	2.65[e]	8[e]
$Hb^{II}A$	-75.9, -63,.9	8-15[e]	1[e]

[a]Budd (1978).   [b]CTT, Chironomus thummi thummi.
[c]Individual chains of HbA as p-hydroxymethylbenzoate adducts.
[d]Gersonde et al. (1972).   [e]Amiconi et al. (1972).
[f]Hoffman and Mangum (1970).   [g]Antonini and Brunori (1971).

VII. CONCLUSIONS

We have attempted here to present a picture of the utility of model compounds for aiding in the analysis of hemoprotein NMR spectra. The above examples demonstrate that the study of models not only provides indications of the likely location and occasional definitive assignment of certain resonances, but also provides the basis for obtaining critical qualitative, and sometimes quantitative interpretations of the protein hyperfine shifts in terms of the electronic/molecular structure within the heme cavity.

Acknowledgment

The writing of this article was supported by a grant from the National Institute of Health, HL 16087.

REFERENCES

Amiconi, G., Antonini, E., Brunori, M., Formanek, H., and Huber, R. (1972). *Eur. J. Biochem. 31,* 52-58.
Antonini, E., and Brunori, M. (1971). "Hemoglobins and Myoglobins in Their Reactions with Ligands." North-Holland, Amsterdam.
Brassington, J. G., Williams, R. J. P., and Wright, P. E. (1975). *JCS Chem. Commun.,* 338-339.
Budd, D. L. (1978). Ph.D. Thesis, University of California, Davis.
Caughey, W. S., and Johnson, L. F. (1969). *JCS Chem. Commun.* 1362-1364.
Cavaleiro, J. A. S., Rocha Gonsalves, A. M. d'A., Kenner, G. W., Smith, K. M., Shulman, R. G., Mayer, A., and Yamane, T. (1974). *JCS Chem. Commun.* 392-393.
Davis, D. G., Lindstrom, T. R., Mock, N. H., Baldassare, J. J., Carache, S., Jones, R. T., and Ho, C. (1971). *J. Mol. Biol. 60,* 101-111.
Dolphin, D., ed. (1978). "The Porphyrins." Academic Press, New York.
Evans, B., Smith, K. M., La Mar, G. N., and Viscio, D. B. (1977). *J. Am. Chem. Soc. 99,* 7070-7072.
Felton, R. H., Owens, G. S., Dolphin, D., Forman, A., Borg, D. C., and Fajer, J. (1973). *Ann. N.Y. Acad. Sci. 206,* 504-514.

Frye, J. S., and La Mar, G. N. (1975). *J. Am. Chem. Soc. 97*, 3561-3562.

Fulton, G. P., and La Mar, G. N. (1976). *J. Am. Chem. Soc. 98*, 2119-2128, and references therein.

Gelin, B. R., and Karplus, M. (1977). *Proc. Nat. Acad. Sci. USA 74*, 801-805.

Gersonde, K., Sick, H., Wollmer, A., and Buse, G. (1972). *Eur. J. Biochem. 25*, 181-189.

Giacometti, G. M., Traylor, T. G., Ascenzi, P., Brunori, M., and Antonini, E. (1977). *J. Biol. Chem. 252*, 7447-7448.

Goff, H. (1977). *J. Am. Chem. Soc. 99*, 7723-7725.

Goff, H., and La Mar, G. N. (1977). *J. Am. Chem. Soc. 99*, 6599-6606.

Goff, H., La Mar, G. N., and Reed, C. A. (1977). *J. Am. Chem. Soc. 99*, 3641-3646.

Hoffmann, R. J., and Mangum, C. P. (1970). *Comp. Biochem. Physiol. 36*, 211-228.

James, T. J. (1975). "Nuclear Magnetic Resonance in Biochemistry." Academic Press, New York.

Jesson, J. P. (1973). *In* "NMR of Paramagnetic Molecules" (G. N. La Mar, W. D. Horrocks, Jr., and R. H. Holm, eds.), pp. 1-52. Academic Press, New York.

Johnson, M. E., Fung, L. W.-M., and Ho, C. (1977). *J. Am. Chem. Soc. 99*, 1245-1250.

Keller, R., Groudinsky, O., and Wüthrich, K. (1973). *Biochim. Biophys. Acta 328, 233-238*.

*Keller, R., Groudinsky, O., and Wü*thrich, K. (1976). *Biochim. Biophys. Acta 427*, 497-511.

Kurland, R. J., Davis, D. G., and Ho, C. (1968). *J. Am. Chem. Soc. 90*, 2700-2701.

Kurland, R. J., Little, R. G., Davis, D. G., and Ho, C. (1971). *Biochemistry 10*, 2237-2246.

La Mar, G. N. (1973). *In* "NMR of Paramagnetic Molecules," (G. N. La Mar, W. D. Horrocks, Jr., and R. H. Holm, eds.), pp. 85-126. Academic Press, New York.

La Mar, G. N., and Viscio, D. B. (1974). *J. Am. Chem. Soc. 96*, 7354-7355.

La Mar, G. N., and Walker, F. A. (1973a). *J. Am. Chem. Soc. 95*, 1782-1790.

La Mar, G. N., and Walker, F. A. (1973b). *J. Am. Chem. Soc. 95*, 6950-6956.

La Mar, G. N., and Walker, F. A. (1978). *In* "The Porphyrins" (D. Dolphin, ed.). Academic Press, New York.

La Mar, G. N., Eaton, G. R., Holm, R. H., and Walker, F. A. (1973a). *J. Am. Chem. Soc. 95*, 63-75.

La Mar, G. N., Horrocks, W. D., Jr., and Holm, R. H., eds. (1973b). "NMR of Paramagnetic Molecules," Chapters 1-4. Academic Press, New York.

La Mar, G. N., Frye, J. S., and Satterlee, J. D. (1976). *Biochim. Biophys. Acta 428*, 78-90.

La Mar, G. N., Bold, T. J., and Satterlee, J. D. (1977a). *Biochim. Biophys. Acta 498,* 189-207.

La Mar, G. N., Del Gaudio, J., and Frye, J. S. (1977b). *Biochim. Biophys. Acta 498,* 422-435.

La Mar, G. N., Budd, D. L., and Goff, H. (1977c). *Biochem. Biophys. Res. Commun. 77,* 104-110.

La Mar, G. N., Overkamp, M., Sick, H., and Gersonde, K. (1978a). *Biochemistry 17,* 352-361.

La Mar, G. N., Viscio, D. B., Gersonde, K., and Sick, H. (1978b). *Biochemistry 17,* 361-367.

La Mar, G. N., Viscio, D. B., Smith, K. M., Caughey, W. S., and Smith, M. L. (1978c). *J. Am. Chem. Soc. 100,* 8085-8092.

La Mar, G. N., Budd, D. B., Gersonde, K., and Sick, H. (1978d). *Biochim. Biophys. Acta 537,* 270-283.

LaMar, G. N., Budd, D. L., Viscio, B., Smith, K. M., and Langry, K. C. (1978e). *Proc. Nat. Acad. Sci. USA 75,* 5755-5759.

Lehmann, H., and Kynoch, P. A. M. (1976). "Human Hemoglobin Variants and Their Characteristics." North-Holland, Amsterdam.

Lemberg, R., and Barrett, J. (1973). "Cytochromes." Academic Press, New York.

McDonald, C. C., Phillips, W. D., and LeGall, J. (1973). *Biochemistry 13,* 1952-1959.

McGrath, T. M., and La Mar, G. N. (1978). *Biochim. Biophys. Acta 534,* 99-111.

Mayer, A., Ogawa, S., Shulman, R. G., and Gersonde, K. (1973). *J. Mol. Biol. 81,* 187-197.

Mayer, A., Ogawa, S., Shulman, R. G., Yamane, T., Cavaleiro, J. A. S., Rocha Gonsalves, A. M. d'A., Kenner, G. W., and Smith, K. M. (1974). *J. Mol. Biol. 86,* 749-756.

Mispelter, J., Momenteau, M., and Lhoste, J. M. (1977). *Mol. Phys. 33,* 1715-1728.

Morishima, I., and Inubushi, T. (1977). *JCS Chem. Commun.,* 616-617.

Morishima, I., Inubushi, T., Neya, S., Ogawa, S., and Yonezawa, T. (1977). *Biochem. Biophys. Res. Commun. 78,* 739-746.

Morrow, J. S., and Gurd, F. R. N. (1975). *CRC Crit. Rev. Biochem. 3,* 221-287.

Ogawa, S., and Shulman, R. G. (1971). *Biochem. Biophys. Res. Commun. 42,* 9-15.

Ogawa, S., Shulman, R. G., and Yamane, T. (1972a). *J. Mol. Biol. 70,* 291-300.

Ogawa, S., Shulman, R. G., Fujiwara, M., and Yamane, T. (1972b). *J. Mol. Biol. 70,* 301-313.

Oldfield, E., and Allerhand, A. (1973). *Proc. Nat. Acad. Sci. USA 70,* 3531-3535.

Peisach, J., Blumberg, W. E., and Adler, A. (1973). *Ann. N.Y. Acad. Sci. 206,* 310-327.

Perutz, M. F. (1976). *Br. Med. Bull. 32,* 195-208.

Perutz, M. F., and Ten Eyck, L. F. (1971). *Cold Spring Harbor Symp. Quant. Biol. 36,* 295-310.

Perutz, M. F., Heidner, E. J., Ladner, J. E., Beetlestone, J. G., Ho, C., and Slade, E. F. (1974). *Biochemistry 13,* 2187-2200, and references therein.

Phillips, W. D. (1973). *In* "NMR of Paramagnetic Molecules" (G. N. La Mar, W. D. Horrocks, Jr., and R. H. Holm, eds.), pp. 421-478. Academic Press, New York.

Redfield, A. G., and Gupta, R. K. (1971). *Cold Spring Harbor Symp. Quant. Biol. 36,* 405-411.

Satterlee, J. D., and La Mar, G. N. (1976). *J. Am. Chem. Soc. 98,* 2804-2808.

Satterlee, J. D., La Mar, G. N., and Bold, T. J. (1977). *J. Am. Chem. Soc. 99,* 1088-1093.

Seybert, D. W., and Moffatt, K. (1976). *J. Mol. Biol. 106,* 895-902.

Sheard, B., Yamane, T., and Shulman, R. G. (1970). *J. Mol. Biol. 53,* 35-48.

Shulman, R. G., Ogawa, S., Wüthrich, K., Peisach, J., and Blumberg, W. E. (1969a). *Science 165,* 251-257.

Shulman, R. G., Wüthrich, K., Yamane, T., Antonini, E., and Brunori, M. (1969b). *Proc. Nat. Acad. Sci. USA 63,* 623-628.

Shulman, R. G., Glarum, S. H., and Karplus, M. (1971). *J. Mol. Biol. 57,* 93-115.

Shulman, R. G., Hopfield, J. J., and Ogawa, S. (1975). *Quart. Rev. Biophys. 8,* 325-420.

Smith, G. M., and Kamen, M. D. (1974). *Proc. Nat. Acad. Sci. USA 71,* 4303-4306.

Smith, K. M., ed (1975). "Porphyrins and Metalloporphyrins." Elsevier, Amsterdam.

Swift, T. J. (1973). *In* "NMR of Paramagnetic Molecules" (G. N. La Mar, W. D. Horrocks, Jr., and R. H. Holm, eds.), pp. 53-83. Academic Press, New York.

Viscio, D. B. (1977). Ph.D. Thesis, University of California, Davis.

Viscio, D. B., and La Mar, G. N. (1978a). *J. Am. Chem. Soc. 100,* 8092-8096.

Viscio, B. and La Mar, G. N. (1978b). *J. Am. Chem. Soc. 100,* 8096-8100.

Walker, F. A., and La Mar, G. N. (1973). *Ann. N.Y. Acad. Sci. 206,* 328-348.

Wilbur, D. J., and Allerhand, A. (1976). *J. Biol. Chem. 251,* 5187-5194, and references therein.

Williams, R. J. P., Wright, P. E., Mazza, G., and Ricard, J. R. (1976). *Biochim. Biophys. Acta 412,* 127-147.

Wüthrich, K. (1969). *Proc. Nat. Acad. Sci. USA 63,* 1071-1078.

Wüthrich, K. (1970). *Struct. Bonding 8,* 53-121.

Wüthrich, K., and Baumann, R. (1973). *Helv. Chim. Acta 56*, 585–596.

Wüthrich, K., and Baumann, R. (1974). *Helv. Chim. Acta 57*, 336–350.

Wüthrich, K., Shulman, R. G., and Peisach, J. (1968a). *Proc. Nat. Acad. Sci. USA 60*, 373–380.

Wüthrich, K., Shulman, R. G., and Yamane, T. (1968b). *Proc. Nat. Acad. Sci. USA 61*, 1199–1206.

Wüthrich, K., Shulman, R. G., Wyluda, B. J., and Caughey, W. S. (1968c). *Proc. Nat. Acad. Sci. USA 62*, 636–643.

Yamazaki, I. (1974). "Molecular Mechanisms of Oxygen Activation" (O. Hayaishi, ed.), pp. 535–558. Academic Press, New York.

MULTINUCLEAR NMR APPROACHES TO THE SOLUTION STRUCTURE OF ALKALINE PHOSPHATASE: ^{13}C, ^{19}F, ^{31}P, AND ^{113}Cd NMR*

Joseph E. Coleman
Ian M. Armitage
Jan F. Chlebowski
James D. Otvos
Antonius J. M. Schoot Uiterkamp

Department of Molecular Biophysics and Biochemistry
Yale University
New Haven, Connecticut

I. INTRODUCTION

Fourier transform NMR has allowed the direct detection of the resonances from individual nuclei contained in protein molecules with molecular weights from 10,000 to 100,000. While sensitivity remains a major limitation of this technique, it has now been possible to detect resonances for individual nuclei from protons to metal ions located within the three-dimensional structure of a native protein at mM concentrations. In the case of ^{1}H and ^{19}F, small sample sizes, a few tenths of a milliliter, partially compensate for the high concentrations required. The chemical shifts, relaxation times (T_1 and T_2), nuclear Overhauser effects, coupling to adjacent nuclei, and paramagnetic relaxation provide a set of parameters with which to determine both static and dynamic aspects of protein structure that are unrivaled by other solution methods.

Alkaline phosphatase of *E. coli* is a particularly good example with which to illustrate the NMR approach to solution structure determination. The enzyme is a dimer of identical subunits containing two catalytic Zn(II) ions as well as sites

Original work from the authors' laboratories was supported by NIH Grants AM 09070-14 and AM 18778-03 and NSF Grant PCM76-82231.

Joseph E. Coleman *et al.*

for additional structural Zn(II) and Mg(II) ions (Anderson *et al.*, 1975; Chlebowski and Coleman, 1976a,b; Bosron *et al.*, 1977). Resolution of individual proton resonances has not been possible due to the large molecular weight of the dimer. ^{13}C, ^{19}F, ^{31}P, and ^{113}Cd NMR on the other hand have been successfully applied to this enzyme, and the results are described in this chapter. ^{13}C has been introduced at the γ position of histidyl residues and ^{19}F at the m position of tyrosyl and p position of phenylalanyl residues to assess changes in the conformation of the amino acid side chains accompanying metal ion and phosphate binding. ^{31}P NMR has been used to assess the chemical nature of the covalent and noncovalent phosphorus-containing intermediates formed by the enzyme. The native Zn(II) ions have been substituted by the NMR sensitive and catalytically active ^{113}Cd(II) ion in order to detect structural alterations at the active center accompanying phosphorylation as reflected in changes of the electron distribution around each metal ion.

A. Stoichiometry of Metal Ion and Phosphate Binding to Alkaline Phosphatase

The amino acid sequence of the alkaline phosphatase monomer is nearly complete and the data on amino acid composition and sequence show each monomer to contain ~425 amino acid residues.[1] The molecular weight calculated from the sequence is ~43,000. Hence the molecular weight of the dimer on which the stoichiometry in this paper is based is 86,000. As shown by equilibrium dialysis studies with ^{65}Zn(II) and ^{60}Co(II), the alkaline phosphatase dimer (concentrations 10^{-5} to 10^{-4}M) binds two nondialyzable metal ions at pH 6.5 (Applebury and Coleman, 1969a). While no precise determination of the formation constant for the enzyme complex with these tightly bound metal ions has been carried out, an early study of Zn(II) ion dissociation (Cohen and Wilson, 1968) and the pH dependency of binding coupled with the assumption that the three histidyl nitrogen ligands have pK_a values near neutral pH (see below), suggest that the pH-independent formation constant for these two Zn(II) ions must be 10^{10} or greater. Dialysis studies using

---

[1]*The phosphorylated serine is residue 99 from the NH$_2$-terminal threonine residue (100 from the N-terminal arginine in the arginine isozyme) in the preliminary numbering of the sequence as it is presently available from the work of R. A. Bradshaw, P. A. Neumann, F. Cancedda, K. Schrifla, J. D. Hecht, and M. J. Schlesinger (personal communication from R. A. Bradshaw).*

Zn(II) chelating agents to compete for enzyme zinc from pH 6.5 to 9.5 confirm these estimates, log K_{Zn} = 11.8 at pH 9.5 and log K_{Zn} = 10.3 at pH 6.5 (Csopak, 1969).

The two tightly bound metal ions are those that occupy identical sites of low symmetry as shown by the d-d transitions of the active Co(II) derivative (Simpson and Vallee, 1968; Applebury and Coleman, 1969b; Taylor et al., 1973). The molar extinction coefficients, the ligand field splitting, and the magnetic circular dichroism of the d-d transitions of the two Co(II) enzymes suggest the coordination geometry to be distorted tetrahedral or five-coordinate (Taylor et al., 1973). In the recent literature these have been termed the catalytic metal ions (Bosron et al., 1977).

At neutral pH and above, the enzyme can bind as many as six ^{65}Zn(II) or ^{60}Co(II) ions per dimer (Applebury and Coleman, 1969a). Vallee and co-workers have shown that if mild isolation procedures are employed, taking particular precautions with the use of EDTA, the protein as isolated contains four Zn(II) and one to two Mg(II) ions per dimer (Anderson et al., 1975; Bosron et al., 1977). The second pair of Zn(II) ions has been termed the structural pair. Since the NMR studies to be described here show the metal ion stoichiometry to influence the phosphate binding stoichiometry as well as the allosteric behavior of this enzyme, it is necessary to define a number of abbreviations indicating the metal ion and phosphate content of the particular species under discussion and these are indicated in Fig. 1.

At the concentrations of enzyme employed for all the studies reported here, ultracentrifugation and differential scanning calorimetry show the apoenzyme to remain a dimer even after undergoing temperature-induced unfolding (Chlebowski and Mabrey, 1977; Falk and Vallee, 1978). As particularly clearly shown by differential scanning calorimetry, the binding of the first two metal ions is cooperative, and at ratios of metal/dimer <2, the two metal dimer and the apoenzyme are the only significant species present (Chlebowski and Mabrey, 1977). The enzyme containing the two catalytic metal ions will be designated Me(II)$_2$AP, with the stoichiometry of additional species designated as shown sequentially in Fig. 1.

The alkaline phosphatase dimer reacts with substrate or binds phosphate in a negatively cooperative manner, i.e., even though two identical active sites must be present (as shown by the 32 Å separation of the metal ions across the two-fold axis; Knox and Wyckoff, 1973), only one can interact with substrate or phosphate at any one instant, as shown by a variety of kinetic and [32$_p$]-binding studies (see Reid and Wilson, 1971, for review). The negative cooperativity must be mediated by conformational changes propagated to the other site across the monomer-monomer interface when ligand binds at the first site.

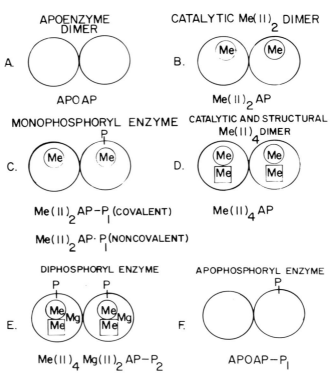

Fig. 1. Schematic representation of molecular forms of alkaline phosphatase. Metal ions displaying a high affinity for
the enzyme are identified (Me in circle) with the "catalytic"
sites. The structures depicted are not intended to suggest the
actual location of the metal ion binding sites with respect to
the intersubunit contact domain or the relative position of the
bound metal ions to one another and the phosphorylation site.

The ^{113}Cd-NMR to be shown here provides evidence for such
changes. While the alkaline phosphatase dimer containing two
catalytic metal ions is shown by ^{31}P NMR to be negatively cooperative at all concentrations, forming only Me(II)$_2$AP-P$_1$ or
Me(II)$_2$AP·P$_1$ (Fig. 1), addition of the structural metal ions
and Mg(II) at mM concentrations of enzyme induces binding of a
second phosphate to form Me(II)$_4$ Mg(II)$_2$AP-P$_2$ (Fig. 1). The
metallophosphatases formed with Mn(II), Co(II), Zn(II), and
Cd(II) containing both two and four metal ions per dimer and
Mg(II) have been completely characterized by NMR and ESR studies and Me(II) will refer to one or another of these four metal ions throughout.

II. ^{31}P NMR OF PHOSPHORUS-CONTAINING INTERMEDIATES OF THE AL-
    KALINE PHOSPHATASE REACTION

A. ^{31}P Chemical Shifts of the Enzyme-Phosphate Complex

A summary of the alkaline phosphatase mechanism is given
by (1).

$$ROP + E \underset{k_{-1}}{\overset{k_1}{\rightleftharpoons}} ROP \cdot E \underset{k_{-2}}{\overset{k_2}{\rightleftharpoons}} E-P \underset{k_{-3}}{\overset{k_3}{\rightleftharpoons}} E \cdot P \underset{k_{-4}}{\overset{k_4}{\rightleftharpoons}} E + P_i \quad (1)$$

with ROH and H_2O entering at k_2 and k_3 steps respectively, and R'OH / R'OP + E branch at the E-P intermediate.

E-P represents the covalent phosphoserine intermediate formed
with Ser 99 in the sequence.[1] At this stage the phosphoryl
group can be readily transferred to an acceptor alcohol R'OH,
if that alcohol carries an amino group on the carbon adjacent
to the hydroxyl function, e.g., ethanolamine or Tris (Reid and
Wilson, 1971). E·P represents the noncovalent Michaelis complex with inorganic phosphate. The enzyme catalyzes the incorporation of ^{18}O from H_2O^{18} into inorganic phosphate at all pH
values (Applebury et al., 1970). At pH values below 5.5 inorganic phosphate can be used to form significant equilibrium
concentrations of E-P. Hence P_i is a substrate for the enzyme
and the reaction can be run from right to left as far as E-P.
At low pH dephosphorylation of E-P, k_3 is rate-limiting, accounting for high equilibrium concentrations of E-P. At alkaline pH where the enzyme is maximally active the major intermediate is E·P (Applebury et al., 1970). Many kinetic studies
have attempted to define the rate-limiting step at alkaline pH,
but a single rate-limiting step has not been identified. Evidence suggested that for oxyphosphates dephosphorylation, k_3,
and dissociation of E·P, k_4, proceeded at rates of comparable
magnitude such that both might contribute to the steady-state
rate (Chlebowski and Coleman, 1974). Dissociation of product
complex, E·P, is apparently the rate-limiting step for the much
more slowly hydrolyzed O-phosphorothioates. Burst kinetics are
observed at alkaline pH (Chlebowski and Coleman, 1974). The
^{31}P-NMR data (see below) suggests that dissociation of product,
P_i, from the enzyme is slow enough to make a significant contribution to the rate limitation (Hull et al., 1976; Chlebowski et
al., 1977). Recent rapid flow kinetics with methylumbelliferyl
phosphate using enzyme freed of contaminating phosphate have
shown a one-mole burst at alkaline pH, compatible with dissociation of E·P as the rate-limiting step (Bale et al., in

press). The pH dependency of the steady-state rate (sigmoid to
high pH) and the shift from high equilibrium concentrations of
E-P to high equilibrium concentrations of E·P is thus explained
by a large increase in the dephosphorylation rate of E-P, $\underline{k_3}$,
as the pH is raised.

The chemical shift values for the ^{31}P-NMR resonances of the
various phosphorus-containing intermediates formed by *E. coli*
alkaline phosphatase from inorganic phosphate are summarized in
Fig. 2. At pH 8, the only detectable form of enzyme-bound phosphate is the noncovalent complex E·P. The resonance for the
Michaelis complex is at ~5 ppm, ~2 ppm further downfield than
the phosphate dianion at this pH (Fig. 2A). At pH 5, the predominant species of enzyme-bound phosphate is the phosphoserine
intermediate E-P, with a resonance appearing at ~8.5 ppm (Fig.
2C).

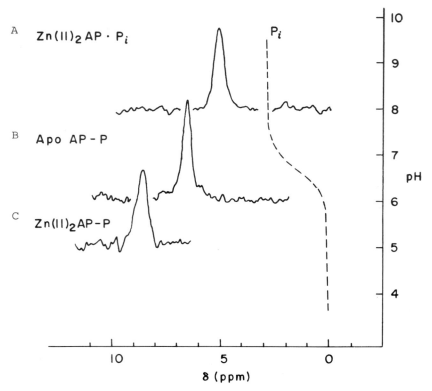

Fig. 2. ^{31}P *NMR spectra of phosphate complexes of*
1-2 × *10^{-3}* $\underline{M}$ *Zn(II) and apoalkaline phosphatase.   Condi-*
tions: 0.01 $\underline{M}$ *Tris, 0.01* $\underline{M}$ *NaOAc, 0.1* $\underline{M}$ *NaCl.   pH given on*
right ordinate.   Dashed line shows the ^{31}P *chemical shift for*
inorganic phosphate.

Phosphate binding to the enzyme as well as phosphorylation and dephosphorylation of Ser 99 are absolutely metal ion dependent. Moreover the relative rates of phosphorylation and dephosphorylation are markedly influenced by the species of first transition or IIB metal ion present at the active site (Applebury et al., 1970). For example, substitution of Cd(II) for the native Zn(II) ion slows both phosphorylation and dephosphorylation by over two orders of magnitude. Dephosphorylation, however, is slowed to a greater degree such that at neutral pH the Cd(II) enzyme is almost completely in the E-P form. Turnover of the Cd(II) enzyme is very slow, but the intermediates appear entirely analogous to those formed by the Zn(II) enzyme. The Cd(II) phosphoryl enzyme can be treated with high concentrations of chelating agents at neutral pH, which rapidly remove the metal ions but leave the phosphoserine intact, forming a species known as the apophosphoryl enzyme. The ^{31}P-resonance for this species occurs at ~6.5 ppm, ~2 ppm upfield from either the Zn(II) or Cd(II) phosphoryl enzyme (Fig. 2B).

Variation in the pH value at which E-P/E·P = 1 from pH 5 for the Zn(II) enzyme to pH 8 for the Cd(II) enzyme, the induction of significant dephosphorylation of the Cd(II) enzyme at pH values above 9, and the metal dependence of the dephosphorylation step have led to the postulate that the mechanism may involve the dephosphorylation of the phosphoserine intermediate E-P, by a metal-coordinated hydroxide (Coleman and Chlebowski, 1979). This would help explain the lability of the phosphoserine in the metal enzyme, its stability in the apoenzyme, and provide two symmetrical enzyme-linked R-O⁻ nucleophiles, one providing for phosphorylation, one for dephosphorylation (Fig. 3). The coordination state of the Zn(II) may be in a state of flux during the enzymatic reaction, since ligand binding is probably accompanied by significant conformational change at the active site. Hence Zn(II) may partici-

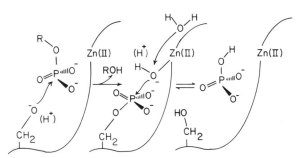

Fig. 3. Molecular mechanism for alkaline phosphatase. (From Coleman and Chlebowski, 1979.)

pate in the initial binding of phosphate (substrate) and may
retain or pick up a solvent ligand when the phosphate is transferred to serine 99.

A preliminary approach to the interpretation of the significance of the chemical shifts observed for the phosphorylated
apo- and metalloalkaline phosphatases can be achieved by comparing the chemical shift values for the enzyme intermediates
to those observed for phosphate monoesters, acyclic diesters,
and five- and six-membered cyclic diesters (Fig. 4). It has
recently been proposed that an empirical correlation can be
made between the chemical shift of the ^{31}P resonance and the
magnitude of the smallest O-P-O bond angle in these phosphate
esters or their corresponding dihedral (torsional) angles ω,
defining the ROPO grouping (Gorenstein *et al.*, 1976; Gorenstein, 1975). Inclusion of the ^{31}P chemical shifts for the
phosphoserine derivatives of the three chemical forms of alkaline phosphatase, Zn(II), Cd(II), and apo, in this correlation
shows that their chemical shifts occur midway between those expected for normal phosphate monoesters and cyclic phosphodiesters. Thus the enzyme phosphomonoesters are abnormal and
correspond to phosphate esters whose minimum O-P-O bond angle
is less than 100°. Hence the enzyme phosphoserine must be distorted. The apparent distortion of the phosphoryl group in the
Zn(II) enzyme may provide the driving force for its rapid turnover. The presence of the metal ion contributes 2-3 ppm to the
downfield chemical shift of the phosphoserine, but does not account for the major fraction of the unusual downfield shift.
This must be largely contributed to by the surrounding protein
structure, since the apophosphoryl enzyme still shows an unusual downfield shift. The 2-3 ppm chemical shift contributed
by the metal ion is larger than observed in simple coordination
to a phosphate dianion.

Such distortion could facilitate the departure of RO$^-$ pa-
ralleling the observation that hydrolysis of the methyl ester
external to the ring occurs 10^7 times faster in methyl ethylene
phosphate than it does in acyclic methyl esters of phosphate.
If a similar effect is present in the enzyme phosphoserine intermediate and is enhanced when the metal ion is present, this
phosphoserine should readily undergo hydrolysis.

The ability to prepare the apophosphoryl enzyme from the
Cd(II) enzyme allows the determination of some interesting comparative properties of the phosphoserine in the metallo- and
apoenzyme derivatives. E-P is stable over a limited pH range
in the metalloenzymes because of rapid dephosphorylation at
high pH [higher for Cd(II) than Zn(II)] and loss of the metal
ion at low pH. The ^{31}P chemical shift data on Zn(II)$_2$AP-P$_1$ and
Cd(II)$_2$AP-P$_1$ over the observable pH range are shown in Fig. 5.
The chemical shift of E-P does not change with pH, although
more and more E·P is appearing as the pH rises. On the other

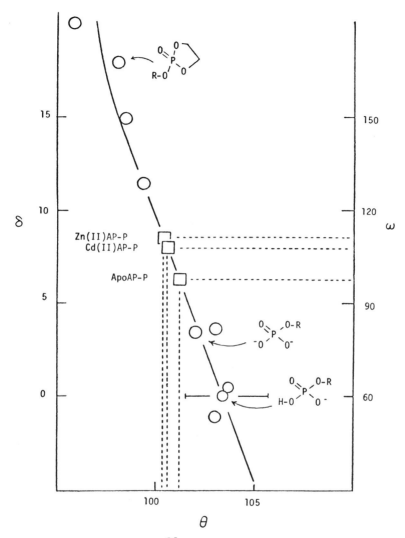

Fig. 4. Plot of the ^{31}P chemical shifts for several model phosphate esters versus the O-P-O bond angle (θ) and torsional angle (ω) (Gorenstein, 1975; Gorenstein et al., 1976). Open squares indicate the observed chemical shift for the various alkaline phosphatase phosphoserine monoesters.

hand, in the apoenzyme the phosphoseryl residue is stable from pH 2 to 10.

The stability of the apophosphoryl enzyme over a large pH range allows a comparison of the change in the ^{31}P chemical

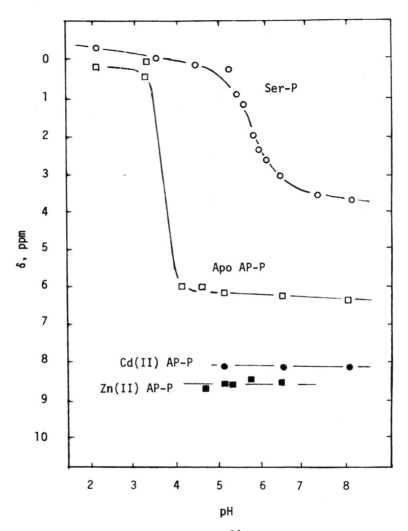

Fig. 5.   pH dependence of the ^{31}P NMR chemical shifts of phosphoserine and phosphoryl alkaline phosphatases (AP-P). Conditions: 0.01 $\underline{M}$ Tris, 0.01 $\underline{M}$ NaOAc, 0.1 $\underline{M}$ NaCl, enzyme concentrations 0.6 × $\overline{10}^{-3}$ $\underline{M}$-2.0 × $\overline{10}^{-3}$ $\underline{M}$; (○) phosphoserine, (□) apophosphoryl alkaline phosphatase, (●) Cd(II)$_2$ phosphoryl alkaline phosphatase, (■) Zn(II)$_2$ alkaline phosphatase.   (From Chlebowski et al., 1976.)

shift of the enzyme phosphoserine to that of free phosphoserine in the pH region of its second pK_a, pH 5 to 6 (Fig. 5).   Phosphoserine shows the normal ^{31}P chemical shift transition from dianion to monoanion with a pK_a of ~5.5. Not only does the ^{31}P resonance of the apophosphoryl enzyme start out at alkaline pH with a chemical shift 2.5 ppm further downfield than a normal dianion, the lack of change in chemical shift down to pH 4 suggests that no protonation has occurred. Below pH 4 there is a rapid change to the chemical shift position expected for protonated phosphoserine (Fig. 5). Previous data show that below pH 4.0 the protein dissociates into monomers and unfolds reversibly to a random polypeptide structure. Since the chemical shift of the enzyme phosphoserine at alkaline pH is further downfield than the dianion of free phosphoserine, it appears reasonable to assign the alkaline form of phosphoserine 99 to the dianion form. Hence the enzyme structure must prevent the protonation of this group until the enzyme unfolds at low pH. A positively charged cage around the phosphate binding site could account for a very low pK_a. Such positively charged groups may participate in phosphate dianion binding and are believed to include at least one arginyl residue (Daemen and Riordan, 1974). Of course, hydrogen bonding of one of the phosphate oxygens in the apoenzyme cannot be ruled out.

Analysis of the relaxation phenomena associated with the phosphorus nucleus of the apophosphoryl enzyme suggest some perhaps unexpected conclusions about the molecular motions that make significant contributions to nuclear relaxation in protein molecules. This analysis is covered briefly in the following section.

B. *Chemical Exchange Contributions to the Nuclear Overhauser Enhancement and Dipolar Relaxation of ^{31}P in Apophosphoryl Alkaline Phosphatase*

Analysis of nuclear spin relaxation mechanisms can lead to detailed descriptions of the motional properties of molecules. Models that include both overall molecular rotation and internal degrees of freedom now exist (Hubbard, 1970; Woessner, 1962). In the case of biopolymers, models that include internal modes are required for the complete interpretation of spin lattice relaxation processes (Oldfield *et al.*, 1975). Analysis of internal modes is usually directed at motions that can be classified as bond isomerizations in which nuclei relax via dipolar interactions with rigidly bonded nuclei on the same rotating group; for example, ^{13}C and ^{1}H relaxation in a methyl group with rotation about its C-X bond.

Interpretation of spin lattice relaxation behavior for some of the heavier nuclei in macromolecules may present a special problem in that there are few rigidly bound nuclei of high magnetic moment within the radius required for effective relaxation. Thus for ^{31}P in a phosphate moiety it is possible that the most effective relaxation may come from hydrogen-bonded protons contributed by protein-bound water molecules or dissociable protons on adjacent amino acid side chains.

^{31}P NMR spectra, proton coupled and proton decoupled, of apophosphoryl alkaline phosphatase compared to the corresponding spectra of the model compound phosphoserine are shown in Fig. 6. Coupling of the ^{31}P nucleus of the enzyme to the β-protons of the seryl residue is clearly present in ApoAP-P as shown by the considerable decrease in linewidth on proton decoupling (Fig. 6A,B). The collapse of this unresolved triplet with proton decoupling is more clearly visualized for the free phosphoserine. From the linewidths of the proton-decoupled spectra, the $\underline{T}_2^*$ values are 0.53 and 0.24 sec for the ^{31}P nucleus of phosphoserine and ApoAP-P, respectively. Determination of $\underline{T}_1$ by both inversion recovery and progressive saturation gave values of 19.0 sec for phosphoserine and 1.5 sec for ApoAP-P (Chlebowski *et al.*, 1976). Using gated decoupling techniques, both the enzyme and the model give rise to an observable nuclear Overhauser enhancement (NOE) as indicated by the loss of amplitude in Fig. 6C compared to Fig. 6B. Neither the $\underline{T}_1$ nor NOE value can be explained using the appropriate equations for dipolar relaxation and assuming a rigid sphere model with the appropriate overall rotational correlation time for the alkaline phosphatase dimer of 7×10^{-8}s (Hull and Sykes, 1975). Such a model would give rise to an NOE value of 1 and a $\underline{T}_1$ of approximately 30 sec. Contributions from other relaxation mechanisms such as chemical shift anisotropy are possible; however, at the field strength employed here, ^{31}P data on other macromolecules suggest that this mechanism will not contribute more than 15% to the relaxation rate and nothing to the NOE. In order to explain the relaxation data assuming that the dipolar relaxation is to the β protons of the seryl residue, it becomes necessary to assume that the ^{31}P-^1H vector under consideration undergoes internal rotation. Using the appropriate equations derived for this case (Doddrell *et al.*, 1972), a correlation time τ_G for internal rotation of $\underline{\leq}1 \times 10^{-9}$ sec is required to account for the relaxation data.

The above analysis of the relaxation of the enzyme phosphorus is rendered largely if not completely incorrect by the finding of a radical change in $\underline{T}_1$ and NOE of the phosphorus of the apoenzyme on changing the solvent from H_2O to D_2O. $\underline{R}_1$ ($^1/\underline{T}_1$) decreases about fourfold and the NOE decreases substantially in D_2O (Table I). Hence more than 80% of the ^{31}P relaxation comes from dipolar interaction with exchangeable protons, not the β protons of serine.

If one again assumes a single isotropic correlation time of 7×10^{-8} sec for alkaline phosphatase, even a full complement of protons hydrogen bonded to the phosphate and modulated at this

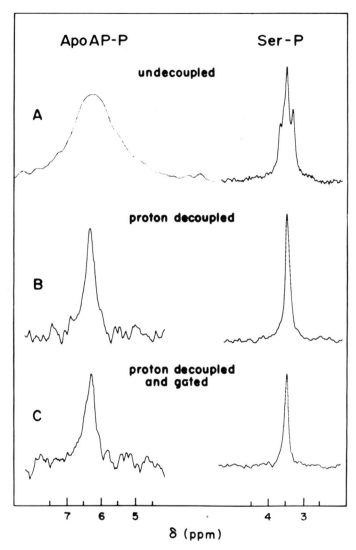

Fig. 6. Effect of proton decoupling on the ^{31}P NMR spectra of phosphoserine and apophosphoryl alkaline phosphatase. (From Chlebowski et al., 1976.)

molecular rotation frequency could not produce the high R_1 and NOE values observed in the H_2O case. As was the case with the β protons of the seryl residue, an internal mode is required that modulates the dipolar interactions at a more effective re-

TABLE I. ^{31}P Relaxation parameters for apophosphoryl alkaline phosphatase (APoAP-P)

R_1 (sec^{-1})		NOE
H_2O	0.67	1.22
D_2O	0.12	1.12

laxation frequency. Such an internal mode could result in two ways, bond rotation or proton exchange. In the case of the enzyme phosphoserine, two possible bond rotations, 1 and 2, can modulate H-P dipolar interactions of the nearest exchangeable protons (Fig. 7). If one treats for type 1 rotation assuming a minimum H-P distance of 2.0 Å for two protons with internal rotation at an angle of 22° for the P-H vector relative to the rotation axis and an overall enzyme rotation time of 7×10^{-8} sec, an internal rotation time constant of 5×10^{-10} sec would be required to account for the observed R_1 and NOE. This very rapid internal rotation is an order of magnitude greater than that reported for individual amino acid side chains in similar systems (Oldfield *et al.*, 1975). Rotation of type 2 of the entire hydrogen-bonded phosphate group would be even less likely to occur.

An alternative means of modulating dipolar interactions at enzyme active sites exists in the form of exchange of acidic

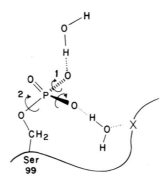

Fig. 7. Alkaline Phosphatase. Possible hydrogen-bonded water molecules at the phosphoryl site of ApoAP-P.

protons (or metal-coordinated water molecules in metalloen-zymes) (Armitage and Prestegard, 1978). Rates on the order of 10^8-10^9 sec^{-1} are in fact reasonable. Estimates for the mean residence time of water molecules at the active sites of man-ganese enzymes are reported to range from 2×10^{-8} to 5×10^{-9} sec (Dwek, 1973). Thus a model may be proposed that considers the modulation of the ^{31}P-H interaction as originating in an exchange of one proton for another of arbitrary spin at approxi-mately the same site.

In the case of apophosphoryl alkaline phosphatase the pro-tons would be carried on H_2O bound in the active site cavity or on amino acid side chains in the cavity. The two protons in Fig. 7, for example, need not move very far in this exchange. An exchange of a proton from a site in which it is covalently bound to a phosphate oxygen (or to a metal-coordinated water oxygen) to a site in which it is hydrogen bonded to the same oxygen would provide an oscillation of sufficient magnitude to be approximated as a jump to infinite distance. This would be accompanied by a 90% reduction in the dipolar interaction of the proton with ^{31}P. Thus a major if not predominant contribu-tion to dipolar relaxation of the ^{31}P of the apophosphoryl en-zyme by exchangeable protons could provide the requisite relaxa-tion process with the required τ_e of $\sim 10^{-9}$ sec. Such an anal-ysis does not rule out internal motion as also being present, but shows that rapidly exchanging protons in the active site cavities of enzymes could provide the dominant dipolar relaxa-tion with a dominating $\tau_e \cong 10^{-9}$ sec without invoking rapid mo-tion of the bound phosphate to explain observed dipolar relaxa-tion and NOE values.

C. Negative Cooperativity of Phosphate Binding to Alkaline Phos-phatase as Detected by ^{31}P NMR

Most rapid-flow kinetic studies at low pH where dephospho-rylation of E-P is the rate-determining step in the alkaline phosphatase reaction have shown a 1 mole burst of alcohol to occur during the rapid presteady state phosphorylation of the enzyme (Ko and Kézdy, 1967; Trentham and Gutfreund, 1968; Fernley and Walker, 1969; Halford et al., 1972; Halford and Schlesinger, 1974; Chlebowski and Coleman, 1974; Chlebowski and Coleman, 1976). Most phosphate binding studies with ^{32}P$_i$ that determine E·^{32}P stoichiometry by equilibrium dialysis have also shown 1 mole of phosphate to be bound per mole of dimer with a dissociation constant of $\sim 10^{-6}$ M (Applebury et al., 1970; Simp-son and Vallee, 1970). Determinations of E-P formation by the enzyme (10^{-5} to 10^{-4} M concentrations) from ^{32}P$_i$ also showed that the Zn(II) enzyme at low pH or the Cd(II) enzyme at pH 6.5 forms only 1 mole of E-P per mole of dimer (Schwartz, 1963;

Lazdunski et al., 1969; Applebury et al., 1970; Lazdunski,
1974; Chlebowski and Coleman, 1976). Some studies have indi-
cated that at high concentrations (10^{-3}-10^{-2} $\underline{M}$) of phosphate,
the binding of a second equivalent of P_i could be detected.
Covalent incorporation of two moles of $\bar{P}_i$ per enzyme dimer has
similarly been reported on incubation of the enzyme at low pH
with elevated concentrations of substrate or phosphate (Lazdunski, 1974). These results are, however, ambiguous, because
of the inherent imprecision in determining exact stoichiometries
in ^{32}P-labeling experiments at high phosphate and relatively low
enzyme concentrations. Observation of a phosphate binding stoichiometry of 2 P_i per enzyme dimer does not necessarily reflect
a loss of negative cooperativity (see below) (Chappelet-Tordo
et al., 1974).

The majority of studies support the widely held belief that
phosphate or substrate interaction with the alkaline phosphatase dimer is negatively cooperative, i.e., that phosphate binding at one of two identical sites must be associated with conformational changes propagated to the second site, which reduces
its affinity for the phosphate dianion. The ^{113}Cd NMR studies
to be presented below show that the two catalytic metal sites,
initially present with identical coordination spheres, are both
altered on phosphorylation of one site as reflected by the
87-ppm difference in the chemical shifts of the two ^{113}Cd(II)
ions. Since neither chemical shift is the same as that of the
^{113}Cd in the unliganded enzyme, a single phosphorylation changes
both sites, compatible with the predicted molecular change in a
negatively cooperative dimer.

1. Stoichiometry of Phosphate Binding to Co(II) and Cd(II)
Alkaline Phosphatases as Measured by ^{31}P NMR

The distinctive chemical shifts of ^{31}P resonances from the
various forms of enzyme-bound phosphate allow the use of ^{31}P
NMR to follow the interaction of phosphate with the enzyme as
a function of the metal ion stoichiometry. In the presence of
apoalkaline phosphatase the ^{31}P resonance for P_i occurs at ~2
ppm, the chemical shift expected for P_i under those conditions
in the absence of enzyme (Fig. 8a) (Chlebowski et al., 1976,
1977). Thus the NMR confirms that no specific interactions of
phosphate occur with the apoenzyme. If one equivalent of Cd(II)
per dimer is added to a solution of apoenzyme containing two
equivalents of P_i per dimer, a new ^{31}P resonance, accounting
for 25% of the phosphate present, appears at the chemical shift
expected for the Cd(II) phosphoryl enzyme (Fig. 8B). Addition
of a second equivalent of Cd(II) causes the E-P resonance to
increase accounting for ~50% of the phosphate present (Fig. 8C).
The other 50% remains at the chemical shift of inorganic phos-

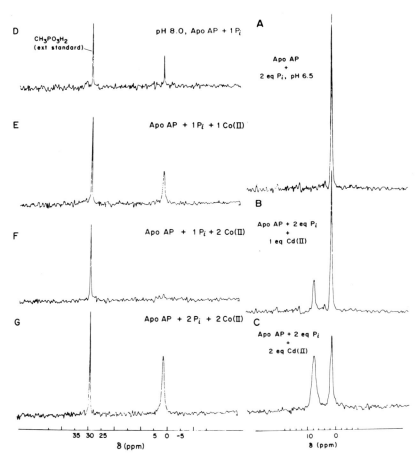

Fig. 8. ^{31}P *NMR spectra of phosphate binding to Co(II) and Cd(II) alkaline phosphatases. Conditions: 0.01 $\underline{M}$ Tris, 0.01 $\underline{M}$ NaOAc, 0.1 $\underline{M}$ NaCl, phosphate (P_i) and metal ion stoichiometries as indicated. Spectra (A)-(C), Cd(II) additions to 1.7 × 10^{-3} $\underline{M}$ apoalkaline phosphatase, pH 6.5; spectra (D)-(G), Co(II) additions to 1.6 × 10^{-3} $\underline{M}$ apoalkaline phosphatase, pH 8.0. Resonance at 29.4 ppm is that of methyl phosphonate, an external standard. (From Chlebowski et al., 1977.)*

phate. Thus Cd(II)$_2$AP forms only 1 mole of phosphoserine, the ^{31}P NMR confirming the negative cooperativity of Cd(II)$_2$AP.

An analogous experiment can be carried out by making use of the paramagnetic relaxation of the bound P_i by the active site Co(II) ions. Addition of Co(II) to the apoenzyme dimer causes a progressive and complete loss of the ^{31}P signal if only one mole of P_i permole of dimer is present (Fig. 8D-F). E·P is the dominant species of the Co(II) enzyme at pH 8.0. Subsequent addition of a second equivalent of P_i to Co(II)$_2$AP·P$_1$ results in the reappearance of a P_i resonance. Thus the second phosphate does not bind near the other Co(II) ion, another demonstration of the negative cooperativity of Me(II)$_2$AP.

2. Formation of Diphosphoryl Alkaline Phosphatase

Since the ^{31}P NMR studies require at least 10^{-3} M concentrations of enzyme and phosphate (up to 4×10^{-3} M concentrations have been used for ^{113}Cd NMR), the NMR method should potentially detect binding of a second phosphate if it occurs at high concentrations. In fact, some early discrepancies on the stoichiometry of P_i binding as determined by ^{31}P NMR are present in the literature (Chlebowski *et al.*, 1976, 1977; Hull *et al.*, 1976). With precise control of the stoichiometry of metal ion in the NMR samples, additional features of the negative cooperativity of phosphate binding have been revealed.

The conclusion from the ^{31}P NMR results on Cd(II)$_2$AP and Co(II)$_2$AP presented above must be that the dimer containing the two catalytic metal ions is strongly negatively cooperative at all concentrations of phosphate. In contrast, at high concentrations of enzyme and phosphate, addition of the pair of structural metals and Mg(II) to form Me(II)$_4$Mg(II)$_2$AP induces the binding of a second phosphate, and in the case of Cd(II) at pH 6.5, the phosphorylation of the second ser 99. The ^{31}P spectra on the addition of P_i to the Cd(II)$_4$Mg(II)$_2$AP is shown in Fig. 9 as a function of titration with P_i compared to the same titration of Cd(II)$_2$AP. In contrast to the latter enzyme, both 1 and 2 moles of P_i induce formation of E-P and it is not until the addition of the third mole of P_i that a resonnance indicating significant free phosphate appears. The signals in Fig. 9C integrate as 2 moles of E-P (δ = 8.1 ppm) and 1 mole of P_i (δ = 0.8 ppm).

All metal ions can be rapidly removed from the diphosphoryl enzyme with retention of most of the phosphoryl groups, thus forming the diphosphoryl apoenzyme ApoAP-P$_2$. The ^{31}P NMR spectrum of this species is shown in Fig. 10B. If ApoAP-P$_2$ is now reconstituted with two ^{113}Cd(II) ions, the phosphoryl resonance splits and over the next 100 hours the high-field portion of the resonances (accounting for about half the phosphate) gradually disappears, and a narrow resonance appears at the chemical shift expected for inorganic phosphate (Fig. 10C-F). Thus the addition of the two catalytic Cd(II) ions to the diphosphoryl apoenzyme causes a clear asymmetry in the phosphoryl sites and the dephosphorylation of one of them. From the data

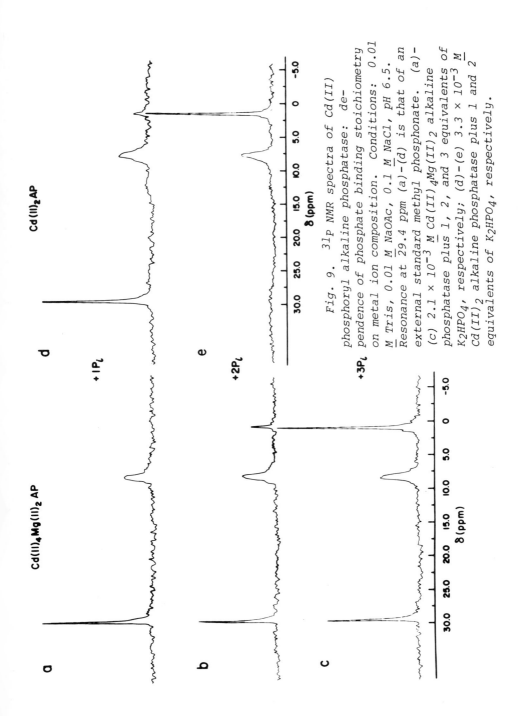

Fig. 9. ^{31}P NMR spectra of Cd(II) phosphoryl alkaline phosphatase: dependence of phosphate binding stoichiometry on metal ion composition. Conditions: 0.01 M Tris, 0.01 M NaOAc, 0.1 M NaCl, pH 6.5. Resonance at 29.4 ppm (a)-(d) is that of an external standard methyl phosphonate. (a)-(c) 2.1×10^{-3} M Cd(II)$_4$Mg(II)$_2$ alkaline phosphatase plus 1, 2, and 3 equivalents of K$_2$HPO$_4$, respectively; (d)-(e) 3.3×10^{-3} M Cd(II)$_2$ alkaline phosphatase plus 1 and 2 equivalents of K$_2$HPO$_4$, respectively.

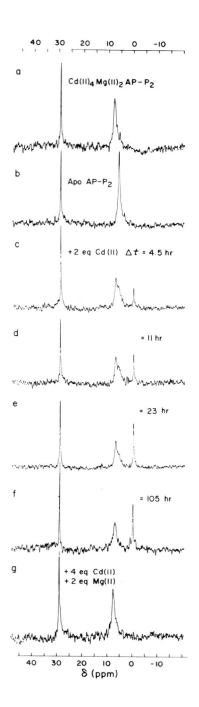

Fig. 10

in Fig. 10C-F, the dephosphorylation rate can be estimated to be $\sim 10^{-5}$ sec^{-1}. This is, in fact, the dephosphorylation rate for $Cd(II)_2AP-P_i$ as will be outlined in Section II,D. Readdition of the structural pair of Cd(II) ions and two Mg(II) ions to the $Cd(II)_2AP \cdot P_1$, which has completely dephosphorylated one of the sites, results in rephosphorylation of the second site to reform $Cd(II)_4Mg(II)_2AP-P_2$ (Fig. 10G).

A hypothesis of negative cooperativity of phosphate binding to E. coli alkaline phosphatase that is in accord with all the above findings can be stated as follows. In the case of the dimer containing two catalytic metal ions, phosphorylation of the first site reduces the binding affinity of the second site for the phosphate (or substrate) dianion by >3 orders of magnitude, since no evidence of the binding of a second phosphate ion is observed even at P_i concentrations $>10^{-2}$ M. Such a structural change could be minor, e.g., the movement of a positively charged residue away from the second site, and need not alter the basic steps of the mechanism. Simply stated, there is no affinity for P_i at the second site of the monophosphate complex of the two metal enzyme. Addition of the structural pair of metal ions and Mg(II) results in an increase in the affinity of P_i at the second site to a detectable level.

The ^{31}P NMR experiments on $Cd(II)_4Mg(II)_2AP$ (Fig. 9) suggest that the dissociation constant for P_i binding at the second site is at least 10^{-4} M. Once formed, E·P is converted to E-P. For the Cd(II) enzyme, dephosphorylation and dissociation of phosphate is normally followed by immediate rebinding of P_i ($k_{-4} = 2 \times 10^7$ $M^{-1}sec^{-1}$) to form E·P, which rephosphorylates the serine residue. Dephosphorylation is extremely slow so that the overall equilibrium lies in favor of E-P. Thus all bound P_i appears as E-P. When only two Cd(II) ions are added back to the apodiphosphoryl enzyme, however, dephosphorylation and phosphate dissociation are not followed by the rebinding of P_i at the second site; strong negative cooperativity is reestablished. Thus the dephosphorylation data of Fig. 10 are adequately described in terms of the rate constants obtained for the monophosphoryl enzyme. The distinction is that for $Cd(II)_2AP-P_2$,

Fig. 10. Demonstration by ^{31}P NMR of the reversible, metal-ion dependent formation of di- and monophosphoryl Cd(II) alkaline phosphatase. Conditions: 0.01 M Tris, 0.01 M NaOAc, 0.1 M NaCl, pH 6.5, 3.2×10^{-3} M alkaline phosphatase, metal ion, and phosphate stoichiometries as indicated. Apodiphosphoryl alkaline phosphatase (ApoAP-P_2) was generated from $Cd(II)_4Mg(II)_2(AP-P_2)$ by dialysis versus orthophenanthroline. Spectra (c)-(f) were taken over the indicated time intervals following addition of two equivalents of Cd(II) to sample (b). Spectrum (g) was generated on addition of Cd(II) and Mg(II) to sample (f) to yield the indicated metal ion stoichiometry. The resonance at 29.4 ppm is that of methyl phosphonate an external standard.

P_i dissociation from the second site is essentially irreversible, while for $Cd(II)_4Mg(II)_2AP-P_2$ it is not.

The correlation of the expression of negative cooperativity and enzyme, phosphate, and metal ion concentration is currently under investigation by $[32_p]$-labeling methods. It is possible that under conditions of low enzyme and phosphate concentration the enzyme remains largely under negative allosteric control. Since the structural pair of metal ions clearly attenuates the negative cooperativity and is considerably less tightly bound (both NMR and ESR suggest they are in relatively facile exchange with free metal ion) (Chlebowski *et al.*, 1976; Weiner *et al.*, 1978), the possibility exists that the metal cations as well as the phosphate anions in the environment may be allosteric modulators of the activity of alkaline phosphatase.

D. *Phosphate Exchange at the Active Site of Alkaline Phosphatase as Measured by ^{31}P NMR*

When an additional P_i per mole of enzyme dimer is added to $Me(II)_2AP-P_1$ or $Me(II)_4Mg(II)_2AP-P_2$ [$Me(II) = Zn(II)$, $Mn(II)$, or $Co(II)$], the resonance for the P_i appears at the chemical shift expected for free inorganic phosphate, but the line is broadened over that in the presence of the apoenzyme. A typical spectrum of $Zn(II)_4Mg(II)_2AP$ in the presence of four moles of P_i per mole of dimer is shown in Fig. 11. The enzyme solution was at pH 5.5, so that the enzyme-bound phosphate was equally partitioned between E·P ($\delta = 3.0$) and E-P ($\delta = 8.2$). The resonances for E·P and one for free P_i are separated by ~125 Hz, indicating that the rate of chemical exchange between E·P and free P_i is clearly slower than the chemical shift difference (~700 $\overline{sec}^{-1}$, pH 5.5). Thus E·P and P_i are in slow exchange relative to their chemical shifts. On the other hand, values of the spin-lattice relaxation time T_1 for E·P and P_i in the presence of excess P_i are nearly equivalent and range from 2 to 4 sec, depending on pH. Thus chemical exchange is fast relative to the relaxation rates, since $\tau_M/\underline{T_1}$ $_{obs} << 1$. Hence the relative concentrations of phosphate species are reflected in the peak intensities.

When the slow exchange limit (on the chemical shift time scale) pertains, it is possible to estimate the lifetimes from the linewidths of the different phosphate species and thus the dissociation rate constant for E·P. As determined from the linewidth of the E·P species for the Zn(II) enzyme (Fig. 11), this dissociation rate is 60 ± 20 sec^{-1} at pH 8 and 100 ± 20 sec^{-1} at pH 5.5 (Chlebowski *et al.*, 1977) based on the exchange contribution (~20 Hz, pH 8; ~32 Hz, pH 5.5) to the linewidth of the E·P resonance in the presence of P_i. A value of 10 to 20 sec^{-1} for dissociation of E·P was reported by Hull *et al.* (1976)

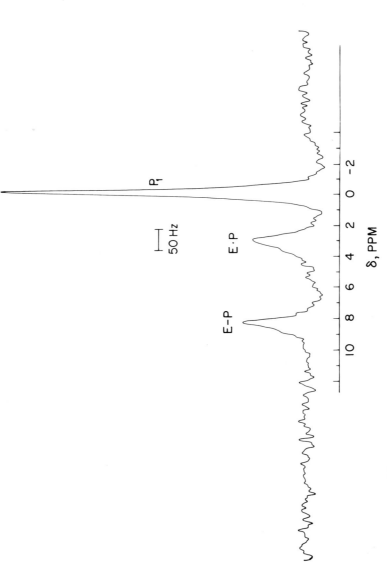

Fig. 11. ^{31}P NMR spectrum of Zn(II) alkaline phosphatase, $Zn(II)_4Mg(II)_2AP$, in the presence of four moles of P_i, pH 5.5.

from similar ^{31}P NMR data. These values are similar to the
range of values, 20 to 50 sec^{-1}, reported for k_{cat} at alkaline
pH (Fernley and Walker, 1969; Halford *et al.*, 1972; Hull *et al.*,
1976) and suggest that dissociation of E·P is the rate-limiting
step in phosphate ester hydrolysis.

The best values for the kinetic constants involved in the
exchange of P_i with both E·P and E-P are summarized in Eq. (2).
These are given for alkaline pH, but in the case of P_i turnover
are relatively pH-independent except for k_3:

$$k_{3(Cd)} = \sim 10^{-5} \ sec^{-1}$$

$$E-P \xrightleftharpoons[\ k_{-3(Zn)} = 0.2 \ sec^{-1}\ ]{\ k_{3(Zn)} = \sim 10^2 \ sec^{-1}\ } E \cdot P \xrightleftharpoons[\ k_{-4(Me)} = \sim 10^7 \ sec^{-1}M^{-1}\ ]{\ k_{4(Me)} = 10-100 \ sec^{-1}\ } E + P_i$$

$$k_{-3(Cd)} = 0.003 \ sec^{-1}$$

$$(2)$$

Measurements of the dissociation constant, K_d for E·P show
little dependence on metal ion; i.e., for Zn(II), Co(II),
Mn(II), and Cd(II), $K_d \cong 10^{-6}$ M (Applebury *et al.*, 1970; Chlebowski and Coleman, 1974). Likewise, the kinetic constant k_4
measured by the NMR linewidth is the same for Zn(II) and Co(II)
(Chlebowski *et al.*, 1977). The resonance of free P_i in the
presence of Co(II)$_2$AP·P_i was shown in Fig. 8 and is similar to
that observed in the presence of Zn(II)$_2$AP·P_i. Hence exchange
broadening occurs via exchange with E·P, but no paramagnetic
broadening is present as expected from the lifetimes involved.
Thus it can be assumed that k_4 is 10-100 sec^{-1} for all Me(II)
derivatives and thus $k_{-4} \cong 10^7$ sec^{-1} M^{-1}. The modulation of
the E·P and P_i resonances by the E·P $\rightleftarrows P_i$ exchange would not be
expected to be observed in the presence of the Cd(II) enzyme,
since at pH 6.5 the overall reaction [Eq. (2)] is governed by
the extremely slow dephosphorylation of E-P and the equilibrium
for enzyme-bound phosphate is very far in favor of E-P^2. When
care is taken to achieve optimal and comparable conditions
(e.g., Figs. 9 and 11), the P_i resonance is narrower in the
presence of the Cd(II) enzyme than in the presence of the Zn(II)
enzyme.

The k_{-3} values for the formation of E-P are directly obtainable from the ^{18}O exchange data for the Zn(II) enzyme, 0.2
sec^{-1} (Applebury *et al.*, 1970; Eargle *et al.*, 1977; Bock and
Cohn, 1978), and the time rate of phosphorylation for the Cd
enzyme, 0.003 sec^{-1}, as measured by ^{32}P incorporation (Applebury *et al.*, 1970). Dephosphorylation by Zn(II) is $\sim 10^2$ sec^{-1}
(Aldridge *et al.*, 1964; Bale, 1978), while by Cd(II) it is

dramatically slower, $\underline{k_3} \simeq 10^{-5}$ sec^{-1} (pH 6.5)[2] (Applebury *et al.*, 1970). The latter number can be directly measured from $E-{}^{32}P \rightleftharpoons {}^{31}P_i$ exchange which gives a value of 3×10^{-5} sec^{-1} (Chlebowski and Coleman, 1976).

Cohn and Hu (1978) have recently shown that ^{18}O exerts an isotopic chemical shift on the ^{31}P NMR resonance of P_i, large enough to resolve the species of phosphate containing 0, 1, 2, 3, or 4 atoms of ^{18}O. Thus the change in relative peak heights of the resulting five-line multiplet can be used to measure enzyme-catalyzed ^{18}O exchange out of $HP^{18}O_4^{2-}$. Bock and Cohn (1978) have followed the exchange out of ^{18}O from $HP^{18}O_4^{2-}$ catalyzed by Zn(II) and Co(II) enzymes and show that the exchange catalyzed by Zn(II) is compatible with the kinetic scheme in which E·P dissociates at least tenfold more rapidly than E-P is reformed (Bock and Cohn, 1978). On the other hand Co(II) catalyzes the exchange of more than one ^{18}O atom per turnover implying that significant rephosphorylation of the serine occurs before E·P can dissociate. In view of the apparent similar magnitudes of $\underline{k_4}$ and $\underline{k_{-4}}$ for the various metal ions this is best explained by the hypothesis that phosphorylation of ser 99 by E·P catalyzed by Co(II) is significantly faster (10-100-fold) than by Zn(II).

III. LIGANDS TO THE CATALYTIC METAL IONS AS DETERMINED BY ^{13}C NMR

Of the five Zn(II) metalloenzymes whose crystal structures have been determined, one (carbonic anhydrase) has three histdyl nitrogens as the ligands from the protein to the metal ion, two have a combination of histidyl nitrogens and carboxylate oxygens (carboxypeptidase, N,N,O; thermolysin, N,N,O; and superoxide dismutase, N,N,N,O; see Chlebowski and Coleman, 1976, for review), and one (alcohol dehydrogenase) has one histidyl and two cysteinyl sulfurs liganded to the catalytic Zn(II) and four cysteinyl sulfurs liganded to the structural Zn(II) (Bränden *et al.*, 1973; Eklund *et al.*, 1974). Thus histidyl nitrogen atoms are prime candidates as ligands to catalytic Zn(II) ions in metalloenzymes. While ESR spectra of $Cu(II)_2AP$ show at least

---

[2]Note that the estimate of the rate constant for the dephosphorylation of the Cd(II) enzyme of <0.01 sec^{-1} given in Chlebowski et al. (1977) (based on 99% E-P) conveys an improper impression of the size of this number, which is actually 10^{-5} sec^{-1} (confirmed by $^{32}P_i \leftrightarrow {}^{31}P_i$ exchange) or an E-P/E·P ratio of ~10^4.

three magnetically equivalent nitrogens to be ligands to the
Cu(II) ion (Taylor and Coleman, 1972) and protection of histidyl residues from photooxidation by Zn(II) (Tait and Vallee,
1966) has suggested histidyl nitrogens as ligands to the Zn(II)
ion in alkaline phosphatase, until recently no direct evidence
for this conclusion had been obtained.

The ^{13}C NMR of proteins carrying histidyl residues ^{13}C-en-
riched at the γ carbon is potentially a powerful probe for the
conformation of the histidyl residues, since the resonance from
the quaternary carbon without directly bonded protons is narrow. The linewidth of the resonance can be additionally improved by substitution of deuterons for the nearby β protons.
In order to explore the role of histidyl residues in the structure and function of alkaline phosphatase, $[\gamma-^{13}C]-\beta,\beta$-dideuter-
ohistidine, first synthesized by Browne *et al.* (1976) and now
commercially available from Merck and Co., has been incorporated
biosynthetically into *E. coli* alkaline phosphatase by using a
histidine auxotroph of *E. coli*.

The ^{13}C NMR spectra of the resulting ^{13}C-enriched alkaline
phosphatase containing three different metal ions, Zn(II),
Cd(II), and Mn(II), are summarized in Figs. 12, 13, and 14
(Coleman *et al.*, 1978; Otvos *et al.*, 1978). In each case the
spectra are those of homogeneous samples containing precise
stoichiometries of the various metal ions. For comparative
purposes the spectra of the various derivatives should be compared to the protein containing the full complement of metal
ions, Zn(II)$_4$Mg(II)$_2$AP. The ^{13}C spectrum of this species is
shown in Figs. 12D and 14A. There are eight well-resolved ^{13}C
resonances for this species. Since there are a total of 16
histidyl residues per dimer, the occurrence of only eight resonances confirms the twofold symmetry of the dimer. The chemical shift range of the resonances of the γ carbons is within the
range observed thus far in nonmetalloproteins, with the exception of resonances 1 and 2, which are downfield of this range,
and resonance 8, which is significantly upfield. Resonances 1,
2, and 8 do not titrate from pH 5 to 10 (Browne *et al.*, 1976).
In the apoenzyme, the chemical shift range contracts to that
observed in nonmetalloproteins. Resonances 1 and 2 move 3 to 5
ppm upfield in the apoenzyme, while resonance 8 moves downfield
~2 ppm (Fig. 12A). These shifts and the titration data strongly suggest that resonances 1, 2, and 8 in the metalloenzyme
arises from the γ carbons of imidazole side chains coordinated
to the metal ion.

Readdition of two Zn(II) ions to the apoalkaline phosphatase restores the original chemical shifts of resonances 1, 2,
and 8 (Fig. 12C). Although Mg(II) causes some shifts in the
$[\gamma-^{13}C]$-histidyl resonances (Fig. 12B), the specific shifts of
resonances 1, 2, and 8 are a function of the addition of the
two catalytic Zn(II) ions. A total of four Zn(II) ions and two

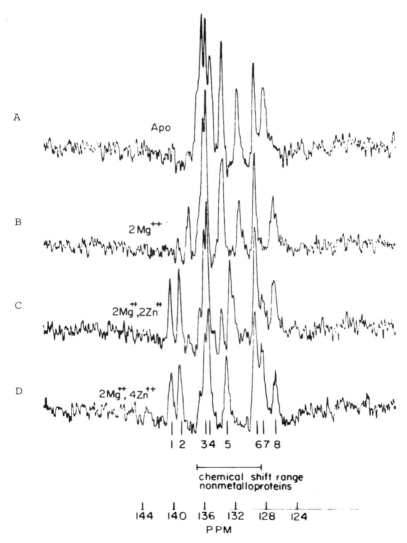

Fig. 12. ^{13}C NMR spectra of $\left[\gamma-^{13}C\right]-\beta,\beta$-dideuterohistidine
alkaline phosphatase. Conditions: 0.01 $\underline{M}$ Tris, 0.01 $\underline{M}$ NaOAc,
0.1 $\underline{M}$ NaCl, pH 7.9, 2.0 × 10^{-3} $\underline{M}$ alkaline phosphatase, metal
ion additions as indicated. Chemical shifts are relative to
tetramethyl silane. (From Otvos et al., 1978.)

Mg(II) ions are required to complete the changes observed in the
$[\gamma-^{13}C]$-histidyl resonances, suggesting additional modulation of

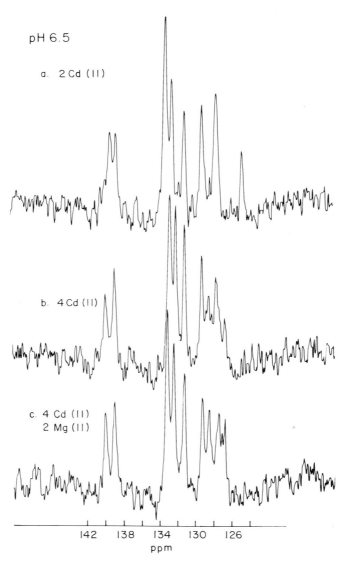

Fig. 13. ^{13}C NMR spectra of Cd(II)$[\gamma-^{13}C]-\beta,\beta$-dideutero-
histidine alkaline phosphatases. Conditions: 0.01 M Tris,
0.01 M NaOAc, 0.1 M NaCl, pH 6.5, 2.0 × 10^{-3} M alkaline phosphatase, metal ion stoichiometries as indicated. (From Otvos et
al., 1978.)

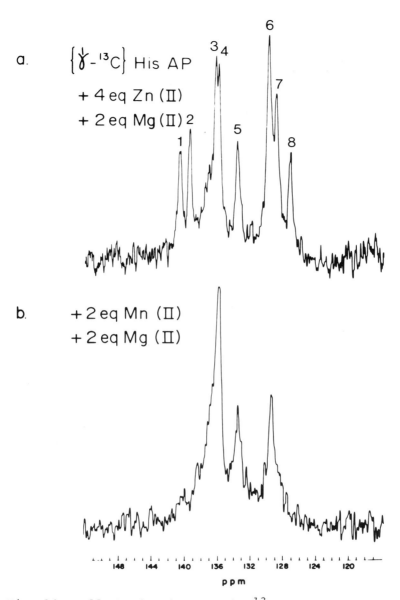

Fig. 14. Effect of Mn(II) on the ^{13}C NMR spectrum of $[\gamma-^{13}C]-\beta,\beta$-dideuterohistidine alkaline phosphatase. Conditions: 0.01 $\underline{M}$ Tris, 0.01 $\underline{M}$ NaOAc, 0.1 $\underline{M}$ NaCl, pH 7.8, 1.9×10^{-3} $\underline{M}$ alkaline phosphatase, metal ion stoichiometries as indicated. (From Otvos et al., 1978.)

the protein structure by "structural" Zn(II) and Mg(II) ions
(Figs. 12D and 13C).

In nonmetalloproteins the most downfield γ carbon resonances
are those from the N_τ-tautomers (i.e., those with the proton on
the τ-nitrogen in the neutral imidazole group). While further
downfield than normal, the chemical shift of the γ carbons of
the "ligand" histidyl residues 1 and 2 almost certainly means
the proton is on the N^τ, hence the metal is coordinated to the
N^π. Resonance 8 must represent a ligand histidyl with the metal coordinated to N^τ.

Addition of two Cd(II) ions to the $[\gamma\text{-}^{13}C]$-histidyl-labeled
apoenzyme at pH 6.5 results in the downfield shift of resonances
1 and 2 and the upfield shift of resonance 8, similar to the
shifts caused by Zn(II) addition (Fig. 13B). Resonance 8, however, is shifted considerably farther upfield by Cd(II) suggesting that while the two catalytic Cd(II) ions coordinate the
same three histidyl residues, there may be significant differences in the length or angle of at least one of the metal-ligand
bonds.

Lastly addition of two Mn(II) ions that generate the solid-
state ESR spectrum characteristic of the two catalytic Mn(II)
ions (Haffner *et al.*, 1974) result in paramagnetic relaxation
of resonances 1, 2, and 8 as would be expected for ligand histidyls. In addition, a fourth histidyl resonance, number 7, is
broadened beyond detection. While this might be a fourth ligand, it may be the histidyl that reacts with mono- and dichlo-
roacetyl-β-glycerophosphate, "substrate analog" reagents, which
have been used to alkylate a histidyl near the active center
(Csopak and Folsch, 1970).

Aside from the major shifts in the ^{13}C resonance accompanying metal ion ligation, there are a number of more subtle
changes in the envelope of the $[\gamma\text{-}^{13}C]$-histidyl residues of the
enzyme as a function of additional variables. The binding of
both Mg(II) and the structural metal ions alter certain resonances, e.g., the chemical shifts of resonances 5 and 7 are
further altered by the binding of the two structural Zn(II) ions
(Fig. 12D). The resonances of the ligands also undergo further
shifts on the addition of the second pair of Cd(II) ions (Fig.
13C). Phosphate addition (not shown here) causes other shifts
in the ^{13}C resonances. Thus ^{13}C NMR of specific carbons in
amino acid side chains is a powerful probe of the conformational
changes associated with the modulation of the catalytic activity
of alkaline phosphatase.

IV. ^{113}Cd NUCLEAR MAGNETIC RESONANCE OF Cd(II) ALKALINE PHOSPHATASE

The inherent sensitivities of most metal nuclei are at least an order of magnitude lower than ^{1}H due in part to the low natural abundance and a spin quantum number $I > 1/2$. Hence metal ion NMR has not found application in the study of metalloenzymes. Recently Armitage et al. (1976, 1978) have shown that NMR of the spin-1/2 isotope of cadmium, ^{113}Cd(II), can, by isotopic enrichment, be made sensitive enough to readily detect satisfactory NMR signals from ^{113}Cd(II) substituted at the Zn(II) sites of a number of metalloenzymes. With a chemical shift range >600 ppm reported for model complexes of cadmium, the nucleus promises to be extremely sensitive to the nature and coordination geometry of its macromolecular ligands. Chemical shifts observed thus far in metalloenzymes have spanned approximately 200 ppm. The extremes of chemical shift for ^{113}Cd-containing proteins now extend from ~600 ppm for metallothionein (Otvos and Armitage, unpublished data) to -130 ppm for one of the sites in concanavalin A (Cardin et al., 1978).

One additional consequence of the large chemical shift range for ^{113}Cd that is not immediately obvious, at least as far as biochemical applications are concerned, is the effect of chemical exchange on the position, amplitude, and lineshape of the ^{113}Cd resonance. The appendix presents some detailed models for chemical exchange as it may apply for the Cd(II) ion in metalloenzymes. Interconverting coordination conformers with lifetimes on the order of 1 to 10^{-5} sec may have pronounced effects on the ^{113}Cd resonance.

Since ^{113}Cd NMR can potentially monitor all metal binding sites simultaneously, it is a particularly powerful method to examine changes in enzyme structure as functions of metal ion and phosphate stoichiometry. Cd(II)$_2$AP is a functional enzyme, but with a turnover number ~100 times smaller than that of the Zn(II) enzyme (Chlebowski and Coleman, 1976). The phosphoserine is formed as an intermediate, but its dephosphorylation is slow enough that the enzyme exists almost totally in the phosphoryl form at neutral pH (Chlebowski and Coleman, 1976; Applebury et al., 1970). Two additional Cd(II) ions and one or two Mg(II) ions also stabilize the enzyme in analogous fashion to the Zn(II) enzyme, although stabilization by Cd(II) does not appear to be as great as by Zn(II) (Coleman and Chlebowski, 1979).

The ^{113}Cd NMR spectrum of ^{113}Cd(II)$_2$AP in the absence of halide counter ions shows a single resonance at 117 ppm (Fig. 15A). This observation leads to the obvious conclusion that the two catalytic Cd(II) ions in the unliganded protein dimer experience identical environments, supporting the X-ray data, which show twofold symmetry between the monomers (see below).

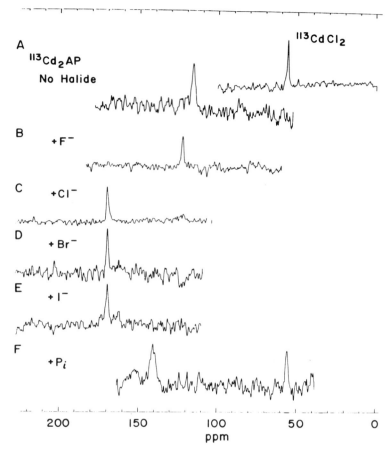

Fig. 15. ^{113}Cd *NMR spectra of* $^{113}Cd(II)_2$ *alkaline phosphatase. Conditions: 0.01 $\underline{M}$ Tris, 0.01 $\underline{M}$ NaOAc, pH 6.0,
3.6 × 10^{-3} $\underline{M}$ $^{113}Cd(II)_2$ alkaline phosphatase. Halide present
at 0.1 $\underline{M}$ as indicated (B-E). Spectrum of $^{113}Cd(II)Cl_2$ (insert
to A) was obtained under identical sampling conditions. (F)
4.5 × 10^{-3} $\underline{M}$ $^{113}Cd(II)_2$ alkaline phosphatase, 4.1 × 10^{-3} $\underline{M}$
K_2HPO_4, pH 6.5. Chemical shifts are relative to 0.1 $\underline{M}$
$Cd(ClO_4)_2$. (From Armitage et al., 1978.)*

Increasing halide concentrations (added as NaCl, NaBr, or NaI)
have a marked stimulatory effect on the activity of alkaline
phosphatase (Reid and Wilson, 1971; Wilson *et al.*, 1964; Nakamura *et al.*, 1978). Addition of increasing concentrations of
halides to $^{113}Cd(II)_2$AP causes a progressive downfield chemical

shift of the cadmium resonance. The chemical shift change at
0.1 M halide is shown and is the same for Cl⁻, Br⁻, and I⁻ (53
ppm), much less for F⁻ (~5 ppm) (Fig. 15) (Armitage et al.,
1978). The direction of the shift to low field with increasing
concentration of these anions is as expected for specific binding to the metal ion and can be fit to a dissociation constant
of ~0.02 M (Fig. 16). The magnitude of this apparent dissociation constant is much larger than those for the cadmium halides,
which range from 10^{-3} M for Cl⁻ to 10^{-6} M for I⁻. The similarity of the chemical shifts for Cl⁻, Br⁻, and I⁻ is also not expected from the chemical shifts of the inorganic salts and thus
one cannot exclude the possibility that the ^{113}Cd(II) chemical
shifts are sensitive to the nonspecific binding of anions to
positive charges on the protein, which would follow a Debye-
Huckel ionic strength function.
    In order to interpret ^{113}Cd(II) resonances completely, it
is necessary to understand the various chemical exchange processes that can modulate the ^{113}Cd resonances in metalloenzymes, which include ligand (solvent or protein) exchange and
intramolecular rearrangements. For example, spin-lattice relaxation measurements and comparison of the integrated intensities of the ^{113}Cd resonances of the above samples with similar
resonances from other ^{113}Cd(II)-substituted metalloenzymes have

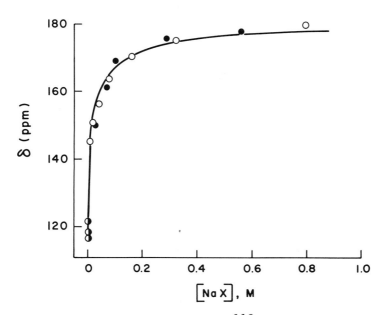

Fig. 16. Chemical shift of the ^{113}Cd resonance of
^{113}Cd(II)$_2$AP as a function of salt concentration. (O) NaCl,
(●), NaBr.

shown that the ^{113}Cd resonance in the unliganded ^{113}Cd(II)$_2$AP
appears to represent only $50 \pm 15\%$ of the total Cd(II). Even
more dramatic reductions in signal amplitude are observed when
additional Cd(II) ions are added to the unliganded Cd(II)$_2$AP.
The latter phenomenon apparently reflects an enhanced exchange
modulation brought about by the metal ions at the secondary
metal binding sites. Such modulation is markedly reduced by
prior phosphorylation of the protein. Hence signal amplitudes
of ^{113}Cd(II) resonances in metalloproteins cannot be directly
compared without additional information on the relaxation and
exchange phenomena applying to a given signal.

The resonance amplitude of the unliganded ^{113}Cd(II)$_2$AP is
best explained by an equilibrium between at least three conformers. The model is only speculative, but provides the minimum number of exchanging sites required to give rise to a
broadened resonance that would escape detection and produce no
exchange broadening in the dominant resonance with which it is
in slow exchange. The calculated spectra for a three-conformer
equilibrium along with the appropriate rate constants that reproduce the observed spectra are shown in the appendix. The
conformers result from a modulation of the two ^{113}Cd(II) ions
at the two fully occupied catalytic sites, where coordination
to the three histidyl ligands is inferred on the basis of the
characteristic ^{13}C resonances of the ligand histidyls in the
[γ-^{13}C]-histidyl-labeled Cd(II)$_2$AP (shown in Fig. 12A). The
slow interconversion of one conformation to a second (1 to 2)
may represent a slow intramolecular exchange of a protein ligand, followed by a more rapid exchange of solvent with one of
these conformers, giving rise to a third species (3). Further
speculation as to the chemical identity of the exchanging intermediates and how this model correlates with the expression
of negative cooperativity, exhibited by the ^{113}Cd(II)$_2$ enzyme,
must await additional experimental data.

On phosphorylation by the addition of one equivalent of
phosphate per dimer, the ^{113}Cd spectrum of Cd(II)$_2$AP shows two
resonances at 142 and 55 ppm (Fig. 15F). Addition of a second
equivalent of phosphate does not alter the chemical shifts of
the ^{113}Cd resonances. Thus phosphorylation at a single site of
the enzyme destroys the symmetric relationship between the metal ion binding sites and alters the environment of both Cd(II)
ions from that present in the absence of phosphate. The two
cadmium resonances in the phosphoryl enzyme are separated by
87 ppm and are relatively narrow. Hence they appear to represent equal populations of states that are not in fast or intermediate exchange with each other or with free Cd(II) ions.
These observations are consistent with the existence of negative
homotropic interactions between subunits. The stoichiometric
binding of phosphate, which induces asymmetry in the initially
symmetrical metal ion sites, is presumably associated with con-

formational changes propagated across the monomer-monomer interface. One possible picture of this process is that phosphorylation of the enzyme freezes out certain conformational isomers present in significant concentrations in the unliganded enzyme dimer. As shown more dramatically by the addition of the structural Cd(II) ions (see below), the amplitude and T_1 values of ^{113}Cd(II) resonances may be a sensitive reflection of conformational flux at enzyme active sites (see appendix). This difference in chemical shifts of the two catalytic ^{113}Cd(II) ions is the first direct evidence that binding of phosphate at one site alters the coordination sphere of the metal ion at the other catalytic site.

The ^{113}Cd resonances in the monophosphoryl Cd(II) enzyme can be used to monitor effects on the structure of the metal sites caused by additional ions that modulate the turnover rate of the enzyme. The ^{113}Cd-NMR spectrum of the monophosphoryl Cd(II)$_2$Mg(II)$_2$AP in 0.01 $\underline{M}$ Tris-0.01 $\underline{M}$ Na acetate is shown in Fig. 17A (Chlebowski et $al.$, 1978). In the absence of Cl$^-$, the two signals occur at 75.6 and 137.5 ppm. If an additional pair of ^{113}Cd(II) ions and an additional equivalent of P$_i$ are added to the monophosphoryl enzyme in the presence of $\overline{Mg}$(II), resonance from additional Cd(II) ions appears at 4 ppm. There is considerable improvement in signal to noise for the signals from the two enzyme-bound Cd(II) ions (Fig. 17B). A second pair of Cd(II) ions or Zn(II) ions stabilizes the protein structure (Coleman and Chlebowski, 1979; Chlebowski and Mabrey, 1977). This may alter certain dynamic processes modulating the active site Cd(II) ions and hence alter their intensities under a constant set of NMR signal collecting conditions.

The T_1 values for the two downfield resonances in Fig. 17B are ~4 sec. Integration of these signals, taking account of the long T_1 values, shows each of the downfield resonances to account for ~1 Cd(II) ion per dimer. The upfield broadened resonance in Fig. 17B has a T_1 of ~1.5 sec, and this resonance also accounts for ~1 Cd(II) ion per dimer. Hence ~25% of the Cd(II) of the 4 Cd(II) enzyme is not visible by NMR, which suggests that the Cd(II) represented by the broadened upfield peak is in relatively rapid exchange with ^{113}Cd(II) at sites on the molecule whose signals are broadened beyond detection by chemical exchange processes. This phenomenon has been documented for other proteins (Schoot Uiterkamp et $al.$, 1978). The magnitude and variation in T_1 values of the various enzyme-bound forms of ^{113}Cd(II) make it impractical to employ pulsing conditions that allow complete recovery of the resonance between pulses. Hence direct comparison of signal amplitudes is not a measure of the amount of Cd(II) represented by each signal.

A differential effect of Mg(II) on the separate ^{113}Cd resonances of the monophosphoryl enzyme can be shown. Addition of Mg(II) (one or two per dimer) to the monophosphoryl ^{113}Cd(II)$_2$AP

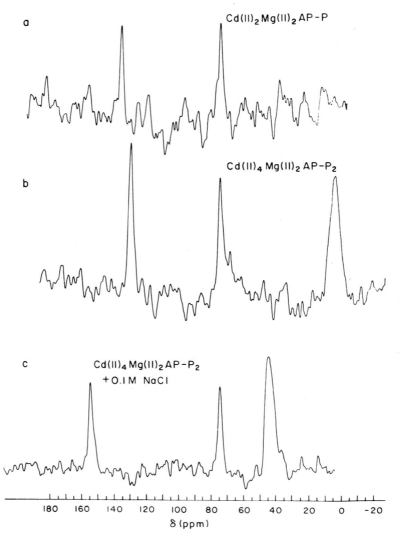

Fig. 17. ^{113}Cd NMR spectra of Cd(II) phosphoryl alkaline
phosphatase. Conditions: 0.01 $\underline{M}$ Tris, 0.01 $\underline{M}$ NaOAc, pH 6.5
(a) 3.6 × 10⁻³ $\underline{M}$ $^{113}Cd(II)_2Mg(II)_2$ phosphorylalkaline phosphatase; (b) and (c), metal ion and phosphate stoichiometry as indicated; (c) sample (b) plus 0.1 $\underline{M}$ NaCl. (From Chlebowski et
al., 1978.)

causes the upfield peak at 55 ppm to move downfield by 20 to
75 ppm (compare Figs. 17A and 15F). A variety of data including

central metal exchange at the downfield, but not upfield peak, leads us to believe that the [113]Cd at the phosphorylated site in the monophosphoryl enzyme is represented by the resonance at 55 ppm [or 75 ppm in the presence of Mg(II)]. Hence Mg(II) may have a rather specific interaction with the phosphorylated site. This might explain some of the ambiguities in functional studies in which some changes in activity seem to require only one Mg(II) (Bosron et al., 1977).

The effect of Cl⁻ on the Cd(II) resonance of the unliganded enzyme was noted above. Chloride also specifically shifts the most downfield resonance in the phosphoryl enzyme further downfield by ~20 ppm as well as interacting with the "free" Cd(II) (Fig. 17). The effect of Cl⁻ on the most downfield resonance of the enzyme is not observed in the absence of Mg(II). The resonance of what is believed to be the Cd(II) at the initially phosphorylated site (~75 ppm) is unaffected, perhaps reflecting further differences in the two sites of the phosphorylated enzyme. As discussed in Section II,C,1, $Cd(II)_4 Mg(II)_2 AP$, at mM concentrations, induces the phosphorylation of the serine at both active sites. The [113]Cd(II) spectrum shown in Fig. 17B is that of the diphosphoryl enzyme.

The [113]Cd chemical shift values for the two catalytic Cd(II) ions of the diphosphorylated $[113]Cd(II)_4 Mg(II)_2 AP-P_2$ (Fig. 17B) remain different, showing that the diphosphoryl enzyme is still asymmetric. The chemical shifts of the two phosphoserines of the dimer do not reflect any large difference in environment, although ^{31}P chemical shifts are not particularly sensitive to structural changes. The phosphoryl group may not in fact be in the inner coordination sphere of the metal ion when it is present as the phosphoserine (see mechanism in Fig. 3).

V. ^{19}F NMR OF FLUORINATED AMINO ACIDS INCORPORATED INTO
    ALKALINE PHOSPHATASE

A large amount of structural and dynamic information can potentially be obtained by ^{19}F NMR studies on bacterial proteins containing m-fluorotyrosine and p-fluorophenylalanine residues that have been incorporated biosynthetically using tyrosine and phenylalanine auxotrophs of E. coli. There are 11 tyrosyl and 8 phenylalanyl residues in each subunit of alkaline phosphatase and active enzymes containing the fluorinated residues are produced by E. coli auxotrophs growing on the fluorinated amino acids. Thus ^{19}F-labeled amino acid residues can provide conformational probes at a total of 38 separate sites in this dimeric enzyme.

A detailed analysis of the ^{19}F chemical shifts and relaxation times of fluorotyrosine-labeled alkaline phosphatase has been carried out by Hull and Sykes (1974, 1975a, 1976; Hull *et al.*, 1976). This was the first application of specific ^{19}F-labeling of a protein and demonstrated the sensitivity of ^{19}F chemical shifts to the chemical environment and its use as a monitor of changes in the enzyme conformation induced by pH, metal ion, solvent composition, and phosphate binding. Hull and Sykes (1974, 1975a,b) developed a detailed analysis of the ^{19}F spin relaxation data for alkaline phosphatase and showed how this data could be used to describe the internal rotation about the C_β-aromatic ring bond and C_α-C_β bond of the 11 resolved tyrosyl residues. Browne and Otvos (1976) have also reported on the ^{19}F NMR of alkaline phosphatase labeled with m-fluorotyrosine and 4-fluorotryptophan. These studies will not be reviewed in detail here, but the sensitivity and resolution possible using ^{19}F NMR probes is illustrated by the ^{19}F NMR of alkaline phosphatase containing perdeuterofluorotyrosyl residues (all aromatic protons were replaced with deuterons) (Fig. 18).

The use of ^{19}F NMR to probe macromolecular structure carries with it certain theoretical limitations. The dependence of the ^{19}F-{^1H} NOE on rotational correlation time has been previously documented, and predicts that for dipolar-dominated relaxation in a macromolecule, $\tau_R \geq 10^{-8}$ sec, proton irradiation will result in the complete loss of the ^{19}F signal as a result of a negative nuclear Overhauser enhancement (NOE) (Hull and Sykes, 1975b). In fact, all but one of the ^{19}F resonances of alkaline phosphatase disappear on proton irradiation (Hull and Sykes, 1975b). The one residue (giving rise to the resonance between -58 and -60 ppm, which is not lost on proton decoupling) may have a significant contribution to its relaxation from internal rotation. Thus the usual improvement in resolution accompanying proton decoupling cannot be realized unless the more time-consuming gated decoupling technique is employed. Furthermore, without proton irradiation, cross relaxation may play an important role, which must be taken into account in the analysis of the relaxation data. Both of these limitations can be partially circumvented by the complete deuteration of the ring protons of m-fluorotyrosine. This can be achieved as for normal tyrosine by the method of Griffiths *et al.* (1976). The ^{19}F resonance of free m-fluorotyrosine appears as a quartet of total width 21 Hz due to coupling to the hydrogens on the ring. On deuteration, this multiplet is reduced to a total width of ~6 Hz. Thus, even for an enzyme with a τ_R of 7×10^{-8} sec (the τ_R appropriate for alkaline phosphatase), improvement in resolution may be expected and is observed experimentally (Fig. 18). The experimental ^{19}F linewidths of the perdeuterated sample approach those calculated by Hull and Sykes (1975b). Separation

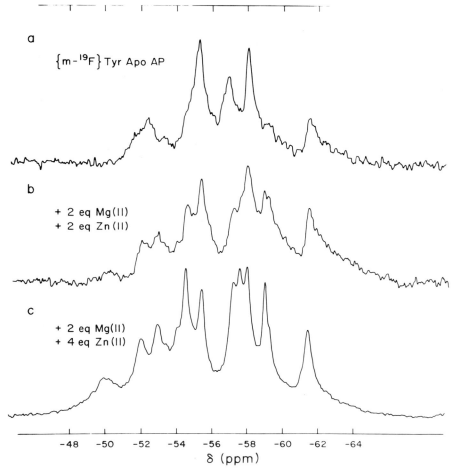

Fig. 18. ^{19}F NMR spectra of deuterated m–F-tyrosine alkaline phosphatase as a function of metal ion content. Conditions same as in Fig. 19. (a) 1.4×10^{-3} M apoalkaline phosphatase, 1.5 hours accumulation; (b) as (a) plus 2 Eq Mg(II), 2 Eq Zn(II), 2.5 hours accumulation; (c) as (a) plus 2 Eq Mg(II), 4 Eq Zn(II), 7.5 hours accumulation.

of the dipolar contributions to the spin-lattice relaxation from those of solvent protons and other protein groups is also simplified. Deuteration of the ring protons has effectively removed the dipolar contribution from these protons and solvent contributions can be determined directly from T_1 measurements performed in D_2O and H_2O. These studies show solvent and exchangeable protons make an appreciable contribution to the relaxation of the fluorine on the tyrosyl residues of alkaline phosphatase.

As described by Hull and Sykes (1974, 1975a,b), removal of
both the catalytic and structural Zn(II) ions and Mg(II) ions
from alkaline phosphatase results in substantial alterations
of a number of the fluorotyrosyl resonances of the enzyme (Fig.
18A). While it is impossible to interpret such changes precisely, a number of the tyrosyl residues obviously sense conformational changes in the structure of the polypeptide chain accompanying removal of the metal ions. The more downfield broader
resonances of the enzyme have been interpreted to reflect residues that are "buried" in the protein structure with more immobilization, more van der Waals contacts and altered dielectric
constants compared to "surface," solvent-exposed residues (Hull
and Sykes, 1974; 1975a,b; 1976). It is possible that the
changes in ^{19}F chemical shifts of the apoenzyme represent a
transition to a more open structure with more solvent access
to certain regions of the polypeptide chain. This may also be
reflected in the more rapid tritium exchange observed for the
apoenzyme than for the metalloenzyme (Brown et al., 1974). Readdition of Zn(II) and Mg(II) to the apoenzyme restores the
chemical shifts of the ^{19}F resonances to those observed in the
native protein (Fig. 18B,C). A total of four Zn(II) ions are
required for the complete change, suggesting that the chemical
shifts and enhanced resolution of the fluorotyrosyl resonances
reflect conformational states induced by both catalytic and
structural Zn(II) ions.

The relative usefulness of ^{19}F-tyrosyl residues and ^{19}F-
phenylalanyl residues as probes of protein structure, at least
in the case of alkaline phosphatase, can be appreciated by comparing Figs. 18 and 19. Resonances from all 11 ^{19}F-tyrosyl
residues are resolved and span ~11 ppm. On the other hand,
resonances from the 8 ^{19}F-phenylalanyl residues are poorly resolved and span only ~7 ppm (Fig. 19). The greater chemical
shift range for ^{19}F-tyrosyl undoubtedly results from the direct
participation of its hydroxyl group in hydrogen bonding, which
makes it more sensitive to the tertiary structure of the protein. No similar involvement of the ^{19}F-phenylalanyl residue
in protein structure is possible, and this gives rise to ~40%
reduction in the chemical shift range. The lack of resolution
in the case of ^{19}F-phenylalanyl is also contributed to by the
increased width of this ^{19}F resonance due to proton coupling
(42 Hz). At least a part of the intense resonance at ~-37 ppm
is due to denatured enzyme. Incorporation of fluorophenylala-
nine has been known to change the conformation of proteins.
Mechanisms responsible for this effect are unclear and require
investigation.

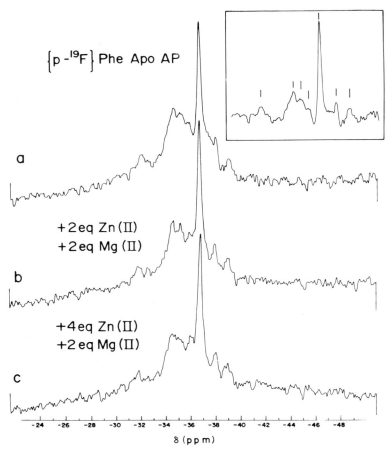

Fig. 19. ^{19}F NMR spectra of p-F-phenylalanine alkaline phosphatase as a function of metal ion content. Conditions: 0.01 M Tris, 0.01 M NaOAc, 0.1 M NaCl, pH 8. (a) 1.5 × 10⁻³ M apoalkaline phosphatase; (b) as (a) plus 2 Eq Mg(II), 2 Eq Zn(II); (c) as (a) plus 2 Eq Mg(II), 4 Eq Zn(II).

VI. SUMMARY AND CONCLUSIONS

While the alkaline phosphatase dimer (MW ~ 86,000) is too large a protein for the application of high-resolution ¹H NMR to the determination of the solution structure, NMR of the heavier nuclei ¹³C, ¹⁹F, ³¹P, and ¹¹³Cd incorporated at strategic locations in the protein has revealed many structural feat-

ures of the molecule, especially near the active center, The
two catalytic metal ions occupy identical active sites, one on
each monomer, coordinated to three histidyl ligands as identified by characteristic metal-ion-dependent chemical shifts of
the resonances of the [γ-^{13}C]-histidyl-labeled protein. The
^{13}C chemical shifts show both Zn(II) and Cd(II) to coordinate
two imidazole rings via the π-nitrogens, while the third is
coordinated via the τ-nitrogen. Paramagnetic broadening by
Mn(II) shows a fourth histidyl to be very near the catalytic
metal ion.

^{113}Cd NMR of ^{113}Cd(II) substituted for the two catalytic
Zn(II) ions of the native enzyme shows the coordination spheres
of both ions to be identical in the unphosphorylated enzyme.
Formation of the phosphoryl enzyme results in chemical shifts
for the two ^{113}Cd(II) ions, which differ by 87 ppm. Thus phosphorylation of one site alters the other site via conformational
changes propagated across the monomer-monomer interface. This
is a direct demonstration of an asymmetric chemical change accompanying phosphorylation and is compatible with the negative
cooperativity of phosphate and substrate binding demonstrated
by this enzyme. ^{31}P NMR of the phosphoryl enzyme shows the
phosphoserine intermediate to have an unusual downfield chemical shift (8 ppm from phosphoric acid), possibly reflecting a
distorted phosphomonoester.

^{13}C NMR of the [γ-^{13}C]-histidyl-labeled protein and ^{19}F NMR
of the enzymes labeled with m-fluorotyrosyl and p-fluorophenyl-
alanyl residues shows that the two catalytic Zn(II) ions, the
two structural Zn(II) ions, and Mg(II) ions all induce significant alterations in protein structure as detected by changes in
chemical shifts of the respective resonances, which must reflect
the structural changes responsible for the modulation of catalytic activity by these ions. The binding of two structural
Cd(II) ions and Mg(II) in addition to the two catalytic Cd(II)
ions induces phosphorylation of the second active site at high
concentrations of phosphate, in contrast to the two Cd(II) enzyme, which remains negatively cooperative at all phosphate concentrations. Despite its size, a remarkably detailed picture of
the active center of alkaline phosphatase is emerging from NMR
techniques alone.

REFERENCES

Aldridge, W. N., Barman, T. E., and Gutfreund, H. (1964). *Bio-
    chem. J. 92,* 23c-25c.
Anderson, R. A., Bosron, W. F., Kennedy, F. S., and Vallee, B.
    L. (1975). *Proc. Nat. Acad. Sci. USA 72,* 2989-2993.

Applebury, M. L., and Coleman, J. E. (1969a). *J. Biol. Chem.* *244*, 308-318.

Applebury, M. L., and Coleman, J. E. (1969b). *J. Biol. Chem.* *244*, 709-718.

Applebury, M. L., Johnson, B. P., and Coleman, J. E. (1970). *J. Biol. Chem. 245*, 4968-4974.

Armitage, I. M., and Prestegard, J. M. (1978) (submitted for publication).

Armitage, I. M., Pajer, R. T., Schoot Uiterkamp, A. J. M., Chlebowski, J. F., and Coleman, J. E. (1976). *J. Am. Chem. Soc. 98*, 5710-5712.

Armitage, I. M., Schoot Uiterkamp, A. J. M., Chlebowski, J. F., and Coleman, J. E. (1978). *J. Mag. Res. 29*, 375-392.

Bale, J. R. (1978). *Fed. Proc. 37*, 1287.

Bock, J. L., and Cohn, M. (1978). *J. Biol. Chem. 253*, 4082-4085.

Bosron, W. F., Anderson, R. A., Falk, M. C., Kennedy, F. S., and Vallee, B. L. (1977). *Biochemistry 16*, 610-614.

Bränden, C. I., Eklund, H., Nordström, B., Bowie, T., Söderlund, G., Zeppezauer, E., Ohlsson, I., and Åkeson, Å. (1973). *Proc. Nat. Acad. Sci. USA 70*, 2439.

Brown, E. M., Ulmer, D. D., and Vallee, B. L. (1974). *Biochemistry 13*, 5328-5334.

Browne, D. T., and Otvos, J. D. (1976). *Biochem. Biophys. Res. Commun. 68*, 907.

Browne, D. T., Earl, E. M., and Otvos, J. D. (1976). *Biochem. Biophys. Res.Commun. 72*, 398-405.

Cardin, A. D., Bailey, D. B., and Ellis, P. D. (1978). *J. Mag. Res.* (in press).

Chappelet-Tordo, D., Iwatsubo, M., and Lazdunski (1974). *Biochemistry 13*, 3754-3762.

Chlebowski, J. F., and Coleman, J. E. (1974). *J. Biol. Chem. 249*, 7192-7202.

Chlebowski, J. F., and Coleman, J. E. (1976a). *J. Biol. Chem. 251*, 1202-1206.

Chlebowski, J. F., and Coleman, J. E. (1976b). *In* "Metal Ions in Biological Chemistry" (H. Sigel, ed.), Vol. 6, pp. 1-140. Dekker, New York.

Chlebowski, J. F., and Mabrey, S. (1977). *J. Biol. Chem. 252*, 7042-7052.

Chlebowski, J. F., Armitage, I. M., Tusa, P. P., and Coleman, J. E. (1976). *J. Biol. Chem. 251*, 1207-1216.

Chlebowski, J. F., Armitage, I. M., and Coleman, J. E. (1977). *J. Biol. Chem. 252*, 7053-7061.

Chlebowski, J. F., Armitage, I. M., Otvos, J. D., and Coleman, J. E. (1978). *J. Biol. Chem.* (in press).

Cohen, S. R., and Wilson, I. B. (1968). *Biochemistry 5*, 904-909.

Cohn, M., and Hu, A. (1978). *Proc. Nat. Acad. Sci. USA 75*, 200-203.

Coleman, J. E., and Chlebowski, J. F. (1979). *Advan. Inorg. Biochem. 1,* 1-66.
Coleman, J. E., Chlebowski, J. F., Otvos, J. D., Schoot Uiterkamp, A. J. M., and Armitage, I. M. (1978). *Trans. Am. Cryst. Assoc. 14,* 17-37.
Csopak, H. (1969). *Eur. J. Biochem. 1,* 186-192.
Csopak, H., and Folsch, G. (1970). *Acta Chem. Scand. 24,* 1025.
Daemen, J. M., and Riordan, J. F. (1974). *Biochemistry 13,* 2865-2871.
Doddrell, D., Glushko, V., and Allerhand, A. (1972). *J. Chem. Phys. 56,* 3683-3689.
Dwek, R. A. (1973). "Nuclear Magnetic Resonance in Biological Systems: Applications to Enzyme Systems." Clarendon Press, Oxford.
Eargle, D. H., Licko, V., and Kenyon, G. L. (1977). *Anal. Biochem. 81,* 186-195.
Eklund, H., Nordström, B., Zeppezauer, E., Söderland, G., Ohlsson, I., Bowie, T., and Bränden, C. I. (1974). *FEBS Lett. 44,* 200-204.
Falk, M. C., and Vallee, B. L. (1978). *Biochemistry* (in press).
Fernley, H. N., and Walker, P. G. (1969). *Biochem. J. 111,* 187-194.
Gorenstein, D. G. (1975). *J. Am. Chem. Soc. 97,* 898-901.
Gorenstein, D. G., Kar, D., Luxon, B. A., and Momii, R. K. (1976). *J. Am. Chem. Soc. 98,* 1668-1673.
Griffiths, D. W., Feeney, J., Roberts, G. C. K., and Burgen, A. S. V. (1976). *Biochim. Biophys. Acta 446,* 479-485.
Haffner, P. H., Goodsaid-Zalduondo, F., and Coleman, J. E. (1974). *J. Biol. Chem. 249,* 6693-6695.
Halford, S. E., and Schlesinger, M. J. (1974). *Biochem. J. 141,* 845-852.
Halford, S. E., Schlesinger, M. J., and Gutfreund, H. (1972). *Biochem. J. 126,* 1081-1090.
Hubbard, P. S. (1970). *J. Chem. Phys. 52,* 563-568.
Hull, W. E., and Sykes, B. D. (1974). *Biochemistry 13,* 3431-3437.
Hull, W. E., and Sykes, B. D. (1975a). *J. Chem. Phys. 63,* 867-880.
Hull, W. E., and Sykes, B. D. (1975b). *J. Mol. Biol. 98,* 121-153.
Hull, W. E., and Sykes, B. D. (1976). *Biochemistry 15,* 1535-1546.
Hull, W. E., Halford, S. E., Gutfreund, H., and Sykes, B. D. (1976). *Biochemistry 15,* 1547-1561.
Knox, J. R., and Wyckoff, H. W. (1973). *J. Mol. Biol. 74,* 533-545.
Ko, S. H. D., and Kézdy, F. J. (1967). *J. Am. Chem. 89,* 7139-7140.

Lazdunski, M. (1974). *Progr. Bioorg. Chem. 3*, 81-140.
Lazdunski, C., Petitclerc, C., Chappelet, D., and Lazdunski, M. (1969). *Biochem. Biophys. Res. Commun. 37*, 744-749.
Nakamura, K.-I., Chlebowski, J. F., and Coleman, J. E. (1978) (in press).
Oldfield, E., Norton, R. S., and Allerhand, A. (1975). *J. Biol. Chem. 250*, 6368-6380.
Otvos, J. D., Coleman, J. E., and Armitage, I. M. (1978). *J. Biol. Chem.* (in press).
Reid, T. W., and Wilson, I. B. (1971). *Enzymes 4*, 373-415.
Schoot Uiterkamp, A. J. M., Armitage, I. M., and Coleman, J. E. (1978). *J. Biol. Chem.* (in press).
Schwartz, J. H. (1963). *Proc. Nat. Acad. Sci. USA 49*, 861-878.
Simpson, R. T., and Vallee, B. L. (1968). *Biochemistry 7*, 4343-4350.
Simpson, R. T., and Vallee, B. L. (1970). *Biochemistry 9*, 953-958.
Tait, G. H., and Vallee, B. L. (1966). *Proc. Nat. Acad. Sci. USA 56*, 1247-1251.
Taylor, J. S., and Coleman, J. E. (1972). *Proc. Nat. Acad. Sci. USA 69*, 859-862.
Taylor, J. S., Lau, C. Y., Applebury, M. L., and Coleman, J. E. (1973). *J. Biol. Chem. 248*, 6216-6220.
Trentham, D. R., and Gutfreund, H. (1968). *Biochem. J. 106*, 455-460.
Weiner, R., Chlebowski, J. F., Armitage, I. M., and Coleman, J. E. (1978). *J. Biol. Chem.* (in press).
Wilson, I. B., Dayan, J., and Cyr, K. (1964). *J. Biol. Chem. 239*, 4182-4185.
Woessner, D. E. (1962). *J. Chem. Phys. 36*, 1-4.

APPENDIX: EFFECTS OF CHEMICAL EXCHANGE ON ^{113}Cd RESONANCES

Most spectroscopists are familiar with the range of rates
and rearrangements that can be studied by NMR. This range is
dictated by the difference in chemical shifts and relaxation
times of the nuclei in the coexisting states. For typical diamagnetic systems, kinetic processes in the range $10^{-1}-10^3$ sec^{-1}
can have an effect on $\underline{T}_2$ relaxation. This range of rates can
be extended by one to two orders of magnitude if the chemical
shift difference $|\omega_{o(A)} - \omega_{o(B)}|$ between exchanging species can
be increased. An example of this is the case in which ligand
exchange involves a site containing a paramagnetic ion. Another
example arises in the NMR studies of metal nuclei, since the extreme sensitivity of the chemical shift of these nuclei to small
changes in electron configuration leads to a large range of
chemical shift values. This becomes important when considering
chemical exchange phenomena for the ^{113}Cd nucleus.

The chemical shift range for common compounds of cadmium is
over 600 ppm. In the ^{113}Cd(II)-substituted metalloenzymes that
we have studied thus far, where the protein ligands are nitrogen and/or oxygen, the observed chemical shifts span approximately 200 ppm. Within a given enzyme a chemical shift range
of approximately 100 ppm for the Cd nucleus in two coexisting
states can be observed (see Section IV).

For a two-site model for exchange between sites or configurations of equal population, Fig. A1 shows the computed spectra
for a range of τ values for the case where $|\omega_{o(A)} - \omega_{o(B)}|$ is
equal to 100 ppm (see legend for details). This exchange pro-
cess may be representative of a fluctuation in the ligand geometry at the metal binding site, ligand exchange, or exchange
of the central metal ion. This latter mechanism is perhaps the
least favorable as a result of the very slow ($<10^{-2}$ sec^{-1}) dissociation rate constant and rapid association rate constant
($>10^4$ sec^{-1}) for metal ions in metalloenzymes. Thus with stoichiometric amounts of metal ion to enzyme, the dominant species
at equilibrium will be enzyme-bound metal ion. Under conditions
where the populations of two exchanging states are not equal,
the maximal amount of broadening of the dominant resonance decreases as the population of the minor component decreases (see
below). Thus this mechanism predicts very little broadening of
the dominant resonance when the population of the minor species
(free metal ion) is negligible. Thus, if we assume a chemical
shift difference of 100 ppm, an exchange of the central metal
ion will not in general explain the failure to detect ^{113}Cd
resonances as is frequently observed for Cd-substituted metalloenzymes. If we consider either of the two more likely exchange
mechanisms, intramolecular rearrangement or ligand exchange, it
is apparent from Fig. A1 that for species lifetimes in the
range $10^{-4} \leq \tau \leq 10^{-2}$ sec significant broadening of the resonance will be observed.

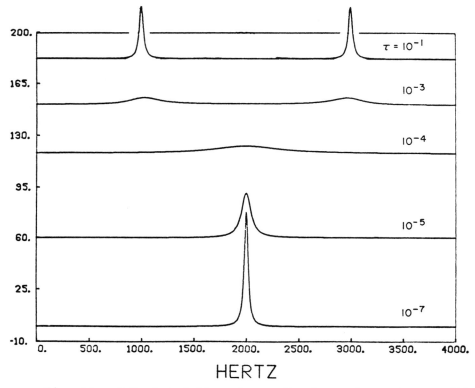

Fig. A1. Calculated line shapes for two-site exchange with a separation in resonance frequency of 2000 Hz and a natural linewidth of 35 Hz. τ values represent the lifetime of the dominant species. Populations in the ratio 50:50.

One need not assume equal populations of exchanging states. The equilibrium constants governing accessible conformers of protein macromolecules may differ from unity significantly. We have therefore calculated the line shapes for the [113]Cd case given above when the populations of exchanging species are 70:30 (Fig. A2) and 90:10 (Fig. A3). This model is sufficient to explain why the minor resonance can elude detection with existing instrumental sensitivity. Nevertheless a 10% fractional population under these conditions can give rise to an increase in linewidth by a factor of ~7 for the dominant resonance when τ values are ~10^−3 sec. Increase in linewidth by a factor of 7' is frequently sufficient to make the observation of the major resonance unlikely with present instrumentation. The unequal population model also predicts much smaller changes in the

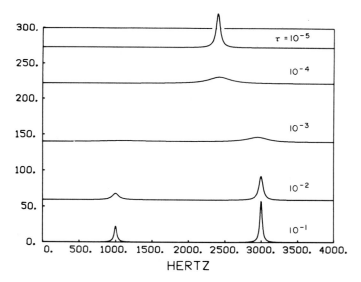

Fig. A2.   Same as Fig. A1 with ratio 70:30.

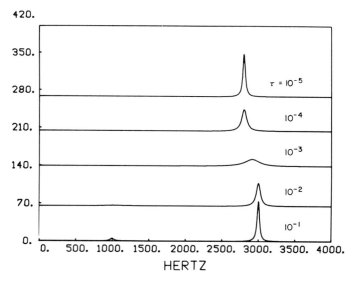

Fig. A3.   Same as Fig. A1 with populations in the ratio
90:10.

chemical shift of the dominant resonance than expected from the
chemical shift difference between the resonances involved.
Thus, one would predict that processes modulating the cadmium
chemical shift will give rise to a clearly observable resonance
only when the site lifetime is in the range $10^{-5} \geq \tau \geq 10^{-2}$ sec.

Exchange broadening of this type is observed for the second
pair of ^{113}Cd(II) ions added to alkaline phosphatase (see Fig.
16). It is seen in even more dramatic fashion in the various
isozymes of carbonic anhydrase where ligand exchange rates vary
greatly between isozymes and are apparently related to the turnover rate of the enzyme (Schoot Uiterkamp et al., 1978). In a
number of other instances, however, reasonably sharp resonances
are observed for ^{113}Cd(II) ions in metalloenzymes, yet the amplitude of the resonance (taking account of relative relaxation
rates) does not appear to account for all the ^{113}Cd(II) even
though by all other analytical criteria the binding sites are
fully occupied. The resonances of the unphosphorylated
^{113}Cd(II)$_2$AP is an example (see Fig. 15). We believe this must
represent a complex equilibrium between several conformers of
the protein that have different chemical shifts for the resonance of the central ^{113}Cd(II) ion. Such equilibria may involve
exchange of monodentate ligands from solution, exchange of a
protein ligand for solvent, or more complex fluctuations in the
conformation of the whole molecule or sections of it. If such
conformational isomerism gives rise to more than two interconverting species, the effects on the shape and amplitude of the
^{113}Cd(II) resonances become complex and depend on the relative
rates of interconversion and the equilibrium concentrations of
the species. An example of the computer simulation of a three-
site equilibrium in which conversion between two of them (1 to
2) is slow, while interchange between the two others (2 and 3)
is fast or intermediate on the chemical shift time scale is
shown in Fig. A4. If the intermediate time scale pertains, then
a sharp but reduced resonance will be observed for site 1 (the
dominant resonance), while resonances for sites 2 and 3 are not
detected because of exchange broadening. Conformational equilibria of this type are one explanation for the varying amplitudes of the dominant ^{113}Cd resonances we have observed in several metalloenzymes, alkaline phosphatase being only one example.
Once the chemical shift difference between the resonances of
protein conformers becomes great enough for the resonances to
be sensitive to rate processes on the time scale of 10^2 to
10^5 sec^{-1}, then dynamic changes in protein structure will certainly influence the resonances. The required chemical shift
range as a function of changing protein conformation is clearly
present in the case of ^{113}Cd(II), since the conformational
change associated with phosphorylation of alkaline phosphatase

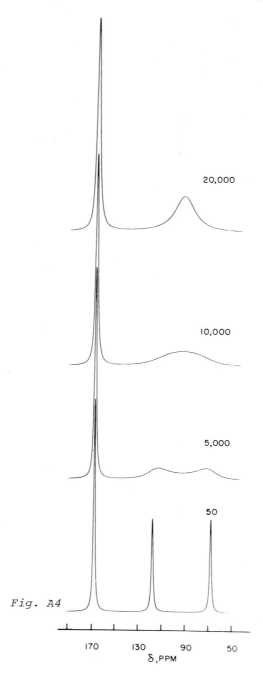

20,000

10,000

5,000

50

Fig. A4

170   130   90   50

δ,PPM

induces a chemical shift change of nearly 100 ppm (see Fig. 15 of text). While independent structural techniques will be required to identify the precise nature of such conformational changes, [113]Cd(II) NMR is potentially remarkably sensitive to the existence of such equilibria.

Fig. A4. [113]Cd(II) resonances expected for a system exchanging between three conformers as a function of the rates of interconversion between them. The spectra have been calculated using the Kubo-Sack matrix approach. The populations of the three states are assumed to be 50% for 1 (the lowfield resonance), 25% for 2 (the middle resonance) and 25% for 3 (the upfield resonance). At slow rates of exchange between all three, ~50 sec^{-1}, three separate resonances are observed (lowest spectrum). The upper three spectra were generated by leaving a slow rate of interconversion, 40 to 60 sec^{-1}, between conformers 1 and 2 or 1 and 3, and increasing the rate of conversion between 2 and 3 to 10^3 sec^{-1} or higher. At intermediate rates, resonances from conformers 2 and 3 are broadened such that they may not be detected at present sensitivity, leaving a dominant narrow resonance representing 50% of the total [113]Cd resonance. The overall rate given on the spectra is a composite number reflecting both the interconversion rates and populations. For the spectrum labeled 10,000, the individual rate constants are 50 sec^{-1} for the interconversion of 1 and 2 or 1 and 3, and ~2500 sec^{-1} between 2 and 3.

CATALYTIC GROUPS OF SERINE PROTEINASES
NMR INVESTIGATIONS

John L. Markley

Biochemistry Division
Department of Chemistry
Purdue University
West Lafayette, Indiana

I. INTRODUCTION

A. *Scope of the Chapter*

Serine proteinases and their inhibitors have been the subjects of a great number of magnetic resonance experiments. This review focuses on structural studies of catalytic groups of the serine proteinases themselves, touches only lightly on studies of inhibitor binding to serine proteinases, and does not include the interesting NMR studies of protein proteinase inhibitors. For the benefit of the reader unfamiliar with the field of serine proteinases, I begin with a brief summary of their properties. A thorough description of the subject is provided in recent reviews (Bender and Killheffer, 1973; Huber and Bode, 1978; Kraut, 1977; Stroud *et al.*, 1977). I then discuss in turn the various serine proteinases that have been investigated by nuclear magnetic resonance spectroscopy. The final section considers general conclusions and unanswered questions.

B. *Historical Perspective*

Serine proteinases comprise an important and ubiquitous class of enzymes. These hydrolases of peptide and ester bonds contain a highly reactive serine residue that is required for enzyme activity. If the serine is blocked chemically, for example, by reaction with diisopropylphosphorofluoridate to form

397

the diisopropylphophoserine derivative (Jansen *et al.*, 1949),
all enzyme activity is lost. Serine proteinases have been
found in virtually all organisms from bacteria to man. In mam-
mals this class of enzyme plays roles as diverse as digestion
(trypsin, chymotrypsin, elastase), blood clotting (thrombin),
complement fixation, and cell fertilization (acrosin) and
growth (see, for example, the articles in Reich *et al.*, 1975).
The first use of the word Enzym was made by Kühne (1877) over
100 years ago in a paper on trypsin. Since then several funda-
mentals of enzymology have emerged in part from studies of serine proteinases. These include the concepts of an active site
comprised of groups involved in covalent catalysis and other
groups involved in specific substrate binding, control of enzyme activity by naturally occurring inhibitors, and activation
and deactivation of enzyme activity by limited proteolysis.

There are two well-known families of serine proteinases:
those related to chymotrypsinogen and those related to subtilisin. The sequences of members of each subfamily are homologous; that is, they can be aligned so that critical residues
fall in the same positions, allowing for gaps and insertions
(De Haen *et al.*, 1975; James *et al.*, 1978). Although members
of the two different subfamilies are not homologous, there is
ample evidence that they share a common mechanism of catalysis
(Kraut, 1977).

C. *Structural Features of Serine Proteinases*

A large number of serine proteinases have been sequenced,
and this remains a fruitful area of research. The sequence provides the fundamental information about each protein and allows
it to be classified according to evolutionary type and mode of
activation (De Haen *et al.*, 1975; James *et al.*, 1978). A number of serine proteinases have been studied in exquisite detail
by X-ray crystallography. These include different forms of
chymotrypsin (Matthews *et al.*, 1967; Cohen *et al.*, 1969; Birktoft and Blow, 1972; Tulinsky *et al.*, 1973; Vandlen and Tulinsky, 1973), chymotrypsinogen (Freer *et al.*, 1970), bovine trypsin (Stroud *et al.*, 1974; Krieger *et al.*, 1974; Fehlhammer and
Bode, 1975; Bode and Schwager, 1975; Bode *et al.*, 1976a), bovine trypsinogen (Bode *et al.*, 1976b; Fehlhammer *et al.*, 1977;
Kossiakoff *et al.*, 1977), the complex between porcine trypsin
and soybean trypsin inhibitor (Kunitz) (Blow *et al.*, 1974;
Sweet *et al.*, 1974), the complex between bovine trypsin and bovine pancreatic trypsin inhibitor (Kunitz) (Rühlmann *et al.*,
1973; Huber *et al.*, 1974, 1975), the complex between bovine
trypsinogen and bovine pancreatic trypsin inhibitor (Kunitz)
(Bode *et al.*, 1978), elastase (Watson *et al.*, 1970; Schotton
and Watson, 1970; Sawyer *et al.*, 1978), proteinase B (Codding

et al., 1974; Delbaere et al., 1975) and proteinase A (see James et al., 1978) from *Streptomyces griseus*, α-lytic proteinase from myxobacter 495 (see James et al., 1978), and subtilisins (Wright et al., 1969; Alden et al., 1970; Drenth et al., 1972). In addition, the binding of small-molecule inhibitors and substrates and the effects of chemical modification have been investigated by difference Fourier analysis (for references, see the review articles cited above). The three-dimensional structures have led to insights concerning mechanisms of catalysis and inhibition, zymogen activation, and enzyme specificity.

D. Mechanisms of Catalysis

The reaction sequence of serine proteinases is understood in great detail. It has been postulated to consist of the symmetrical series of steps shown in Fig. 1 (Kraut, 1977).

The key role of serine-195 of chymotrypsin as a nucleophile in the attack of substrates was elucidated by studies of inhibitors such as diisopropylphosphorofluoridate mentioned above. The identity of the reactive serine was established by enzymatic cleavage and isolation of the organic phosphorus labeled peptide (Oosterbaan et al., 1955). The catalytic function of the critical serine was confirmed by trapping an acetyl enzyme intermediate of substrate hydrolysis (Oosterbaan and van Adrichem, 1958). Enzymatic degradation of the acetyl enzyme established that serine-195 in the chymotrypsinogen sequence (Hartley, 1964; Hartley and Kauffman, 1966; Meloun et al.,

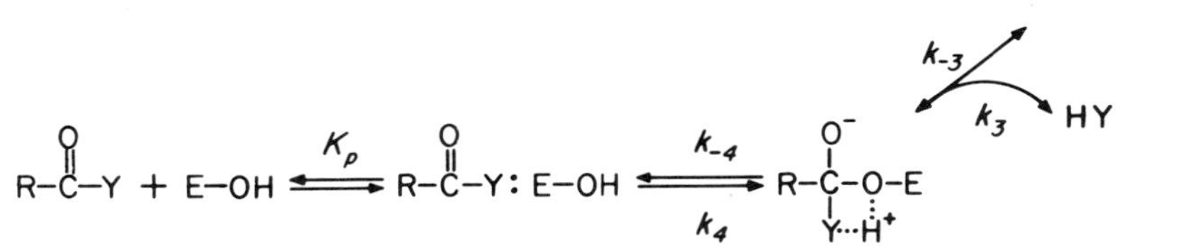

Fig. 1. Reaction sequence postulated for serine proteinases. (From Kraut, 1977.)

1966) is the acylated group.[1] Chemical modification by photo-
oxidation (Weil *et al.*, 1953; Ray and Koshland, 1960) and by
acylation with site-specific reagents (Schoellman and Shaw,
1962; Ong *et al.*, 1964) demonstrated that a histidine is also
required for enzyme activity. A critical histidine residue had
been suspected on the basis of pH-activity profiles that reveal
that serine proteinases are activated with a pK_a around 7 (see
Hess, 1971).

The first high-resolution X-ray structure of a serine proteinase, that of tosyl chymotrypsin (Matthews *et al.*, 1967),
demonstrated that the side chains of histidine-57 and serine-
195 are adjacent in the active site and further that the side
chain of residue 102 is hydrogen bonded to the imidazole of his-
tidine-57 (Blow *et al.*, 1969). The discovery of these three
residues led to the important "charge relay" hypothesis concerning their function.

Blow and co-workers (1969) correctly predicted that the original sequences of chymotrypsin and trypsin contained errors
and that residue 102 is an aspartate rather than an asparigine.
As originally formulated (Blow *et al.*, 1969), the charge relay
was envisioned as a mechanism for the generation of an alkoxide
ion at serine-195, which would explain its high reactivity:

$$(1)$$

---

[1]*The chymotrypsinogen numbering system is used throughout
to specify residues in this and related serine proteinases.*

The triad of catalytic side chains has come to be referred to as the "charge relay system." Kraut (1977) recently has suggested that the term charge relay system be used in a more limited sense to describe the structure that binds the proton that is shuttled back and forth between the reactive serine and the leaving or entering group in the transition states.

The charge relay hypothesis generated a large number of both theoretical and experimental studies that tested the ideas involved. The existence of an alkoxide ion in solution as shown in Eq. 1 is unlikely because of the extremely high pK_a' of a serine hydroxyl. The original charge relay hypothesis has been modified to include a concerted transfer of two protons, which accompanies attack of the substrate by the serine-195 (Hunkapiller $et\ al.$, 1973):

$$
\begin{array}{c}
\text{His}^{57} \qquad\qquad\qquad \text{Ser}^{195}\\[4pt]
\text{Asp}^{102} - \text{C} \overset{O}{\diagup} \quad \overset{\text{His}^{57}}{\bigg\langle}\qquad\qquad \overset{\text{Ser}^{195}}{|}\\
\text{O} \cdots \text{H-N} \quad\text{N} \cdots \text{H} - \text{O}\quad\text{C} = \text{O}\\
\text{H-N}\\
R_2
\end{array}
$$

(2)

$$
\begin{array}{c}
\text{His}^{57} \qquad\qquad \text{Ser}^{195}\\
\text{Asp}^{102} - \text{C} \overset{O}{\diagup}\\
\text{OH} \cdots \text{N} \quad\text{N-H} \cdots \text{O} - \text{C} - \text{O}^-\\
\text{HN}\\
R_2
\end{array}
$$

Here the essential feature is the relay of charge from the histidine imidazole to the aspartate. Proponents of this hypothesis have emphasized that the proton transfer from histidine-57 to aspartate-102 is important for catalysis because it provides a significant decrease in the free energy of the transition state (Hunkapiller $et\ al.$, 1973; Koeppe and Stroud, 1976). The idea is that the free energy of a neutral histidine-aspartate is lower than that of a histidine cation-aspartate

anion pair given their environment in the enzyme. This conclusion now appears unlikely since there are additional experimental data discussed below that indicate that the pK_a of his-
tidine-57 is greater than that of aspartate-102 and remains so
in the transition state.

The major specificity of the reaction is determined by
residues of the proteinases that make up the binding pocket for
the side chain of the substrate residue that will become the
new carboxyl terminal after hydrolysis. Both the size of the
pocket and its charge are important. For example, chymotrypsin
differs from trypsin in that residue 189 at the bottom of the
substrate specificity pocket is a neutral serine instead of a
negatively charged aspartate (Hartley, 1970). Chymotrypsin has
a specificity for large hydrophobic side chains on the residue
preceding the cleavage point, whereas trypsin is specific for
the positively charged residues arginine and lysine. In elastase, the specificity pocket is occluded by the bulky groups
valine-216 and threonine-226 (Schotton and Watson, 1970),
whereas in chymotrypsin and trypsin these residues are glycines.

Other residues are thought to participate in catalysis.
These include the peptide N-H groups of glycine-193 and serine-
195, which are thought to stabilize the oxyanions generated in
the two tetrahedral transition states (Robertus et al., 1972)
and the groups that bind the peptide chain in subsites on both
sides of the cleavage site (see Kraut, 1977).

E. Precursor Molecules and Zymogen Activation

Many mammalian serine proteinases are produced by cells as
inactive precursors called preenzymes or zymogens. The zymogens are converted to activated enzymes by limited proteolysis
near the amino terminal. The activation peptide may be lost
as in the case of trypsin or may be retained by a disulfide
linkage as in the case of π-chymotrypsin.

One of the goals in the field of serine proteinases has
been to determine the mechanism of zymogen activation: how a
modification as parsimonious as cleavage of a single peptide
bond can increase the catalytic activity a million-fold. The
picture provided by the first comparison of X-ray structures of
a zymogen (chymotrypsinogen; Freer et al., 1970; Robertus et
al., 1972) and its activated enzyme (chymotrypsin; Matthews et
al., 1967) indicated that the zymogen lacks both a developed
binding pocket and the correct orientation of one of the hydrogen bonding groups (glycine-193) that stabilizes the oxyanion
transition state.

These changes also have been reported to accompany activation of trypsinogen to trypsin (Kossiakoff et al., 1977; Fehlhammer et al., 1977; Bode et al., 1978). The above studies in-

dicated no significant differences in the positioning of the
catalytic triad (serine-195, histidine-57, aspartate-102) between enzymes and zymogens. This view has been questioned recently since studies of refined X-ray structures indicate that
the hydrogen bond between the serine-195 O$^\gamma$ and the histidine-
57 N$^{\varepsilon 2}$ is strained or more probably absent in enzymes but present in zymogens (Matthews *et al.*, 1977; Birktoft *et al.*, 1976).

An additional hypothesis concerning zymogen activation
holds that zymogens have a flexible structural domain that becomes rigid on activation. The idea has arisen from "blurring"
observed in high-resolution X-ray diffraction electron density
maps of trypsinogen (Felhammer *et al.*, 1977). However, similar
"blurring" was not reported for the earlier X-ray structure of
chymotrypsinogen (Freer *et al.*, 1970).

The key event in the zymogen-to-enzyme conversion in trypsin, chymotrypsin, and related proteinases is the cleavage of
a single peptide bond. This generates a new amino terminal
with an IleVal or a ValVal- sequence (Zwilling *et al.*, 1975).
The new positively-charged amino terminus forms an ionic interaction with the carboxylate of aspartate-194 adjacent to the
active site. This interaction triggers a reorganization of hydrogen bonds, leading to a conformational change and the generation of increased activity. Hence, trypsinogen and similar
zymogens are regulated enzymes that possess a built-in effector
in the form of the -LysIleVal- sequence (Huber and Bode, 1978),
which is activated by limited proteolysis of the -LysIle- bond.

It has been shown that exogenous isoleucylvaline can bring
about the same conformational change under certain conditions.
There appears to be some kind of communication between the iso-
leucylvalyl binding site and the specificity pocket, since the
conformational change is observed only if the specificity pocket
is occupied (Bode and Huber, 1976; Bode *et al.*, 1978). Chymotrypsin loses activity if the critical amino terminal is blocked
by acetylation (Oppenheimer *et al.*, 1966; Ghelis *et al.*, 1967).
The inactivation of chymotrypsin at high pH is thought to be a
consequence of the deprotonation of the amino group of the amino
terminal isoleucine-16, which would abolish its interaction with
aspartate-194 (Himoe *et al.*, 1967).

The bacterial serine proteinase α-lytic proteinase does not
contain the normal IleVal or ValVal amino terminal (Olson *et
al.*, 1970). The critical interaction with aspartate-194 is provided instead by the side chain of arginine-138 (James *et al.*,
1978). No zymogen has been found for α-lytic proteinase or related bacterial serine proteinases.

II. NUCLEAR MAGNETIC RESONANCE STUDIES

A. *Chymotrypsinogen Family*

1. *α-Lytic Proteinase*

The enzyme α-lytic proteinase from the soil bacterium *Sorangium sp.* (Myxobacter 495) has the distinction of being the serine proteinase investigated by the widest range of NMR experiments: ^{1}H-NMR in ^{1}H$_2$O (Robillard and Shulman, 1974a; Westler and Markley, 1979) and ^{2}H$_2$O (Westler and Markley, 1979), ^{13}C-NMR (Hunkapiller et al., 1973; 1975) and ^{15}N-NMR (Bachovchin and Roberts, 1978) of enriched analogs, and ^{31}P-NMR of the diisopropylphosphoryl derivative (Porubcan et al., 1979).

The first NMR study of the enzyme was the elegant selective ^{13}C enrichment experiment of Hunkapiller et al. (1973). They chose α-lytic proteinase for their study of the active site histidine because this enzyme has only one histidine residue (obviating assignment problems), is stable toward self-proteolysis (permitting long-term signal averaging), and is of bacterial origin and can be produced in high yield (facilitating isotopic substitution). L-Histidine labeled with 90% ^{13}C in $C^{\varepsilon 1}$ was synthesized and fed to the bacteria that produce the enzyme.[2] Undecoupled ^{13}C-NMR spectra of α-lytic proteinase labeled at the $C^{\varepsilon 1}$ position of histidine-57 and of the unlabeled control are shown in Fig. 2a,b. The difference spectrum (Fig. 2c) clearly shows the presence of a signal from the labeled carbon $C^{\varepsilon 1}$ split to a doublet by the directly bonded $C^{\varepsilon 1}$-H. The spectrum provides information in the form of the carbon chemical shift, the $^1J_{CH}$ coupling constant, and the line width. The pH dependence of the chemical shift gave a calculated pK_a' of 6.8 (Fig. 3). The titration curve is similar to that observed for histidine $C^{\varepsilon 1}$

---

[2] *Because of differences in the numbering of the histidine ring in the chemical and biochemical literature, the crystallographic convention is utilized here: $C^{\varepsilon 1}$ is $C(2)$ in the usual biochemical convention; $N^{\delta 1}$ is N_π; and $N^{\varepsilon 2}$ is N_τ. Abbreviations used are DFP, diisopropylphosphorofluoridate; DIP, diisopropylphosphoryl; BCtg, bovine chymotrypsinogen A; BCtr, bovine chymotrypsin A; α-LP, α-lytic proteinase; TMS, $(CH_3)_4Si$; DSS, 3-trimethyl-silyl-1 propanesulfonic acid sodium salt (2,2-dimethyl-2-silapentane-5-sulfonate). The symbol pH* is used to indicate the uncorrected pH meter reading of ^{2}H$_2$O solutions obtained in ^{1}H$_2$O. Notation used follows the "Recommendations for the Presentation of NMR Data for Publication in Chemical Journals," Pure Appl. Chem. 45, 219 (1976).*

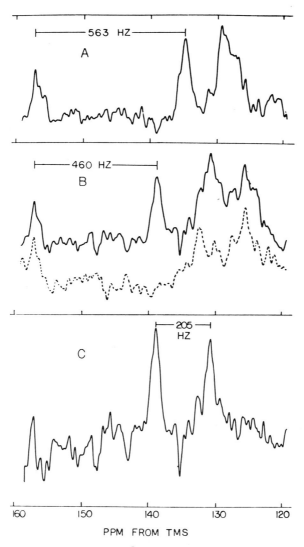

Fig. 2. Measurement of $^1J_{CH}$ for the histidine $^{13}C^{\varepsilon}1-^1H$ of α-lytic proteinase at pH 5.98: (a) Proton-decoupled ^{13}C NMR spectrum selectively ^{13}C enriched α-lytic proteinase, 50,000 transients; (b) (——) proton-decoupled ^{13}C-NMR spectrum of natural abundance α-lytic proteinase, 250,000 transients; (c) difference spectrum obtained by computer subtraction of the natural abundance spectrum (- - -) of (b) from the ^{13}C enriched spectrum (-) of (a). Enzyme samples were 5-6 mM in 0.2 M KCL in a 12 mm NMR tube, 34°C. (From Hunkapiller et al., 1973.)

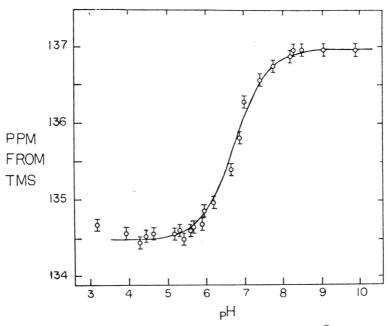

Fig. 3. Chemical shift of the histidine $^{13}C^{\varepsilon 1}$ NMR peak of α-lytic proteinase as a function of pH. The solid line is a theoretical titration curve calculated using a pK_a' of 6.75 and Δδ of 2.46 ppm. (From Hunkapiller et al., 1973.)

in peptides and proteins. Hunkapiller *et al.* (1973) discovered, however, that the pH dependence of $^1J_{CH}$ did not fit that expected for a histidine with a pK_a' of 6.8. Their studies of imidazole in model compounds demonstrated that the magnitude of $^1J_{CH}$ depends on the protonation state of the ring. The $^1J_{CH}$ for the cation falls in the range 218-222 Hz, whereas that for the neutral imidazole falls in the range 204-209 Hz. These values hold for 4-methylimidazole in both dioxane and water; hence the coupling constant appears not to be affected by the dielectric properties of the solvent. The values of $^1J_{CH}$ for the enzyme at three pH values are shown in Table I. At pH 8.2, $^1J_{CH}$ is 205 ± 3 Hz as expected for a neutral imidazole. However, at pH 5.2, which is well below the pK_a' derived from the chemical shift (Fig. 3), $^1J_{CH}$ is still 205 ± 3 Hz. At pH 3.3 Hunkapiller *et al.* (1973) found three pairs of peaks, two representing charged species ($^1J_{CH}$ = 222 ± 3 and 218 ± 3 Hz) and one representing an uncharged species ($^1J_{CH}$ = 208 ± 3 Hz). The peak having chemical shift 134.05 ppm ($^1J_{CH}$ = 222 Hz) was sharper than the others and was assigned to denatured protein.

TABLE I.   Chemical shift and coupling constant values of the histidine $C^{\varepsilon}1$ of α-lytic proteinase.[a]

pH	^{13}C NMR chemical shift ppm (±0.12) from Me_4Si	$^1J_{CH}$ Hz (±3)
8.2	136.95	205
5.2	134.57	205
3.3[b]	134.81	208
	134.05	222
	132.46	218

[a]*From Hunkapiller et al. (1973).*
[b]*Three doublets were found at this pH.*

Since the other two peaks had similar areas, they concluded that the pK_a' of histidine-57 is near 3.3. The inflection in the chemical shift with the pK_a' of 6.8 was attributed to an electrostatic effect attending the protonation of aspartate-102.

As critics of this experiment have pointed out (Robillard and Shulman, 1974a; Egan et al., 1977) the ^{13}C chemical shift may be, in this case, a more reliable indicator of the charge on histidine-57 than the coupling constant. Precise measurement of $^1J_{CH}$ was hampered by the background of natural abundance resonances and by the large line widths of the components of the histidine doublet (around 30 Hz) compared to the small change reported in the coupling constant (13 Hz).

Hunkapiller et al. (1973) attempted unsuccessfully to resolve a low-field ^1H-NMR peak in solutions of α-lytic proteinase in ^1H_2O analogous to the N-H peak discovered earlier by Robillard and Shulman (1972) and assigned to the proton hydrogen bonded between histidine-57 and aspartate-102. Robillard and Shulman (1974a) subsequently published a spectrum of the low-field N-H of α-lytic proteinase at pH 3.0 and compared it with the low-field peaks of other serine proteinases (Fig. 4). Since a single peak was observed at pH 3.0 with a chemical shift similar to that expected for a protonated histidine (Robillard and Shulman, 1972; Patel et al, 1972; Griffin et al., 1973), Robillard and Shulman proposed that the simplest explanation of the data would be a pK_a' of histidine-57 well above 3 (Robillard and Shulman, 1974a). We recently reinvestigated the low-field N-H of α-lytic proteinase at 3°C using the "2-1-4" pulse sequence (Redfield et al., 1975) and found that the peak observed

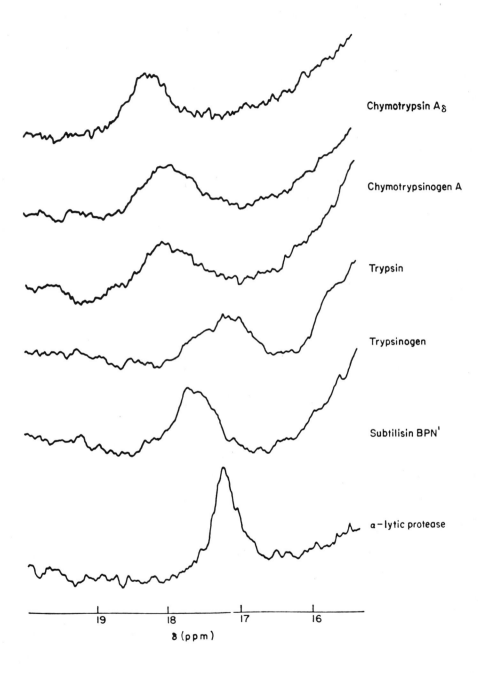

Chymotrypsin A$_\delta$

Chymotrypsinogen A

Trypsin

Trypsinogen

Subtilisin BPN'

α-lytic protease

19   18   17   16

δ (ppm)

by Robillard and Shulman (1974a) titrates with pH with a pK_a' around 6.9 (Westler and Markley, 1079). This titration curve shown in Fig. 5 is similar to that reported for chymotrypsin (Robillard and Shulman, 1972, 1974a) and discussed in the next section. According to the interpretation of Robillard and Shulman (1972) the inflection is to be assigned to histidine-57. The absence of a perturbation at low pH suggests that the pK_a' of aspartate-102 is less than 3.0. Curiously, the low field NH peak is present with enzyme samples that have been lyophilized at pH 4 but not with samples lyophilized at pH 6 (both procedures yield fully active enzyme). This may explain the disagreement between Hunkapiller et al. (1973) and Robillard and Shulman (1974a).

In an attempt to resolve the conflict between the ^{13}C-NMR (Hunkapiller et al., 1973) and low-field ^1H-NMR (Robillard and Shulman, 1972, 1974a) studies, we also have investigated the ^1H-NMR resonance of the $C^{\varepsilon 1}$-H of histidine-57 of α-lytic proteinase in ^2H_2O (Westler and Markley, 1078, 1979).

^1H-NMR titration studies of the model compounds, $N^{\delta 1}$-CH_3 histidine (I) and $N^{\delta 2}$-CH_3 histidine (II) (Westler and Markley, 1979), lead to the following conclusions concerning the $C^{\varepsilon 1}$-H ^1H-NMR chemical shift:

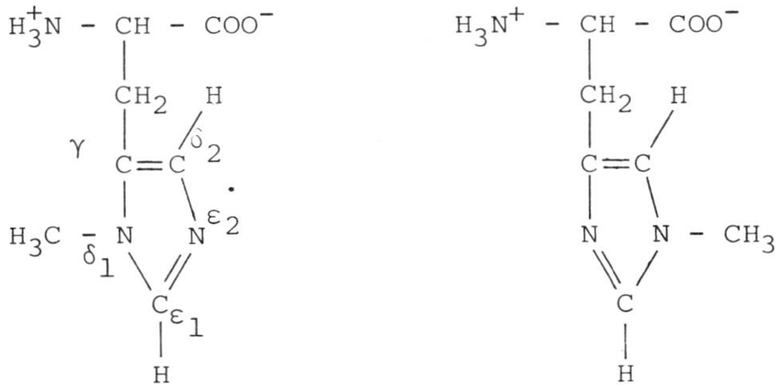

(I)   (II)

Fig. 4.   220 MHz ^1H NMR spectra of enzyme samples in ^1H_2O solutions. The concentration of enzyme in each sample was 3 to 4.5 m$\underline{M}$. All spectra were recorded at 3°C and are an average of 500 scans each. The specific data for each enzyme sample are as follows: bovine chymotrypsin, in 0.25 $\underline{M}$ NaCl, pH 3.5; bovine chymotrypsinogen, in 0.25 $\underline{M}$ NaCl, pH 3.2; bovine trypsin in 0.1 $\underline{M}$ CaCl$_2$, pH 3.6; bovine trypsinogen, in 0.1 $\underline{M}$ CaCl$_2$, pH 0.3; subtilisin, in 0.1 $\underline{M}$ N-acetyl-L-tryptophan, pH 6.0; α-lytic proteinase, in 0.001 $\underline{M}$ HCl, pH 3.0. (From Robillard and Shulman, 1974a.)

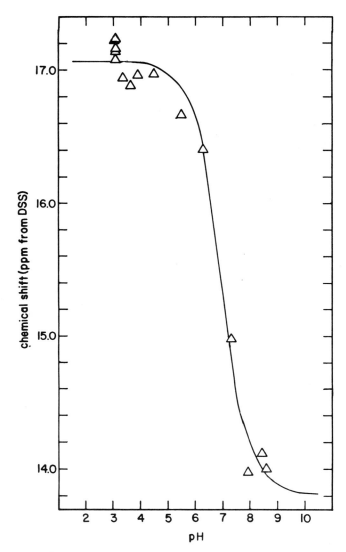

Fig. 5. ^1H-NMR titration curve (at 360 MHz) of the low-field N-H peak of α-lytic proteinase in ^1H_2O. Experimental points were obtained with 2 m$\underline{M}$ α-lytic proteinase 90/10 (v/v) $^1H_2O/^2H_2O$, 3^OC using the 2-1-$\overline{4}$ pulse sequence (Redfield et al., 1975) with 250-500 transients. The solid line is a theoretical titration curve with pK_a' 6.9. This peak is observed with samples that have been lyophilized at pH 4 but not with samples that have been lyophilized at pH 6. The titration curve is not reversible in that the peak disappears when the pH is lowered from 8 to 3.

(i) The chemical shift of the $C^{\varepsilon 1}$-H is perturbed less than 0.1 ppm by titration of either the amino or carboxyl groups of (I) and (II). (ii) The chemical shift of the $C^{\varepsilon 1}$-H depends to a small extent on the tautomeric form of the imidazole. The chemical shift of the $C^{\varepsilon 1}$-H of (I) is 0.10 ppm downfield from that of (II). The imidazole titration shift is 1.04 ppm for (I) and 1.07 ppm for (II). (iii) The imidazole pK_a' values are 6.8 for (I) and 6.1 for (II) (Reynolds et al., 1973; Westler and Markley, 1979). It appears safe to assume that the methyl group resembles a proton closely enough (Reynolds et al., 1973) so that these results may be used to predict the titration curve of histidine-57 based on the pK_a' values of Hunkapiller et al. (1973) (Model A: $pK_{a\ Asp}' = 6.8$; $pK_{a\ His}' = 3.3$), or based on the low-field N-H results of Fig. 5 (Model B: $pK_{a\ His}' = 6.9$; $pK_{a\ Asp}' < 3.0$). The calculated titration curves are shown in Fig. 6. Model A neglects the influence of the change in charge of aspartate-102, which is expected to be small. Proton chemical shifts are sensitive, in addition, to the presence of nearby anisotropic groups (such as aromatic rings) and to changes in local dielectric constant. However,

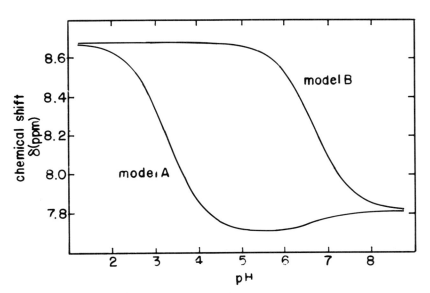

Fig. 6. Two models showing ^{1}H NMR titration curves predicted for the histidine C^{1}-H peak of α-lytic proteinase: model A is based on a pK_a' of 3.3 for histidine-57 and a pK_a' of 6.8 for aspartate-102; model B is based on a pK_a' of 6.9 for histidine-57. The predicted chemical shifts were derived from pH titration studies of N-methyl histidines. (From Westler and Markley, 1979.)

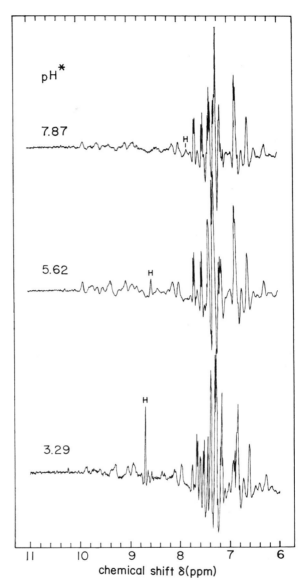

Fig. 7. Resolution-enhanced pulse-Fourier-transformed 360 MHz ^1H NMR spectra of the low-field region of pre-exchanged α-lytic proteinase. Chemical shifts are measured from internal 2,2-dimethyl-2-silapentane sulfonic acid sodium salt. Resolution enhancement to remove broad peaks was carried out in the computer by multiplication of the first 200 data points in the time domain with an increasing linear function ranging from zero to one. The intensity of peak H is decreased as it broadens

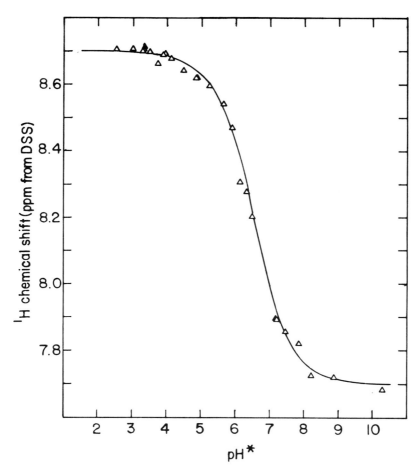

Fig. 8. ^1H NMR titration curve (at 360 MHz) of the histidine $C^{\varepsilon 1}$-H peak of α-lytic proteinase. The triangles are the experimental points, and the solid line is a theoretical curve with pK_a' value of 6.5 ± 0.1. (From Westler and Markley, 1979.)

at high pH as a consequence of the mathematical manipulation of the data to increase resolution. pH* is the meter reading obtained in ^2H_2O solutions from a glass electrode standardized in H_2O buffers. Sample conditions: 30 mg protein/0.5 ml of 0.2 M KCl in ^2H_2O, 30°C. (Westler and Markley, unpublished.)

the above environmental factors are assumed to remain constant between pH 5 and 10.5 since there are no changes in ORD spectra of α-lytic proteinase in this range (Patterson and Whitaker, 1969).

The aromatic region of the ^{1}H-NMR spectrum of α-lytic proteinase is shown in Fig. 7. The titrating peak labeled H has been resolved in samples that have been lyophilized from ^{2}H$_2$O and dissolved in ^{2}H$_2$O. Resolution is improved by prior exchange of the protein in 3 M guanidine hydrochloride in ^{2}H$_2$O followed by ultrafiltration to desalt the protein (Westler and Markley, 1979) or simply by an incubation for one hour at pH 10.5 in ^{2}H$_2$O followed by lyophilization at pH 4 (Westler and Markley, 1979). None of these procedures degrades the specific activity of the enzyme. The pH dependence of peak H is given in Fig. 8. The curve may be fit to a single pK_a' of 6.5.[3] A minor perturbation of the peak may be present at low pH around 4.2.

The titration curve of Fig. 8 resembles that of Model B but not that of Model A of Fig. 6. Hence we conclude that the pK_a' of histidine-57 is 6.5 and is higher than that of aspartate-102. In order to fit the ^{1}H-NMR results to the pK_a' values of Hunkapiller et al. (1973), one would need to invoke (i) a major change in the environment of histidine-57 at pH 6.5 that is not expected from solution and X-ray studies of α-lytic proteinase and related serine proteinases (Paterson and Whitaker, 1969; Vandlen and Tulinsky, 1973) and (ii) environmental effects at low pH that deshield the $C^{\epsilon 1}$-H of the neutral histidine but not the $C^{\epsilon 1}$-H of the positively charged histidine so that the histidine titration with pK_a' 3.3 becomes "silent."

Hunkapiller et al. (1973) reported three species at pH 3.3. In some samples we observe two histidine peaks at low pH (separated by 0.05 ppm at pH 3.3). One state is favored by preincubation of the enzyme at low pH and the other by preincubation at high pH. The environment of histidine-57 appears to be slightly different in the two states as evidenced by the small difference in the chemical shift, but the pK_a' value of histidine-57 is nearly the same for both states (Westler and Markley, 1979).

Bachovchin and Roberts (1978) have carried out a definitive ^{15}N-NMR study of histidine-57 of α-lytic proteinase. Two derivatives were prepared by growing histidine auxotrophs of Myxobacter on defined media containing either [^{15}N$^{\delta}$1]-L-histidine or [^{15}N$^{\delta}$1,^{15}N$^{\epsilon}$2]-L-histidine. The spectra of the derivatives

---

[3]The current pK_a' value of 6.5 (Westler and Markley, 1978b) is higher than that reported earlier in an abstract (Westler and Markley, 1979); in the previous study the pH electrode was calibrated with a buffer solution later found to be one pH unit below the nominal value of 7.00.

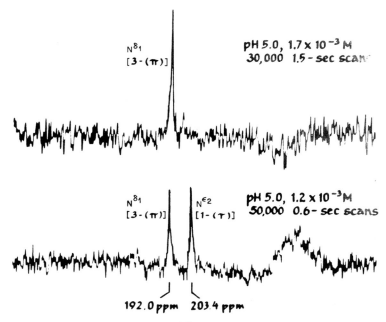

Fig. 9. Proton-coupled ^{15}N NMR spectra, at 18.2 MHz, 26°C, of α-lytic proteinase analogs produced by a histidine auxotroph of Myxobacter 495. Top spectrum: α-lytic proteinase obtained from a culture grown on [99% $^{15}N^{\delta}1$]-L-histidine. Bottom spectrum: α-lytic proteinase obtained from a culture grown on [95% $^{15}N^{\delta}1$, 95% $^{15}N^{\epsilon}2$]-L-histidine. The chemical shift values are referenced to external HNO_3. (From Bachovchin and Roberts, 1978.)

are shown in Fig. 9, and the titration curves in Fig. 10. The fact that the titration shift is larger for $^{15}N^{\epsilon}2$ than for $^{15}N^{\delta}1$ indicates that a proton is being added to the $N^{\epsilon}2$ with a pK_a' of around 7.0 ± 0.1 at 26°C. The $N^{\epsilon}2$ peak could not be resolved at pH values between 6.5 and 9.5.

Results of ^{31}P-NMR studies of the diisopropylphosphoryl derivatives of α-lytic proteinase (Porubcan et al., 1979) are shown in Fig. 11. The inflection with pK_a' 7.9 is assigned to protonation of histidine-57 since the pK_a agrees with that (7.8) obtained from the ^{1}H-NMR titration curve of the $C^{\epsilon}1$-H of histidine-57 of the same derivative (also given in Fig. 11).

The ^{1}H-, ^{15}N-, and ^{31}P-NMR studies appear to be in general agreement that the pK_a of histidine-57 must be higher than that of aspartate-102. The difference between the pK_a' value of 7.0 obtained in the ^{15}N-NMR experiment (Bachovchin and Roberts,

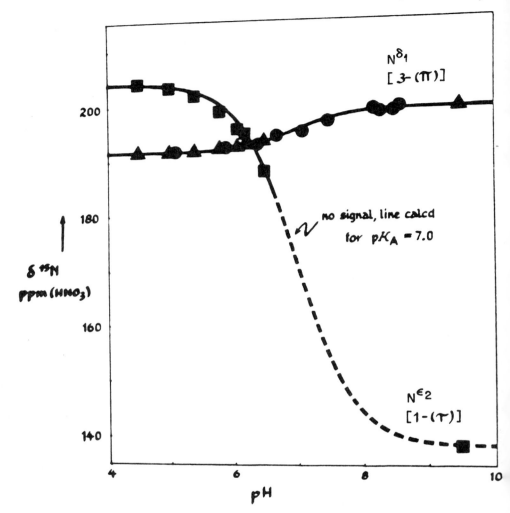

Fig. 10. Dependence of the ^{15}N NMR chemical shifts of the ^{15}N-enriched histidine nitrogens in α-lytic proteinase (see Fig. 9 for spectra) as a function of pH. ●, ^{15}N enriched at N^δ (N_3 or π); ▲, ■, enriched at $N^\delta 1$ and $N^\epsilon 2$ (N_1 or τ), respectively. (From Bachovchin and Roberts, 1978.)

1978) and the pK'_a value of 6.5 obtained in the ^1H-NMR study of the histidine $C^{\epsilon 1}$-H (Westler and Markley, 1979) may be attributed to differences in experimental conditions. The former experiment was carried out in 0.1 M KCl in ^1H_2O at 26°C, and the latter in 0.2 M KCl in ^2H_2O at 30°C. The pK'_a of histidine-57 is expected to be temperature dependent with the pK'_a increas-

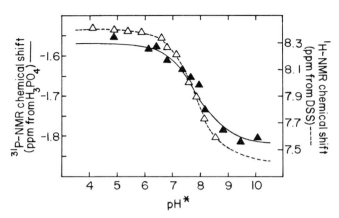

Fig. 11. Titration curves of diisopropylphosphoryl α-lytic proteinase. △ ^1H NMR peaks (360 MHz) assigned to the histidine $C^\varepsilon l$-H; (---) theoretical titration curve with pK'_a 8.1 ± 0.1. ▲ ^{31}P-NMR peaks (40.5 MHz) assigned to the trialkylphosphate; (-) theoretical titration curve with pK'_a 7.9 (From Westler and Markley, 1979.)

ing with decreasing temperature (Hanai, 1976). Using the thermodynamic parameters of Hanai (1976) obtained for chymotrypsin, if the pK_a of histidine-57 is 6.5 at 30°C, the pK'_a should be around 7.1 at 3°C. This is close to the value of 6.9 obtained from the titration curve of the low-field N-H at 3° in H_2O (Westler and Markley, 1979).

Hunkapiller et al. (1975) made use of α-lytic proteinase enriched with ^{13}C in the $C^\varepsilon l$-H of histidine-57 to investigate the effect of binding a peptide aldehyde inhibitor, N-acetyl-L-Ala-L-Pro-L-Alaninal. In the complex, the lower pK'_a of the catalytic triad obtained from $^1J_{CH}$ was greater than 4.6 as compared to 3.3 in the free enzyme (Hunkapiller et al., 1973). The chemical shift of the $C^\varepsilon l$ of histidine-57 showed a cooperative transition with a pH_{mid} of 6.25. The curve was purported to represent two interacting groups with pK'_a values 7.0 and 5.0. A cooperative curve can be obtained, however, only if two or more groups having similar pK'_a values interact.

The chemical shift data probably indicate that the pK'_a of one of the components of the catalytic triad (histidine-57) is lowered from the value of 6.8 in the free enzyme to pH_{mid} value around 6.25 in the complex. The second pH_{mid} near 6.25 could correspond to protonation of the anion of the inhibitor hemiacetal.

2. Chymotrypsin and Chymotrypsinogen

Chymotrypsin and chymotrypsinogen contain two histidine residues. In addition to histidine-57 at the catalytic site, a histidine located at position 40 plays a role in zymogen activation (Freer et al., 1970; Kraut, 1971; Wright, 1973). In the zymogen, histidine-40 is hydrogen bonded to aspartate-194, which is adjacent to the active site (Freer et al., 1970). Upon activation, the newly formed α-amino group of isoleucine-16 displaces histidine-40 and forms a salt linkage with aspartate-194, which is considered critical for maintaining the enzyme in its active configuration. Histidine-40 changes its orientation and hydrogen bonds instead to the carbonyl of glycine-193 (Matthews et al., 1967).

Robillard and Shulman (1972, 1974a) carried out extensive studies of a low-field titrating, N-H peak observable at low temperature (3-5°C) in solutions of chymotrypsinogen and chymotrypsin in ^1H_2O Their spectra and titration curves are reproduced in Fig. 12. A continuous titration curve was observed with the zymogen. However, with the enzyme, it was not possible to resolve the peak within the main pH transition, and a smaller transition appeared at pH 5.5. The exchangeable nature of the peak was established by showing that it is absent in samples dissolved in ^2H_2O. The N-H peak was assigned by chemical modification to the proton hydrogen-bonded between histidine-57 and aspartate-102, and the assignment was substantiated by inhibitor binding experiments (Robillard and Shulman, 1974b). Histidine-57 was alkylated by a chloromethyl ketone reagent forming z-Gly-Leu-Phe-chymotrypsin A. As shown in Fig. 13, the pK_a of the transition is raised from 7.5 to 8.4, and its magnitude is much smaller than expected for deprotonation of the histidine. Robillard and Shulman (1974a) suggested that the transition at pH 8.4 represents dissociation of a tetrahedral intermediate complex stable at lower pH in which there is a covalent bond between the O^γ of serine-195 and the peptide carbonyl of the inhibitor.

The low-field N-H was not detected with solutions of tosyl chymotrypsin A_δ (Robillard and Shulman, 1974a), perhaps because the aspartate-102-histidine-57 hydrogen bond is weakened in that derivative (Sigler et al., 1968; Henderson et al., 1971).

Fig. 12. (a) 220 MHz ^1H NMR spectra of the low-field N-H peak of bovine chymotrypsin A_δ in ^1H_2O. Enzyme concentration was 3 to 4 mM in buffered solutions of ionic strength 0.5, 3°C. (b) Comparison of the titration curves of the low-field N-H of chymotrypsin A_δ (□) shown above and of bovine chymotrypsinogen A (×). The dashed line represents a theoretical titration curve with a pK_a' of 7.5. (From Robillard and Shulman, 1974a.)

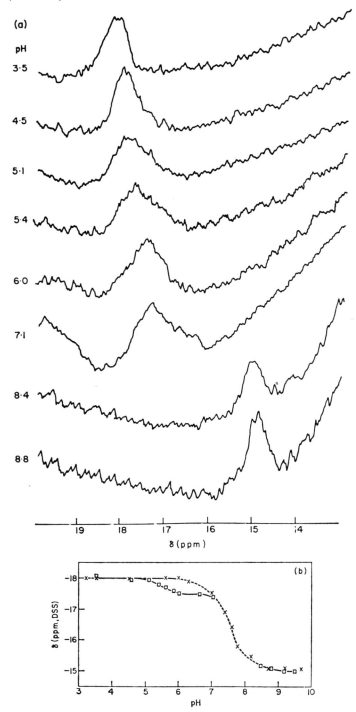

(a)

pH

3·5

4·5

5·1

5·4

6·0

7·1

8·4

8·8

δ (ppm)

(b)

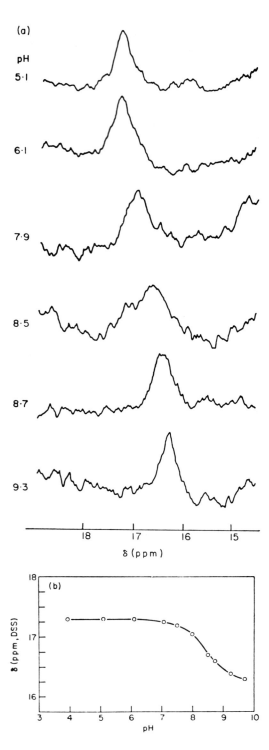

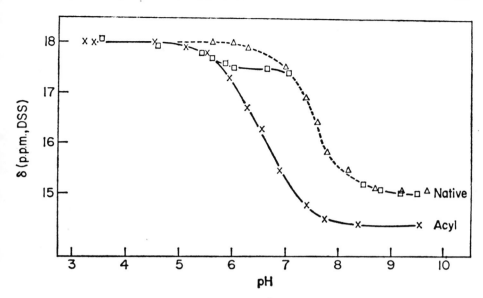

Fig. 14. A comparison of the ^1H NMR titration curves for the low-field N-H peaks of bovine chymotrypsinogen A ( ) and chymotrypsin A$_\delta$ ( ) and the acylated derivative N^2-acetylalanyl-N^1-benzoylcarbamyol-chymotrypsin A$_\delta$. (From Robillard and Shulman, 1974a.)

Another acyl enzyme derivative of serine-195 was found that yielded the low-field peak: the titration curve of N-(N-acetylalanyl)-N-benzoylcarbamoyl-chymotrypsin A$_\delta$ is compared with the titration curves of the native enzyme and zymogen in Fig. 14. The pK_a' of the major transition is lowered from 7.5 to 6.5, and the minor transition at pH 5.5 is absent in the acyl derivative. There can be no histidine-57-serine-195 hydrogen bond in acyl chymotrypsin. Hence the difference in chemical shift of 0.6 ppm between the N-H of the native and acyl enzyme at pH 9.5 (Fig. 14) was taken as an indication that the hydrogen bond is present in the native enzyme and further that its presence adds a small positive charge to histidine-57 and a corresponding negative charge to O$^\gamma$ of serine-195. Robillard and Shulman (1972, 1974a) reached the following conclusions concerning the

Fig. 13. (a) 220 MHz ^1H NMR spectra at 16°C of the alkylated derivative z-Gly Leu Phe chymotrypsin A. Enzyme concentrations were approximately 2 mM in buffered solutions of ionic strength 0.4. (b) Titration curve of the resonances presented in (a). (From Robillard and Shulman, 1974a.)

active site: (i) the hydrogen-bonded triad exists and is similar in both the zymogen and enzyme; (ii) the pK_a' of histidine-
57 is 7.5 in both enzyme and zymogen; (iii) histidine-57 and
serine-195 are only slightly polarized, in contrast with the
view that the charge relay produces appreciable amounts of alkoxide ion (Blow *et al.*, 1969).

In subsequent papers, Robillard and Shulman (1973, 1974b)
extended their investigations to enzyme inhibitor complexes.
They found that binding of the following inhibitors interferes
with the normal titration of histidine-57: N-acetyl-L-tryptophan, 2-phenylethylborate, benzene boronic acid, and bovine
pancreatic trypsin inhibitor (Kunitz) (BPTI).   The pH-independent chemical shifts of the low-field N-H in these complexes
are given in Table II.   The X-ray structures of subtilisin:aro-
matic boronic acid complexes have been elucidated (Matthews *et
al.*, 1975).   Based on these, Robillard and Shulman (1974b) proposed the model shown in Fig. 15 to explain these NMR results.
They established credence for this model by showing that the
two aromatic benzene boronic acid subtilisin complexes have
identical pH-independent chemical shifts (17.2 ppm), which are
the same as that of the 2-phenylethyl boronic acid complex of
chymotrypsin A_δ.   The fact that the chemical shift of the low-
field N-H is significantly different in the aromatic boronic
acid complexes as compared to the complex with BPTI is important.   Recently it has been pointed out that histidine-57 has
a different orientation in the subtilisin aromatic boronic com-

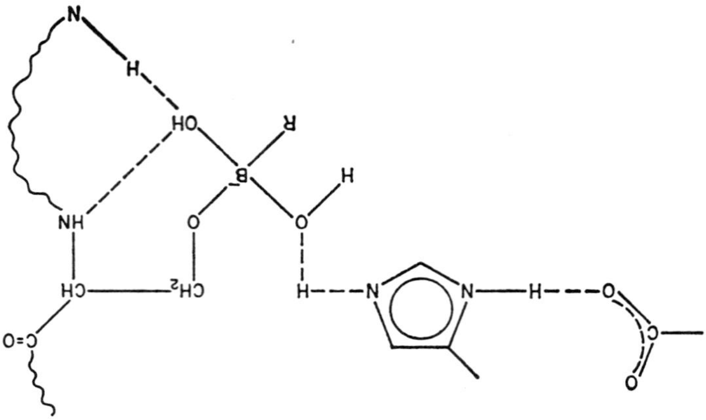

*Fig. 15.   Schematic representation of the active site re-
gion of chymotrypsin complexed with boronate inhibitors based
on a combination of NMR and X-ray crystallographic data.   (From
Robillard and Shulman, 1974b.)*

TABLE II.   *Chemical shifts of the low-field N-H assigned
to the proton hydrogen-bonded between aspartate-102 and histi-
dine-57 in complexes of chymotrypsin A$_\delta$ and chymotrypsinogen A
with various inhibitors*[a]

Inhibitor	Chemical shift, ppm from internal DSS
Chymotrypsin A$_\delta$	
N-acetyl-L-tryptophan	17.5
borate	15.9
benzene boronic acid	16.3
2-phenylethyl boronic acid	17.2
bovine pancreatic trypsin inhibitor (Kunitz)	14.9
Chymotrypsinogen A	
borate	16.0
benzene boronic acid	16.1
2-phenylethyl boronic acid	16.1

[a]*From Robillard and Shulman (1974b).*

plexes, where the $N^{\epsilon 2}$ is directed toward the location of the
substrate's leaving group (Matthews *et al.*, 1975, 1977), and in
the bovine trypsin:BPTI complex, where the $N^{\epsilon 2}$ is directed toward the Ser-195 O^γ (Rühlmann *et al.*, 1973). The chemical shift
change may reflect this different orientation. Since the chemical shift of the low-field N-H in the chymotrypsin A$_\delta$:BPTI complex is similar to that of free chymotrypsin A$_\delta$ at high pH, Robillard and Shulman (1974b) concluded that histidine-57 is
neutral in the complex.

Another NMR approach to the active site of chymotrypsin and
chymotrypsinogen has involved investigations of the histidine
$C^{\epsilon 1}$-H by ^{1}H-NMR in ^{2}H$_2$O. Early attempts to resolve the histidine $C^{\epsilon 1}$-H peaks of chymotrypsin and chymotrypsinogen were un-
successful (Bradbury and Wilairet, 1967; Bradbury *et al.*, 1971)
probably because of interference from slowly exchangeable N-H
resonances. The N-Hs that obscure the histidine peaks may be
removed by pre-exchange in ^{2}H$_2$O (Ibañez *et al.*, 1976) as shown
in Fig. 16. Two peaks are resolved that titrate with pH as
shown in Fig. 17 (Markley and Ibañez, 1978). In contrast with
the studies of the low-field N-H of these proteins (Robillard
and Shulman, 1972, 1974a), it was possible to take the solutions
to low pH and follow additional transitions. Since signals are
present from both histidines in the molecule one obtains information about both histidine-57 and histidine-40. The latter is

also of interest because of its role in zymogen activation.
However, before this information can be utilized one must assign the individual NMR peaks to the respective histidines.

The peak assignments of chymotrypsinogen (Fig. 17a) were
based on ^{1}H-NMR spectra (Markley and Ibañez, 1978) and later
confirmed by ^{31}P-NMR spectra (Porubcan et al., 1979) of the DIP
derivative. Reeck et al. (1977) demonstrated that one can resolve the ^{31}P-NMR signals from DIP-chymotrypsinogen and DIP-
chymotrypsin (Fig. 18). We recently found that the ^{31}P chemical shift is pH dependent (Porubcan et al., 1979) and yields
a titration curve that resembles the ^{1}H-NMR titration curve of
one of the histidine $C^{\varepsilon}1$-^{1}H peaks of DIP-chymotrypsinogen (Markley and Ibañez, 1978).[4] The two curves are compared in Fig. 19.
The ^{31}P peak has inflections at 3.3 and 7.5, and the ^{1}H peak
has inflections at 3.1 and 7.6. As shown in Fig. 20, the
$C^{\varepsilon}1$-H peak that has transitions matching the ^{31}P peak is altered in the DIP derivative, whereas the other $C^{\varepsilon}1$-H peak follows that of the native zymogen. Based on the X-ray structure
of DIP-trypsin (Stroud et al., 1974; see also Kraut, 1977), the
phosphate is expected to be within hydrogen-bonding distance of
histidine-57 but considerably farther away from histidine-40.
Hence we reach the following conclusions: (i) The inflections
in the ^{31}P peak reflect changes occurring at histidine-57.
(ii) Peak DIP-BCtg-H1 (Figs. 19 and 20) corresponds to histi-
dine-57 of the DIP derivative. (iii) Peak BCtg-H1 corresponds
to histidine-57 of the native zymogen. (iv) By difference,
peaks BCtg-H2 and DIP-BCtg-H2 correspond, respectively, to his-
tidine-40 of native and DIP-chymotrypsinogen. (v) From the direction and magnitude of the ^{1}H-NMR titration shifts, the higher pK_a' affecting histidine-57 in both native and DIP-chymotrypsinogen is assigned to protonation of the histidine itself, and
the lower pK_a' is assigned to an environmental perturbation probably related to protonation of aspartate-102 (Porubcan et al.,
1978; Markley and Ibañez, 1978).

[4]D. D. Mueller (personal communication) and co-workers have
observed the pH dependence of these peaks.

Fig. 16. Comparison of 250 MHz correlation ^{1}H-NMR spectra
of bovine chymotrypsinogen A before (spectrum a) and after
(spectra b-d) exchange procedures (see text) simplify the histidine $C^{\varepsilon}1$-H region of the spectrum and permit resolution of
peaks BCtg-H1 and BCtg-H2 corresponding to individual histidine
residues. Sample conditions: 25 mg protein/0.5 ml 0.5 M KCl
in [2]H_2O containing 6 M guanidinium chloride at pH* 3.0 for 12
hours at 25°C and then desalted. (d) Sample first exchanged
as in spectrum b by heating in [2]H_2O, then incubated in 2 M guanidinium chloride in [2]H_2O at pH* 8.0 for 21 days at 25°C, and
finally desalted. (From Markley and Ibañez, 1978.)

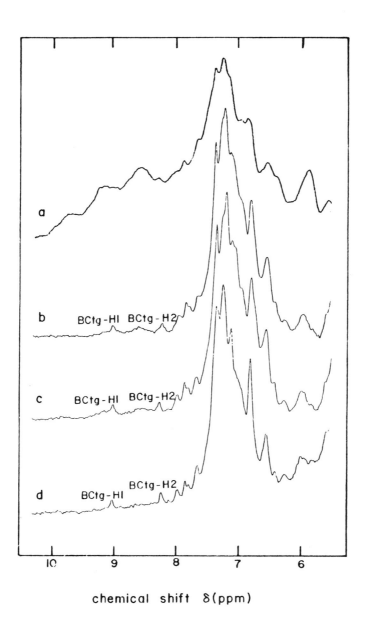

Fig. 16

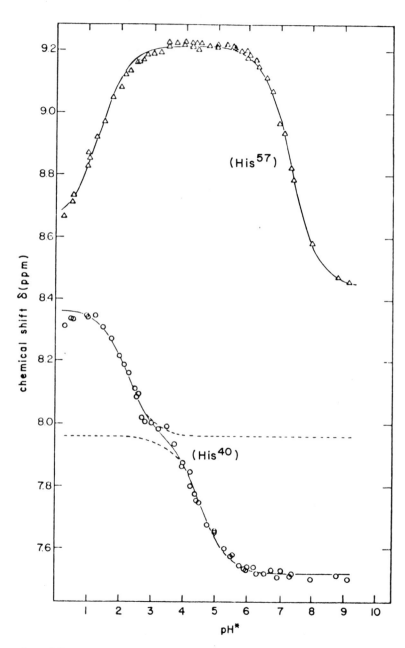

Fig. 17a

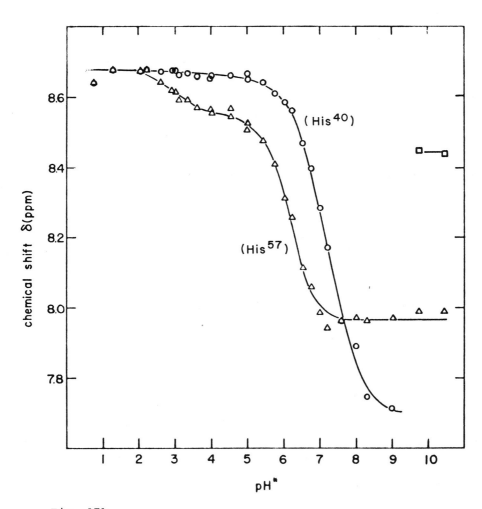

Fig. 17b

Fig. 17. (a) ^{1}H-NMR titration curves (at 250 MHz) of histidine $C^{\varepsilon}1$-H peaks of bovine chymotrypsinogen A, BCtg (pre-exch). Both peaks BCtg-H1 ($\triangle$) assigned to histidine-57 and BCtg-H2 ($\bigcirc$) are affected by two transitions. (b) ^{1}H-NMR titration curves (at 250 MHz) of the histidine $C^{\varepsilon}1$-H peaks of bovine chymotrypsin A_{α}, $BCtr_{\alpha}$ (pre-exch). Peak $BCtr_{\alpha}$-H1 ($\bigcirc$) assigned to histidine-40 is affected by a single transition; peak $BCtr_{\delta}$-H2 ($\triangle$) assigned to histidine-57 is affected by two transitions.

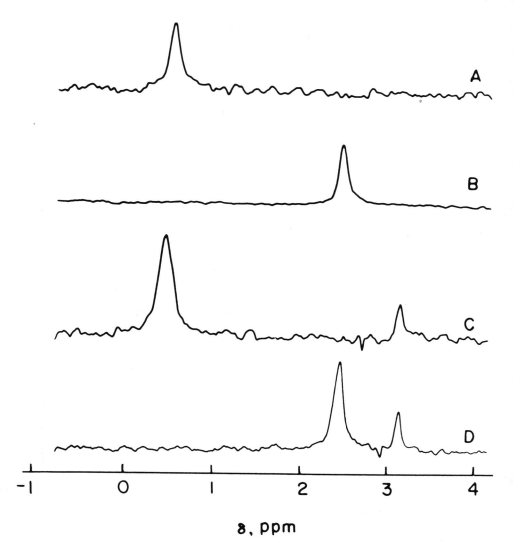

δ, ppm

Fig. 18. Proton noise-decoupled ^{31}P NMR spectra (40.5 MHz) of the phosphorus peaks of bovine diisopropylphosphoryl-chymotrypsin A_α and diisopropylphosphoryl-chymotrypsinogen A in 0.25 M KCl, 0.01 M Tris-Cl, pH 8.0 (at 25°C). (A) Diisopropylphosphoryl-chymotrypsin A_α, 10.4 mg/ml, 32000 pulses; (B) diisopropylphosphoryl chymotrypsinogen A, 54.5 mg/ml, 8000 pulses; (C) trypsin-treated diisopropylphosphoryl-chymotrypsinogen 10.7 mg/ ml (total), 48,000 pulses; and (D) diisopropylphosphoryl-neochymotrypsinogen produced by the reaction of TLCK-treated chymotrypsin A_α on diisopropylphosphoryl-chymotrypsinogen, 20.3 mg/ml, 16,000 pulses. (From Reeck et al., 1977.)

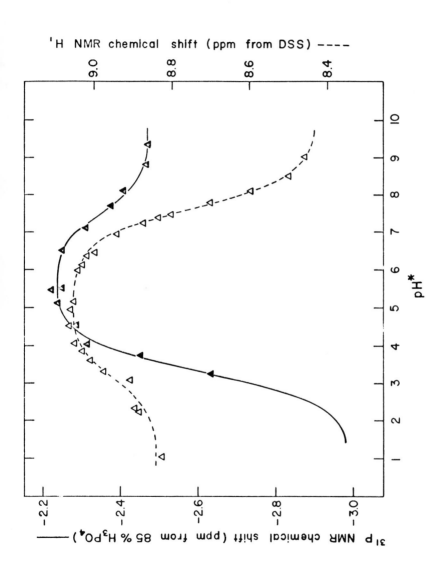

Fig. 19. Titration curves of bovine diisopropylphosphoryl-chymotrypsinogen A. (△) [1H] NMR peaks (250 MHz) assigned to the histidine-57 $C^{\epsilon 1}$-H; (---) theoretical titration curve with pK_a' 3.3 and 7.5; (▲) [31P] NMR peaks (40.5 MHz) assigned to the trialkylphosphate ester; (-) theoretical titration curve with pK_a' 3.3 and 7.5. (From Porubcan et al., 1979.)

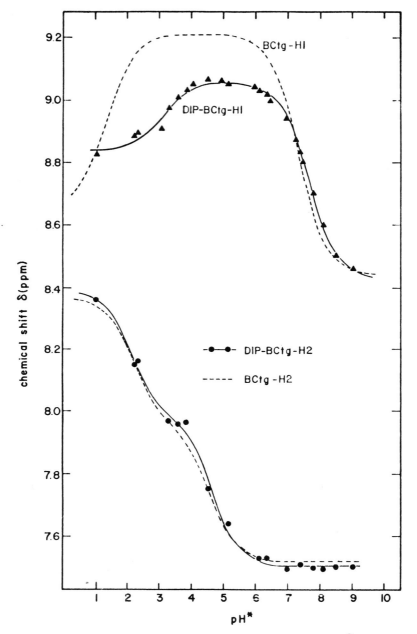

Fig. 20. The pH dependence of the histidyl $C^{\epsilon 1}$-H ^{1}H-NMR peaks (at 250 MHz) of diisopropylphosphoryl bovine chymotrypsinogen A. Titration curves of free chymotrypsinogen A (---) are shown for comparison. Sample conditions: 25 mg protein /0.5 ml 0.5 $\underline{M}$ KCl in ^2H_2O, 31°C. (From Markley and Ibañez, 1978.)

The peak assignments of chymotrypsin A_α (Markley and Ibañez, 1978)[5] were based on ^{1}H-NMR titration curves of the BPTI complex (Fig. 21). In this complex one $C^\epsilon 1$-H peak (BCtr$_\alpha$:BPTI-Ha) was found to titrate as in the native enzyme, and the other (BCTr$_\alpha$: BPTI-Hb) gave a pH-independent chemical shift from pH 4 to 8.5 (Ibañez et al., 1976; Markley and Ibanez, 1978). The positions of histidine-40 and histidine-57 in the chymotrypsin:BPTI complex should be similar to the corresponding residues in the trypsin:BPTI complex, whose structure is known from X-ray diffraction analysis (Rühlmann et al., 1973). In the latter complex, histidine-57 lies in the contact region between the two proteins and has both ring nitrogens hydrogen-bonded, whereas the ring of histidine-40 lies just beyond the contact region and appears unaffected by complexation. Hence the nontitrating peak in Fig. 21 is assigned to histidine-57 and the normally titrating peak to histidine-40. Below pH 3 the nontitrating peak disappears and a new peak appears, which follows the low pH limb of the titration curve of histidine-57. This result is consistent with the well-known low pH transition that weakens the binding constant of the complex (Engel et al., 1974). The nontitration of histidine-57 in this complex is also expected from Robillard and Shulman's studies (1974b) of the low-field N-H of this complex.

Because the two inflections that affect the $C^\epsilon 1$-H of histidine-57 have the same sense, the assignment of one of these to the protonation of histidine-57 is less certain than in the case of the zymogen, where the inflection that ran counter to a normal histidine titration shift could be eliminated as an external environmental perturbation. We chose to assign the larger titration shift (higher pK_a') to the protonation of histidine-57 itself (Markley and Ibañez, 1978) for the reasons outlined earlier in the discussion of α-lytic proteinase. This choice appears to be substantiated by the ^{15}N results with α-lytic proteinase (Bachovchin and Roberts, 1978). The lower pK_a' was assigned to an external perturbation, probably the protonation of aspartate-102.

---

[5]The early ^{1}H NMR studies of chymotrypsin were carried out using the rapid trypsin activation product of chymotrypsinogen at room temperature, and it was assumed that the histidine $C^\epsilon 1$-H peaks corresponded to chymotrypsin A_δ (Ibañez et al., 1976). Recent NMR studies of homogeneous chymotrypsin A_α and chymotrypsin A_δ indicate that the histidine peaks are visible with the former but not the latter form of the enzyme (Markley and Ibañez, 1978). The failure to resolve the peaks with chymotrypsin A_δ probably indicates exchange broadening, although the mechanism is not understood at present.

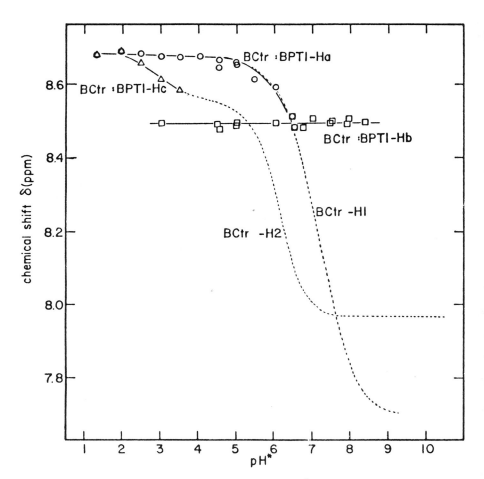

Fig. 21. The pH dependence of the ^{1}H-NMR peaks (at 250 MHz) assigned to the two histidyl residues of bovine chymotrypsin A_α in the complex with bovine pancreatic trypsin inhibitor. Titration curves of the histidyl $C^\varepsilon 1$-H peaks in bovine chymotrypsin A_δ alone (---). Because of interfering peaks from the complex, peak BCtr:BPTI-Ha could not be traced at chemical shifts upfield of $\delta = 8.4$. (From Markley and Ibañez, 1978.)

The assignments and NMR titration parameters of chymotrypsinogen and chymotrypsin are summarized in Table III. Unlike the results of the low-field N-H study (Robillard and Shulman, 1972, 1974a), the ^{1}H-NMR (Markley and Ibañez, 1978) and ^{31}P-NMR (Porubcan *et al.*, 1979) data indicate that there are major dif-

TABLE III. Transitions affecting the histidine $C^{\varepsilon 1}$-H NMR peaks of bovine chymotrypsinogen A and chymotrypsin A_α[a]

Histidine NMR Peak	Residue assignments	pK_a' value	NMR chemical shift (ppm from DSS)		
			δ low pH	δ high pH	$\Delta\delta$ low-high
Chymotrypsinogen A					
BCtg-H1[b]	Asp[102]	1.36 ± 0.03	8.65 ± 0.01	0.21 ± 0.03	-0.56 ± 0.04
	His[57]	7.33 ± 0.02	9.21 ± 0.03	9.44 ± 0.01	0.77 ± 0.04
BCtg-H2[b]	Asp[194] [c]	2.29 ± 0.06	8.35 ± 0.01	7.96 ± 0.01	0.39 ± 0.02
	His[40]	4.57 ± 0.04	7.95 ± 0.01	7.42 ± 0.01	0.39 ± 0.02
Chymotrypsin A_α					
BCtr$_\alpha$-H2[b]	Asp[102] [c]	3.17 ± 0.23	8.64 ± 0.02	8.50 ± 0.02	0.14 ± 0.04
	His[57]	6.78 ± 0.04	8.50 ± 0.02	7.71 ± 0.01	0.79 ± 0.03
BCtr$_\alpha$-H1	His[40]	7.30 ± 0.02	8.63 ± 0.01	7.62 ± 0.02	1.01 ± 0.02

[a] From Markley and Ibañez (1978). Values for chymotrypsin A_α have been revised according to recent data obtained at 360 MHz (Ibañez and Markley, unpublished).

[b] Two pK' values calculated simultaneously, assuming Hill coefficients of unity.

[c] In these cases, there may be some electron transfer from Asp[102] to His[57]; see Section III.

ferences in the environment and pK_a' of histidine-57 in the zymogen and enzyme. The ^{31}P-NMR data (Reeck et al., 1977; Porubcan et al., 1978) indicate in addition that there are differences in the environment of the phosphate. The ^{31}P peak is less shielded in the DIP-enzyme than in the DIP-zymogen (Reeck et al., 1977; Porubcan et al., 1979). The change in the ^{31}P chemical shift was explained by fewer hydrogen bonds to the phosphate in the zymogen than in the enzyme (Reeck et al., 1977). The X-ray structure of DIP-trypsinogen (R. M. Stroud, personal communication), however, reveals a more profound change in the environment of the phosphate in the DIP-zymogen as compared with the DIP-enzyme.

The recent investigation of refined X-ray data confirms that the hydrogen bond between aspartate-102 and histidine-57 is conserved in both chymotrypsinogen and chymotrypsin (Matthews et al., 1977). This may explain why the chemical shift of the low-field N-H is the same in the zymogen and enzyme (Robillard and Shulman, 1972, 1974a). However, the hydrogen bond between serine-195 and histidine-57 appears to be present in the zymogen but not the enzyme (Matthews et al., 1977). This structural change may account for the changes in the pK_a values of histidine-57 and aspartate-102 observed in studies of the $C^{\varepsilon 1}$-H (Markley and Ibañez, 1978). Since the chemical shift of the low-field N-H is the same in both enzyme and zymogen (Robillard and Shulman, 1974a), this NMR peak does not provide information about possible polarization of serine-195.

The discrepancies in pK_a' values obtained by ^{1}H-NMR of the low-field N-H and of the histidine $C^{\varepsilon 1}$-H (Table III) probably result from differences in experimental conditions. The studies of the low-field N-H (Robillard and Shulman, 1974a) were carried out in ^{1}H$_2$O at 3°C whereas the studies of the $C^{\varepsilon 1}$-H were performed in ^{2}H$_2$O at 31°C. Hanai (1976) has reported thermodynamic parameters for histidine-57 of chymotrypsin derived from proton release studies of inhibitor binding. These indicate that the pK_a' of histidine-57 in ^{1}H$_2$O is 7.2 at 3°C and 6.6 at 31°C. These values are in reasonable agreement with Robillard and Shulman's pK_a' of 7.5 at 3°C in ^{1}H$_2$O and Markley and Ibañez's pK_a' of 6.8 at 31°C in ^{2}H$_2$O. Thermodynamic parameters are unavailable for histidine-57 of chymotrypsinogen so that a similar comparison cannot be made.

The changes in titration behavior of histidine-40 (Fig. 16) are consistent with structural information. In chymotrypsinogen, the imidazole of histidine-40 is located in a hydrophobic environment adjacent to the carboxylate of aspartate-194 (Freer et al., 1970). The $C^{\varepsilon 1}$-H NMR peak is shielded as expected for a nonpolar environment, and the two inflections represent the addition of protons to histidine-40 itself and to aspartate-194

(Markley and Ibañez, 1978). In chymotrypsin, histidine-40 is more exposed and is hydrogen-bonded to the carbonyl of glycine-193 (Birktoft and Blow, 1972; Tulinsky et al., 1973). The $C^{\varepsilon 1}$-H NMR peak has a more normal chemical shift and a titration curve with a single pK_a' (Markley and Ibañez, 1978).

Gorenstein and Findlay (1976) reported the existence of two slowly interconverting ^{31}P NMR peaks in spectra of DIP chymotrypsin A_{α}. They speculated that the two peaks represent two isomers in slow exchange: the lower-field signal representing the normal serine O^{γ} phosphorylated ester and the higher-field signal representing a histidine-57 imidazole N-phosphoramide. We also detect a second peak in spectra of DIP-chymotrypsin A_{δ} after "aging" in solution (Porubcan et al., 1979). Since we do not see the second peak in other DIP serine proteinases, we see no basis for invoking an imidazole N-phosphoramide. The two peaks may be related to the conversion between DIP-chymotrypsin A_{α} and DIP-chymotrypsin A_{δ} (Gladner and Neurath, 1954; Massey and Hartley, 1956; Corey et al., 1965).

3. Trypsin and Trypsinogen

Trypsins and trypsinogens from the cow and pig have been the subjects of ^{1}H-NMR studies in $^{2}H_2O$. Bovine trypsin and its zymogen contain three histidine residues; porcine trypsin and its zymogen contain four histidine residues. The positions of these residues in the protein sequences are listed in Table IV.

TABLE IV. Comparison of the bovine chymotrypsinogen and bovine trypsinogen numbering systems for various residues discussed

Bovine chymotrypsinogen[a]	Bovine trypsinogen[b]	Porcine trypsinogen[c]
histidine-40	histidine-29	histidine-29
histidine-57	histidine-46	histidine-46
phenylalanine-71	aspartate-59	histidine-59
asparagine-91	histidine-79	histidine-79
aspartate-102	aspartate-90	aspartate-90
serine-195	serine-183	serine-183

[a]Hartley and Kauffman (1966), Blow et al. (1969). [b]Walsh and Neurath (1964). [c]Hermodson et al. (1973).

The histidine assignment problem is compounded in these proteins. Thus far, assignments have been made only for histidine-57. A further complication has been the inability to resolve certain histidine peaks of bovine trypsin and trypsinogen. No histidine $C^{\epsilon 1}$-H peaks have been resolved in samples of bovine trypsin (Porubcan *et al.*, 1978); and only the peak corresponding to histidine-57 has been resolved in spectra of bovine trypsinogen (Porubcan *et al.*, 1978). Addition of calcium ion or various buffers has not led to the appearance of the missing peaks. The most probable cause of their disappearance is exchange broadening. It may be possible to resolve the peaks at higher or lower temperature (all the work has been at 31°C).

Titration curves of the four $C^{\epsilon 1}$-H peaks of porcine trypsinogen (Porubcan *et al.*, 1978) are shown in Fig. 22. The corresponding ^{1}H-NMR titration curves of porcine trypsin (Markley and Porubcan, 1976) are presented in Fig. 23. The assignment of peak PTg-Hl to the $C^{\epsilon 1}$-H of histidine-57 was based on the studies of DIP-trypsinogen summarized in Table V. Only one of the histidine titration curves of porcine trypsinogen (PTg-Hl) is altered appreciably in the DIP derivative (Porubcan *et al.*, 1978). The pK_a of this ^{1}H-NMR peak is equal (within experi-

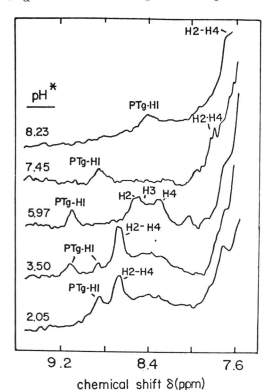

Fig. 22a

chemical shift δ(ppm)

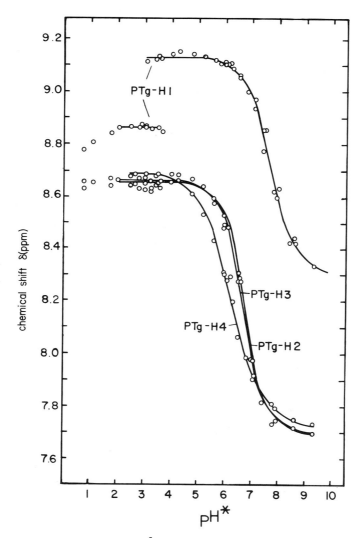

Fig. 22.   (a) 250 MHz ^{1}H-NMR spectra of the histidyl $C^\varepsilon 1$-H region of pre-exchanged porcine trypsinogen (PTg) at various pH* values; 25 mg of protein/0.5 ml of 0.5 M KCl in H_2O, 31°C. Note that at high pH all of the peaks broaden significantly, making distinct resolution of peaks H2-H4 difficult.   (b) ^{1}H-NMR titration curves of the histidyl $C^\varepsilon 1$-H peaks of porcine trypsinogen (PTg) from (a). Peak PTg-Hl is affected by two transitions and is assigned to histidine-57 of the active site. Peaks H2-H4 are not assigned to specific histidines. The pK' values are listed in Table VI. (From Porubcan et al., 1978.)

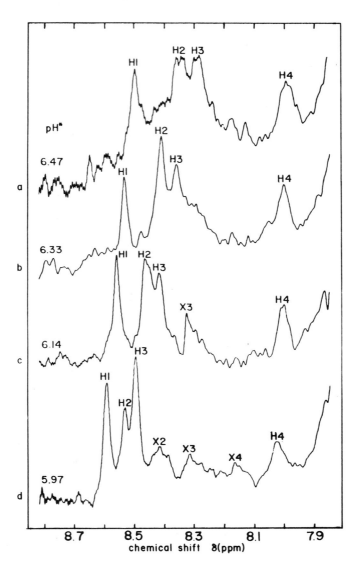

Fig. 23. (a) ^{1}H-NMR spectra at 250 MHz of the histidine $C^{\varepsilon 1}$-H region of porcine trypsin (preexchanged) obtained with 12.5 min of signal averaging using 0.3 ml of 2 m$\underline{M}$ trypsin in 0.5 $\underline{M}$ KCl in ^2H_2O, 31°C. Peaks H1, H2, H3, and H4 correspond to the four histidine residues of porcine β-trypsin. Peak H4 is from histidine-57. X2, X3, and X4 are unassigned nontitrating peaks.

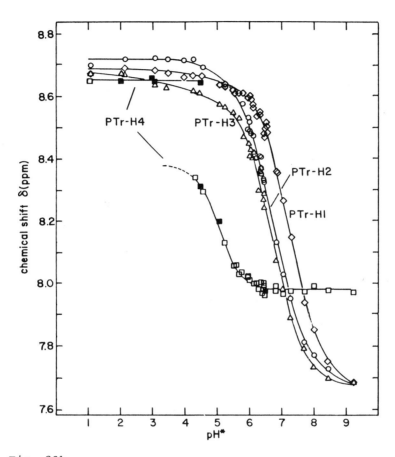

Fig. 23b

(b) ^1H-NMR titration curves (at 250 MHz) of the four histidine residues of porcine -trypsin Hl, ($\Diamond$); H2, ($\bigcirc$); H3, ($\triangle$); H4 ($\square$). Peak H4 is assigned to histidine-57. ($\square$) Chemical shifts of peak H4 obtained using trypsin preexchanged in N-H peaks; ($\blacksquare$) additional data obtained for peak H4 from samples subjected to a subsequent exchange in D_2O while bound to a Sepharose-STI support. (Modified from Markley and Porubcan, 1976.)

TABLE V. DIP-trypsinogen and DIP-trypsin. Comparison of ^1H-*NMR parameters for the* $C^{\varepsilon 1}$-*H of histidine-57 and* ^{31}P-*NMR parameters for the phosphorus*

	pH_{mid} values from	
	^1H-NMR[a]	^{31}P-NMR[b]
Porcine DIP-trypsinogen	7.67 ± 0.02[a]	7.42 ± 0.02[b]
Porcine DIP-trypsin	7.36 ± 0.02	7.31 ± 0.05

[a]*From Porubcan et al. (1978).* [b]*From Porubcan et al. (1979).*

mental error) to that of the ^{31}P-NMR peak of porcine DIP-tryp-sinogen. Similar experiments were used to assign the $C^{\varepsilon 1}$-H peak of histidine-57 in bovine trypsinogen and bovine DIP-tryp-sinogen (Porubcan *et al.*, 1978).

The original assignment of peak H4 in spectra of porcine trypsin to histidine-57 (Markley and Porubcan, 1976) was based on three independent experiments: (i) the disappearance of the peak in TLCK trypsin; (ii) the altered titration behavior of the peak in the BPTI:trypsin complex; and (iii) the disappear-ance of peaks H1-H3 following exchange of trypsin in ^2H_2O while bound to Sepharose-linked soybean trypsin inhibitor Kunitz. We have established recently (J.-G. Vanecek, M. A. Porubcan, and J. L. Markley, unpublished; see Porubcan, 1978) by a tri-tium exchange experiment that the disappearance of peaks H1-H3 cannot be the result of deuterium exchange because the exchange kinetics are too slow. This experiment was prompted by the tritium exchange experiments of Krieger *et al.* (1974), who found that all the histidines $C^{\varepsilon 1}$-H of benzamidine-inhibited bo-vine trypsin exchange at abnormally slow rates. These recent results invalidate the exchange experiment. The reason peaks H1-H3 disappear has not been elucidated. We continue to have faith in the assignment of peak H4 to histidine-57 on the basis of the other two assignment experiments (Markley and Porubcan, 1976) and on the basis of new investigations of DIP-trypsins. The ^1H-NMR titration curves of porcine DIP-trypsin indicate that only peak H4 is altered in the derivative. The higher pK_a' of the ^1H-NMR peak is 7.3 (Porubcan, 1978), which is iden-tical to the pK_a' (7.3) of the ^{31}P-NMR peak of porcine DIP-trypsin (Porubcan *et al.*, 1979). See Table V.

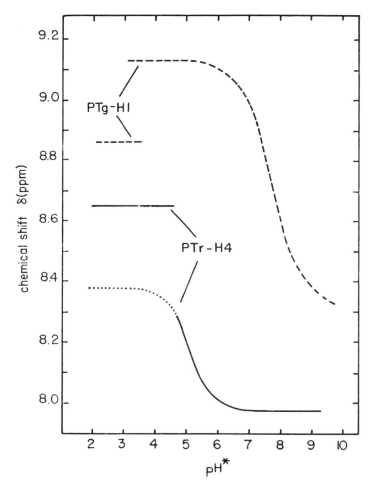

Fig. 24. Comparison of the titration behavior of histidine-57 (from Figs. 22 and 23) in porcine trypsin (PTr-H4, solid line) and porcine trypsinogen (PTg-H1, dashed line) showing the presence of discontinuous curves at low pH and the differences in pK' values in the zymogen and enzyme. (From Porubcan et al., 1978.)

As in the case of bovine chymotrypsinogen and chymotrypsin, the ^{1}H-NMR titration curves of the $C^\epsilon 1$-H of porcine trypsinogen and trypsin indicate that the environment and pK_a' of histidine-57 are significantly different in the zymogen and activated enzyme. The titration curves of the peaks assigned to $C^\epsilon 1$-H of histidine-57 are compared in Fig. 24. The titration parameters for these peaks are included in Table VI. In analogy with chymotrypsinogen, the higher pK_a' of the zymogen (7.7) is assigned to protonation of histidine-57 itself, whereas the lower pH_{mid} (3.2) is assigned to an external environmental perturbation, which probably includes the protonation of aspartate-102 (Porcuban et al., 1978a). Likewise, in the enzyme, the higher pK_a' (5.0) is assigned to histidine-57 and the lower pH_{mid} (4.5) to aspartate-102 (Markley and Porubcan, 1976).

The two laboratories that have obtained X-ray structures of trypsin and trypsinogen did not detect significant differences in the catalytic triad between zymogen and enzyme (Fehlhammer et al., 1977; Kossiakoff et al., 1977). Thus the structural changes leading to the changes in pK_a' values must be subtle.

It is noteworthy that the lower transition in porcine trypsinogen and trypsin is slow on the NMR time scale whereas the higher transition is rapid. The interpretation for this is that the $N^{\epsilon 2}$ of histidine-57 is accessible to solvent, whereas in order to protonate aspartate-102 there must be movement of a protein group, probably the imidazole of histidine-57 (Markley and Porubcan, 1976).

An important difference between the chymotrypsinogen/chymotrypsin titration curves (Fig. 17) and the trypsinogen/trypsin titration curves (Figs. 22 and 23) is the fact that only one titration curve is altered in the latter pair. Although peaks have not been assigned to histidine-40 in trypsins or trypsinogens, the titration curves indicate that (i) the environment of histidine-40 must be similar in both porcine trypsinogen and trypsin, (ii) histidine-40 must not interact significantly with aspartate-194 in porcine trypsinogen, and (iii) the environment of histidine-40 must be more exposed in porcine trypsinogen than in bovine chymotrypsinogen. The X-ray results for bovine trypsinogen are divided on these points: one laboratory reports an interaction between histidine-40 and aspartate-102 (Fehlhammer et al., 1977) and one does not (Kossiakoff et al., 1977).

The titration curves of the $C^{\epsilon 1}$-H peaks assigned to histidine-57 are quite similar in the three zymogens studied (Fig. 25). A significant feature of these curves is the deshielding of the $C^{\epsilon 1}$-H proton by 0.6 ppm as compared with a normal histidine. This deshielding is removed as a consequence of the low pH transition affecting histidine-57 (Porubcan et al., 1977, 1978). The existence of high-resolution X-ray structures of trypsin and trypsinogen (Bode and Schwager, 1975; Fehlhammer et al., 1977) permitted an investigation into the origin of the deshielding interaction (Porubcan et al., 1978). The only group close enough to the $C^{\epsilon 1}$-H of histidine-57 to cause its deshielding in trypsinogen is the carbonyl of serine-124. The distance and the angle between this carbonyl and the imidazole ring are different in trypsinogen (Fehlhammer et al., 1977) and

TABLE VI. *Comparison of the H-NMR titration behavior of peaks assigned to histidine-57 C^{ε_1}-H in serine proteinases[a]*

| | First protonation | | | Second protonation | |
	δ high pH (ppm from DSS)	pK'_1	$\Delta\delta_1$	pK'_2	$\Delta\delta_2$
Bovine chymotrypsinogen A[b]	8.44	7.3	0.77	1.4	-0.56
Bovine trypsinogen[c]	8.36	7.7	0.68	1.8	-0.43
Porcine trypsinogen[c]	8.31	7.7	0.81	3.4[g]	-0.28
Bovine chymotrypsin A_α[h]	7.92	6.8	0.79	3.2	0.14
Porcine trypsin[d]	7.94	5.0	0.41	4.5[g]	0.28
α-Lytic Proteinase[e]	7.75	6.5	0.88	[f]	

[a] All data were obtained at 31°C in ^2H_2O.
[b] From Markley and Ibañez (1978).
[c] From Porubcan et al. (1978).
[d] From Markley and Porubcan (1976).
[e] From Westler and Markley (1979).
[f] The pK' of Asp[102] is probably less than 3.
[g] This transition appears to be cooperative in hydrogen ion; hence the value is a pH_{mid} rather than a pK_a'.
[h] Ibañez and Markley, unpublished.

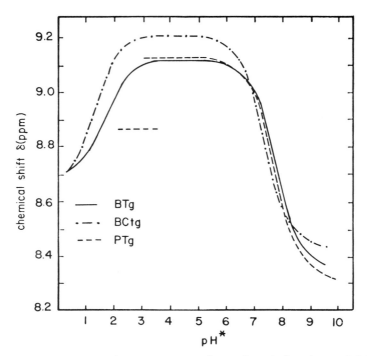

Fig. 25. Comparison of the titration behavior of histidine-57 ($C^\varepsilon 1$-H NMR peaks) in bovine trypsinogen (——), porcine trypsinogen (- - -) (Porubcan et al., 1978), and bovine chymotrypsinogen (- —— -) (From Markley and Ibañez, 1978.)

in trypsin (Bode and Schwager, 1975) (see Fig. 26). Calculations of the environmental shift $\Delta\delta_{env}$ based on the anisotropy of the carbonyl double bond reported by Mital and Gupta (1970) indicated that $\Delta\delta_{env}$ is 0.7 ppm for bovine trypsinogen and essentially zero for bovine trypsin (Porubcan et al., 1978a). The agreement with the experimental chemical shift difference of 0.6 ppm between zymogen and enzyme is most satisfactory. Similar calculations based on the X-ray coordinates of chymotrypsinogen and chymotrypsin did not provide quantitative agreement with NMR results, probably because of greater errors in these structures. Qualitatively, however, there is an analogous difference in the environment of the $C^\varepsilon 1$-H of histidine-57 between chymotrypsinogen and chymotrypsin.

Calculations of the contribution of the serine-214 carbonyl to the chemical shift of the $C^\varepsilon 1$-H of histidine-57 in the BPTI: bovine trypsin complex, based on the X-ray structure of Huber et al. (1974), indicate that the peak should be shielded by

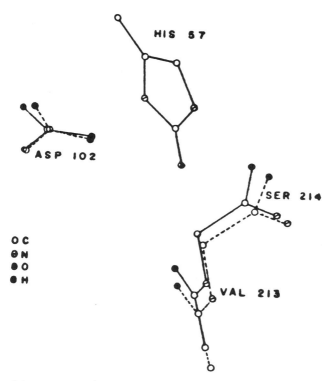

Fig. 26. Comparison of the three-dimensional structure around histidine-57 in trypsinogen (____) and trypsin (---). The coordinates were rotated to superimpose the histidine rings. Differences in the orientation of the serine-214 carbonyl provide a mechanism for the chemical shift differences of the histidine $C^{\varepsilon}1$-H in trypsinogen compared with trypsin. The X-ray coordinates used were from Fehlhammer et al. (1977) and Bode and Schwager (1975). (From Porubcan et al., 1978.)

0.3 ppm compared to native trypsin (Porubcan et al., 1978a). The titration curve of the $C^{\varepsilon}1$-H of histidine-57 in the BPTI: porcine trypsin complex (Markley and Porubcan, 1976) is shown in Fig. 27b. The chemical shift of histidine-57 in the complex is pH independent and lies 0.3 ppm downfield of the neutral histidine of the free enzyme. If this shift on complex forma- tion is corrected by $\Delta\delta_{env} = 0.3$ ppm, the total shift is 0.6 ppm. Since this downfield shift is even greater than the transition in the free enzyme assigned to protonation of histidine-57 itself (0.4 ppm), we have concluded that histidine-57 is positively charged in the BPTI:trypsin complex. This result could be in error only if we have overlooked interactions in the complex that contribute a selective deshielding interaction of 0.6 ppm.

Trypsinogen also forms a stable complex with BPTI and gives the titration curve for histidine-57 shown in Fig. 27a. Similar results have been obtained with BPTI complexes with bovine and porcine trypsinogen (Porubcan *et al.*, 1978). The chemical shift of histidine-57 is pH independent in the BPTI:trypsinogen complex and has the same chemical shift as in the BPTI:trypsin complex. The pH independence implies that histidine-57 is excluded from solvent in the complex. The similarity of chemical shift of the $C^{\epsilon 1}H$ of histidine-57 in BPTI complexes with tryp-

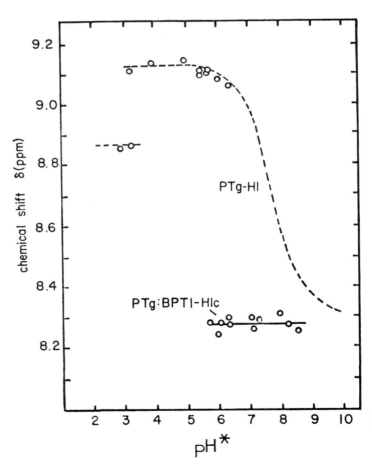

Fig. 27a

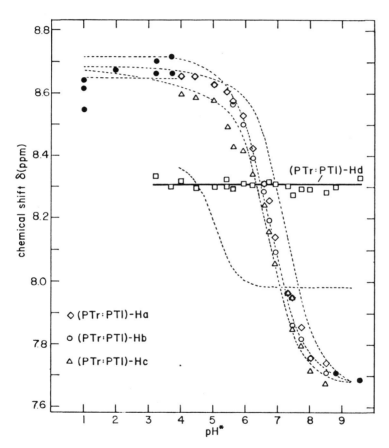

Fig. 27b

Fig. 27. ^1H *NMR titration curves (250 MHz) of the histi-*
dine $C^{\varepsilon}1$-H peaks resolved in complexes between bovine pancreatic
trypsin inhibitor (Kunitz) and (a) porcine trypsinogen or (b)
porcine trypsin. Peaks corresponding to histidine-57 in the
complexes are labeled. (Adapted from Markley and Porubcan,
1976, and Porubcan et al., 1978.)

sinogen and trypsin suggests that the structures of the complexes are identical as far as histidine-57 is concerned.   The
X-ray results for BPTI complexes with bovine trypsinogen and
trypsin indicate that binding of BPTI induces a conformational
change in trypsinogen leading to an active site structure that
resembles that of trypsin (Bode *et al.*, 1976, 1978; Bode and
Huber, 1976).   If the structures around histidine-57 are similar, the NMR results suggest that histidine-57 is positively
charged in the BPTI:trypsinogen complex.

B. Other Serine Proteinases

1. Subtilisin

Robillard and Shulman (1974a) found a low-field N-H peak
spectra of subtilisin at low pH analogous to the N-H peak observed for members of the chymotrypsinogen family (Fig. 3).
They did not carry out pH titration study of the free enzyme.
The chemical shift of the peak was pH independent in complexes
with 2-phenylethylboronic acid and benzene boronic acid (Robillard and Shulman, 1974b).

2. Staphylococcal Proteinase

^{1}H-NMR titration curves for histidines of staphylococcal
proteinase have been reported (Markley *et al.*, 1975).   The spectra were analyzed on the basis of an amino acid composition that
indicated three histidine residues per molecule (Drapeau *et al.*,
1972).   Although more than three $C^{\varepsilon 1}$-H peaks were observed at certain pH values, this was attributed to structural heterogeneity
in the sample (Markley *et al.*, 1975).   The recent sequence of
Staphylococcal proteinase (Drapeau, 1978) establishes that there
are six histidines per molecule; hence the split peaks probably
correspond to individual histidines.   The NMR spectroscopy of
this enzyme needs to be reinvestigated in light of the sequence
data.   Since none of the histidines gave biphasic titration
curves similar to histidine-57 of trypsin and chymotrypsin it
was concluded that staphylococcal proteinase differs substantially from these enzymes (Markley *et al.*, 1975).   This conclusion appears to be confirmed by the sequence (Drapeau, 1978).

III. CONCLUSIONS AND FUTURE NEEDS

The experiments described above illustrate the way in which chemical, X-ray crystallographic, and spectroscopic studies of proteins complement one another. Chemical experiments provide protein sequences, reaction rates, and identify critical amino acid residues. X-ray crystallography provides detailed three-dimensional structures of enzymes, enzyme-inhibitor complexes, and chemically modified derivatives. Spectroscopic techniques give information about reactivities of individual groups, their chemical environments, and dynamic properties.

The problem of making NMR assignments has been appreciated for several years. The rigor of NMR assignments in proteins ranges from iron-clad assignments based on selective isotopic enrichment and comparison of proteins differing in single amino acid replacements, to less definitive assignments based on specific chemical modification of the residues being assigned, to indirect assignments based on chemical modifications farther removed and on properties deduced from X-ray structures. There can be no ambiguity in the NMR assignments of α-lytic proteinase labeled at histidine-57 with ^{13}C (Hunkapiller et al., 1973) or ^{15}N (Bachovchin and Roberts, 1978). Selective isotopic exchange also is one of the best techniques for assigning histidine $C^{\varepsilon}1$-H peaks (Markley, 1975). Unfortunately, attempts to use this technique for assignments in chymotrypsin and trypsin and their zymogens have not been successful (Porubcan, 1978; Markley and Ibañez, 1978). The histidine residues of these serine proteinases exchange much more slowly (Krieger et al., 1976) than those of other proteins studied.[6]

After NMR assignments have been made, the NMR parameters still must be interpreted in terms of chemical events. The NMR investigations of the active sites of serine proteinases demonstrate that certain nuclei are more responsive than others to

---

[6]*After this review was submitted, a very interesting report was made of parallel tritium exchange and ^{1}H-NMR studies of the histidine $C^{\varepsilon}1$-Hs of the trypsinlike enzyme from Streptomyces erythreus [Miyamato, K., Matsuo, H., and Narita, K. (1978), VIIth Int. Conf. Magn. Resonance Biol. Syst., Nara Japan, Abstr. A14]. This enzyme has four histidines. Histidine-57 has a very slow but measurable tritium exchange rate in this enzyme. All four ^{1}H-NMR peaks were assigned, and titration studies demonstrated that histidine-57 has a pK_a' of 6.7. The ^{1}H-NMR titration curve of the peak assigned to histidine-57 has a shallow protonation shift similar to that of histidine-57 of porcine trypsin (Markley and Porubcan, 1976) and bovine chymotrypsin (Markley and Ibañez, 1978).*

various processes of interest. Ideally, the NMR studies should
include nuclei from all catalytic groups. For example, the key
functional group in catalysis is the hydroxyl of serine-195.
It would be of great advantage to monitor this group directly
rather than infer its properties from data concerning the neighboring histidine. Likewise the pK_a' of aspartate-102 has been
deduced thus far only from its effects on histidine-57.

Three of the interpretations based on NMR data of serine
proteinase have been controversial: (i) the pK_a' of histidine-
57; (ii) the similarity or difference of the catalytic triad in
enzymes and zymogens; (iii) the charge on histidine-57 in enzyme-inhibitor complexes. This confusion reflects uncertainties
in the field of serine proteinases in general.

The protonation scheme for the aspartate-102-histidine-57
couple is shown in Fig. 28. Conventional wisdom held that the
pK_a' of histidine-57 should be higher than that of aspartate-102
and that species $(EH^{-,+})$ should predominate over species (HE).
The experiments of Robillard and Shulman (1972, 1974a) rein-
forced this view. The recent ^{15}N-NMR results with α-lytic proteinase (Bachovchin and Roberts, 1978) provide overwhelming
support for a histidine-57 pK_a' near 7. Our ^{1}H-NMR investigations of the $C^{\varepsilon}1$-H of histidine-57 in α-lytic proteinase, trypsin, chymotrypsin, and their zymogens also are in agreement
(Markley and Porubcan, 1976; Westler and Markley, 1979;
Markley and Ibañez, 1978; Porubcan et al., 1978). In addition,
the titration curves of the mammalian serine proteinases provide pK_a' values (albeit indirect) for aspartate-102 that are
lower than those for histidine-57. In several X-ray structures
of serine proteinases, anions from the environment have been observed to bind in the neighborhood of histidine-57 at low pH but
not at neutral pH (Shotton and Watson, 1970; Tulinsky et al.,
1973; Delbaere et al., 1975; Huber and Bode, 1977). These results have been interpreted as indicating a pK_a' for histidine-
57 near neutrality.

The two experiments that support a higher pK_a' for aspartate-
102 than for histidine-57 (Hunkapiller et al., 1973; Koeppe and
Stroud, 1976) need to be reinvestigated. It should be possible
to obtain more precise values for the pH dependence of the
$^{13}C^{\varepsilon}1$-$C^{\varepsilon}1$-^{1}H coupling constant $(^{1}J_{CH})$ of histidine-57 of [13]C
enriched proteinases by means of ^{1}H-NMR spectroscopy of
the $C^{\varepsilon}1$-^{1}H rather than by ^{13}C-NMR spectroscopy of the $^{13}C^{\varepsilon}1$.
Several molecular orbital calculations of the catalytic triad
also support the reversal of pK_a' values (Amidon, 1974; Scheiner
et al., 1975; Kitayama and Fukutome, 1976; Beppeu and Yomosa,
1977). These calculations have overestimated the hydrophobic
character of histidine-57 and have neglected dipoles from surrounding protein and solvent, which would stabilize the ion
pair, species $(EH^{-,+})$ in Fig. 28.

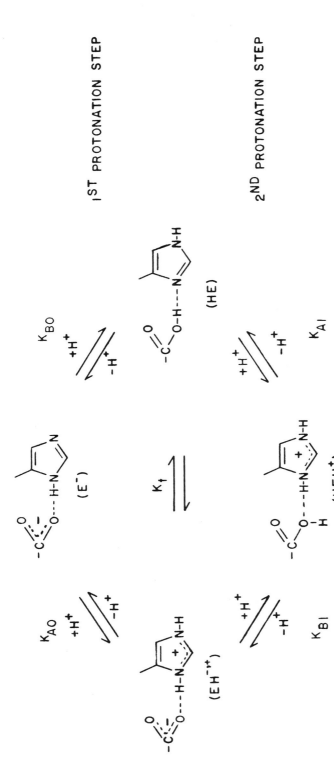

Fig. 28. Protonation scheme for the active site residues histidine-57 and aspartate-102 of serine proteinases. (From Markley and Ibañez, 1978.)

Two kinds of experiments indicate differences in the catalytic groups in serine proteinase zymogens and enzymes: (i)
^{1}H-NMR studies of the $C^{\epsilon}1$-H of histidine-57 (Ibañez, 1978); Porubcan et al., 1978) and (ii) ^{31}P-NMR studies of DIP derivatives
Reeck et al., 1977; Porubcan et al., 1978b). The differences
include changes in the pK_a' values of histidine-57 and aspar-
tate-102, and the environment of histidine-57. The strength
of the aspartate-102 histidine-57 hydrogen bond apparently is
not affected since the chemical shift of the low-field N-H
assigned to this bond is the same in chymotrypsinogen and
chymotrypsin (Robillard and Shulman, 1972, 1974a).

The ^{31}P-NMR data from DIP derivatives are summarized in
Fig. 29. The chemical shifts of the zymogens lie in the region
between -2.2 and -3.0 ppm, whereas the chemical shifts of mammalian enzymes lie between 0 and 1.3 ppm. It is interesting to
note that the ^{31}P chemical shift of α-lytic proteinase, which
does not share the same activation mechanism as the mammalian

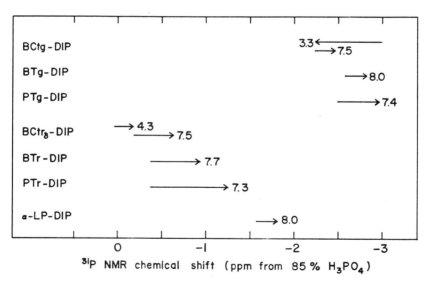

Fig. 29. Summary of ^{31}P NMR titration data of diisopropylphosphoryl derivatives of serine proteinases. The arrows represent the extent of titration shifts; and the numbers indicate
pK_a' values obtained from fitting the data to theoretical titration curves. Abbreviations used: DIP, diisopropylphosphoryl;
BCTg, bovine chymotrypsinogen A; BTg, bovine trypsinogen; PTg,
porcine trypsinogen; BCtr$_α$, bovine chymotrypsin A$_α$; BTr, bovine
trypsin; PTr, porcine trypsin; α-LP, α-lytic proteinase. (From
Porubcan et al., 1979.)

enzymes (James *et al.*, 1978), is intermediate between these.
If this pattern holds up for other examples, ^{31}P chemical shifts
of DIP derivatives may be useful eventually for classifying enzymes and zymogens.

There is general agreement that the titration behavior of
histidine-57 is altered in complexes with certain inhibitors.
Hisditine-57 does not titrate in stable complexes between chymotrypsin or subtilisin and benzene boronic acid (Robillard and
Shulman, 1972, 1974b). Histidine-57 also does not titrate in
the complex between chymotrypsin and BPTI (Robillard and Shulman, 1974b; Markley and Ibañez, 1978) or in complexes of trypsin or trypsinogen with BPTI (Porubcan *et al.*, 1978; Markley,
1978). Histidine-57 is nontitrating in STI complexes with trypsin (Markley, 1978). The chemical shifts of the low-field N-H
in protein proteinase inhibitor: serine proteinase complexes
are catalogued in Table VII.

The charge state of the histidine-57 imidazole in complexes
with protein proteinase inhibitors is unclear because of conflicting experimental ^{1}H-NMR evidence. The low-field N-H peak
in these complexes (Robillard and Shulman, 1974b; Markley,
1978) has a pH independent chemical shift in the range 13.6 to
14.9 ppm (Table VII) that is similar to the chemical shift of
the low-field N-H of chymotrypsin A_δ at high pH (14.9 ppm),
where histidine-57 presumably is deprotonated (Robillard and
Shulman, 1974a). On the other hand, the $C^\epsilon 1$-^{1}H peak assigned to
histidine-57 has a pH-independent chemical shift in complexes
that is similar to that of the singly protonated histidine-57
aspartate-102 couple, in which histidine-57 appears to have a positive charge of at least one-half. The dilemma cannot be resolved by reference to the X-ray structures of these complexes
alone, because the locations of protons are not resolved. (The
question of the charge on histidine-57 is separate from whether
or not the proteinase Ser^{195}-O^Δ inhibitor peptide carbonyl bond
is tetrahedral.) Provided that the ^{1}H-NMR assignments are correct, the conflicting data imply that either the proton hydrogen-bonded between aspartate-102 and histidine-57 or the histi-
dine-57 $C^\epsilon 1$-H has an altered environment in the complexes compared to the free enzyme, and that the change in environment af-
fects the chemical shift so as to mimic or obscure the titration
shift.

^{1}H-NMR studies of zymogens (Markley and Ibañez, 1978; Porubcan *et al.*, 1978a) reveal that the $C^\epsilon 1$-H of histidine-57 is subject to a large environmental shift. However, this shift can
be accounted for by anisotropy calculations (Porubcan *et al.*,
1978) based on the X-ray structures of trypsin and trypsinogen. Similar calculations for the BBTI:trypsin complex (Porubcan *et al.*, 1978) indicate that the $C^\epsilon 1$-H of histidine-57
should be shielded by 0.3 ppm compared to the free enzyme.
This even strengthens the case for a positive charge on histi-

TABLE VII. Chemical shift of the extreme low-field H-NMR peak in various proteinase:protein proteinase inhibitor complexes at pH 7

	Low-field peak chemical shift (ppm from DSS)
Complex	
Bovine chymotrypsin A :BPTI[a]	14.9
Bovine chymotrypsin A :BPTI[b,c]	14.7
Bovine chymotrypsinogen A:BPTI[b,c]	14.7
Bovine chymotrypsin A :STI[b,d]	14.5
Bovine trypsin:BPTI[b,c]	13.6
Bovine trypsinogen:BPTI[b,c]	13.6
Bovine trypsin:STI[b,d]	13.8
Porcine trypsin:BPTI[b,c]	14.2
Porcine trypsin:STI[b,d]	14.4
Porcine anhydrotrypsin:BPTI[b,c]	14.1

[a]*From Robillard and Shulman (1974a); in water, $3^{o}C$, 220 MHz, time-averaged continuous wave spectrum.*
[b]*From Markley (1978); in 0.5 $\underline{M}$ CCl $30^{o}C$.*
[c]*Data obtained by correlation spectroscopy at 360 MHz.*
[d]*From 250 MHz correlation spectra.*

dine-57 in the complex. The argument could be in error only if an important deshielding interaction has been overlooked.

Recent data (Markley, 1978) indicate that the low-field N-H is exchange-broadened in free proteinases but not in complexes between proteinases and proteinase inhibitors. Near $0^{o}C$ the low-field NH peak of porcine trypsinogen is very broad and has an intensity of 0.5 ± 0.2 protons at pH 3.0 or 7.0. Both of these pH values are well removed from the pH_{mid} values of transitions involving histidine-57 and aspartate-102. It was proposed that at low temperature two conformational forms of the enzyme are "frozen out" on the NMR time scale (lifetime of species: 0.5-2.0 msec). The low-field N-H studied previously corresponds to one of these forms. It was pointed out (Markley, 1978) that if the other form of the enzyme is the one stabilized on protein proteinase inhibitor binding, then the chemical shift of the sharp peak of the complex might not be readily interpretable in terms of the charge state of histidine-57. Additional experiments are required to clarify these results.

The titration curve of histidine-57 of free porcine trypsin is not affected by saturation of the enzyme with the competitive inhibitor benzamidine (Porubcan, 1978). Krieger *et al.*

(1974) suggested on the basis of their X-ray results that the side chains of aspartate-102 and histidine-57 are altered on benzamidine binding. Since such changes should result in differences in the chemical shift of pK_a' of histidine-57 that were not observed (Porubcan, 1978), the NMR results agree with the subsequent X-ray analysis of Bode and Schwager (1975) in which the positions of the surrounding main and side chains are not perturbed by benzamidine binding.

   In the discussion above it has been assumed that either species (EH$^{-,+}$) or species (HE) would predominate as the first protonation product. In this case, one can ignore either the right or left side of the protonation scheme (Fig. 26) and can interpret the NMR results in terms of two pK_a' values. We suggested in our initial study of porcine trypsin that the full scheme might be required in order to explain the abnormally small protonation shift of the $C^{\varepsilon 1}$-H of histidine-57 (0.4 ppm). The small shift could result from the presence of an appreciable amount of species (HE). Since the protonation shifts of chymotrypsin (0.79 ppm) (Ibañez and Markley, unpublished) and α-lytic proteinase are still larger (0.7 ppm) (Westler and Markley, 1978b), the trend could be explained by decreasing values for K_t, where $K_t = $ [HE]/[EH$^{-,+}$]. The evidence for this speculation is weak since the observed differences in protonation shifts could be explained equally well by environmental differences affecting the $C^{\varepsilon 1}$-H chemical shift. Investigations of the histidine ring ^{15}N-^{1}H coupling constants may provide reliable values for K_t.

   The protonation shifts of the histidine-57 $C^{\varepsilon 1}$-H in all three zymogens studied are uniform, and their magnitudes are relatively normal. Hence the simple protonation scheme E$^-$ + H$^+$ $\rightleftarrows$ EH$^{-,+}$ + H$^+$ $\rightleftarrows$ HEH$^+$ is sufficient to explain the results.

   One may speculate about the significance of the pK_a' values obtained for histidine-57. In all the species studied to date, histidine-57 has a higher pK_a' in the zymogen than in the corresponding active enzyme. Hence at pH values near neutrality, correspondingly more of the enzyme than the zymogen is in the catalytically active form. This could contribute to the observed increased activity of the enzyme. Also in the enzymes studied, formation of the DIP-derivative invariably leads to an increase in the pK_a' of histidine-57. If, as suggested by Kraut (1977), DIP derivatives are models for the transition state, this indicates that the basicity of histidine-57 increases on approaching the transition state. Having a lower pK_a' in the free enzyme ensures that more of the molecules will be in the catalytically active, unprotonated form. The increase in pK_a'

of histidine-57 as the transition state is approached means
that the imidazole develops a higher affinity for the proton
to be removed from serine-195.

The pK_a' around pH 7 obtained from steady-state kinetics
generally has been attributed to protonation of the catalytic
triad in the free enzyme. The pK_a' of chymotrypsin at 25°C and
ionic strength 0.1 obtained from k_{cat}/K_m vs. pH for chymotrypsin in 6.80 ± 0.03 with only one exception (Fersht, 1977).
The pK_a' of histidine-57 of chymotrypsin from NMR studies at
360 MHz is 6.78 ± 0.04 (Ibañez and Markley, unpublished).
Note that this value is higher than that (6.1) reported by
Markley and Ibañez, 1978. At 250 MHz the histidine peak was
confused with an aromatic peak at high pH. We do not have a
satisfactory explanation for the low pK_a' of 5.0 obtained for
histidine-57 of porcine trypsin (Markley and Porubcan, 1976).

It is hoped that future studies will answer some of the
questions that remain about serine proteinases. Additional experiments are required to interpret the temperature dependence
of low-field N-H peaks (Markley, 1978) and the disappearance of
the ^{15}N peak of [^{15}N]-histidine-57 α-lytic proteinase at intermediate pH values (Bachovchin and Roberts, 1978). Explanations
also are required for the absence of all of the $C^{\varepsilon 1}$-H peaks from
bovine trypsin (Porubcan, 1978), bovine chymotrypsin A_δ (Markley and Ibañez, 1978) and one from DIP-chymotrypsin.

More details are needed in order to understand how the proton transferred from serine-195 to histidine-57 is shuttled onto
the leaving group. Does this transfer require movement of the
protein, and is such movement rate limiting (Satterthwait and
Jencks, 1974)? How does water bind to the acyl-enzyme intermediate?

NMR experiments may be useful in studies of the evolution
of the active site of serine proteinases and for studies of additional important members of the chymotrypsin family. ^1H-NMR
investigations of hydrogen bonds may also prove useful for investigations of the change in hydrogen bonding on activation of
zymogens to enzymes (Fehlhammer et al., 1977).

ACKNOWLEDGMENTS

I thank Drs. W. W. Bachovchin and J. D. Roberts for supply-
ing a preprint of their article on ^{15}N NMR of α-lytic proteinase and Dr. R. M. Stroud for advance information concerning the
X-ray structure of DIP-trypsinogen.

REFERENCES

Alden, R. A., Wright, C. S., and Kraut, J. (1970). *Phil. Trans. Roy. Soc. (London) B257,* 119-124.
Amidon, G. L. (1974). *J. Theor. Biol. 46,* 101-109.
Bachovchin, W. W., and Roberts, J. D. (1978). *J. Am. Chem. Soc. 100,* 8041-8047.
Bender, M. L., and Killheffer, J. V. (1973). *CRC Crit. Rev. Biochem. 1,* 149-199.
Beppeu, Y., and Yomosa, S. (1977). *J. Phys. Soc. Japan 42,* 1694-1700.
Birktoft, J. J., and Blow, D. M. (1972). *J. Mol. Biol. 68,* 187-240.
Birktoft, J. J., Kraut, J., and Freer, S. T. (1976). *Biochemistry 15,* 4481-4485.
Blow, D. M. (1976). *Acc. Chem. Res. 9,* 145-152.
Blow, D. M., Birktoft, J. J., and Hartley, B. S. (1969). *Nature 221,* 337-340.
Blow, D. M., Janin, J., and Sweet, R. M. (1974). *Nature 249,* 54-57.
Bode, W., and Huber, R. (1976). *FEBS Lett. 68,* 231-235.
Bode, W., and Schwager, P. (1975). *J. Mol. Biol. 98,* 693-717.
Bode, W., Fehlhammer, H., and Huber, R. (1976a). *J. Mol. Biol. 106,* 325-335.
Bode, W., Schwager, P., and Huber, R. (1976b). *In* "Proteolysis and Physiological Regulation," pp. 43-76. Academic Press, New York.
Bode, W., Schwager, P., and Huber, R. (1978). *J. Mol. Biol. 118,* 99-112.
Bradbury, J. H., and Wilairet, P. (1967). *Biochem. Biophys. Res. Commun. 29,* 84-89.
Bradbury, J. H., Chapman, B. E., and King, N. L. R. (1971). *Int. J. Prot. Res. 3,* 351-356.
Codding, P. W., Delbaere, L. T. J., Hayakawa, K., Hutcheon, W. L. B., James, M. N. G., and Jurášec, L. (1974). *Can. J. Biochem. 52,* 208-220.
Cohen, G. H., Silverton, E. W., Matthews, B. W., Braxton, H., and Davies, D. R. (1969). *J. Mol. Biol. 44,* 129-141.
Corey, R. B., Battfay, O., Brueckner, D. A., and Mark, F. G. (1965). *Biochim. Biophys. Acta 94,* 535-545.
DeHaen, C., Neurath, H., and Teller, D. C. (1975). *J. Mol. Biol. 92,* 225-259.
Delbaere, L. T. J., Hutcheon, W. L. B., James, M. N. G., and Thiessen, W. E. (1975). *Nature 257,* 758-763.
Drapeau, G. R. (1978). *Can. J. Biochem. 56,* 534-544.
Drapeau, G. R., Boily, Y., and Houmard, J. (1972). *J. Biol. Chem. 247,* 6720-6725.
Drenth, J., Hol, W. G. J., Jansonius, J. N., and Koehoek, R. (1972). *Eur. J. Biochem. 26,* 177-181.

Egan, W., Shindo, H., and Cohen, J. S. (1977). *Ann. Rev. Biophys. Bioeng. 6*, 408.

Engel, J., Quast, U., Heumann, H., Krause, G., and Steffen, E. (1974). *In* "Proteinase Inhibitors," pp. 412/419 (Fritz, H., Tschesche, H., Greene, L. J., and Truscheit, E., eds.). Springer-Verlag, New York.

Fehlhammer, H., and Bode, W. (1975). *J. Mol. Biol. 98*, 683-692.

Fehlhammer, H., Bode, W., and Huber, R. (1977). *J. Mol. Biol. 111*, 415-438.

Fersht, A. R. (1977). "Enzyme Structure and Mechanism," pp. 146-147. W. H. Freeman and Co., San Francisco.

Fink, A. (1976). *Biochemistry 15*, 1580-1586.

Freer, S. T., Kraut, J., Robertus, J. D., Wright, H. T., and Nguyen-Huu-Xuong (1970). *Biochemistry 9*, 1997-2009.

Ghelis, C., Labouesse, J., and Labouesse, B. (1967). *Biochim. Biophys. Res. Commun. 29*, 101-106.

Gladner, J. A., and Neurath, H. (1954). *J. Biol. Chem. 206*, 911-924.

Gorenstein, D. G., and Findlay, J. B. (1976). *Biochem. Biophys. Res. Commun. 72*, 640-645.

Griffin, J. H., Cohen, J. S., and Schechter, A. N. (1973). *Biochemistry 12*, 2096-2099.

Hanai, K. (1976). *J. Biochem. (Tokyo) 79*, 107-116.

Hartley, B. S. (1964). *Nature 201*, 1284-1287.

Hartley, B. S. (1970). *Phil. Trans. Roy. Soc. (London) B257*, 77-87.

Hartley, B. S., and Kauffman, D. L. (1966). *Biochem. J. 101*, 229-231.

Henderson, R., Wright, C. S., Hess, G. P., and Blow, D. M. (1971). *Cold Spring Harbor Symp. Quant. Biol. 36*, 63-69.

Hermodson, M. A., Erisson, L. H., Neurath, H., and Walsh, K. A. (1973). *Biochemistry 12*, 3146-3153.

Hess, G. P. (1971). *Enzymes 3*, 213-248.

Himoe, A., Parks, P. C., and Hess, G. P. (1967). *J. Biol. Chem. 242*, 919-929.

Huber, R., and Bode, W. (1977). *In* "NMR in Biology," pp. 1-31 (Dwek, R. A., Campbell, I. D., Richards, R. E., and Williams, R. J. P., eds.). Academic Press, New York.

Huber, R., and Bode, W. (1978). *Acc. Chem. Res. 11*, 114-122.

Huber, R., Kukla, D., Bode, W., Schwager, P., Bartels, K., Deisenhofer, J., and Steigemann, W. (1974). *J. Mol. Biol. 39*, 73-101.

Huber, R., Bode, W., Kukla, D., Kohl, U., and Ryan, C. A. (1975). *Biophys. Struct. Mech. 1*, 189-201.

Hunkapiller, M. W., Smallcombe, S. H., Whitaker, D. R., and Richards, J. H. (1973). *Biochemistry 12*, 4732-4743.

Hunkapiller, M. W., Smallcombe, S. H., and Richards, J. H. (1975). *Org. Magn. Reson. 7*, 262-265.

Ibañez, I. B., Porubcan, M. A., and Markley, J. L. (1976). *Fed. Proc., Fed. Am. Soc. Exp. Biol. 35*, Abs. 545.

James, M. N. G., Delbaere, L. T. J., and Brayer, G. D. (1978). *Can. J. Biochem. 56,* 396-402.

Jansen, E. F., Nutting, M. D. F., Jang, R., and Balls, A. K. (1949). *J. Biol. Chem. 179,* 189-199.

Kitayama, H. P., and Fukutome, H. (1976). *J. Theor. Biol. 60,* 1-18.

Koeppe, R. E. II, and Stroud, R. M. (1976). *Biochemistry 15,* 3450-3458.

Kossiakoff, A. A., Chambers, J. L., Kay, L. M., and Stroud, R. M. (1977). *Biochemistry 16,* 654-664.

Kraut, J. (1971). *Enzymes 3,* 165-183.

Kraut, J. (1977). *Annu. Rev. Biochem. 46,* 331-358.

Krieger, M., Kay, L. M., and Stroud, R. M. (1974). *J. Mol. Biol. 83,* 209-230.

Kühne, W. (1877). *Verhl. Naturhut.-Med. Ver. Heidelberg 1,* 194-198.

Markley, J. L. (1975). *Acc. Chem. Res. 8,* 70-80.

Markley, J. L. (1978). *Biochemistry 17,* 4648-4656.

Markley, J. L., and Finkenstadt, W. R. (1975). *Biochemistry 14,* 3562-3566.

Markley, J. L., and Ibañez, I. B. (1978). *Biochemistry 17,* 4627-4640.

Markley, J. L., and Porubcan, M. A. (1976). *J. Mol. Biol. 102,* 487-509.

Markley, J. L., Finkenstadt, W. R., Dugas, H., Leduc, P., and Drapeau, G. R. (1975). *Biochemistry 14,* 998-1005.

Massey, V., and Hartley, B. S. (1956). *Biochim. Biophys. Acta 21,* 361-367.

Matthews, B. W., Sigler, P. B., Henderson, R., Blow, D. M. (1967). *Nature 214,* 652-656.

Matthews, R. A., Alden, L. A., Birktoft, J. J., Freer, S. T., and Kraut, J. (1975). *J. Biol. Chem. 250,* 7120-7126.

Matthews, D. A., Alden, R. A., Birktoft, J. J., Freer, S. T., and Kraut, J. (1977). *J. Biol. Chem. 252,* 8875-8883.

Meloun, B., Kluh, I., Kostha, V., Morávek, L., Púsik, Z., Vaněcek, J., and Keil, B. (1966). *Biochim. Biophys. Acta 130,* 543-546.

Mital, R. L., and Gupta, R. R. (1970). *Z. Phys. Chem. 243,* 121-126.

Olson, M. O. J., Nagabhushan, N., Dzwiniel, M., Smillie, L. B., and Whitaker, D. R. (1970). *Nature 228,* 438-442.

Ong, E. B., Shaw, E., and Schoellman, G. (1964). *J. Am. Chem. Soc. 86,* 1271-1272.

Oosterbaan, R. A., and van Adrichem, M. E. (1958). *Biochim. Biophys. Acta 27,* 423-425.

Oosterbaan, R. A., Kunst, P., and Cohen, J. A. (1955). *Biochim. Biophys. Acta 16,* 299-300.

Oppenheimer, H. L., Labouesse, B., and Hess, G. P. (1966). *J. Biol. Chem. 241,* 2720-2730.

Patel, D., Woodward, C., and Bovey, F. (1972). *Proc. Nat. Acad. Sci. USA 69,* 599–602.

Paterson, G. M., and Whitaker, D. R. (1969). *Can. J. Biochem. 47,* 317–321,

Porubcan, M. A. (1978). Ph.D. thesis, Purdue University.

Porubcan, M. A., Ibañez, I. B., and Markley, J. L. (1977). *Fed. Proc., Fed. Am. Soc. Exp. Biol. 36,* Abs. 2586.

Porubcan, M. A., Neves, D. E., Rausch, S. K., and Markley, J. L. (1978). *Biochemistry 17,* 4640–4647.

Porubcan, M. A., Ibañez, I. B., Westler, W. M., and Markley, J. L. (1979). To be published.

Ray, W. J., Jr., and Koshland, D. E., Jr. (1960). *Brookhaven Symp. Biol. 13,* 135–150.

Redfield, A. G., Kunz, S. D., and Ralph, E. K. (1975). *J. Magn. Resonance 19,* 114–117.

Reeck, G. R., Nelson, T. B., Paukstelis, J. V., and Mueller, D. D. (1977). *Biochem. Biophys. Res. Commun. 74,* 643–649.

Reich, E., Rifkin, D. B., and Shaw, E. (eds.) (1975). "Proteases and Biological Control." Cold Spring Harbor Laboratory.

Reynolds, W. F., Peat, I. R., Freedman, M. H., and Lyerla, J. R., Jr. (1973). *J. Am. Chem. Soc. 95,* 328–331.

Robertus, J. D., Kraut, J., Alden, R. A., and Birktoft, J. J. (1972). *Biochemistry 11,* 4293–4303.

Robillard, G., and Shulman, R. G. (1972). *J. Mol. Biol. 71,* 507–511.

Robillard, G., and Shulman, R. G. (1973). *Ann. N.Y. Acad. Sci. 222,* 220–225.

Robillard, G., and Shulman, R. G. (1974a). *J. Mol. Biol. 86,* 519–540.

Robillard, G., and Shulman, R. G. (1974b). *J. Mol. Biol. 86,* 541–558.

Rühlmann, A., Kukla, D., Schwager, P., Bartels, K., and Huber, R. (1973). *J. Mol. Biol. 77,* 417–436.

Satterthwait, A. C., and Jencks, W. P. (1974). *J. Am. Chem. Soc. 96,* 7018–7031.

Sawyer, L., Shotton, D. M., Campbell, J. W., Wendell, P. L., Muirhead, H., Watson, H. C., Diamond, R., and Ladner, R. C. (1978). *J. Mol. Biol. 118,* 137–208.

Scheiner, S., Kleier, D. A., and Lipscomb, W. N. (1975). *Proc. Nat. Acad. Sci. USA 72,* 2606–2610.

Schoellman, G., and Shaw, E. (1962). *Biochem. Biophys. Res. Commun. 7,* 36–40.

Schotton, D. M., and Watson, H. C. (1970). *Nature 225,* 811–816.

Sigler, P. B., Blow, D. M., Matthews, B. W., and Henderson, R. (1968). *J. Mol. Biol. 35,* 143–164.

Stroud, R. M., Kay, L. M., and Dickerson, R. E. (1974). *J. Mol. Biol. 83,* 209–230.

Stroud, R. M., Kossiakoff, A. A., and Chambers, J. L. (1977). *Ann. Rev. Biophys. Bioeng. 6,* 177–193.

Sweet, R. M., Wright, H. T., Janin, J., Chothia, C. H., and Blow, D. M. (1974). *Biochemistry 13*, 4212-4228.

Tulinsky, A., Vandlen, R. L., Morimoto, C. N., Mani, N. V., and Wright, L. H. (1973). *Biochemistry 12*, 4185-4192.

Vandlen, R. L., and Tulinsky, A. (1973). *Biochemistry 12*, 4193-4200.

Walsh, K. A., and Neurath, H. (1964). *Proc. Nat. Acad. Sci. USA 52*, 884-889.

Watson, H. C., Schotton, D. M., Cox, J., and Muirhead, H. (1970). *Nature 225*, 806-811.

Weil, L., James, S., and Buchert, A. R. (1953). *Arch. Biochem. Biophys. 46*, 266-278.

Westler, W. M., and Markley, J. L. (1978). *Fed. Proc., Fed. Am. Soc. Exp. Biol. 37*, Abs. 2882.

Westler, W. M., and Markley, J. L. (1979) (to be submitted).

Wright, H. T. (1973). *J. Mol. Biol. 79*, 1-11.

Wright, C. S., Alden, R. A., and Kraut, J. (1969). *Nature 221*, 235-242.

Zwilling, R., Neurath, H., Ericsson, L. H., and Enfield, D. L. (1975). *FEBS Lett. 60*, 247-249.

^{31}P NMR IN LIVING TISSUE:
THE ROAD FROM A PROMISING
TO AN IMPORTANT TOOL IN BIOLOGY

D. G. Gadian
G. K. Radda
R. E. Richards
P. J. Seeley

Department of Biochemistry
University of Oxford
Oxford

I. INTRODUCTION

In this chapter we shall examine the use of NMR in the study of intact living tissue and organs. We shall, in particular, concentrate on ^{31}P NMR although we shall also briefly examine the possibility of using other nuclei. Many of the methods overlap with those used in studying cell suspensions, and this material is covered in a separate chapter (Shulman, 1979).

It is appropriate to ask why we and others have initially chosen to concentrate on ^{31}P as the most suitable nucleus for observation in living tissues:

(1) Since the ^{31}P nucleus has spin 1/2, it is not subject to electric quadrupole relaxation. Narrow resonances can therefore be obtained.

(2) Although the sensitivity of the ^{31}P nucleus is only about 1/15 that of the proton, the spectra observed are much simpler and therefore far more easily assigned and interpreted. In Fig. 1, for example, we compare the high-resolution proton spectrum of an important metabolite, glucose 6-phosphate, with its phosphorus spectrum. Even for a simple molecule like this, the proton spectrum is rather complex. If one imagines that in a living tissue the number of different molecules from which signals will be obtained is very large, whereas the different phosphate compounds are far more limited in range, it is clear that the phosphorus nucleus has significant advantages.

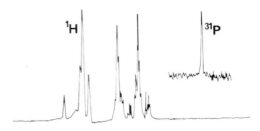

Fig. 1. ^{1}H and ^{31}P NMR spectra of glucose 6-phosphate.
The total width of each spectrum is 5 ppm.

(3) ^{31}P NMR spectra exhibit a wide range of chemical shifts
(~30 ppm for biological phosphates) and in Fig. 2 we have
grouped the shifts of a variety of compounds of biological interest. Also the chemical shift of the signal of a given phosphorus compound is a sensitive indicator of the environment of
the nucleus. For example, the signals obtained from phosphate
compounds in different states of ionization occur at different
frequencies, while those of mixtures give signals at intermediate values, because of fast chemical exchange. Thus one may
simply titrate phosphate compounds through their pK values and
read off from the frequencies of the signals the state of ionization of molecules, and hence the apparent pH of their environment (Fig. 3). Other interactions, like the chelation of
metal ions by phosphates, can also result in easily recognizable
spectra. One such interaction is illustrated in Fig. 4, where
we show the phosphorus resonances of ATP in the absence and
presence of added magnesium ions, as demonstrated some years
ago by Cohn and Hughes (1962). One can not only deduce how magnesium is bound to ATP, but also use the information to find out
what proportion of the ATP in a given solution is complexed to
magnesium.

(4) Another feature of ^{31}P NMR is that the phosphorus relaxation times are sensitive to the nature and mobility of the
phosphate environment. For example, Cohen and Burt (1977) have
studied the interaction of magnesium ions with phosphocreatine
and have deduced from measurements of relaxation times and the
nuclear Overhauser effect that the concentration of free magnesium ions in intact muscle is 3 mM.

Binding to macromolecules reduces mobility and therefore
alters the relaxation times. An illustration of this is given
by an experiment that was designed to establish the stereochemical specificity of binding of the inhibitor glucose 6-
phosphate to glycogen phosphorylase. Glucose 6-phosphate is
present in two anomeric forms that differ only in the chirality
of a single optical centre:

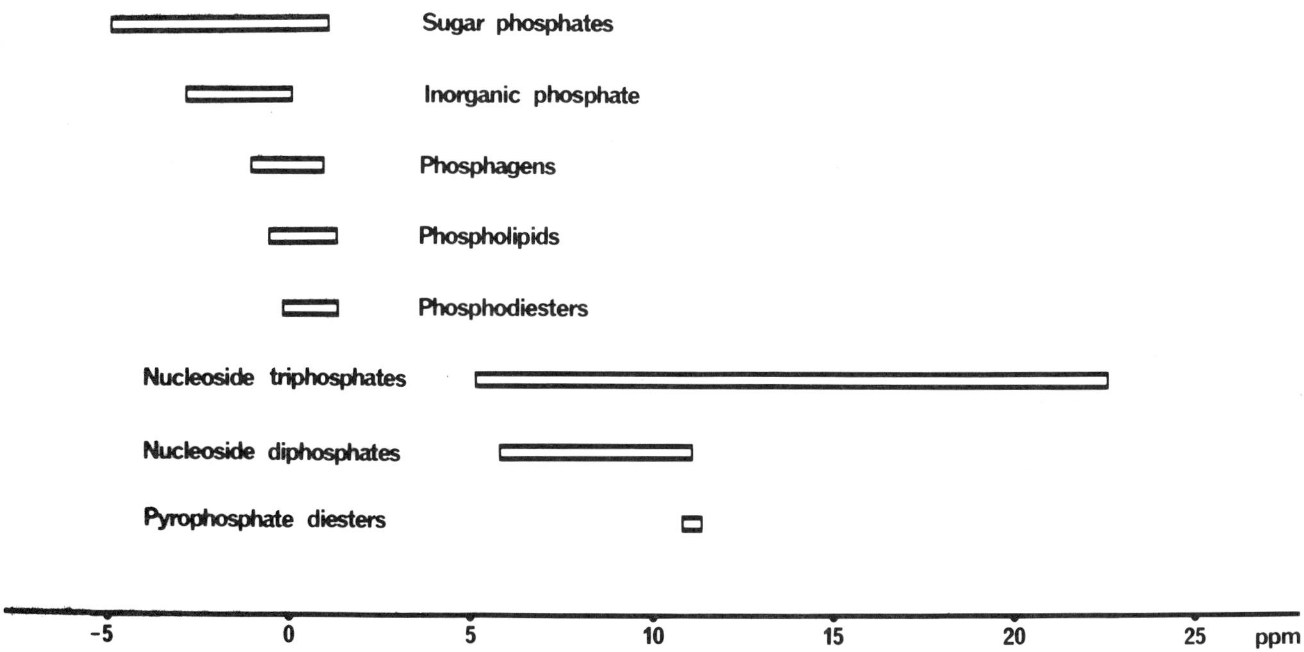

Fig. 2. Chemical shift ranges for groups of biological phosphates. The scale is referenced to 88% phosphoric acid at 0 ppm.

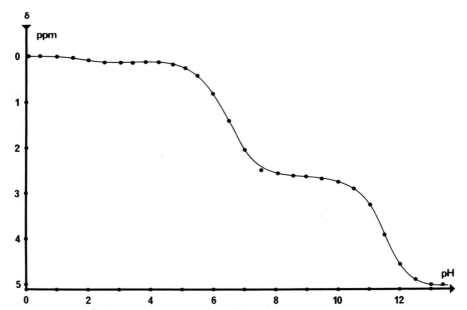

Fig. 3. *Relation between* ^{31}P *resonance frequency and pH for inorganic phosphate (temperature, 24°C; ionic strength, ≃120 mM).*

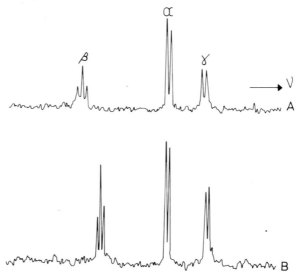

Fig. 4. ^{31}P *NMR spectra of ATP and its complex with magnesium ion. (A) 5 mM ATP, pH 7.30. (B) 5 mM ATP + 7.5 mM magnesium chloride, pH 7.30. The spectral widths are each 36 ppm.*

The α- and β-anomers of glucose 6-phosphate.

The two forms of glucose 6-phosphate are slowly interconvertible in solution (Bailey *et al.*, 1968) and are always in an
equilibrium mixture. Because of the small stereochemical difference between the two forms, two signals are obtained from
glucose 6-phosphate (Battersby and Radda, 1976). This is thus
an example of <u>slow</u> exchange between interconverting forms of a
phosphate group. In principle, the signals could be assigned
to the α and β anomers on the basis of further NMR experiments,
but since there is good chemical evidence that the proportion
of α:β is 40:60, one can assign the observed resonances on this
basis (Fig. 5). The simple question we intended to answer was
which of these two anomers could bind to phosphorylase, or indeed if there was any specificity at all. Taking a mixture of
the two isomers and adding increasing amounts of enzyme to the
solution, one can see (Fig. 5) that only the signal from the
α anomer is broadened. The glucose 6-phosphate inhibitor site
of phosphorylase is thus highly specific for this isomer. In
addition, the experiment illustrates that when a small molecule
binds to a macromolecule the signal is considerably broadened
because of the reduction in the rate of molecular tumbling.
The sugar phosphate is, in this example, exchanging rapidly between free and bound states. In principle, line broadening can
be used to learn something about interactions of metabolites
with enzyme in tissues.
    Note that there is a wide variation in the rate and extent
of motion of phosphorus-containing molecules in tissues and
therefore in their signal linewidths. Thus the high-resolution
spectra of tissues reveal the resonances of the relatively mobile compounds such as ATP and phosphocreatine; membrane phospholipids and DNA produce resonances that are often too broad
to make a significant contribution.
    (5) Apart from technical considerations, there are impor-
tant biological reasons for choosing ³¹P as our primary target
for observation. The dynamic steady state of living systems is
maintained by a continuous input of energy that is used to drive
synthetic reactions and to do different forms of work (mechanical, osmotic, electrochemical, etc.). The overall efficiency
of energy utilization in living systems requires that the indi-

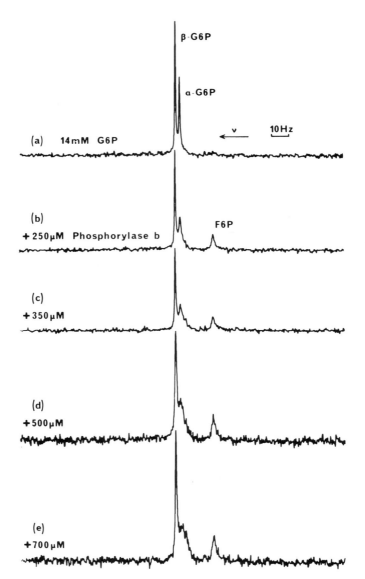

Fig. 5. *The interaction of glucose 6-phosphate with gly-*
cogen phosphorylase b. Phosphorylase was added in increasing
concentration to 14 mM glucose 6-phosphate in 50 mM triethanol-
amine hydrochloride, 100 mM potassium chloride, 1 mM EDTA, pH
7.0. Each spectrum was accumulated from the application of
2000 rf pulses (70°) applied at 4.1 sec intervals. The arrow
shows the direction of increasing frequency (Battersby and
Radda, 1976).

vidual energy conversion processes--of which there are many--
should be as efficient as possible. At the same time, because
of the varying demands imposed on the living organism by the
changing environment, the energy utilization must be strictly
controlled. Sudden or large energy demands are often met from
storage systems (e.g., glycogen in muscle) whereas energy consumption is almost completely shut down in some situations
(e.g., dormancy and hibernation).

Phosphate compounds are involved in the energetics of a
living organism in three ways. (1) The favorable free energy
of hydrolysis of organic phosphates like ATP, phosphocreatine,
and phosphoenolpyruvate makes these compounds the primary energy currency in many processes, particularly in view of their
kinetic stability in the absence of catalysts. (2) In some
cases phosphoenzyme intermediates can be formed from ATP in a
nearly isoenergetic process so that most of the energy available
from ATP breakdown is retained in the macromolecule-phosphate.
This may involve a phosphate covalently linked to a group on
the enzyme or an enzyme product complex in which noncovalent interactions of the enzyme with the product or within the protein
itself provide a way of retaining the energy in the intermediate. (3) Efficient energy conversion often involves highly
organized systems, like membranes where phospholipids have an
essential role in the maintenance of structure and possibly in
the control of energy coupling. The understanding of the
structural and dynamic features of these membranes is prerequisite to the elucidation of the mechanism of energy coupling at
the molecular level.
The importance of the phosphate group in biology is illustrated by the sequence of reactions involved in the anaerobic
catabolism of glucose (glycolysis). The first step is the
phosphorylation of glucose to glucose 6-phosphate, at the expense of breaking the terminal pyrophosphate linkage of ATP.
Further down the sequence a second ATP is used for phosphorylation (fructose 6-phosphate → fructose 1,6-diphosphate). Thus
each intermediate in glycolysis will retain at least one phosphate group, only to be removed at the end of the sequence
phosphoenolpyruvate → pyruvate. It is interesting to speculate
that evolution learned to use this trick in order to confer on
the small substrate a convenient charged group that would aid
its stereospecific recognition by the appropriate enzyme molecules.

II. INTERPRETATION OF TISSUE SPECTRA

In 1974, we reported that useful ^{31}P spectra can be obtained from systems varying in complexity from solutions of purified enzymes to intact tissue (Hoult *et al.*, 1974). We demonstrated that in the ^{31}P NMR spectrum of an intact, relaxed Vastus lateralis muscle, freshly excised from the hind leg of a rat, the resonances of at least four of the major metabolites can be observed. In a typical spectrum (Fig. 6) the resonances could be assigned to the phosphates of ATP, phosphocreatine, inorganic phosphate, and sugar phosphate (mainly glucose 6-phosphate). We also concluded that the frequency of the inorganic phosphate resonance defined the "apparent pH" of its environment (7.1 in the specific example given) and that the frequencies of the ATP peaks corresponded to those for the Mg-ATP complex at the same pH as the inorganic phosphate. We also noted the relatively large breadth of the inorganic phosphate signal compared to that of phosphocreatine, and speculated that this might indicate a distribution of pH within the muscle or even compartmentation of inorganic phosphate. As expected, "aging" of the muscle resulted in the initial breakdown of phosphocreatine followed by that of ATP, i.e., NMR could follow the time courses of these relatively slow metabolic changes. We also showed that during 160 minutes of ischemia the intracellular pH dropped by about one unit. Similar results were later reported by Burt *et al.* (1976a) for other muscle types. These observations established the basis for subsequent work.

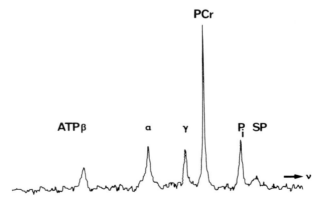

Fig. 6. ^{31}P *NMR spectrum of a rat Vastus lateralis muscle (1 g). Collection time, 13.4 minutes; temperature, 15°C, spectrum width, 38 ppm; PCr, phosphocreatine; P_i, inorganic phosphate; SP, sugar phosphate. The muscle was immersed in Locke solution.*

In order to relate the NMR observations to function, it is important to choose muscles and organs that can be maintained under controlled physiological conditions. The preparations we have used include amphibian skeletal muscle and perfused mammalian heart and kidney, but before describing how we study these tissues, it is necessary to discuss the following problems: (A) assignment of resonances, (B) calibrations for quantitative measurements, (C) calibration of pH determinations, and (D) relaxation times and linewidths.

A. The Problem of Assignments

The initial assignment of the observed resonances requires the accurate measurement of chemical shifts (see Section III), knowledge of these shifts for a range of expected metabolites (see Fig. 2 and Appendix 1), and an understanding of the factors that could influence, for a given compound, its resonant frequency. Of these three criteria, the last can lead to problems because the cellular environment of a given metabolite is not known with any certainty. Indeed, one would wish to use the chemical shift of a metabolite already identified as an indication of its "state" in the cell.

In many tissue spectra the assignment of the resonances of ATP, phosphocreatine, and inorganic phosphate is unambiguous because these signals are well separated from those of other compounds known to be present at high concentrations in the tissue. It is much harder to assign the resonances in the phosphate monoester region because many compounds of this type are present in tissues. Tentative assignments can be confirmed by conventional freeze-extraction of the tissue. The levels of metabolites in the extracts can be measured by biochemical methods and compared with values deduced from NMR measurements on the whole tissue.

Additional information can be obtained from NMR spectra of the extracts, which give narrower resonances than those observed in intact tissue spectra. The most definitive method of assignment using extracts is to record their spectra as a function of pH, before and after addition of expected compounds. If the additional signal always coincides with the resonance of interest in the extract, this should provide additional evidence that the assignment is indeed correct.

An interesting example of how biochemical methods can be combined with NMR observations to assign unknown resonances comes from the detection of "new" ³¹P resonances in several muscle types. The Chicago group reported resonances in the "phosphodiester region" (~-0.5 to 0.5 ppm) for the Northern frog leg muscle, rabbit soleus, and beef heart (Burt *et al.*,

1976b), and we found such a resonance in the red semitendinosus
muscle from the rabbit (Seeley *et al.*, 1976), and for frog and
toad gastrocnemius (Dawson *et al.*, 1977a).

Combination of the NMR data followed by extraction and various chemical and chromatographic identification procedures led
to the assignment of the major phosphodiester resonance in several muscles as that from glycerol 3-phosphoryl-choline (Burt *et
al.*, 1976b; Seeley, 1975). More recently another component,
L-serine-ethanolamine phosphodiester, was identified in a similar way in the pectoralis muscle of chickens with hereditary
muscular dystrophy (Chalovich *et al.*, 1977). Since several
phosphodiesters have been observed, the presence of a signal in
this region of the spectrum must always be followed up by extraction and chemical identification.

It is interesting that the concentration of these phosphodiesters in muscle varies widely from species to species. In
fact, their occurrence was first reported in 1954 (Roberts and
Lowe, 1954) and was widely investigated in the following 15
years (for references, see Chalovich *et al.*, 1977) before they
became forgotten. While NMR may well help in elucidating their
functional role, at present it can only give us a quick and
analytically accurate estimate of these compounds in different
tissues.

B. *Quantitation--Determination of Concentration*

1. *Relative Concentrations*

Relative concentrations of metabolites can be determined
from the areas of their resonances. The areas are proportional
to the quantities of metabolites within the volume enclosed by
the radiofrequency (rf) coils, only if the rf pulses are applied at time intervals much greater than the spin-lattice relaxation times (T_1) of the resonances. However, in order to
optimize the signal-to-noise ratio of the spectra, pulses are
normally applied at time intervals approximating to the T_1
values of the resonances. Under these conditions, the areas
of the individual resonances are reduced by factors determined
by their T_1 values. These "saturation" factors vary with pulse
angle and interval, and with magnetic field. They may also depend on the type and state of the tissue, and it is therefore
essential to determine them under the appropriate conditions.
For frog gastrocnemius muscles at $4°C$ (Dawson *et al.*, 1977a),
in experiments performed at 129 MHz in which $70°$ rf pulses are
repeated every 2 sec, the relative concentrations of ATP,
phosphocreatine, sugar phosphates, inorganic phosphate, and
the phosphodiesters are arrived at by multiplying their resonance areas by 1.0, 1.25, 1.5, 1.8, and 1.3, respectively.

2. Absolute Concentrations

Quantitative calibration in absolute terms is complicated
by the facts that the exact volume that contributes signal cannot in general be determined and that the sample contains both
tissue and an unknown amount of buffer solution. For these
reasons, special calibration methods must be employed. The
exact method will depend on the particular tissue under examination.

Calibrated spectra have been obtained from frog sartorii,
which can be maintained in a steady "resting" state for sufficiently long periods (Dawson et al., 1977a). A resting spectrum for frog sartorii was obtained in a normal tris-buffered
Ringer's solution to confirm that the inorganic phosphate level
was low and thus that the muscles were in a steady resting
state. A spectrum was then obtained while circulating Ringer's
solution containing 10.0 mmol liter^{-1} phosphate buffer. During
the latter spectral accumulation (Fig. 7A) the ratio of phosphocreatine to inorganic phosphate remained constant, confirming that phosphate penetrates only slowly into the muscle fibers (Stella, 1928). Finally the muscles were removed and a
further spectrum (Fig. 7B) was obtained. The concentration of
phosphocreatine within the muscle fibers was estimated from

$$[PCr] = 10\underline{x}/[\underline{z} - \underline{y} + \underline{x}(P_i/PCr)_{resting}] \quad \text{mmol liter}^{-1}$$

using the symbols shown in Fig. 7. The last term is a correction for the internal P_i of the resting muscle based on the
first accumulation. This is not shown in Fig. 7, but the P_i
was small, with $P_i/PCr = 0.054$. From Fig. 7, $\underline{x} = 3.5$, $\underline{y} = 5.1$,
$\underline{z} = 6.4$, and so $[PCr] = 23.5$ mmol (liter fiber)$^{-1}$.

For many purposes, the ratio of concentrations provides a
sufficient basis for interpretation.

C. Calibration of pH Measurements

The measurement of intracellular pH involves both theoretical and practical difficulties. The two most commonly used
techniques are (i) insertion of a pH-sensitive microelectrode
into the cell, and (ii) analysis of the distribution of weak
acids or bases. These and other methods have well-documented
advantages and have been critically reviewed (Waddell and
Bates, 1969; Cohen and Iles, 1975).

^{31}P NMR provides an alternative method of determining intracellular pH, as first shown by Moon and Richards (1973). In
principle, any resonance whose frequency is sensitive to pH can
be used, but in practice the resonance of inorganic phosphate

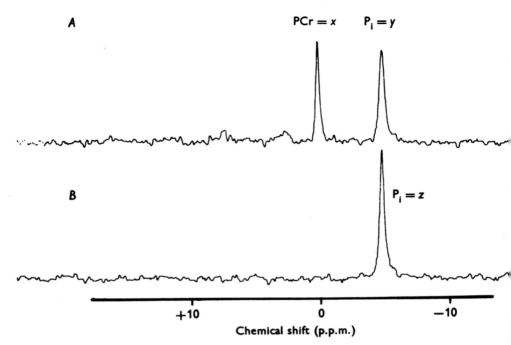

Fig. 7. *Absolute calibration of peak areas in frog sar-*
torious. Spectrum A: Four frog sartorii in phosphate buf-
fered (10 mmol liter^{-1}) Ringer solution. Spectrum B: Buffer
alone after removal of muscle. 238 scans at 16 sec intervals.
$\underline{x}$, $\underline{y}$, and $\underline{z}$ are relative peak areas. (From Dawson et al.,
1977a.)

is generally the most suitable because it is readily observable
in the majority of spectra, and because P_i has a pK_a in the pH
region of interest.

Inorganic phosphate exists mainly as HPO_4^{2-} and $H_2PO_4^-$ around
neutral pH. In the absence of chemical exchange, these two
species would give rise to two resonances separated from each
other by about 2.3 ppm. In solution, however, the two species
exchange with each other rapidly ($\sim10^9$–10^{10} sec^{-1}), and as a
result, the observed spectrum consists of a single resonance
the frequency of which is determined by the relative amounts of
two species. The frequency of the signal as a function of pH
gives a "standard" pH curve. One must ensure that other factors (e.g., chelation by metal ions) produce negligible shifts
in frequency. Control experiments have demonstrated that the
effects on the inorganic phosphate resonance of likely variation
in ionic strength or metal-ion binding within tissues are very
small. For a calibration under physiological conditions see
Fig. 8.

Chemical shift (ppm)

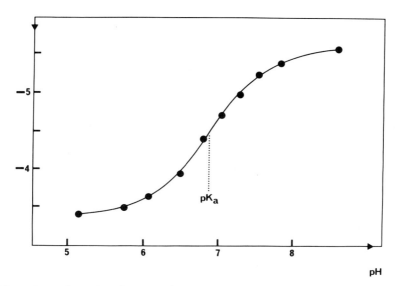

Fig. 8. The chemical shift of inorganic phosphate as a function of pH under physiological conditions of ionic strength and at 4°C. The filled circles are experimental points; the continuous line is drawn from the equation $pH = 6.88 + \log_{10} (\delta - \delta_1)/(\delta_2 - \delta)$, where δ is the observed chemical shift and δ_1 and δ_2 are the shifts of $H_2PO_4^{1-}$ and HPO_4^{2-}, -3.35 and -5.6, respectively. (The chemical shift scale is zeroed on the phosphocreatine resonance.)

The width of the P_i resonance may also reflect the distribution of pH within a tissue volume and allows limits to be placed on the rates at which such distributions are averaged. This is an additional advantage of ³¹P NMR over other methods, which yield only an average value for intracellular pH.

What are the problems in converting an observed resonance frequency of an ionizable phosphate group to a measurement of intracellular pH?

1. Chemical Shift Measurements--Use of a Lock and of Standards

The first necessity is to measure chemical shifts accurately. It is standard practice to use a field-frequency lock, in order to counteract magnetic field drifts, and to ensure that chemical shifts are (at least in principle) reproducible from one experiment to another. A deuterium resonance is frequently used for the lock, and in many experiments it is convenient simply to add the required amount of D_2O to the sample, providing an "internal" lock. In some cases, however, it may be impossible or impractical to do this, and an "external" lock, provided, for example, by a capillary tube containing D_2O, may then be required. It is essential to be able to correlate the results obtained with the two types of lock.

When using an internal D lock, it is necessary to consider any effects the nature of the sample may have on the D shift, for such effects will be reflected in the observed ^{31}P shifts; in fact, the observed shift will be equal to the true ^{31}P shift minus the D shift.

Within the pH range 3-9, the D resonance of 10% D_2O in water shifts by less than 0.02 ppm, which is within the experimental error of most measurements. Therefore, the use of an internal rather than external lock has a negligible effect on the ^{31}P pH titrations. However, the chemical shift of 10% D_2O in water increases linearly with KCl concentration, the magnitude of the shift being 0.16 ppm per molar KCl (Gadian, 1974). For studies performed at high ionic strength, it is therefore necessary to apply a correction to the observed ^{31}P shift.

Many of these problems are avoided by the use of internal or external phosphate standards. A convenient external standard is that used by Burt et al. (1976a): methylene diphosphonic acid. Phosphocreatine (Dawson et al., 1977a) and phosphodiester signals are suitable internal references for tissues. Internal standards, when available, are recommended (as long as all the appropriate controls have been performed) because the standard experiences precisely the same field as that experienced by all the compounds being measured. Addition of a standard to a capillary tube or to the buffer surrounding a tissue could perhaps lead to errors if the field in this region differs slightly from that in the tissue as a result of magnetic susceptibility effects, etc.

Glonek and van Wazer (1974) have criticized the use of 85% phosphoric acid as a standard on the basis of its large linewidth (~5 Hz). An additional disadvantage of phosphoric acid is that it has a different magnetic susceptibility from aqueous solutions. This causes chemical shifts measured relative to this compound to differ according to whether the sample is aligned parallel or perpendicular to the magnetic field, and hence according to whether a superconducting or electromagnet is used. Thus, for example, in the Oxford laboratory where a superconducting magnet is used, the phosphocreatine resonance at neutral pH is 2.35 ppm upfield of 85% phosphoric acid, whereas Burt et al. (1976a), using an electromagnet, quote a value of 3.2 ppm. Care should therefore be taken when expressing shifts relative to this standard.

2. Knowledge of the pK Values

It is imperative to know the pK of the compound used for the determination of pH under the particular experimental or cellular conditions. In several publications this fact has not been recognized. Some time ago, we (Gadian, 1974) carried out a series of studies to determine the influence of several factors on the pK values of some of the most common phosphate-containing metabolites.

Several titrations are shown in Fig. 9 and the data together with the calculated pK_a values are collected in Appendix 1.

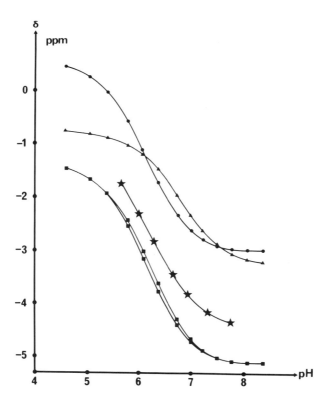

Fig. 9. pH titrations of inorganic phosphate and phosphate esters. ●, Glucose 1-phosphate; ▲, inorganic phosphate; ★, inosine 5'-phosphate; ■, glucose 6-phosphate. Signals from both α and β anomers of glucose 6-phosphate were resolved near the pK_a. Temperature, 24°C; ionic strength, ≈120 mM.

The effects of phosphate concentrations on the chemical shifts and pK_a values are in general negligible. We have done titrations on solutions containing several metabolites at different concentrations. The solution of highest concentration contained 50 m$\underline{M}$ inorganic phosphate, 50 m$\underline{M}$ ATP, and 50 m$\underline{M}$ glycerol 2-phosphate, while that of lowest concentration contained 5 m$\underline{M}$ inorganic phosphate, 5 m$\underline{M}$ glucose 6-phosphate, 5 m$\underline{M}$ ATP, and 5 m$\underline{M}$ phosphocreatine. The inorganic phosphate titrations for these two solutions agreed very closely. There was a constant difference of about 0.04 ppm, which could be attributed to a small shift in the D resonance in the solution of high ionic strength. The glucose 6-phosphate and phosphocreatine titrations followed exactly those obtained at higher concentrations (20-40 m$\underline{M}$). The chemical shifts of 50 m$\underline{M}$ ATP were, however, considerably different from those of 5 m$\underline{M}$ ATP, presumably due to the increased stacking of ATP molecules at the higher concentration.

The effect of high ionic strength (1.6 $\underline{M}$ KCl) on the chemical shifts of inorganic phosphate, glucose 6-phosphate, and phosphocreatine is shown in the titration curves of Fig. 10. There is very little shift in the terminal frequencies of the resonances, but the pK_a values of all the compounds are, as expected, significantly lowered by the increase in ionic strength. This addition of KCl shifts these phosphate resonances indirectly by changing the degree of protonation of the phosphate groups.

The pH titrations of ATP (see Figs. 11 and 12) in the presence of KCl show extensive shifts in the terminal frequencies, in addition to the shifts reflecting changes in the pK_a values. These terminal shifts can be interpreted in terms of the weak binding of K^+ ions to ATP (Dawson *et al.*, 1969).

Some of the results are summarized in Table I are are detailed in Appendix 1.

Lack of knowledge about the exact ionic conditions (and about binding to macromolecules; see below) within a cell or tissue puts some uncertainty on the derived pH values, and in general we feel that it is unwise to quote pH values to better than ±0.1 pH unit unless the pK_as have been determined under the specific cellular conditions. In cases where one wants to know precisely the magnitude of a pH gradient, such determination is essential (see, for example, Section IV,C,2).

The effect of temperature on the pK values also has to be considered. In the range 20-40°C, the ^{31}P resonances of AMP, NAD^+, ATP, phosphocreatine, and inorganic phosphate shift downfield with increasing temperature by between 0.005 and 0.015 ppm K^{-1} (at pH 6.6). For most purposes such a shift can be ignored.

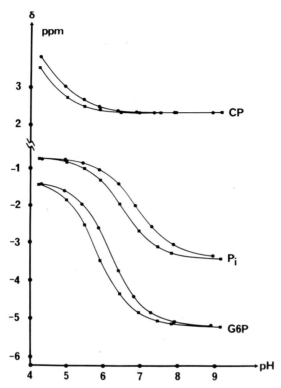

Fig. 10.   The effect of ionic strength on phosphate resonance frequencies.   ●, ionic strength of approximately 100 m$\underline{M}$;
■, +1.6 $\underline{M}$ KCl.   Glucose 6-phosphate curves are the means of
the two anomer frequencies.   Temperature, 24°C.

TABLE I.   Some pK_a Values Determined From the Variation of
Phosphorus Chemical Shifts with pH[a]

	Phosphate Counterions only	1.6 $\underline{M}$ KCl	6 m$\underline{M}$ MgCl$_2$
Compound			
Glucose 6-phosphate	6.15	5.7	−
Inorganic phosphate	6.8	6.5	−
ATP			
α	6.5	6.3	5.4
β	6.6	5.9	4.9
γ	6.6	5.95	5.15

[a]All phosphate concentrations 5 m$\underline{M}$.

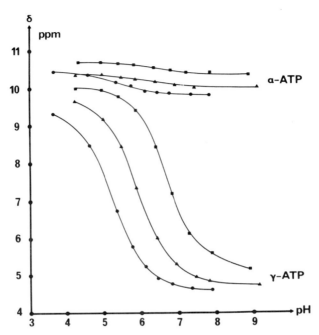

*Fig. 11. The effects of ions on the phosphorus resonance
frequencies of the* α *and* γ *phosphates of ATP.* ■ , *ionic
strength approximately 120 mM;* ▲ , *+1.6 M KCl;* ● , *+6 mM
MgCl₂. The ATP concentration was 5 mM. Temperature, 22°C.*

3. Other Interactions

   a. Metal Ions. Reference has already been made to the interaction of phosphocreatine with Mg^{2+} and of ATP with K^+.
Here we shall discuss the interaction of ATP with a paramagnetic (Mn^{2+}) and a diamagnetic (Mg^{2+}) divalent cation. Until
the availability of spectrometers with the present sensitivities, much of the ^{31}P NMR work had to be carried out at relatively high concentrations of nucleotide (0.2-0.3 M). Thus in
the early work, the effects of high ionic strength (due to the
ionized ATP) and stacking could not be avoided. Nevertheless
many interesting observations on the effect of complexing with
Mg^{2+} (Cohn and Hughes, 1962), paramagnetic ions (Sternlicht *et
al.*, 1965a,b) and monovalent cations have been reported. Here
we shall illustrate some of the possibilities for studying me-
tal-phosphate interactions with reference to more recent work
in our laboratory.

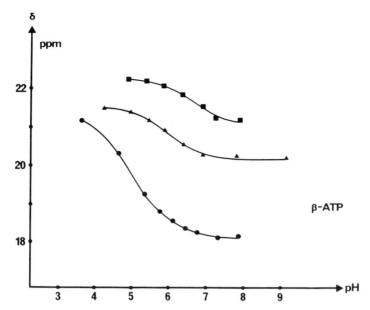

Fig. 12. The effects of ions on the phosphorus resonance frequencies of the β phosphate of ATP. Conditions and symbols as for Fig. 11.

Mn^{2+} - ATP. Study of the interaction between Mn^{2+} and ATP may prove to be useful for cellular systems as there are ways of introducing this cation into cells. In solution this interaction has been investigated in detail by Brown et al. (1973) following the original studies of Cohn and Hughes (1962) and Sternlicht et al. (1965a,b). The measurements of the spin lattice relaxation rates of the phosphorus nuclei in ATP as a function of temperature in the presence of Mn^{2+} were interpreted in terms of the dipolar interaction between the electron and nuclear spins. The relaxation data indicate that the Mn^{2+} ion forms a complex with ATP that has an average lifetime of 6.5 μsec at 25°C and that the ion is somewhat closer to the β and γ phosphorus atoms than the α atom. These data, however, and other data obtained at high concentrations of ATP are not necessarily applicable to tissue studies where nucleotide concentrations are considerably lower (~5 mM).

Mg^{2+} - ATP. Since ATP is often present in the cell as its Mg^{2+} complex, several studies have been performed on the effect of Mg^{2+} ions on the ATP resonances. pH titrations of 5 mM ATP in the absence of Mg^{2+} ions, and in the presence of 6 mM $MgCl_2$

are shown in Figs. 11 and 12. It may be seen that the effects
of Mg^{2+} are qualitatively similar to those produced by addition
of high concentrations of KCl to ATP solutions. In particular,
at physiological pH, Mg^{2+} ions produce large and differential
shifts of the three ATP resonances, the β and γ phosphate signals being most affected. The pK_{a4} of ATP is lowered by about
1.5 units on coordination to Mg^{2+}. The titration curves for
Ca-ATP are similar (Gadian, 1974; Colman and Gadian, 1976).

    The studies we describe were done at concentrations compa-
rable to those found in tissues, and do not necessarily correlate with earlier studies in the literature using higher nucleotide concentrations.

    b. Interactions with Macromolecules. The binding of phos-
phates to enzymes within the cell may well produce chemical
shifts that would affect the absolute value of the measured pH.
It is not easy to exclude this possibility completely, but
there are good arguments against it. First, the pH measured by
^{31}P NMR in several muscles is close to that obtained by direct
electrode measurements: for example, in frog sartorii (Dawson
et al., 1977a) and in the giant barnacle muscle cell (Fig. 13).
Second, in cases where the pH values could be derived from the
chemical shifts of both P_i and glucose 6-phosphate, good agreement was obtained. Third, the natural linewidth of the P_i
resonance in the rat Vastus lateralis muscle is only 2 Hz
(Seeley *et al.*, 1976), suggesting that the interaction with macromolecules in the cell is relatively weak. Nevertheless, in
some situations, particularly when one aims at using the measured pH in a precise and quantitative sence, some uncertainty
in the absolute value must remain.

D. Relaxation Times and Linewidths

    Since information about the metabolite interactions within
the cell can be obtained from measurements of relaxation times
and linewidths (Seeley *et al.*, 1976; Cohen and Burt, 1977), it
is important to have some understanding of the factors that
govern their values. We shall discuss here the mechanisms of
relaxation that are particularly relevant to the phosphorus
nucleus.

    First, the ^{31}P nucleus has spin 1/2. It therefore has
spherical electrical symmetry and there is no relaxation via
quadrupolar processes. Second, because the phosphorus nucleus
has a small magnetic moment and is often separated from other
magnetic nuclei by oxygen atoms, the dipolar relaxation of the
^{31}P nucleus is relatively inefficient. This means that the
contribution of dipolar coupling to the relaxation times may
be small, even when molecular motion is slow, as is often the
case in biological systems.

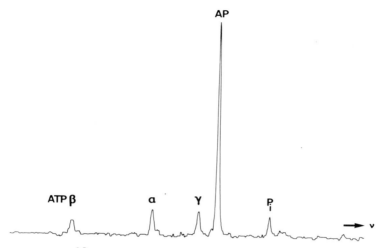

Fig. 13. ^{31}P NMR spectrum of 20 barnacle muscle fibers. *The fibers were suspended in a buffer containing 500 m$\underline{M}$ Na$^+$, 12 m$\underline{M}$ K$^+$, 12 m$\underline{M}$ Ca^{2+}, 20 m$\underline{M}$ Mg^{2+}, 576 m$\underline{M}$ Cl$^-$, and 20 m$\underline{M}$ Tris- maleate, pH 7.27 at 20°C. The spectrum was accumulated by ap- plying 1600 rf pulses at intervals of 2 sec. AP, arginine phosphate; spectrum width, 38.7 ppm.*

    There are other mechanisms that are particularly important for transverse relaxation of phosphorus. One arises from modu- lation of chemical shifts by nuclear exchange between two (or more) chemical environments. For example, the β and γ phosphate resonances of a solution containing 5 m$\underline{M}$ ATP and 5 mM MgCl$_2$ are broadened as a result of exchange between Mg-ATP and the small amount of free ATP that is present. It is likely that this mechanism provides an important contribution to the ATP line- widths in tissue spectra. A second mechanism involves aniso- tropy of the chemical shift. Our earlier results on the contri- bution of this mechanism to ^{31}P relaxation in phospholipids (Berden *et al.*, 1974) have been extended to other phosphate derivatives. We measured the linewidths of several phosphates in 95% glycerol (Table II) at two magnetic field strengths. Be- cause of the high viscosity of the glycerol solutions, the reso- nances were sufficiently broad for us to be able to ignore the effects of field inhomogeneity. There is, for all of the com- pounds except the phosphonium salt, a strong field-dependence of the linewidths, which could arise from only two mechanisms. The first involves nuclear exchange between two chemical envi- ronments, which for the studies presented here can be identified as differing states of ionization of the phosphate groups. For inorganic phosphate, the theoretical contribution of exchange to the linewidth is only about 0.2 Hz at 129 MHz and 0.01 Hz at

TABLE II.   Chemical Shift Anisotropy: Phosphate Line-widths at 36 and 129 MHz[a]

Compound	Linewidth 36 MHz (±1 Hz)	Linewidth 129 MHz (±2 Hz)	Ratio[b,c]
ATP			
α	16	100	6.3
β	13	120	9.2
γ	16	135	8.4
Glucose 6-phosphate	10	82	8.2
Inorganic phosphate	4	30	7.5
NADH	15	77	5.1
p-Nitrophenol phosphate	5	25	5.0
Phosphocreatine	8	32	4.0
Pyridoxal 5'-phosphate	18	80	4.4
Tetramethyl phosphonium iodide	11	13	1.2
Trimetaphosphate	5	30	6.0

[a]*Conditions: Phosphates were dissolved in 20 mM EDTA and the pH adjusted to 7.0 ± 0.1.   Glycerol was then added to the aqueous solution to 95% (w/w).   Spectra were recorded at 22°C with an external field-frequency lock of deuterium oxide.*
[b]*(Linewidth at 129 MHz)/(linewidth at 36 MHz).*
[c]*(Ratio of frequencies)[2] is 12.6.*

36.4 MHz. Furthermore, the strong field dependence of line-widths is observed also for NADH, phosphocreatine and trimeta-phosphate, none of which have pK_a values in the neutral pH range used in the experiments. Thus chemical exchange does not contribute significantly to the linewidths, and the field dependence must arise from the alternative mechanism, anisotropy of the phosphorus chemical shift.

The contribution of chemical shift anisotropy to the line-widths is given by

$$\frac{1}{T_2} = \gamma^2 B_0^2 \, a\sigma_0^2 \left[\frac{2}{15} \tau_R + \frac{1}{10} \frac{\tau_R}{1 + \omega_0^2 \tau_R^2} \right]$$

where γ is the gyromagnetic ratio of the nucleus; B_0 the flux
density of the magnetic field; $_a\sigma_0^2 = {}_a\sigma_x^2 + {}_a\sigma_y^2 + {}_a\sigma_z^2$, in which
$_a\sigma_x$, $_a\sigma_y$, $_a\sigma_z$ are the anisotropic parts of the chemical shift
tensor; ω_0 the resonant frequency; and τ_R is the rotational
tumbling time, which can be estimated using the known viscosity
of 95% glycerol.

   Our data indicate that $_a\sigma_0^2$ is of the order of 100 ppm for
common phosphates. These estimates are ratified by low-reso-
lution spectra of polycrystalline phosphate samples (Seeley,
1975), and by the elegant experiments of Kohler and Klein
(1976) on electronic distribution about the phosphorus nucleus.
The only phosphorus compound in Table II that fails to show a
field dependence in linewidth is that with a symmetric distribution of charge about the phosphorus nucleus--tetramethyl
phosphonium iodide.

   A consequence of chemical shift anisotropy is that it can
very much reduce the gain in sensitivity and resolution to be
expected from increase in the magnetic field. If this mechanism is dominant, the linewidth will increase according to B_0^2.
Since the frequency range of the spectrum increases according
to B_0, the resolution will actually decrease on increasing the
magnetic field. The best resolution will be obtained at a
field such that chemical shift anisotropy accounts for half the
linewidth. This effect is more important for studies involving
the binding of ligands to macromolecules than for those on intact tissue, for which the observed metabolites are relatively
mobile. Indeed, our studies on phosphorylase (Busby et al.,
1975) were performed at 36.4 MHz in preference to 129 MHz, because of the excessively large linewidths observed at the higher
frequency.

   Additional contributions to relaxation can arise from spin-
rotation, and from modulation of scalar coupling by means of
chemical exchange. These mechanisms have been discussed by
Cohen and Burt (1978) in relation to their studies on the interaction of Mg^{2+} ions with phosphocreatine.

   The resonances in tissue spectra typically have T_1 values
in the region 0.5-5 sec and linewidths ranging from about 5 to
100 Hz. However, few interpretations have been made of the observed values (but see Seeley et al., 1976; Busby et al., 1978;
Cohen and Burt, 1977), largely because little is still known
about the precise details of phosphorus relaxation. A further
consequence of the failure of the dipolar mechanism to dominate
phosphorus relaxation is small Overhauser enhancements observed
on hydrogen decoupling phosphate spectra, typically 20-30%
(Seeley, unpublished) at 36.4 MHz.

   The possibility arises that local variations in magnetic
susceptibility within the tissue produce line broadening that
cannot be corrected for. In the case of frog gastrocnemius

muscles, we have observed a linewidth for the phosphocreatine
resonance of 5.5 Hz at 73.8 MHz. Therefore, in this tissue,
magnetic susceptibility variation can generate, at most, a line
broadening of 0.075 ppm.

III. THE ATTAINMENT OF STABLE PHYSIOLOGICAL CONDITIONS IN THE SPECTROMETER

   If we are to make meaningful observations on metabolites
in relation to function it is essential to maintain the tissue
under well-defined physiological conditions within the spectrometer. The need for this, together with the constraints imposed by design of the magnet, the probe, and the size of the
sample tube to some extent determine the choice of appropriate
tissue. It is not surprising therefore that the first muscle
examined under physiological conditions was the frog sartorius.
Examination of perfused organs--heart and kidney--rapidly followed. We shall now briefly describe our own experimental set-
up for these three systems.

A. *Skeletal Muscle*

   The choice of the experimental preparation involves a compromise. Due to the insensitivity of the technique one wishes
to work with as large a sample, and thus as large a signal, as
possible. However, in order to keep the muscle in a physiological condition, it must be adequately oxygenated. In the experiments by Dawson *et al.* (1977a,b) oxygen was delivered by
diffusion from the bathing medium surrounding the muscles; this
can only be accomplished with very thin muscles separated by an
appreciable volume of fluid. In addition, it is highly desirable to choose muscles that function at low temperatures since
the rates of biochemical processes, including oxygen consumption, are then decreased (Hill, 1965). Many of the experiments
were done on frog sartorii from *Rana temporaria* at $4^{O}C$. These
muscles are only about 1 mm thick (by approximately 5 mm wide
and 30 mm long) and so it is relatively easy to supply them
with oxygen by diffusion. $4^{O}C$ was chosen rather than $0^{O}C$,
which is usual in physiological experiments on amphibian
muscles, in order to provide a margin of safety against freezing. Even though one can put four frog sartorii in an 8 mm
inner diameter sample tube, only about one-third of the cross-
sectional area of the NMR tube is occupied by muscle. For certain experiments, therefore, pairs of small gastrocnemii, either
from *R. temporaria* or from the toad *Bufo bufo* were used. These

muscles are roughly circular in cross section, having a diameter of about 4 mm at the widest point and weighing about 100
mg each.   These gastrocnemii fill up appreciably more of the
sample space, but are not as well oxygenated.   In experiments
in which anaerobic metabolism is studied, larger gastrocnemii
(400-500 mg) may be used.

1. Design of Chamber

The glass NMR sample tube was converted into an experimental chamber as shown in Fig. 14.   The support system is constructed entirely of Teflon, glass, and epoxy resin, as these
materials do not appreciably disturb either the homogeneity of
the magnetic field or the tuning of the rf coil.

The muscle holder consists of a central glass capillary
tube to which are attached two Teflon bobbins.   One of these
serves as the top to the experimental chamber, and the other,
a circular disk with two slots cut out, is fixed to the capillary tube approximately 1 mm from the bottom.   The ends of the
muscles are inserted into these slots so that small attached
pieces of bone form stops below the bobbin.   Cotton tied to
the tendon at the other end of the muscle is threaded through
holes in the top of the chamber and tied over the force transducer, which is fitted onto another Teflon support above.   The
muscles are held in a vertical position parallel to the central
glass tube and the magnetic field.   The capillary tube serves
an additional purpose, being filled with a solution of KCl in
D_2O.   The KCl (134 g/kg D_2O) is present to prevent the D_2O from
freezing at the temperatures of around 4°C at which the experiments are conducted.   (The freezing point of pure D_2O is
3.81°C.)   The resonance from deuterium is used for the field-
frequency lock and for optimizing the spatial homogeneity of
the magnetic field.

2. Stimulation and Recording

The muscles are stimulated electrically via two axially located platinum wires, one threaded through the Teflon top and
one sealed into the bottom of the chamber, and the resulting
force development is recorded by a transducer consisting of two
silicon strain gauges forming a bridge circuit bonded above and
below a narrow phosphor-bronze strip attached across a circular
Teflon support (Fig. 14).

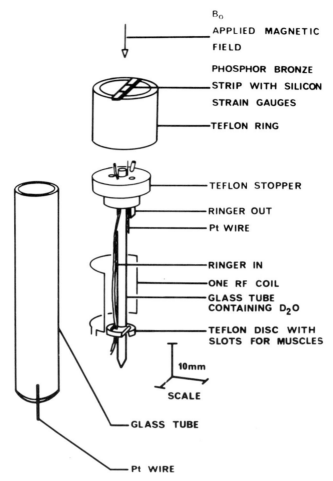

*Fig. 14.   Design of experimental chamber for superfusion
of amphibian muscles.   The volume in which the NMR measurement
is made is defined by the two single-turn rf coils, only one
of which is shown in the diagram.   (From Dawson et al., 1977a.)*

3. Superfusion by Oxygenated Ringer's Solution

Bubbling air or oxygen through the experimental chamber
would result in severe disturbance of magnetic homogeneity;
therefore, a recirculating perfusion system was chosen as the
method of oxygenation. Approximately 200 ml of Ringer's solution was bubbled with 100% oxygen outside the spectrometer, and
this was pumped to and from the chamber by a peristaltic pump.

B. Perfused Heart

The requirements of perfusion are quite conventional. The
heart needs an uninterrupted supply of oxygenated, buffered,
bubble-free perfusion fluid at a controlled temperature. We
use Krebs-Henseleit solution (95% oxygen; 5% carbon dioxide
filtered to 0.45 μm) for most experiments and run the Langendorff heart at 37°C. Since oxygen tension and temperature of
the perfusion medium are thus considerably greater than those
of the surroundings, the apparatus has been constructed almost
entirely out of thermostatted glass tubing to minimize diffusion losses. (Short, flexible joints between glass units are
made by non-toxic 'Tygon' surgical plastic; Fig. 15.) The
first problem in interfacing the perfusion and NMR techniques
is a manipulative one. After conventional aortic cannulation
of the heart adjacent to the magnet and the initial perfusion
period, the organ must be transferred into the magnet without
interruption of flow to the coronary arteries. As a consequence, the perfusion apparatus must have spatial flexibility,
and the length of the circuit (from oxygenated reservoir to
heart) is about three times that of a normal Langendorff system. Flexible plastic tubing joining the glass units allows
the thermostatted column (T_{a2}) to move through space as the
probe and sample are inserted into the bottom-loading magnet.
Oxygen loss over this rather long perfusion circuit is mini-
mized by its glass construction. Slight cooling of fluid
(<1°C) is corrected just before its entry to the probe by heat
exchanger T_b. The narrow plastic tube that conveys medium
through the probe is warmed, as is the sample tube, by a rapid
flux of thermostatting air controlled by a bimetallic sensor
and associated electronics.

The metal cannula customarily used for Langendorff perfusion has been replaced by glass to avoid absorption of rf radiation. Dark paint is fused into the tips of these glass cannulae to render them visible through the aortic wall and thus
reduce the risk of damaging the aortic valve during cannulation.
Signal-to-noise has been optimized by filling the rf coil as
much as possible with the heart and by minimizing field inhomogeneities. The beating heart is immersed in perfusion medium
and is maintained just free (by a fraction of a millimeter) of
the walls of the sample tube (8 mm inside diameter) and the
reference capillary (0.8 mm outside diameter). Hearts in even
light contact with the sample tube had visible ischemia patches
and impaired function; those organs subjected to yet higher surface pressures failed to beat regularly. Immersion of the heart
under a small height of buffer (about 2 cm) made no impact on
its achievement of a functional steady state and caused no additional edema over periods up to 5 hours, but reduced linewidths

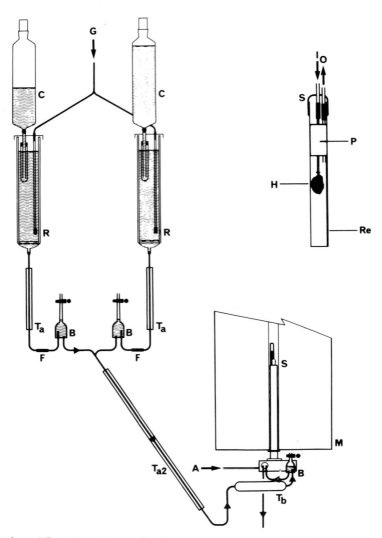

Fig. 15. Heart perfusion apparatus used for NMR measurements. Glassware is maintained at constant temperature by rapid flow of thermostatted water. The right-hand reservoir and bubble trap are auxiliaries used in experiments requiring more than one perfusion solution. G, gas input (95% O_2, 5% CO_2); C, constant head device; R, reservoir; T_a, T_{a2} thermostatted glass tubes; T_b, thermostatted glass helix; F, flow control; B, bubble trap; A, input for thermostatting air; M, superconducting magnet; S, sample tube; I, aortic input for perfusion fluid; O, output for conary flow; P, Teflon support post for cannula; H, heart; Re, fine reference capillary. (From Seeley et al., 1977.)

of intracellular phosphate signals. This experimental set-up
has enabled us to maintain rat and mouse hearts for periods in
excess of 5 hours.

C. Technical Considerations

The modifications to probes that are necessary for studying tissue should be designed in such a way that signal-to-
noise and field homogeneity are degraded as little as possible.

1. Signal-to-Noise

As discussed by Hoult and Richards (1976) and Hoult and
Lauterbur (1979), conductors positioned close to the rf coil
can degrade signal-to-noise. The tissue and bathing medium
have finite conductivity, and therefore generate noise. Since
only the tissue produces signal, it is best to minimize the
proportion of the surrounding medium. It is essential to re-
place the metal cannulae normally used for perfusion by cannulae made of insulating material such as glass. Also, any
electrodes that are required for electrical stimulation should
be kept as far from the rf coil as possible, and it has been
found necessary, for experiments on skeletal muscle, to place
rf stoppers in the stimulation circuit immediately above and
below the sample chamber (Dawson et al., 1977a). It was also
necessary for these experiments to insert rf filters and decoupling capacitors into the transducer and stimulation leads
to prevent leakage of any stray radiation from the laboratory
into the probe.

2. Field Homogeneity

In many experiments, the presence of perfusion tubes and
electrical leads makes sample spinning difficult. In the absence of spinning, it is particularly important to keep field
inhomogeneities to a minimum, and the following precautions
should be taken.

(a) The sample should be positioned as symmetrically as
possible within the sample tube.
(b) The additional material required for perfusion, electrical stimulation, etc., should be as small as possible and
as far from the rf field as possible.
(c) Bubbles in the sample perturb field homogeneity, and
therefore any oxygenation should be performed as far from the
rf coil as possible, preferably outside the spectrometer.

Navon et al. (1977) have described an alternative method for dealing with bubbles, whereby RF pulses are timed to avoid periods when bubbles are present.

IV. BIOCHEMICAL AND PHYSIOLOGICAL RESULTS

Barely three years after the original report that ^{31}P NMR spectra can be obtained from intact muscle (Hoult et al., 1974), many laboratories have taken up the challenge to develop the techniques to a stage where it will become an accepted tool in biological studies on whole tissue. Several of the earlier publications inevitably duplicated and extended the previous observations and it has been gratifying to see the broad agreement between different groups. The road from promise to achievement, however, is a long one and is full of obstacles that must be overcome. First, it is necessary to show that this new method can reproduce well established and accepted observations. Once this is done, it is important to demonstrate that NMR can provide information that is more precise and hopefully also novel, in the sense that no other present method can be used to study the same problem. Finally, one would wish to show that the method can also lead to conceptual advances. In our view, several of these obstacles have been overcome, and in the following discussion we illustrate this with reference to studies on different tissues.

A. Skeletal Muscle

So far the most detailed studies that have been reported on a functioning skeletal muscle preparation are those of Dawson et al. (1977a,b), who studied sartorii and gastrocnemii from the frog and toad. The extensive work of the Chicago group has been mainly concerned with the comparison of different muscles and 'diagnosis' of diseased states (Burt et al., 1977).

1. Resting Muscles and Their Extracts

^{31}P NMR spectra of resting amphibian muscles are shown in Fig. 16. These spectra are remarkably consistent from one set of muscles to another: the only features that vary are the areas of resonances in the phosphodiester region. As first shown by Burt et al. (1976a) the toad muscle contains a particularly large amount of phosphodiester.

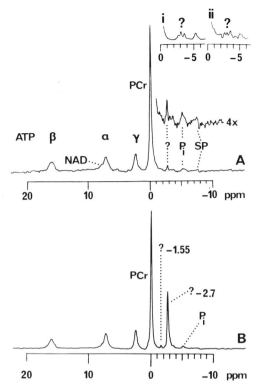

*Fig. 16. Spectra of resting amphibian muscles. (A) Four
frog sartorii; average of 10,000 scans at 2 sec intervals.
The short length of record on the right-hand side of the PCr
peak has been enlarged twice vertically to show fine detail.
The unidentified peaks (?) at -2.7, -3.05, and -3.6 ppm vary
in size (though not in frequency) from experiment to experiment (insets i and ii). SP, sugar phosphate. (B) Two toad
gastrocnemii; average of 6000 scans at 4 sec intervals. Spectral widths, 38.8 ppm; temperature, 4°C. (From Dawson et al.,
1977a.)*

The frequencies of the ATP resonances confirm that this
substance is almost wholly (>95%) bound to Mg^{2+} (cf. Hoult *et
al.*, 1974; Burt *et al.*, 1976a).

The internal pH of the resting frog sartorius muscle determined from the position of the P_i resonance peak (Fig. 16)
is 7.5. Muscle pH is dependent upon several factors, including
the pCO_2 and the pH of the external medium as well as mechanical activity and metabolic state. Tris-buffered Ringer's solution was used in these experiments at a pH of 7.7; the pCO_2 was

negligible. The ^{31}P NMR results agree reasonably well with
those obtained in other muscles under similar conditions using
microelectrodes (Aickin and Thomas, 1976) or the distribution
of DMO (5.5 dimethyl-2,4-oxazolidinedione) (Roos, 1975).

Spectra from extracts of frog sartorii and toad gastrocnemii are shown in Fig. 17. The resonances are much sharper
than those from intact muscle and it is thus possible to resolve the triplet structure of the ATP β peak and the doubling
of the α and γ peaks, which result from spin-spin coupling between neighboring phosphorus nuclei. The results from extracts
are in every way consistent with those from intact muscles.
As Fig. 17 shows, exactly the same resonances are seen as in
intact muscles (Fig. 16), though the ATP peak in the extract
are shifted because the ATP is no longer complexed to Mg^{2+} and
the solution is at a different pH. The ratios PCr/βATP and
PCr/P_i are 8.3 and 12, agreeing almost exactly with what is
found by chemical analysis of extracts (Table III). As in intact muscle, there is an unidentified peak at -2.7 with a barely perceptible one at -3.65 ppm. However, the inset shows that
three peaks (-2.7, -3.1, -3.7) have been seen in other extracts,
inviting direct comparison with Fig. 16.

It is important to note that whereas total extractable metabolite is measured by chemical methods, the NMR resonance linewidth is dependent upon molecular mobility, so that highly immobilized compounds give rise to broad resonances that are not
always observed. Broad resonances can be seen (e.g., from
phospholipids and DNA) under different conditions of data collection. In the frog muscle spectra the linewidths suggest that
the resonances arise almost exclusively from compounds that are
mobile and are therefore probably free in solution.

2. Stimulated Muscles

The effect of a single prolonged contraction on the spectra
observed from toad gastrocnemius is shown in Fig. 18. Application of a single tetanic stimulation lasting 35 sec to a pair
of resting toad gastrocnemius muscles produces the expected
changes; phosphocreatine falls to about half its resting level,
and there are large increases in the amount of P_i and sugar
phosphates. The ATP level remains constant.

A simple experiment was reported by Burt et al. (1977), who
followed "isometric caffeiene contracture" in frog gastrocnemius at 31ºC. After 45 minutes of a contraction, phosphocreatine and ATP were exhausted while P_i and sugar phosphate were
increased markedly. Control muscle showed much smaller changes
during this period.

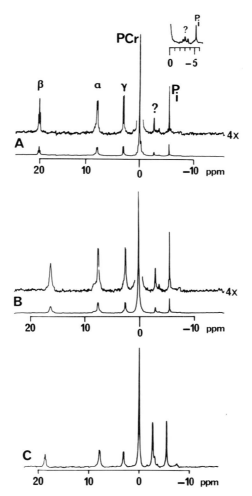

Fig. 17.   *Spectra of extracts from resting amphibian mus-*
cles.   (A) Frog sartorii, magnesium ion chelated by EDTA.
There are unidentified peaks at -2.7 and -3.7 ppm.   The inset
spectrum, from a different batch of frog sartorii, shows an ad-
ditional peak at -3.1 ppm.   (B) Frog sartorii extract plus 30
mM̲ magnesium chloride.   (C) Toad gastrocnemii.   There are un-
identified peaks at -1.6, -2.7, -3.05, and -3.6 ppm.   The mus-
cles were quick-frozen and then extracted in 1.25 mM̲ EDTA in
50% methanol, pH 7.6 for 4 days at -30°C.   (From Dawson et al.,
1977a.)

Both of these experiments demonstrate that it is possible to
measure metabolic responses by NMR, as was foreseen in 1974.   A
more detailed picture of muscle energetics is obtained by observation of recovery from short contractions.

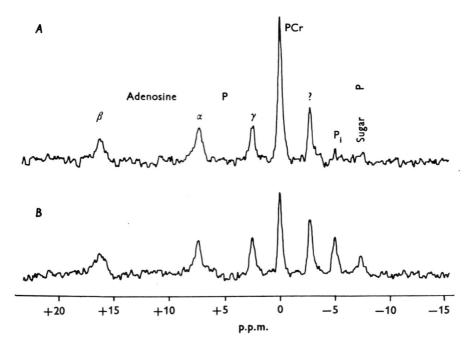

Fig.18. The effect of prolonged stimulation on the ³¹P spectrum of toad gastrocnemius. Spectrum A was obtained on a pair of small gastrocnemii in oxygenated Ringer solution at 4°C. The muscles were then stimulated for 35 sec, the perfusion was turned off, and spectrum B was accumulated; 200 scans at 2 sec intervals. (From Dawson et al., 1977a.)

   a. Contractions and Recovery. To follow such events one must use "gated NMR." For this, the NMR computer is used to synchronize the accumulation of data with the electrical stimulation of the muscle. A given experiment begins at time 0 with a pulse from the computer, which after a present delay (typically 1 sec) triggers the stimulator and the muscles contract. The rf pulses commence 2 sec from time 0 and are repeated every 2 sec throughout the course of the experiment. The first m scans are stored in bin 1, the next m scans in bin 2, etc., until n bins (mn scans) have been accumulated. A trigger pulse is again provided and the cycle recommences (Fig. 19). The process can be repeated many times to build up the required signal-to-noise ratio. By these means, although experiments may last many hours, the kinetics of reactions can be followed with a time resolution set by the minimal interval between successive bins, that is, 2 sec.

TABLE III. Relative Concentration of Phosphate Compounds in Resting Frog Sartorious[a]

	NMR	Chemical	P
$\dfrac{PCr}{\beta\,ATP}$			
$\bar{x}$	6.74	8.14	
S.E.	0.309	0.744	NS
n	6	18	
$\dfrac{PCr}{P_i}$			
$\bar{x}$	16.02	13.10	
S.E.	1.58	3.03	NS
n	6	17	

[a]*The chemical analyses on perchloric acid extracts of resting muscles were kindly made available by Drs. Curtin and Woledge. They are from the control muscles of their recent paper (1977). The NMR estimate of ATP is based on the area of the β peak since this peak is unique to ATP and would not be altered by the presence of ADP or NAD. Rough estimates of actual concentrations may be made by assuming that the resting PCr content is 27 mml kg⁻¹. (See Dawson et al., 1975.)*

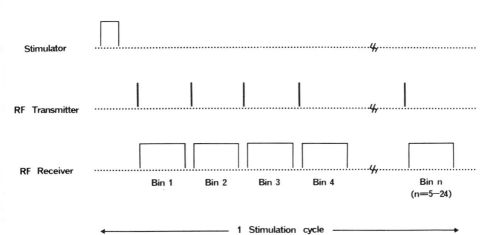

Fig. 19. The synchronization scheme used in a gated NMR experiment.

Recovery from long (25 sec) contractions repeated every 56 minutes is illustrated in Fig. 20. Scans were accumulated in

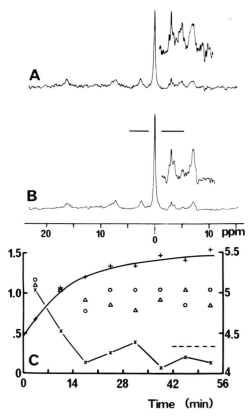

Fig. 20. *Recovery of frog sartorius from long contrac-tions. Four frog sartorious muscles were repeatedly stimulated for 25 sec every 56 minutes and spectra accumulated into eight bins of 7 minutes each. Spectrum A was accumulated from 0-7 minutes after the contraction and spectrum B was obtained throughout the last 28 minutes of recovery. The horizontal line at the Phosphocreatine resonance in spectrum B indicates the height of this peak in spectrum A. Scans were made at 2 sec intervals. The muscles gave seven virtually identical re-sponses, during each of which the tension fell from 1.22 to 0.47 N. Graph (C) shows how phosphocreatine (+), ATP (o), P_i (x), and sugar (P) varied throughout the eight bins. The ordinates show the resonance peak areas as multiples of the mean area for the β ATP peak. The right-hand scale applies to phosphocreatine. The exponential curve drawn through the phos-phocreatine points has a $\underline{T}_{1/2}$ of 9.1 minutes. (From Dawson et al., 1977a.)*

eight bins, each of 7 minutes duration. Spectra from the first
bin (Fig. 20A) and the last four bins are shown (Fig. 20B).
Figure 20C gives the variations in metabolite concentrations
with time after contraction. About 20% of the phosphocreatine
($\sim$5 mmol liter^{-1}) is broken down as a result of a 25 sec contraction, and this is approximately equalled by the increase in
concentration of P_i. This result agrees with the observation
that about 1% of phosphocreatine is broken down per second of
contraction (Gilbert et al., 1971). The concentrations of phosphocreatine and P_i recover with a half-time of approximately 10
minutes. The same value for the half-time of recovery of phosphocreatine was reported by Dydynska and Wilkie (1966) when
studying 30 sec contractions of the same muscle under similar
conditions. Kushmerick and Paul (1976b) also found a half-time
of about 10 minutes for recovery of phosphocreatine and P_i as
well as for recovery oxygen consumption in frog sartorii undergoing 25 sec contractions at 0°C. These observations again
confirm the validity of the NMR measurements. It is of interest to note the unexpected maintenance of sugar phosphates at
a high and constant level throughout the recovery period following long contractions. High levels of sugar phosphates are
not observed in resting muscles; such an event is generally associated with anoxia and muscle deterioration. Yet, in this
particular series of experiments, the muscles were not depleted of ATP and phosphocreatine during the latter half of the
recovery period nor was their mechanical response declining.

The spectra accumulated <u>during</u> the contractions of frog sartorius or toad gastrocnemius stimulated for 1 sec every 2 minutes were not in any way different from those accumulated during the recovery interval between contractions (Dawson et al.,
1977a). This result again agrees with previous biochemical
work, as the changes in phosphocreatine would only be about 1%.
The experiment also shows that the presence of an electrical
impulse does not degrade the spectra.

Spectra illustrating recovery from short contractions are
shown in Fig. 21. In these experiments toad gastrocnemii were
stimulated for 1 sec every 2 minutes.

Spectrum A was obtained during the first 16 sec following
stimulation and spectrum B represents the last 32 sec of recovery. In order to reveal small differences between the two
spectra, record A was subtracted from record B in the computer.
The difference, scaled up four-fold, is shown in Fig. 21C, and
indicates the breakdown and subsequent resynthesis of ~10% of
the steady-state phosphocreatine content.

The chemical shift of the negative P_i peak in the difference spectrum (Fig. 21C) corresponds to a pH of 7.3, whereas
the average pH as obtained from the P_i resonance in Fig. 21A
and B is 7.1. Perhaps there is a pH shift immediately after
contraction that is too small to alter the observed average

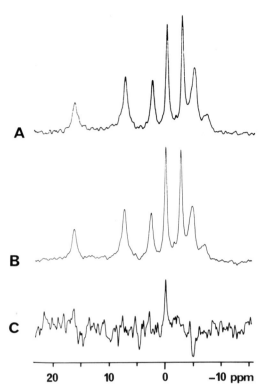

Fig. 21. Spectra from brief, repeated contractions of toad muscle, 1 sec every 124 sec. The spectra are the sum of five separate experiments, each on a pair of gastrocnemii. There was a total of 244 contractions; scans were taken every 2 sec. (A) First 16 sec following relaxation (1952 scans). (B) Last 32 sec of recovery before the next contraction (3904 scans). (C) Difference spectrum (B - A) (ordinate ×4). (From Dawson et al., 1977a.)

pH; alternatively, it is possible that the P_i liberated during contraction is in a more alkaline environment than the average for the muscle.

b. Poisoned Muscles. A standard biochemical approach in the study of complex pathways is to inhibit specifically a particular enzyme in the pathway. Recent experiments on frog gastrocnemius muscles using iodoacetate to inhibit glycolysis and cyanide to prevent oxidative metabolism have indicated the possibility of measuring the rates of reactions in limited parts of a metabolic sequence using ^{31}P NMR.

Experiments performed under anaerobic conditions have been particularly illuminating, and here we shall briefly discuss aspects of these experiments that are of general interest.

(i) As a result of lactic acid formation, the pH of muscles falls during the recovery period following contractions. The pH change can be determined by NMR, and if the buffering capacity of the muscles is known, it is possible to estimate the amount of lactic acid that is formed (Dawson *et al.*, 1978). It is thereby possible to determine the extent to which glycolysis proceeds under a variety of conditions, and in this way to assess what factors control the rate of glycolysis.

(ii) It is often desirable to know the concentration of ADP that is free in solution, rather than bound to the myofibrils. Unfortunately, the resonances of free ADP overlap with the α and γ resonances of ATP and are often too weak to be detectable. However, it is possible to estimate free ADP by use of the creatine kinase equilibrium, which may be written

$$\underline{K} = \quad ATP \quad Cr \ / \ ADP \quad PCr \quad H^+$$

where the concentrations are those free in solution. H^+ , phosphocreatine , and ATP can be determined from the NMR spectra, creatine can be calculated if the total amount of creatine + phosphocreatine is known, and there are values for K_{eq} in the literature. Knowledge of these quantities enables an estimate to be made of free ADP .

(iii) Recently, we have been studying the problem of fatigue in anaerobic frog muscles undergoing repeated isometric contractions at $4^{\circ}C$ (Dawson *et al.*, 1978). From our ³¹P NMR studies, we have concluded that the decline in isometric force development is more likely to be linked to changes in the levels of measured biochemical substances than to any changes in excitatory conduction. We have also found that the force development as the muscles fatigue is proportional to the rate at which ATP is hydrolyzed. This result suggests that the hydrolysis of ATP in each cross-bridge cycle produces a fixed mechanical impulse (force × time), and that the economy of ATP hydrolysis remains unchanged.

3. *Other Aspects of Muscle Metabolism*

We have discussed how by choice of a particular preparation we can strike a successful compromise between the constraints imposed by the NMR measurements and the demands of physiology and obtain information about a function like muscle contraction. This is only one area where ³¹P NMR can be useful. It was demonstrated (Burt *et al.*, 1977; Seeley *et al.*, 1976) that the technique has some "diagnostic" value in that different muscles

can easily be compared. This led the Chicago group to compare
normal and diseased muscles and ourselves to the notion of using NMR to study compartmentation of metabolites.

    a. Comparison of Normal and Diseased Muscle. Burt et al.
(1977b) reported significant differences between the spectra
of normal and diseased human quadriceps muscles. Of particular interest is that the muscles with Duchenne dystrophy do not
contain glycerol 3-phosphorycholine (Chalovich et al., 1978)
and the total P content is also generally lower. Occasionally
Duchenne muscles, especially those with no evidence for X-
linked origin, contained nearly normal total phosphate levels,
though glycerol 3-phosphorylcholine was still absent (Fig. 22).
These observations, together with studies on dystrophic chicken
muscles (Chalovich et al., 1977), encourage the belief that ^{31}P
NMR will improve our understanding of diseased tissue as well
as that of a normally functioning one.

    b. Compartmentation of Metabolites. It has been consistent-
ly observed that the linewidth of the inorganic phosphate signal of phosphorus NMR spectra of mammalian and amphibian skeletal muscle is greater than that for the phosphocreatine signal
(Seeley et al., 1976; Busby et al., 1978; Dawson et al., 1977a).
Yet the phosphorus signals from an aqueous solution containing
both P_i and PCr have the same linewidth, which is determined by
the homogeneity of the $\underline{B_0}$ field. Since the frequency of the
inorganic phosphate signal is sensitive to pH, it seemed possible that a distribution of hydrogen ion concentration within the
muscle volume might account for the width of the signal from
muscle P_i. The pK_a of phosphocreatine is about 4.6 and its
signal would not therefore be affected by pH variations near
neutrality. An alternative explanation (namely, immobilization
of inorganic phosphate by binding to macromolecules) did not
adequately account for the multicomponent P_i signals observed
during metabolic rundown of anoxic muscles (Seeley et al.,
1976). The pH gradient hypothesis had the virtue of simplicity.
    Values for NMR signal linewidth contain contributions from
both the intrinsic linewidth of the resonance and from field
inhomogeneity. Application of a sequence of rf pulses allows
the intrinsic linewidth to be determined. Figure 23 shows some
of the spectra obtained from application of such a pulse sequence (Carr and Purcell, 1954; Meiboom and Gill, 1958) to a rabbit semitendinosus muscle. The transverse relaxation rate of
the glycerol 6-phosphorylcholine signal is considerably less
than that of the P_i signal. $\underline{T_2}$ values are given in Table IV.
Intrinsic linewidths of inorganic phosphate and phosphocreatine
resonances are identical (2 Hz), and much less than those de-

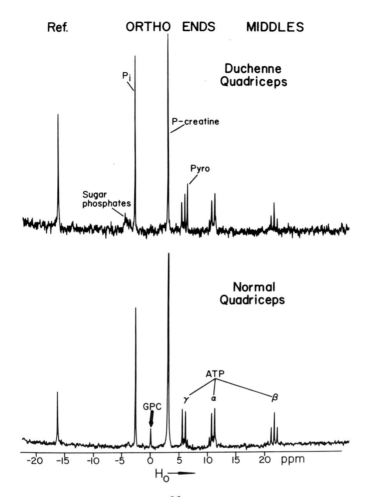

Fig. 22.   Comparison of ³¹P NMR spectra of PCA extracts of human quadriceps muscle: Duchenne dystrophy (upper) and normal (lower).   GPC indicated by the arrow in the normal extract is missing in the extract of the Duchenne muscle.   The peak labelled Pyro in the latter extract corresponds to pyrophosphate, at 7.0 ppm.   (Reproduced by permission from Chalovich et al., 1978.)

termined from spectra accumulated using a simple pulse train (50–200 and 20 Hz, respectively). These data are consistent with the pH distribution hypothesis; i.e., the inorganic phosphate resonance is an envelope of slightly staggered 2 Hz wide signals, which are broadened by field inhomogeneity of about 17 Hz (Seeley, 1975).

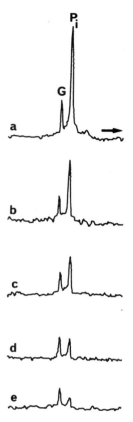

Fig. 23. T_2 determinations for glycerophosphorycholine and inorganic phosphate signals of rabbit semitendinosus muscle. Data were collected using a Carr-Purcell-Meiboom-Gill pulse sequence. The spectra are the transforms of the terminal free induction decay for several refocusing periods: (a) 0, (b) 20, (c) 40, (d) 80, (e) 120 msec. There were 40 scans per spectrum. Interpulse period, 2 msec; sweep width, 7.8 ppm; temperature, $22^{o}C$; G, glycerophosphorylcholine; P_i, inorganic phosphate.

Comparison of T_2 values in Table IV gives additional information on the intracellular environment. The lower T_2 values of P_i and PCr signals compared with that of glycerophosphorylcholine may be interpreted qualitatively in terms of weak binding of both inorganic phosphate and phosphocreatine to macromolecules. The intrinsic T_2 of glycerophosphorylcholine itself gives an upper limit of ~1 P for the viscosity of the interior

TABLE IV. $\underline{T_2}$ Valves Derived From Application of Carr-Purcell-Meiboom-Gill Sequence to Rabbit Semitendinosus Muscle

Inorganic phosphate	200 ± 50
Glycerophosphorylcholine	1800 ± 100
Phosphocreatine	200 ± 50
ATP (α, β, and γ)	10 ± 5

of the muscle cell, although this value is greatly in excess of the cytoplasmic viscosity expected from measurements of ionic (Kushmerick and Podolsky, 1969) and water (Cleveland et al., 1976) diffusion.

These observations led us to examine the effects of perturbing the intracellular pH using an acetate buffer (Busby et al., 1978). Figure 24 shows the spectra of Vastus lateralis muscle that had been bathed in acetate buffer, pH 5.2, for 20 minutes before being placed in the spectrometer. The interesting feature is the increasing amplitude of a signal X of chemical shift -1.05 ppm. We conclude (Busby et al., 1978) that inorganic phosphate in the muscle is present in two types of solution, one at the usual muscle pH (~7.0) and another at about pH 5.8. The time course of the appearance of the "low pH" pool suggests that its magnitude is not dependent upon the extent to which acetate has penetrated the muscle volume macroscopically.

Acetate experiments were also carried out on muscles treated with 2-deoxy-glucose or adrenaline to enhance the intensity of sugar phosphate signals (2-deoxyglucose 6-phosphate and glucose 6-phosphate, respectively). Low and high pH pools were also observed for these phosphate esters, though no evidence could be found for such heterogeneity in phosphocreatine and ATP (Busby et al., 1978). It is not as yet possible to locate the two pools of inorganic and sugar phosphates in defined regions of the muscle structure, though we favor an interpretation based on compartments within individual muscle cells rather than one based on cell heterogeneity. We have tentatively assigned the two phosphate pools to the sarcoplasm and sarcoplasmic reticulum (Busby et al., 1978), but this suggestion depends on diffusion kinetics for acetate for which there are at present no experimental data. Nevertheless we feel that this type of experiment can make a substantial contribution to our knowledge of the organization of the muscle cell.

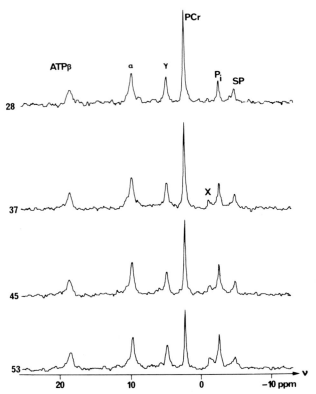

Fig. 24. ^{31}P NMR spectra of a rat Vastus lateralis muscle.
*Data were collected after 20 minutes immersion of the muscle in
an acetate buffer, pH 5.2, at 15°C. Each spectrum is the accumulation of 200 scans at 2 sec intervals. Numbers to the left
of the spectra refer to the time midpoints of each accumulation
in minutes. X is assigned to inorganic phosphate at low pH.
SP, sugar phosphate. (From Busby et al., 1978.)*

B. *Heart*

   1. *Nonperfused*

     Studies on thin skeletal muscles such as frog sartorii place
considerable limitations on the NMR measurements as only small
volumes of tissue can be kept in good physiological condition.
In contrast, perfused hearts of almost any size can be kept
functioning for many hours, and it is the size of the NMR sample
tube that limits the experiments. The interest in studying
heart metabolism led us to examine the possibility of observing
^{31}P NMR spectra from small rat hearts.

In our initial experiments, we observed NMR spectra from a small rat heart that was rapidly cooled after removal from the animal (Gadian *et al.*, 1976). At 4°C the heart no longer beats and energy utilization is consequently substantially reduced. Figure 25 shows the spectrum of such a preparation. The peaks were assigned as for skeletal muscle and the values of the integrals could be measured after about half an hour of accumulation. A notable feature of these integrals is that the three signals corresponding to ATP are in the ratio $\beta:\alpha:\gamma$ 0.5:1.5:0.9. This implies that the γ peak contains a contribution from ADP and that the α peak, in addition to ADP, also contains resonances from other pyrophosphates, possibly NAD and similar molecules.

The pH of the resting heart muscle is ~7. After the temperature of the NMR sample cavity was raised to 30°C and about 10 minutes was allowed for equilibration and for the stoppage of intermittent contractions of the heart, spectra were recorded by accumulating sets of 200 scans successively (Fig. 26). These spectra show a steady increase in the inorganic phosphate level and the rundown of the energy store of the cardiac tissue. The significant result is that the inorganic phosphate resonance progressively shifts to lower frequency, indicating that the intracellular acidity has dropped to pH 6 in approximately 15 minutes.

We have argued (Gadian *et al.*, 1976) that since the inorganic phosphate signals from the hypoxic ischaemic states are significantly shifted from those of normoxic tissue, ³¹P NMR provides a diagnostic tool for abnormal metabolic conditions associated with cardiac infarct. Tissue acidosis is a measure of increased glycogenolysis and the consequential lactate pro-

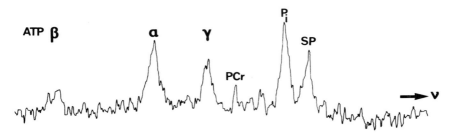

Fig. 25. *³¹P NMR spectrum of a 170 mg rat heart. The heart was rapidly excised from a small anaesthetized rat and immersed in Locke solution maintained at 4°C. The spectrum was accumulated by applying 1040 60° rf pulses at 2 sec intervals. Spectrum width, 38.8 ppm; SP, sugar phosphate. (From Gadian et al., 1976.)*

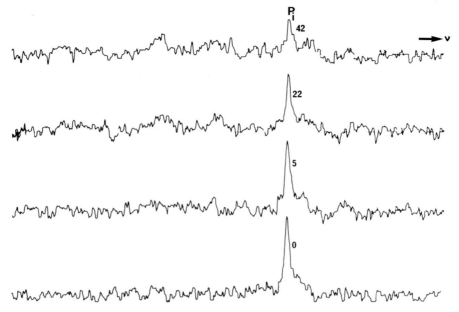

*Fig. 26. Spectra of a 200 mg rat heart at various times af-
ter warming to 30°C. A freshly excised rat heart was chilled
to 4°C for initial data collection (see legend to Fig. 25).
It was then warmed to 30°C, equilibrated for 10 minutes at this
temperature and the spectra collected in successive 6.7 minute
(200 scan) blocks. The numbers are the relative frequencies
of the maximum ordinate of the inorganic phosphate signal in
Hz. Spectrum width, 38.8 ppm. (From Gadian et al., 1976.)*

duction that may well be responsible for tissue damage in cardiac ischemia. To evaluate the validity of such possibilities
it is necessary to have a direct comparison between a normal
functioning heart under physiological conditions and the dam-
aged system.

2. Perfused

a. *Initial Observations.* It is not surprising that the
possibility of working on perfused hearts has occurred to several groups. Jacobus and co-workers (1977) reported data on
perfused rat hearts that demonstrated the feasibility of the
experiment and arrived at similar, though not identical conclusions to our own (Garlick *et al.*, 1977). In their first paper,
Jacobus *et al.* (1977) showed that the intracellular pH in their

preparation dropped from 7.4 to 5.7 during total ischemia and
that after reperfusion only partial recovery of the original
metabolite levels was obtained. The pH after reperfusion remained low (6.1). The pH change observed by us was much smaller (see below) (1 pH unit) although we agreed on the initial
value of 7.4. Unlike Jacobus et al. (1977) we did not observe
the loss of cellular phosphate during perfusion and in ische-
mia-reperfusion sequences the original phosphocreatine and ATP
levels were almost entirely restored. It is interesting that
the level of sugar phosphate after reperfusion remains high and
this is reminescent of the observation in skeletal muscle after
extensive contraction. Perfusion of rat hearts in small NMR
tubes (12 mm, Jacobus et al., 1977; 8 mm inner diameter, Garlick et al., 1977) is very demanding, and the observed spectra,
pH values, and recovery experiments are good indicators of the
physiological status of the heart preparation. In the experiments we reported (Garlick et al., 1977) we could keep the
heart in a functional steady state for periods in excess of 5
hours. A typical spectrum is shown in Fig. 27. It is worth
noting that in the early stages of our investigations hearts
began to fail after about an hour and the levels of the various
metabolites observed by NMR changed throughout the experiment
(Fig. 28). We noted initial synthesis of phosphocreatine (Fig.
28i-iii), but after about 30 minutes inorganic phosphate started
to build up again and the phosphocreatine was slowly depleted
(Fig. 28iv-v). In a good preparation, the phosphocreatine/ATP
ratio is both high and constant.

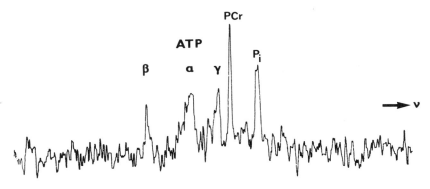

Fig. 27.   ³¹P NMR spectrum of a 160 mg rat heart perfused
by the Langendorff technique at 37°C. The heart was continuously supplied with Krebs-Henseleit buffer (95% oxygen, 5% carbon dioxide) containing 11 mM glucose. The perfusion pressure
was 80 cm water. 1000 60° rf pulses were applied at 4 sec intervals. Spectrum width, 77.5 ppm. (From Garlick et al.,
1977.)

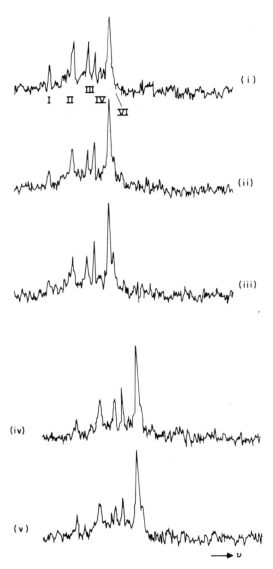

Fig. 28. Consecutive ^{31}P NMR spectra of a Langendorff-per-
fused 150 mg rat heart. Each spectrum is the sum of 200 scans
at 2 sec intervals. Phosphocreatine (peak IV) is being synthesized in spectra (i) to (iii), but is subsequently broken down
in spectra (iv) and (v). Spectra are all 77.5 ppm wide. I,
β-ATP; II, α-ATP + pyrophosphate diesters; III, γ-ATP + β-ADP;
IV, phosphocreatine; VI, inorganic phosphate (and sugar phosphates). Perfusion details in legend to Fig. 27.

The values of relative concentrations observed in perfused
rat hearts and their perchloric acid extracts are given in
Table V. Accuracy of these ratios again requires a knowledge
of the reduction in signal intensities due to saturation effects, i.e., partial saturation of some signals does occur when
spectra are collected with 2 sec intervals between pulses. Accurate concentration ratios are given by integration of "4 sec
spectra" (Table V).

The concentration ratios for perfused heart are in broad
agreement with those for perchloric acid extracts. The β-ATP
signal is taken as the concentration standard for ATP since the
α and γ signals are overlapped by other resonances (see below).
The high value of phosphocreatine/ATP (~1.8) and low inorganic
phosphate/ATP (~1.8) are indices of efficient cardiac respiration. These values remained constant throughout the perfusion
period and are therefore--along with heart and coronary flow
rates--a further indication of a functional steady state of the
heart. The values are also similar to those obtained from biochemical determinations on freeze-clamped extracts (unpublished
observations). A marked discrepancy occurs for "α-ATP"/β-ATP
ratios: values are consistently higher for hearts from young
animals. This age dependence effect in "α-ATP"/β-ATP has also
been observed in rat skeletal muscle (unpublished observations)
and is accounted for by additional signals overlapping that of
α-ATP, at least partly NAD/NADH. "γ-ATP"/β-ATP ratios are also
greater than unity. The majority of the extra γ signal is from
β-ADP and resolved signals from this nucleotide are visible in
extract spectra (Garlick et al., 1977).

b. Ischemia and Recovery. We have followed the time course
of metabolic changes during ischemia and recovery on mouse
hearts (Garlick et al., 1978). For example, Fig. 29 shows that
cessation of flow of the oxygenated buffer results in a rapid
decrease in phosphocreatine, slower changes in the ATP pool,
and a progressive shift of the inorganic phosphate signal to
lower frequencies. This is shown in Fig. 29A-D, in which individual spectra were collected over successive periods of 200
sec. The progressive shift in intracellular pH during global
ischemia is plotted in Fig. 30.

Recently Taylor et al. (1977) also estimated that the intracellular pH of rat hearts decreased from 7.45 to 6.4 during ischemia although they have not commented on the discrepancy between these values and those previously reported (Jacobus et
al., 1977). The same authors (Hollis et al., 1977) quote a pH
transition from 7.4 to 6.4 for global ischemia in rabbit heart.

Spectra collected during the initial stages of recovery
from a 15 minute ischemic period on a mouse heart are shown in
Fig. 31. There is a rapid and unexpected resynthesis of phos-

TABLE V. Integrals of Phosphorus Signals Relative to β-ATP

	β-ATP	α-ATP	γ-ATP	Phospho-creatine	Inorganic phosphate	Sugar phosphate and nucleotide monophosphate[a]
160 mg rat hearts direct observation						
4 sec pulse intervals	1	3.6 ± 0.5	1.7 ± 0.3	1.9 ± 0.3	1.8 ± 0.3[b]	n.o.
2 sec pulse intervals	1	1.6 ± 0.3	1.1 ± 0.2	1.4 ± 0.3	1.6 ± 0.3	n.o.
800 mg rat hearts 3% perchloric acid extraction						
4 sec pulse intervals	1	1.5 ± 0.2	1.2 ± 0.2	1.7 ± 0.3	2.2 ± 0.3	0.9 ± 0.2

[a] n.o., not observable.
[b] Corrected for inorganic phosphate in Krebs-Henseleit buffer.

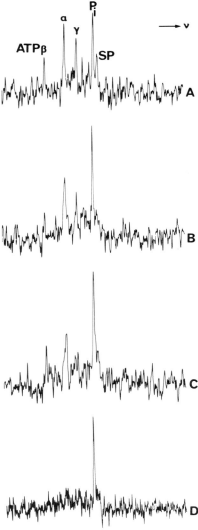

Fig. 29. ³¹P NMR spectra of a mouse heart in total global
ischaemia. Spectra A-D represent successive 200 sec periods
after stopping the flow of perfusion fluid. The inorganic
phosphate signal both increases in intensity and shifts to lower frequency. Spectrum width, 77.5 ppm; temperature, 37°C.

phocreatine and a return of the cellular pH to its value prior
to ischemia, whereas recovery of the nucleotide pool is relatively slow. Subsequently the phosphocreatine level relaxes to
its steady-state value (Garlick et al., 1978; Battersby et al.,
1977).

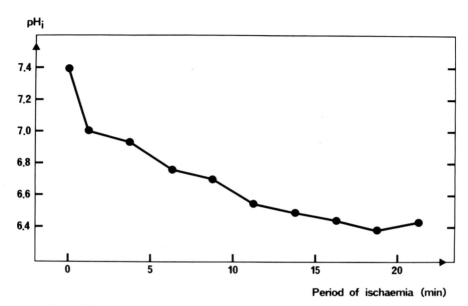

Fig. 30. *The time course of the change in intracellular pH with total global ischemia for a single mouse heart.*

The improved signal-to-noise expected from use of larger bore magnets, and consequent removal of many spatial constraints with larger sample tubes, will lead to many extensions of the method, not least the examination of species variation. The improved sensitivity of such a system has already been used by Hollis and co-workers (1977) in an elegant demonstration of regional ischemia. Ligation of the left anterior descending coronary artery increased the size of the inorganic phosphate signal, but did so by generating intensity at a frequency representing a relatively acid pH (Fig. 32).

C. Other Tissues

Since the first report of Moon and Richards (1973) that high-resolution ^{31}P NMR spectra could be recorded from small molecules in erythrocytes, a variety of cellular systems have been studied (Shulman, this volume). In the same way, once it had been established that tissue preparations could also be examined (Hoult *et al.*, 1974), the measurements could be extended to almost any isolated tissue or organ preparation. Thus ^{31}P NMR spectra have been recorded from whole adrenal glands (Radda, 1975), kidney (Sehr *et al.*, 1977), developing

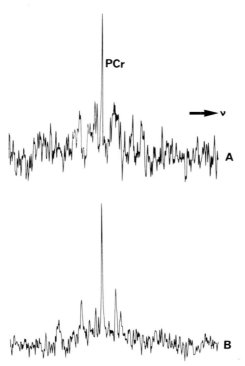

Fig. 31.   Spectra of a mouse heart recovering from 15 minutes of total global ischemia.   A and B represent successive 200 sec periods after reflow.   The major feature is the rapid resynthesis of phosphocreatine.   Perfusion details are given in the legend to Fig. 27.   Spectrum width, 77.5 ppm; temperature, 37°C.

frog embryos (Colman and Gadian, 1976), brain, and liver (unpublished observations).   It is no longer sufficient merely to record a spectrum; the system must be kept under controlled physiological conditions.   It is only then that one can expect NMR to give meaningful results.

1. Kidney

Our experiments on isolated kidneys have to date been concerned with establishing conditions under which we could follow, by ³¹P NMR, the depletion of metabolites during warm and cold ischemia and their recovery following blood perfusion using a circuit with a live anesthetized "assist" animal (Sehr et al., 1977; Seeley et al., 1977).   Such an experiment is briefly described below.

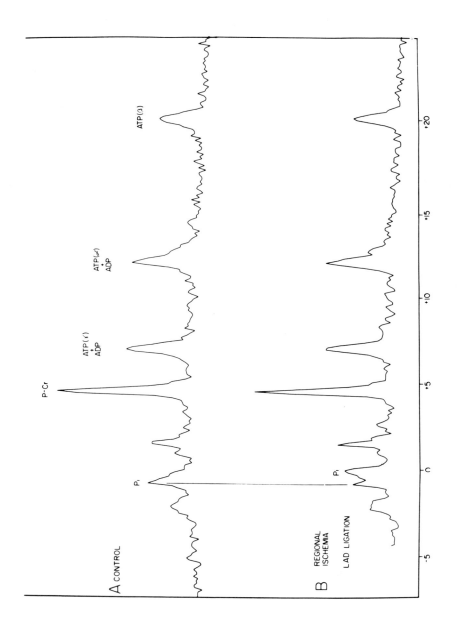

Fig. 32. ^{31}P NMR spectrum (72.9 MHz) of a perfused rabbit heart. Each trace represents the Fourier transformed average of 150 transients requiring 5 minutes total accumulation time. Chemical shifts are expressed in ppm relative to a solution of 0.2 M H_3PO_4 in 15% HClO$_4$ contained in a 1 mm diameter capillary tube. Positive values indicate shifts to a higher field than the reference. (A) Fully perfused heart; (B) same heart and same conditions as in (A) but spectrum obtained immediately after ligation of the left anterior descending coronary artery. The bar line calibrates the location of the control phosphate peak. This figure is reproduced by permission from Hollis et al. (1977).

The kidney was removed from a young rat and was then chilled to 4°C. The collection of ^{31}P NMR spectra was started within 4 minutes of nephrectomy. The initial spectrum shows ATP, a small amount of phosphocreatine, and some other components (Fig. 33). A metabolite that corresponds in frequency to a phosphodiester is very abundant in the kidney. The level of AMP is also much higher in renal tissue than in the other tissues we have examined. The subsequent spectra in Fig. 33 show how, even at 4°C, the energy pool of renal tissue is gradually depleted. A difference between renal tissue and muscle is that changes in cellular pH are smaller in the former case. This is consistent with the relatively low glycogen content and the slow rates of glycolysis in renal tissue at 4°C.

When an isolated ischemic kidney is linked up to the blood circulation of an anesthetized animal, after a period the original ATP level is reestablished (Fig. 34) even if during the blood perfusion a further period of ischemia is allowed to take place. Kidneys perfused in this way produce urine, and it is therefore possible, using this model, to quantitate metabolic effects of procedures involved in renal transplantation.

2. The Adrenal Gland

One system for which ^{31}P NMR has given us a new insight into the biochemical mechanisms involved is the adrenal medulla. This tissue contains the specialized adrenomedullary :chromaffin cells" that are primarily involved in the storage and release of adrenaline. Catecholamine is stored at high concentrations (0.55 $\underline{M}$), together with ATP (0.125 $\underline{M}$) and some acidic proteins in membrane-limited vesicles inside the chromaffin cells (Smith and Winkler, 1972). A ^{31}P NMR spectrum of a whole adrenal gland is shown in Fig. 35. A notable feature of this spectrum is that the ATP resonances are similar to those in the spectra of isolated storage granules (Fig. 35) and that the chemical shifts of the α and β phosphates of ATP are unusual. Detailed studies (Ritchie, 1975; Casey *et al.*, 1977) showed that this spectral pattern is a result of the interaction of ATP with the positively charged catecholamines within the granules (Njus *et al.*, 1978) and that the intragranular pH is around 5.6. These results represent the first direct observation of ATP in a whole tissue within a well-characterized compartment. The abnormally low pH inside the isolated storage vesicles is maintained within the adrenal gland; a result that could not have been concluded from other methods.

^{31}P NMR has been used to prove that the membrane-bound ATPase of isolated granules is an inwardly directed proton pump. Figure 36 shows an experiment in which we have isolated granules, observed the ^{31}P NMR spectrum, and added ATP and Mg^{2+} on

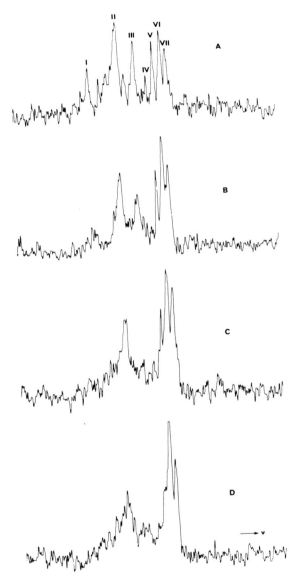

Fig. 33.   ³¹P NMR spectra of a nonperfused rat kidney at 4°C. The kidney was excised, cooled for 10 sec in ice-cold Krebs buffer, and then transferred to the thermostatted NMR tube. Spectra were recorded at 129 MHz in the Fourier transform mode by applying 200 70° rf pulses at 2 sec intervals. Spectral accumulation was started 1 (A), 15 (B), 41 (C), and 57 (D) min after the onset of ischemia. I, β-ATP; II, α-ATP;

the outside. The resonances derived from the ATP inside the
granules and those from ATP outside can be clearly distinguished
because of the different environmental conditions. The external
ATP is being hydrolyzed, as is demonstrated by the gradual decrease of the ATP_{out} resonances and the concomitant increase in
the inorganic phosphate signal. At the same time, there is no
change in the magnitude of the ATP_{in} signals, but the position
of the $\gamma\text{-}ATP_{in}$ peak shifts in a manner that indicates that the
pH inside the granules has decreased by about half a pH unit
(Casey *et al.*, 1977) (Fig. 36). The drop in the intragranular
pH is only observed in the presence of permeable anions, like
chloride, but not when the anion is sulfate or when there are
no anions present (e.g., in a sucrose medium).

NMR has assisted here in the examination of the medium requirements for proton movement, indicating that the proton
translocation is electrogenic (Casey *et al.*, 1977). This proton translocation has been linked to active catecholamine up-
take into the storage granules (Njus and Radda, 1978).

3. *Developing Embryos*

A rather different application of ^{31}P NMR was reported by
Colman and Gadian (1976): examination of the early development
of *Xenopus laevis* (the South African clawed toad). The spectra,
taken of various stages of development from unfertilized egg to
tadpole, are shown in Fig. 37. These, in conjunction with bio-
chemicaldata, led to several interesting conclusions. First,
the nucleoside triphosphate levels remained relatively constant
at about 3.5-4.5 μmol/embryo, at least until the "spontaneous
movement" stage. Second, the inorganic phosphate content of the
embryo increased markedly throughout the course of development.
Third, as expected, the major resonance observed in the early
stages (that corresponding to the yolk phosphoprotein (phosvitin
and lipovitellin) gradually disappears, the protein phosphate
peak disappearing at a greater rate than the lipid phosphate
peak. In fact, while the total phosphorus content of the embryo determined chemically (~200 μmol/embryo) remains constant
during development, the total amount of phosphorus observed by
NMR decreases by about 40%. This NMR loss must be accounted
for by the synthesis of highly immobilized phosphorus-containing
molecules, which cannot just be newly formed nucleic acid. Finally, the internal pH of embryos where the inorganic phosphate
is observed is 6.8 ± 0.2.

*III, γ-ATP; IV, phosphocreatine; V, unidentified; VI, inorganic
phosphate; VII, nucleotide monophosphate plus some sugar phosphates. Spectrum width, 77 ppm. (From Sehr et al., 1977.)*

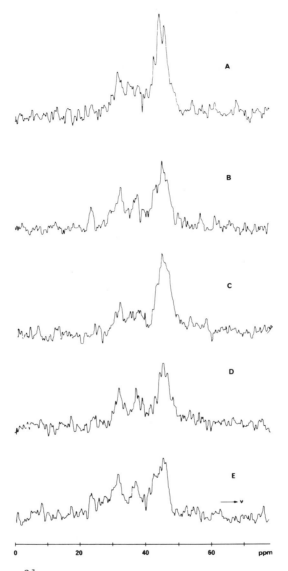

Fig. 34. ^{31}P NMR spectra of a single, blood-perfused kidney. The kidney was excised and rinsed in ice-cold buffer. There was then a 15 minute period of <u>cold</u> ischemia and spectrum (A) was started together with blood perfusion. Data accumulation for spectrum (B) was initiated after 51 minutes of perfusion. After a total of 133 minutes the pump was turned off. Spectrum (C) was started after 12 minutes of <u>warm</u> ischemia. Reperfusion began as adata collection for (C) cane to an end. Spectra (D) and (E) were begun after 12 and 24 minutes of reperfusion, respectively. (Details of NMR data collection are given in the legend to Fig. 33.) (From Sehr et al., 1977.)

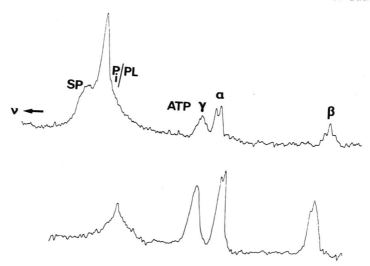

Fig. 35. ^{31}P *NMR spectra of a porcine adrenal gland and of porcine chromaffin granules. The upper trace is the spectrum of the adrenal gland, which was immersed in Tyrode's solution, pH 7.0. The chromaffin granules were suspended in buffered 0.3 $\underline{M}$ sucrose. Spectra were accumulated by applying 2000 70° pulses at 2 sec intervals. Each spectrum is 34.4 ppm wide. Temperature, 16°C; PL, phospholipids; SP, sugar phosphates, mainly AMP.*

V. THE USE OF OTHER NUCLEI

Our reasons for examining the phosphorus nucleus were given above. Much information could certainly be obtained if it were possible to measure and assign high-resolution tissue spectra of other nuclei. The feasibility of such measurements has been demonstrated for both protons (Daniels *et al.*, 1976; Brown *et al.*, 1977) and ^{13}C (Schaefer *et al.*, 1975).

Daniels *et al.* (1976) used the convolution difference technique (Campbell *et al.*, 1975) to enhance resolution in proton NMR spectra of adrenal glands. Considerable additional spectral simplification can be achieved using $\underline{T_1}$ and $\underline{T_2}$ pulse sequences. In this way, signals of highly mobile molecules (e.g., adrenaline and ATP in the adrenal gland) can be observed. The $\underline{T_2}$ pulse sequence has been elegantly applied to the study of red cell metabolism (Brown *et al.*, 1977) and Daniels *et al.* (1977) have surveyed a range of systems using these techniques.

The low natural abundance of the C^{13} isotope (1.1%) makes enrichment of samples in this nuclide a valuable technique for

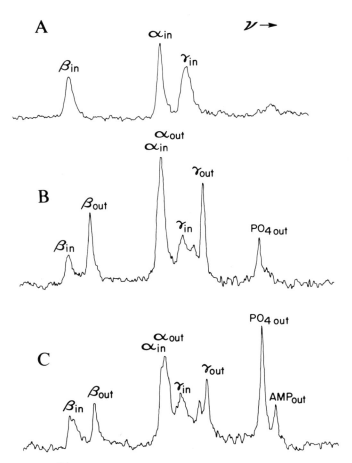

Fig. 36. ^{31}P NMR spectra of chromaffin granule samples.
(A) Chromaffin granules suspended in 0.8 ml of 120 m$\underline{M}$ KCl
+ 40 m$\underline{M}$ MES + 185 m$\underline{M}$ sucrose, pH 6.44.   (B) Chromaffin granule
suspension 1-5 minutes after the addition of 0.1 ml of 100 m$\underline{M}$
ATP + 100 m$\underline{M}$ MgSO$_4$ (pH 7).   (C) Chromaffin granule suspension
34-38 minutes after the addition of MgATP.   Temperature, 25°C;
protein concentration, 80 mg ml^{-1}.   The spectra were collected
by applying 256 rf pulses at 1 sec intervals.   The suffixes
"in" and "out" distinguish phosphates internal and external to
the chromaffin granule, respectively.   It is possible to resolve the signals indicated because of differences in the pH
and ionic composition of the internal and external media.
(From Casey et al., 1977.)

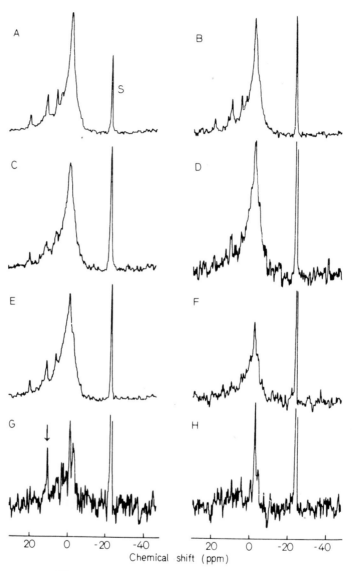

Fig. 37. ^{31}P *NMR spectra of developing embryos of the South African clawed toad Xenopus laevis. (A) Unfertilized eggs (600); (B) midblastula, stage 8 (349); (C) late gastrula, stage 11 (285); (D) postneurula, stage 23 (160); (E) spontaneous movement, stage 26 (384); (F) heart-beat, stage 33/34 (218); (G) swimming tadpole, stage 41 (106); (H) feeding tadpole, stage 46 (83). Figures in parentheses above refer to the number of embryos contributing to spectra. The indicated peak*

both sensitivity enhancement and spectrum simplification. Schaefer *et al.* (1975) have used this method in the study of plant metabolism. In their experiments, fresh soybean ovules were exposed to $^{13}CO_2$. The rates of lipid synthesis and sugar metabolism could be qualitatively estimated from the spectra. Distribution of ^{13}C label in the lipid chains gave specific information about the extent to which glucose is metabolised via the phosphogluconate pathway. This work sets lines of development for ^{13}C NMR examination of tissue and also demonstrate the tractability of botanical material for detailed NMR work.

VI. POSTSCRIPT

In the past four years the study of cellular metabolism by NMR in whole organs and organisms has become well established (Radda and Seeley, 1979). Initially, for the reasons outlined above, the ^{31}P nucleus has proved to be the most convenient and useful in such studies. But it is now quite clear that *in vivo* measurements on other nuclei (e.g., ^{13}C or ^{1}H) are also possible. Inevitably, whenever a new method is introduced to investigate a problem like cellular metabolism that has been extensively explored, one is bound to ask if the method can give new information or merely confirm and slightly extend previous knowledge. In our view, apart from the nondestructive analytical aspect of the method, NMR measurements contain two types of information not available in other measurements. First, since the resonances are sensitive to the chemical environment of the nucleus observed we can study inside the cell interactions, compartmentation, the nature of the cellular environment (e.g., pH), binding, and in general heterogeneity in the distribution of molecular species. Second, we can follow the rates of individual metabolic events in several ways. During changing metabolic conditions the levels of different metabolites can be followed simultaneously as a function of time provided that the changes are slower than the times required for the accumulation of the signals. When periodic variations occur (e.g., in muscle

in spectrum (G) (10 ppm) was not obtained in two subsequent experiments; however, in these additional experiments the nearest developmental stage examined was stage 44. Apart from this possible exception, spectra have been extremely reproducible. The spectra are the sum of three 20 minute (200 scan) accumulations. S, external standard, tetramethylphosphonium iodide. (From Colman and Gadian, 1976.)

contraction and oscillations) the NMR measurements can be gated
and hence faster rates can be measured. But even in the <u>steady</u>
<u>state</u> metabolic fluxes are measurable by the technique of satu-
ration transfer as first reported for the ATPase reaction in
E. coli (Brown *et al.*, 1977) and later applied for the creatine
kinase catalyzed reaction in skeletal and cardiac muscle
(Brown *et al.*, 1978). The possibility of introducing spatial
resolution into these various measurements by one of the methods
of zeugmatography is also real.
    Even without sophisticated imaging techniques, metabolites
in selected organs of live animals should be observable as indeed has been shown recently for mouse brain (Chance *et al.*,
1978).

ACKNOWLEDGMENTS

    P.J.S. is a British Heart Foundation Fellow. The authors'
own work described in this article was supported by the Science
Research Council, British Heart Foundation, and NIH (Grant HL
18708-0251).

REFERENCES

Aickin, C. C., and Thomas, R. C. (1976). *J. Physiol. London*
    260, 25-26P.
Bailey, J. M., Fishman, P. H., and Pentchev, P. G. (1968). *J.*
    Biol. Chem. 243, 4827-4831.
Battersby, M. K., and Radda, G. K. (1976). *FEBS Lett. 72*, 319-
    322.
Battersby, M. K., Garlick, P. B., Seeley, P. J., Sehr, P. A.,
    and Radda, G. K. (1977). *Proc. Missouri Symp. 177, Cellular*
    Function and Molecular Structure (P. F. Agris, R. N. Loepp-
    ky, and B. Sykes, eds.), pp. 175-193. Academic Press, New
    York.
Berden, J. A., Cullis, P. R., Hoult, D. I., McLaughlin, A. C.,
    Radda, G. K., and Richards, R. E. (1974). *FEBS Lett. 46*,
    55-58.
Brown, F. F., Campbell, I. D., Henson, R., Hirst, C. W. J., and
    Richards, R. E. (1973). *Eur. J. Biochem. 38*, 54-58.
Brown, F. F., Campbell, I. D., Kuchel, P. W., and Rabenstein,
    D. C. (1977). *FEBS Lett. 82*, 12-16.
Brown, T. R., Ugurbil, K., and Shulman, R. G. (1977). *Proc. Nat.*
    Acad. Sci. USA 74.

Brown, T. R., Gadian, D. G., Garlick, P. B., Radda, G. K.,
    Seeley, P. J., and Styles, P. (1978). *In* "Frontiers of Bio-
    logical Energetics: Electrons to Tissues" (L. Dutton, J.
    Leigh, and A. Scarpa, eds.). Academic Press, New York.
Burt, C. T., Glonek, T., and Bárány, M. (1976a). *J. Biol. Chem.*
    251, 2584-2591.
Burt, C. T., Glonek, T., and Bárány, (1976b). *Biochemistry 15*,
    4850-4853.
Burt, C. T., Glonek, T., and Bárány, M. (1977). *Science 195*,
    145-149.
Busby, S. J. W., Gadian, D. G., Radda, G. K., Richards, R. E.,
    and Seeley, P. J. (1975). *FEBS Lett. 55*, 14-17.
Busby, S. J. W., Gadian, D. G., Radda, G. K., Richards, R. E.,
    and Seeley, P. J. (1978). *Biochem. J. 170*, 103-114.
Campbell, I. D., Dobson, C. M., Williams, R. J. P., and Xavier,
    A. V. (1973). *J. Magn. Res. 11*, 172-181.
Carr, H. Y., and Purcell (1954). *Phys. Rev. 94*, 630-638.
Casey, R. P., Njus, D., Radda, G. K., and Sehr, P. A. (1977).
    Biochemistry 16, 972-977.
Chalovich, J. M., Burt, C. T., Cohen, S. M., Glonek, T., and
    Bárány, M. (1977). *Arch. Biochem. Biophys. 182*, 683-689.
Chalovich, J. M., Burt, C. T., Danon, M. J., Glonek, T., and
    Bárány, M. (1978). *Ann. New York Acad. Sci. 317*, in press.
Chance, B., Radda, G. K., Seeley, P. J., Silver, I., Nakase, Y.,
    Bond, M., and McDonald, G. (1978). *Proc. Nat. Acad. Sci.*
    USA, in press.
Cleveland, G. G., Change, D. C., Hazlewood, C. F., and Ror-
    schach, H. E. (1976). *Biophys. J. 16*, 1043-1053.
Cohen, R. D., and Iles, R. A. (1975). *Crit. Rev. Clin. Lab.*
    Sci. 6, 101-143.
Cohen, S. M., and Burt, C. T. (1977). *Proc. Nat. Acad. Sci.*
    USA 74, 4271-4275.
Cohn, M., and Hughes, T. R. (1962). *J. Biol. Chem. 237*, 176-
    181.
Colman, A., and Gadian, D. G. (1976). *Eur. J. Biochem. 61*, 387-
    396.
Curtin, N. A., and Woledge, R. C. (1977). *J. Physiol. London*
    270, 455-471.
Daniels, A. J., Krebs, J., Levine, B. A., Wright, P. E., and
    Williams, R. J. P. (1977). *In* "NMR in Biology" (R. A. Dwek,
    I. D. Campbell, R. E. Richards, and R. J. P. Williams,
    eds.), pp. 277-287. Academic Press, New York.
Daniels, A., Williams, R. J. P., and Wright, P. E. (1976). *Na-
    ture 261*, 321-322.
Dawson, J., Gower, D., Kretzschmar, M. K., and Wilkie, D. R.
    (1975). *J. Physiol. London 254*, 41-42P.
Dawson, R. M. C., Elliott, D. C., Elliott, W. H., and Jones,
    K. M., eds (1969). "Data for Biochemical Research." Claren-
    don Press, Oxford.

Dawson, M. J., Gadian, D. G., and Wilkie, D. R. (1977a). *J. Physiol. London 267,* 703-735.

Dawson, M. J., Gadian, D. G., and Wilkie, D. R. (1977b). *In* "NMR in Biology" (R. A. Dwek, I. D. Campbell, R. E. Richards, and R. J. P. Williams, eds.), pp. 289-322. Academic Press, New York.

Dawson, J., Gadian, D. G., and Wilkie, D. R. (1978). *Nature, 274,* 861-866.

Dydynska, M., and Wilkie, D. R. (1966). *J. Physiol. London 184,* 751-769.

Gadian, D. G. (1974). Ph.D. Thesis, Univ. of Oxford, England.

Gadian, D. G., Hoult, D. I., Radda, G. K., Seeley, P. J., Chance, B., and Barlow, C. (1976). *Proc. Nat. Acad. Sci. USA 73,* 4446-4448.

Garlick, P. B., Radda, G. K., Seeley, P. J., and Chance, B. (1977). *Biochem. Biophys. Res. Commun. 74,* 1256-1262.

Garlick, P. B., Seeley, P. J., Battersby, M. K., and Radda, G. K. (1978). *Fed. Eur. Biochem. Soc. 11th Meeting, Copenhagen 1977 42* "Regulatory Mechanisms of Carbohydrate Metabolism" (V. Esmann, ed.), pp. 297-302, Pergamon Press, Oxford and New York.

Gilbert, C., Kretzschmar, K. M., Wilkie, D. R., and Woledge, R. C. (1971). *J. Physiol. London 218,* 163-193.

Glonek, T., and van Wazer, J. R. (1974). *J. Magn. Res. 13,* 390-391.

Hill, A. V. (1965). "Trails and Trials in Physiology. "Edward Arnold, London.

Hollis, D. P., Nunnally, R. L., Jacobus, W. E., and Taylor IV G. J. (1977). *Biochem. Biophys. Res. Commun. 75,* 1086-1091.

Hoult, D. I., and Lauterbur, P. C. (1979). *J. Magn. Res.,* in press.

Hoult, D. I., and Richards, R. E. (1976). *J. Magn. Res. 24,* 71-85.

Hoult, D. I., Busby, S. J. W., Gadian, D. G., Radda, G. K., Richards, R. E., and Seeley, P. J. (1974). *Nature 252,* 285-287.

Jacobus, W. E., Taylor IV, G. J., Hollis, D. P., and Nunnally, R. L. (1977). *Nature 265,* 756-758.

Kohler, S. J., and Klein, M. P. (1976). *Biochemistry 15,* 967-973.

Kushmerick, M. J., and Paul, R. J. (1976). *J. Physiol. London 254,* 711-724.

Kushmerick, M. J., and Podolsky, R. J. (1969). *Science 166,* 1297-1298.

Lam, Y.-F., and Kotowycz, G. (1977). *Can. J. Chem. 55,* 3620-3630.

Meiboom, S., and Gill, D. (1958). *Rev. Sci. Instrum. 29,* 688-691.

Moon, R. B., and Richards, J. H. (1973). *J. Biol. Chem. 248,* 7276-7278.

Navon, G., Ogawa, S., Shulman, R. G., and Yamane, T. (1977). *Proc. Nat. Acad. Sci. USA 74*, 888-891.

Njus, D., and Radda, G. K. (1978). *Biochim. Biophys. Acta 463*, 219-244.

Njus, D., Sehr, P. A., Radda, G. K., Ritchie, G. A., and Seeley, P. J. (1978). *Biochemistry, 17*, 4337-4343.

Radda, G. K. (1975). *Phil. Trans. Roy. Soc. London B. 272*, 159-171.

Radda, G. K., and Seeley, P. J. (1979). *Ann. Rev. Physiol., 41*, 749-769.

Ritchie, G. A. (1975). Ph.D. Thesis, Univ. of Oxford, England.

Roberts, E., and Lowe, I. P. (1954). *J. Biol. Chem. 211*, 1-12.

Roos, A. (1975). *J. Physiol. London 249*, 1-25.

Schaefer, J., Stejskal, E. O., and Beard, C. F. (1975). *Plant Physiol. 55*, 1048-1053.

Seeley, P. J. (1975). Ph.D. Thesis, Univ. of Oxford, England.

Seeley, P. J., Busby, S. J. W., Gadian, D. G., Radda, G. K., and Richards, R. E. (1976). *Biochem. Soc. Trans. 4*, 62-64.

Seeley, P. J., Sehr, P. A., Gadian, D. G., Garlick, P. B., and Radda, G. K. (1977). *In* "NMR in Biology" (R. A. Dwek, I. D. Campbell, R. E. Richards, and R. J. P. Williams, eds.), pp. 247-275. Academic Press, London.

Sehr, P. A., Radda, G. K., Bore, B. J., and Sells, R. A. (1977). *Biochem. Biophys. Res. Commun. 77*, 195-202.

Shulman, R. G. (1979). *In* "Biological Applications of Magnetic Resonance" (R. G. Shulman, ed.). Academic Press, New York.

Smith, A. D., and Winkler, H. (1972). *In* "Catecholamines--Handbook of Experimental Pharmacology" (H. Blaschko and E. Muscoll, eds.), Vol. 33, pp. 538-617. Springer, Berlin.

Stella, G. (1928). *J. Physiol. London 66*, 19-31.

Sternlicht, H., Shulman, R. G., and Anderson, E. W. (1965a). *J. Chem. Phys. 43*, 3123-3132.

Sternlicht, H., Shulman, R. G., and Anderson, E. W. (1965b). *J. Chem. Phys. 43*, 3133-3143.

Taylor, G. J., Jacobus, W. E., Hollis, D. P., Nunnally, R. L., and Weisfeldt, M. L. (1977). *Clin. Res. 25*, 257A.

Waddell, W. J., and Bates, R. G. (1969). *Physiol. Rev. 49*, 285-329.

APPENDIX 1

The ^{31}P Chemical Shifts of Biological Phosphorus Compounds,
Measured as a Function of pH

The resonant frequency of a solution of 85% phosphoric acid
containing about 5 mM EDTA was determined at 20°C using an external ^{2}D lock, and all shifts were measured relative to the
shift of this standard solution. The samples were cylindrical,
with their axis parallel to the field of a 7.5 T superconducting
magnet. The values given below will therefore be smaller by
about 0.8 ppm than values obtained with sample axes perpendicular to the main field (see Section II,C,1).
The concentrations of compounds were in the range 5-50 mM.
All solutions containing 1 mM EDTA, and all measurements were
made at 20°C.
No decoupling was used, and for coupled lines the mean
shift is quoted.

(1) Inorganic phosphate, pK 6.8

pH	4.26	4.91	5.40	5.84	6.36	6.80	7.26	7.89	8.88
Chemical shifts	-0.84	-0.86	-0.92	-1.05	-1.46	-2.01	-2.65	-3.05	-3.31

(2) 5 mM ATP, no added metal ions, pK 6.6 (from γ resonance), 6.5 (α), 6.6 (β)

pH	4.26	4.91	5.40	5.84	6.36	6.80	7.26	7.89	8.88
Chemical shifts									
γ	9.91	9.91	9.74	9.30	8.38	7.17	6.13	5.64	5.21
α	10.51	10.53	10.55	10.53	10.47	10.38	10.32	10.34	10.26
β	22.13	22.23	22.19	22.09	21.89	21.51	21.25	21.29	20.94

(3) 5 mM ATP + 10 mM MgCl2 + 180 mM KCl, pK 4.85 (γ), 4.85 (α), 4.8 (β)

pH	2.99	4.05	4.86	5.30	5.84	6.66	7.16	7.68	8.65
Chemical shifts									
γ	9.85	9.11	7.30	6.16	5.33	4.90	4.82	4.82	4.78
α	10.55	10.45	10.23	10.11	9.98	9.92	9.92	9.92	9.92
β	21.62	21.11	19.97	19.31	18.78	18.50	18.46	18.44	18.42

(4) 50 mM ATP, no added metal ions, pK 6.35 (γ), 6.5 (α), 6.25 (β)

pH	4.02	5.05	5.42	5.72	6.12	6.43	6.80	7.23	7.58	8.24
Chemical shifts										
γ	9.76	9.56	9.28	8.87	8.00	7.19	6.30	5.60	5.34	5.11
α	10.40	10.42	10.40	10.36	10.32	10.28	10.23	10.18	10.22	10.21
β	21.91	21.87	21.76	21.66	21.44	21.23	20.98	20.81	20.77	20.72

(5) 5 mM ADP, no added metal ions, pK 6.8 (α), 6.8 (β)

pH	4.12	5.39	5.80	6.18	6.67	6.98	7.26	7.77	8.89
Chemical shifts									
α	10.55	10.55	10.49	10.46	10.36	10.28	10.21	10.19	10.10
β	9.98	9.83	9.60	9.17	8.11	7.38	6.73	6.15	5.73

(6) 5 mM ADP + 10 mM MgCl$_2$, pK 5.25 (α), 5.25 (β)

pH	3.07	3.97	4.45	4.95	5.46	5.98	6.46	7.12	7.72
Chemical shifts									
α	10.57	10.53	10.40	10.15	9.97	9.52	9.41	9.36	9.30
β	10.00	9.62	9.30	8.45	7.06	6.05	5.58	5.34	5.26

(7) AMP and IMP (resonances overlap throughout titration), pK 6.25

pH	5.36	5.94	6.29	6.63	7.01	7.34	7.71
Chemical shift	-1.43	-2.23	-2.91	-3.52	-4.00	-4.25	-4.40

(8) Phosphocreatine, pK 4.6

pH	4.26	4.91	5.40	5.84	6.36	6.80	7.26	7.89	8.88
Chemical shifts	3.80	2.99	2.59	2.46	2.39	2.35	2.33	2.35	2.35

(9) NADH

pH	3.60	5.08	5.68	6.04	6.62	7.22	7.52	8.24
Chemical shifts	10.43	10.45	10.45	10.45	10.45	10.45	10.45	10.47

(10) NAD$^+$

pH	4.99	5.76	6.25	6.86	7.44	7.89	9.72
Chemical shifts	10.64	10.64	10.64	10.62	10.62	10.64	10.64

(11) Inorganic pyrophosphate, pK 6.35

pH	4.18	5.14	5.46	5.86	6.54	6.92	8.36
Chemical shifts	9.90	9.70	9.53	9.15	7.98	7.49	6.75

(12) Glucose 1-phosphate, pK 6.2

pH	4.59	5.07	5.42	5.81	6.09	6.39	6.76	7.01	7.24	7.52	7.80	8.10	8.40
Chemical shift	0.45	0.27	-0.03	-0.59	-1.12	-1.74	-2.34	-2.64	-2.82	-2.92	-2.97	-2.97	-2.99

(13) Glucose 6-phosphate, pK 6.2

pH	4.59	5.07	5.42	5.81	6.09	6.39	6.76	7.01	7.24	7.52	7.80	8.10	8.40
Chemical shifts													
α	-1.47	-1.66	-1.92	-2.50	-3.04	-3.63	-4.31	-4.65	-4.91	-5.06	-5.10	-5.12	-5.14
β	-1.47	-1.66	-1.97	-2.56	-3.12	-3.76	-4.42	-4.73	-4.91	-5.06	-5.10	-5.12	-5.14

(14) Fructose 6-phosphate, pK 6.2

pH	4.59	5.07	5.42	5.81	6.09	6.39	6.76	7.01	7.24	7.52	7.80	8.10	8.40
Chemical shift	-1.20	-1.36	-1.63	-2.14	-2.64	-3.25	-3.84	-4.09	-4.31	-4.42	-4.46	-4.47	-4.49

(15) Fructose 1,6-diphosphate, pK 5.95 for 1-P and 6.15 for 6-P

pH	5.68	6.00	6.28	6.65	6.94	7.32	7.77
Chemical shifts							
1-P	-2.10	-2.73	-3.26	-3.82	-4.17	-4.47	-4.68
6-P	-2.02	-2.52	-3.00	-3.58	-3.95	-4.30	-4.55

(16) 2-Phosphoglycerate, pK 6.65

pH	4.99	5.76	6.25	6.51	6.86	7.13	7.44	7.76	7.98	8.24	9.72
Chemical shifts	-0.95	-1.31	-1.89	-2.39	-3.07	-3.48	-3.91	-4.13	-4.21	-4.27	-4.37

(17) 3-Phosphoglycerate, pK 6.4

pH	4.58	5.34	5.88	6.17	6.42	6.74	7.12	7.48	7.79	8.79	9.80
Chemical shift	-1.25	-1.52	-2.06	-2.56	-3.08	-3.46	-4.21	-4.51	-4.63	-4.74	-4.76

(18) Glycerol 1-phosphate, pK 6.25

pH	4.02	5.05	5.42	5.72	6.12	6.43	6.80	7.23	7.58	8.24
Chemical shifts	-1.44	-1.65	-1.92	-2.31	-3.03	-3.65	-4.25	-4.69	-4.82	-4.94

(19) Glycerol 2-phosphate, pK 6.25

pH	4.02	5.05	5.42	5.72	6.12	6.43	6.80	7.23	7.58	8.24
Chemical shifts	-0.93	-1.12	-1.41	-1.80	-2.56	-3.21	-3.86	-4.31	-4.46	-4.59

(20) Glyceraldehyde 3-phosphate, pK 6.1

pH	5.08	5.68	6.04	6.36	6.62	6.89	7.22	7.52	7.81	8.24	10.85
Chemical shifts	-1.82	-2.18	-2.83	-3.50	-3.95	-4.44	-4.71	-4.88	-4.98	-5.02	-5.06

(21) Dihydroxyacetone phosphate, pK 5.9 (hydrated) and 5.95 (free)

pH	5.08	5.68	6.04	6.36	6.62	6.89	7.22	7.52	7.81	8.24	10.85
Chemical shifts											
hydrated	-1.69	-2.69	-3.56	-4.27	-4.67	-5.06	-5.23	-5.34	-5.38	-5.40	-5.42
free	-1.39	-2.25	-3.05	-3.63	-3.95	-4.23	-4.37	-4.44	-4.48	-4.50	-4.50

(22) Phosphoenolpyruvate, pK 5.9

pH	4.58	5.34	5.88	6.17	6.42	6.74	7.12	7.48	7.79	8.79	9.80
Chemical shifts	3.42	2.80	1.89	1.33	0.91	0.57	0.34	0.21	0.17	0.18	0.13

(23) Pyridoxal phosphate, pK 6.05

pH	4.12	4.71	5.07	5.44	5.88	6.27	6.56	6.86	7.20	7.66	7.99	8.65
Chemical shifts	-0.90	-1.03	-1.18	-1.50	-2.25	-2.97	-3.50	-3.88	-4.18	-4.39	-4.46	-4.52

Titrations in the Presence of 1.6 $\underline{M}$ KCl, which Illustrate the Effects of High Ionic Strength

(24) Inorganic phosphate, pK 6.5

pH	4.24	4.97	5.44	5.87	6.40	6.91	7.42	7.80	9.13
Chemical shifts	-0.80	-0.85	-1.04	-1.38	-1.98	-2.76	-3.15	-3.25	-3.40

(25) Phosphocreatine

pH	4.24	4.97	5.44	5.87	6.40	6.91	7.42	7.80	9.13
Chemical shifts	3.49	2.68	2.45	2.37	2.34	2.32	2.30	2.35	2.32

(26) Glucose 6-phosphate, pK 5.7

pH	4.24	4.97	5.44	5.87	6.40	6.91	7.42	7.80	9.13
Chemical shifts	-1.48	-1.87	-2.57	-3.51	-4.38	-4.92	-5.09	-5.15	-5.22

(27) 5 mM ATP, pK 5.95 (γ), 6.3 (α), 5.9 (β)

pH	4.24	4.97	5.44	5.87	6.40	6.91	7.42	7.80	9.13
Chemical shifts									
γ	9.30	8.85	8.06	6.92	5.72	4.90	4.61	4.58	4.44
α	9.98	9.98	9.92	9.87	9.81	9.74	9.72	9.76	9.70
β	21.21	21.11	20.91	20.61	20.26	20.04	19.96	20.00	19.92

HIGH-RESOLUTION ^{31}P AND ^{13}C
NUCLEAR MAGNETIC RESONANCE STUDIES OF
ESCHERICHIA COLI CELLS *in vivo*

K. Ugurbil
R. G. Shulman
T. R. Brown

Bell Laboratories
Murray Hill, New Jersey

I. INTRODUCTION

High-resolution nuclear magnetic resonance (NMR) is now
solidly established as a valuable method of studying biological
problems. The advances that made this possible are based upon
earlier investigations of small molecules and the subsequent
applications to larger and more complicated molecules of biological origin. Continuing this flow toward complexity, NMR
spectroscopy, predominantly of ^{31}P nuclei, has recently been
employed in studying a variety of problems in intact cells and
tissue. The first ^{31}P measurements on cells were made by Moon
and Richards (1973), who showed that the 40 MHz spectrum of
erythrocytes allowed one to observe the two ^{31}P peaks of 2,3-
diphosphoglycerate (2,3-DPG) as well as two peaks from intracellular and extracellular inorganic orthophosphate (P_i); it
was possible from the chemical shifts of these resonances to
distinguish the intracellular and extracellular pHs, which differed slightly. A subsequent study on erythrocytes by Henderson *et al.* (1974) showed that incubation with inosine and phosphoenol pyruvate (PEP) resulted in an increase in the 2,3-DPG
levels, which could be observed in the NMR spectrum. At about
this time the high-frequency (129 MHz) ^{31}P NMR spectrometer at
Oxford was used to detect and monitor the levels of P_i, adenosine triphosphate (ATP), and creatine phosphate in muscle preparations under different conditions such as during "aging" and

ischemia (Hoult *et al.*, 1974). Similar observations were also
made on different muscles in Chicago (Burt *et al.*, 1976).

Our own early studies on suspensions of yeast cells showed
that large values of ΔpH, where $\Delta pH = pH^{in} - pH^{ex}$, could be observed in yeast cells when the external pH was changed (Salha-
ny *et al.*, 1975). It was also possible to follow glucose metabolism in Erlich ascites tumor cells (Navon *et al.*, 1977a) and
in several other mammalian cells in suspension (Navon *et al.*,
1978). A substantial part of our effort has been concentrated
on the canonical bacterium *Escherichia coli* (Navon *et al.*,
1977b; Brown *et al.*, 1977; Ogawa *et al.*, 1978; Ugurbil *et al.*,
1978a), which possess numerous advantages as a system of study.
These include the extensive biochemical information that exists
on this organism and the wealth of mutants available, which can
be used to test various points. On the other hand, *E. coli*
provides a strong challenge to the NMR investigator because of
its rapid metabolism. The rates of anaerobic glycolysis and of
respiration of *E. coli* are fast, particularly when compared to
mammalian cells where the rates are one to two orders of magnitude slower. Since the signal-to-noise ratio in an NMR experiment is proportional to the square root of the data accumulation
time, it is more difficult to get good signal-to-noise spectra
from *E. coli* samples than from other cells. Hence problems
once solved for *E. coli* are solved *a fortiori* for other cellular suspensions.

Our ^{31}P NMR studies on suspension of *E. coli* cells have
predominantly dealt with problems in bioenergetics, particularly as influenced by metabolism. Recently, we have shown that
in vivo high-resolution ^{13}C NMR spectra of isotopically enriched metabolites, when compared to ^{31}P NMR spectra on the
same samples, offer considerably better resolution and approximately equal sensitivity (Ugurbil *et al.*, 1978b). Using ^{13}C
NMR we have extended our investigations on *E. coli* to include
glucose transport and catabolism. In this chapter we review
and discuss in detail our ^{31}P and ^{13}C NMR results on *E. coli*
cells.

II. METHODS

NMR studies are generally limited by signal-to-noise ratios
and by resolution. In the study of intact cells these problems
are particularly severe. We have tried to improve the signal-
to-noise ratio in several ways.

First we have increased the total amount of metabolites in
the NMR sample. High cellular concentrations (generally
~5×10^{11} cells/ml for *E. coli*) have been used, where the intra-

cellular volume was approximately 1/4 of the total volume. At
these concentrations the suspensions are still fluid enough to
stir and aerate as discussed below. Sample tubes were 10 mm
in diameter with sample volumes of ~1.5 ml. We have also
closely controlled the cell harvesting and sample preparation
procedures so as to maximize the intracellular concentrations
of metabolites. Details of the harvesting procedure are given
separately (Ugurbil *et al.*, 1978a). After harvesting and final
resuspension at densities suitable for NMR, the cells were always used within approximately 2 hours, thereby avoiding aging.
Our *E. coli* samples, prepared in this way, synthesized high
levels of intracellular ATP ($\approx$5 m$\underline{M}$) when supplied with a carbon
source. On the other hand, the ATP levels were considerably
lower in cells allowed to consume their internal pools of metabolites prior to the NMR experiments by, for example, harvesting
at high temperatures or prolonged waiting before use. The high
ATP levels led to higher concentrations of many metabolites in
the properly prepared samples. Without these controlled preparations, it would not have been possible to do the experiments
discussed here.

Second, we tried, in several ways, to increase the time
available for data accumulation. Generally we worked at 20°C
to decrease metabolic rates slightly and used high substrate
levels. It is well known from *in vitro* studies that nucleotide
phosphates can control the rates of certain glycolytic enzymes.
Hence in the properly harvested cells, which develop high nucleotide triphosphate and low nucleotide diphosphate levels,
metabolic rates were slower, thereby allowing longer accumulation times. Thus, we see that the signal-to-noise ratios depend upon the sample preparation.

Finally, all spectra were measured with a sensitive Bruker
HX-360 NMR spectrometer operating at 145.7 and 90.5 MHz for ^{31}P
and ^{13}C nuclei, respectively. With the above-mentioned *E. coli*
samples this spectrometer allows us to obtain an ~7:1 signal-to-
noise ratio in 10 minutes of data accumulation for a phosphate
metabolite that has an intracellular concentration of ~1 m$\underline{M}$ and
whose resonance linewidth (full width at half-height) is ~50 Hz.

III. ^{31}P NMR SPECTRA AND ASSIGNMENTS

In this section we show ^{31}P NMR spectra of *E. coli* suspensions and show how resonances can be assigned to metabolites.

Figure 1 illustrates the intact cell ^{31}P NMR spectrum of
an anaerobic suspension of freshly harvested *E. coli* cells before and after the addition of glucose. Each spectrum was accumulated for ~2 minutes. The cells were grown aerobically us-

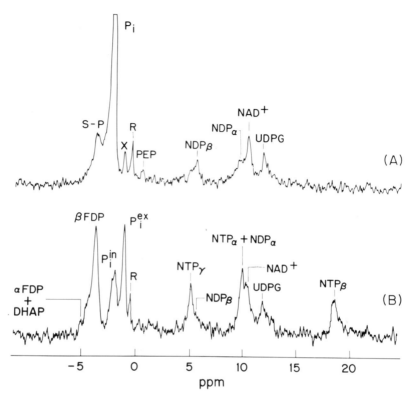

*Fig. 1.   145.7 MHz ^{31}P NMR spectra of anaerobic E. coli
cells at 20°C.   (A) E. coli cells in suspension after harvesting and prior to addition of glucose.   (B) Same sample as in
(A), 4 to 6 minutes after addition of 50 mM glucose.   Each
spectrum consists of 400 scans obtained using a repetition time
of 0.34 sec and a 45 degree rf pulse.   E. coli, strain MRE 600,
were grown aerobically at 37°C using M9 minimal medium supplemented with 20 mM glucose and 0.1 mM CaCl$_2$.   Cells were harvested at midlog phase of growth as described in Ugurbil et al.
(1978a).   The recovered pellets were washed twice and resuspended in an equal volume of 10 mM Na$_2$HPO$_4$, 10 mM KH$_2$PO$_4$, 85 mM
NaCl, 50 mM MES, 100 mM PIPES (pH 7.3).   This gives a suspension
with a cell density of ~5 × 10^{11} cells/ml.   The suspensions were
kept in ice bath until ~15 minutes prior to their use, at which
point they were warmed up to 20°C.   Peak R stems from 0.1% orthophosphoric acid in 1 M HCl contained in a concentric 3 mm
capillary.   X, Unidentified resonance.   PEP, Phosphoenolpyruvate; UDPG, Uridine diphosphoglucose.   Chemical shifts are referred to 85% orthophosphoric acid.*

ing glucose as the carbon source. In midlog phase of growth
they were collected, washed, and resuspended in a phosphate-
containing medium (specified in the figure legend) that lacked
nitrogen and carbon sources. 50 m$\underline{M}$ glucose was added to the
suspension immediately after the spectrum shown in Fig. 1A was
obtained. The resonances were assigned to specific metabolites
on the basis of chemical shifts and their dependence upon pH.
This information was obtained by titrating cellular extracts
that were prepared by the conventional perchloric acid digestion
method (Weibel *et al.*, 1974; Saez and Lagunas, 1976). pH titration data, however, are insufficient for a definite identification because the pKs often depend on salt concentration that is
difficult to control in extracts. Consequently, after tentative assignments were made from pH titration data, they were
checked by adding to the extracts small quantities of the purified metabolites. Figure 2A,B illustrates ^{31}P NMR spectra of
extracts obtained from *E. coli* suspensions under the conditions
of Fig. 1A,B, respectively. Clearly, the resolution in these
spectra is much better; while the linewidths are approximately
50 to 90 Hz in the intact cell spectra, they range from 3 to

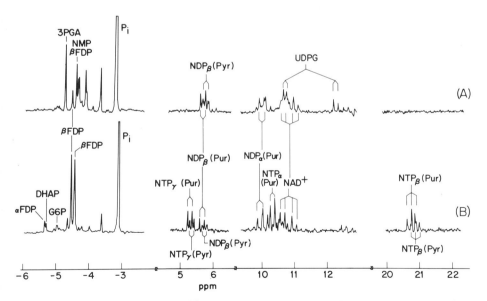

*Fig. 2. 145.7 MHz ^{31}P NMR spectra at 20°C of E. coli cell
extracts: (A) before glucose addition; (B) during glycolysis
in the presence of glucose. Experimental conditions for cell
growth and suspension were identical to those in Fig. 1. Perchloric acid extracts were prepared as described in Navon et al.
(1977a). The solution pH was (A) 8.2 and (B) 8, respectively.*

6 Hz in the proton-decoupled extract spectra. Titration curves
for the major metabolites detected in the extract from glycolysing *E. coli* cells are shown in Fig. 3.

In the absence of an external carbon source, the intact
cell spectrum shows resonances stemming from α and β phosphates
of nucleoside diphosphates (NDP), nicotinamide adenine dinucleotide (NAD), uridine diphosphoglucose (UDPG), P_i, phosphoenol

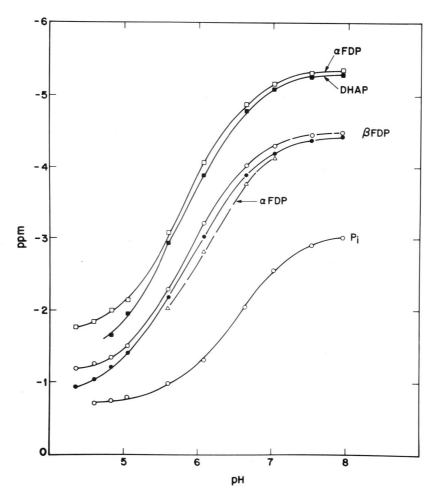

*Fig. 3. Titration curves for the phosphomonoester and or-
thophosphate resonances at 20°C of an extract prepared from
glycolysing anaerobic E. coli cells. The cell extract was different from the sample used in Fig. 2B. However, it was prepared under identical conditions and the whole cell spectra
were very similar.*

pyruvate (PEP), and phosphomonoesters such as sugar phosphates (S-P). The broad phosphomonoester peak in the intact cell spectrum (Fig. 1A) is resolved into several distinct peaks in the extract spectrum (Fig. 2A). Four of these peaks have been assigned to 3-phosphoglycerate (3-PGA), nucleoside monophosphates (NMP) (the different nucleoside monophosphates are not resolved from each other), the two phosphates of β-fructose 1,6-diphosphate (FDP), and the one phosphate of α-FDP. At this pH the 6-phosphate of α-FDP has a chemical shift that coincides with one of the two β-FDP resonances (see Fig. 3). The NDP α and β phosphate resonances each split into two distinct doublets in the extract spectrum. The two doublets have been assigned to purine and pyrimidine nucleoside diphosphates. However, the individual purine or pyrimidine resonances are not clearly resolved even at 145.7 MHz.

In the presence of glucose, the spectrum of glycolysing cells (Fig. 1B) and the corresponding extract spectrum (Fig. 2B) show extensive changes. The α, β, and γ phosphate resonances of nucleoside triphosphates (NTP) are now detectable in the intact cell spectrum at 10.0, 18.7, and 5.0 ppm, respectively. Note that the NTP$_\beta$ peaks appears at ~20.9 ppm in the extract spectrum of the glycolysing cells (Fig. 2B), as opposed to 18.7 ppm in the intact cell. The difference is attributable to Mg^{2+} binding and indicates that in our cells essentially all of the NTP is Mg^{2+} bound. The extract spectrum also shows that the NTP pool is heterogeneous, containing contributions from purine and pyrimidine nucleoside triphosphates. In the phosphomonoester region, resonances have been assigned to dihydroxyacetone phosphate (DHAP), glucose 6-phosphate (G6P), and FDP. In this region ~90% of the overall intensity is from FDP. An intracellular concentration of ~13 mM is calculated for this metabolite from the NMR peak intensity, which was corrected for T_1 effects (Ugurbil *et al.*, 1978a); for this calculation the intracellular volume was assumed to be 1/2 of the total pellet volume (Padan *et al.*, 1976).

The chemical shifts of the resonances stemming from titratable metabolites can be used to determine intracellular pH. For example, the two intense peaks at -1.28 and -2.24 ppm in Fig. 1B stem from the external and internal P_i and reflect the existence of a pH difference across the *E. coli* membrane during glycolysis; the extract spectrum of these cells contains a single P_i resonance (Fig. 2B). P_i is the most suitable metabolite for the purpose of measuring pH because it is usually found in large quantities in *E. coli* as well as other cells. Furthermore its pK, which is higher than those of the phosphomonoesters by 0.5 to 0.8 units, is closer to the physiological pH range. The pK of P_i is salt dependent, becoming less so at high salt concentrations. *E. coli* are known to accumulate 200 mM K^+. The medium in which the *E. coli* cells were suspended for the NMR experiments contained an approximately equal amount of Na^+. Therefore, we have used the titration curve for P_i measured in the E. coli suspension medium to calculate intracellular pH from the chemical shifts of the internal P_i peak. The estimated accuracy of these pH determinations is ~0.1 pH units, which is determined not by the accuracy of measuring peak position, but by the uncertainties in the pK.

IV. BIOENERGETICS

As can be seen from the spectra shown in Fig. 1, the sensitivity for ^{31}P nuclei at 145.7 MHz would allow us to follow the time dependence of metabolite concentrations and pH gradients with 2 minute time resolution. The ability to measure NTP, NDP, and P_i levels while simultaneously determining the intracellular and extracellular pH values make ^{31}P NMR especially suitable for studying aspects of bioenergetics in microorganisms.

A. *Chemiosmotic Hypothesis*

Numerous experiments performed on mitochondria, chloroplasts, bacteria, and their vesicles have demonstrated the presence of coupling among the redox chain, ATP levels, and ion concentration gradients (for reviews see Racker, 1970; Harold, 1972, 1977; Rottenberg, 1975). Oxidation of electron donors such as succinate by the multienzyme electron transport chain can be used to synthesize ATP as well as to accumulate ions and substrates against large concentration gradients. Hydrolysis of ATP has also been shown to support ion accumulation and other energy-linked functions such as transhydrogenation and reverse electron transport (Harold, 1977). One explanation of the coupling among these energy-linked functions is the chemiosmotic hypothesis, first suggested by Peter Mitchell in 1961 (Mitchell, 1961, 1966, 1968; West and Mitchell, 1974). According to this hypothesis in its present form, membranes of micro-organisms and organelles (such as mitochondria and chloroplasts) are impermeable to protons. Substrate oxidation by the redox chain translocates protons across the membrane generating a proton concentration gradient and an electrical potential ($\Delta\psi$) as indicated schematically in the upper part of Fig. 4. The transmembrane electrical potential and the proton concentration gradients together constitute a proton electrochemical potential difference given by the expression

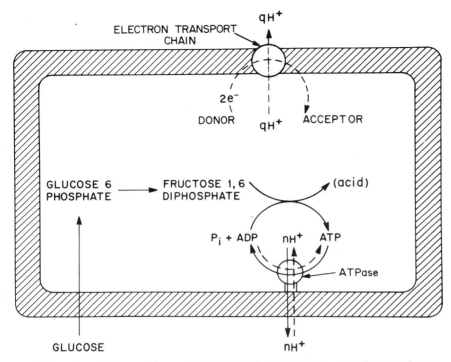

Fig. 4. Schematic representation of the chemiosmotic hypothesis relevant to the NMR experiments performed on E. coli.

$$\Delta\mu_{H^+} = F\ \Delta\psi - 2.3RT\ \Delta pH \qquad\qquad (1)$$

where F is the Faraday constant. $\Delta\mu_{H^+}$ is also postulated to be reversibly and tightly coupled to the reaction ADP + P_i ⟷ ATP by the membrane-bound proton adenosine triphosphatase (ATPase) (bottom of Fig. 4). During net synthesis of ATP, the ATPase translocates protons across the membrane so as to collapse the $\Delta\mu_{H^+}$. Net hydrolysis of ATP by the ATPase, on the other hand, is used to generate a $\Delta\mu_{H^+}$ independently of substrate oxidation. The tight coupling mediated by the ATPase between the phosphorylation of ADP and $\Delta\mu_{H^+}$ requires that

$$\Delta G_{ATPase} + n\ \Delta\mu_{H^+} = 0 \qquad\qquad (2)$$

where ΔG_{ATPase} is the Gibbs free energy for the reaction ADP + P_i ⟷ ATP and n is the number of protons translocated per ATP synthesized. Thus, unlike the earlier chemical coupling scheme, which proposed the existence of a high-energy chemical intermediate between electron transport and ATP synthesis

(Harold, 1972), the chemiosmotic hypothesis assigns this role to $\Delta\mu_{H^+}$ in oxidative phosphorylation and transport. Clearly, the demonstration that ΔpH and $\Delta\psi$ do in fact exist and the quantitative correlation of these with the intracellular concentrations of ATP, ADP, and P_i is critical for this hypothesis.

E. coli provides an excellent system for investigating the proposed relationship between ATP, ΔpH, and electron transport because, as a facultative anaerobe, it can synthesize the ATP required for cellular processes either by fermentation or by respiration in the presence of an electron acceptor. The aspect of the chemiosmotic hypothesis relevant to the NMR experiments on *E. coli* that are presented here are illustrated in Fig. 4. The cells used in our experiments have all been grown aerobically on M9 minimal medium (Roberts *et al.*, 1963) supplemented with either glucose or succinate as the carbon source. Consequently, they are capable of respiration using O_2 as the terminal electron acceptor, thereby creating a $\Delta\mu_{H^+}$ that can be coupled to ATP synthesis (Fig. 4). In addition, *E. coli* are constitutive for glycolysis and glucose transport enzymes; therefore, regardless of the growth conditions, they will catabolize glucose anaerobically to generate ATP via substrate level phosphorylation, which as shown in Fig. 4 can be hydrolyzed to generate a $\Delta\mu_{H^+}$ some part of which will be expressed as a ΔpH.

B. pH Gradients under Aerobic and Anaerobic Conditions

Numerous studies have inferred the existence of pH gradients in microorganisms (Harold, 1977). Quantitative determination of ΔpH and $\Delta\psi$ have been made in *E. coli* cells (Padan *et al.*, 1976) and in their vesicles (Ramos *et al.*, 1976) during respiration. pH gradients have also been measured in the strict anaerobe *S. feacalis* during glycolysis (Harold *et al.*, 1970). Techniques developed and used for these measurements have previously been reviewed (Rottenberg, 1975); determination of the pH gradients are generally based on the distribution of weak acids or bases between the cytoplasmic and outer phases. Although our measurements on aerobic *E. coli* have agreed with those previously obtained by dimethyloxazolidedione (DMO) distribution (Padan *et al.*, 1976) the explicitness of the pH determinations by ^{31}P NMR avoids questions sometimes raised about DMO measurements such as its binding within the cell, the existence of compartmentalization, and the possibility of a distribution of pH values among the cells. In this article, we confine ourselves to the discussion of results obtained with ^{31}P NMR.

Figure 5 illustrates a sequence of ^{31}P NMR spectra obtained as a function of time after glucose addition to an anaerobic suspension of succinate-grown cells. In the absence of glucose,

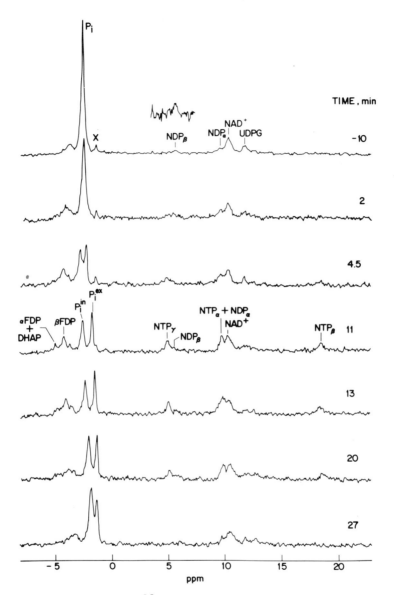

Fig. 5. 145.7 MHz ^{31}P NMR spectra of succinate-grown E. coli cells as a function of time from glucose addition under anaerobic conditions. Cells were grown, harvested, and prepared for
NMR as specified in the Fig. 1 legend, except succinate was used
as the carbon source in the growth medium. At time zero, glucose
was added to the NMR sample to a final concentration of 25 mM.
Each spectrum represents the sum of 200 scans obtained with 90°
pulses and 0.68 sec repetition rate. Sample temperature was
20°C. X, the same unidentified resonance shown in Fig. 1.

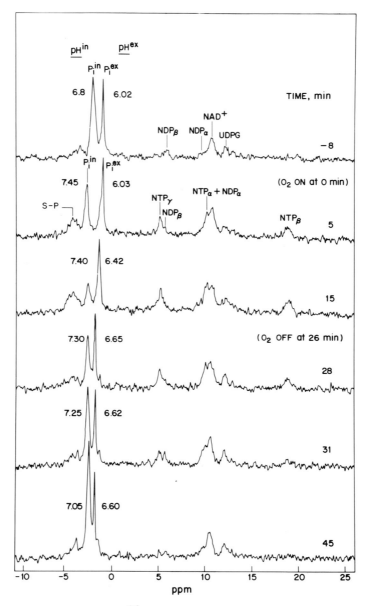

Fig. 6. 145.7 MHz ^{31}P NMR spectra of succinate-grown E. coli cells at 20°C before, during, and after oxygenation. Suspensions for NMR were prepared as in Fig. 5 except the resuspension medium contained 100 mM MES and 50 mM PIPES and its pH was adjusted to 6.4. NMR samples contained 75 mM succinate. Oxygen was introduced by continuous bubbling at the rate of 25 cm³/min. Each spectrum consists of 200 free induction decays obtained using a repetition time of 0.68 sec and 90° pulses.

a single intense P_i resonance is detectable at -2.55 ppm showing that the intra- and extracellular pHs are both 7.1. Shortly after glucose is added, peaks from three distinct phosphomonoesters and the NTP α, β, and γ phosphates become detectable. Two of the phosphomonoester resonances are assigned to FDP and DHAP as indicated in Fig. 5, while the third small peak at slightly higher field is still unidentified. These cells clearly establish a ΔpH during <u>anaerobic</u> glycolysis. The single P_i resonance broadens 2 minutes after glucose addition and splits into two distinct peaks by 4.5 minutes. The internal P_i intensity decreases as the sugar phosphates and NTP are synthesized. At the same time the external P_i resonance shifts upfield, indicating a drop in the external pH as the acidic end products of glycolysis are formed. There is an ~10 minute long "steady state" of glycolysis characterized by constant levels of phosphomonoester peaks and a continuous upfield drift of the external P_i peak. When glucose is exhausted, and subsequently the concentration of FDP decreases, acid production ceases as judged by the position of the external P_i resonance. The internal P_i resonance, however, shifts upfield as the transmembrane pH gradient (alkaline inside), established during glucose catabolism, collapses in the absence of an energy source.

Oxygenation of the cells can be accomplished by bubbling O_2 through the cellular suspensions. In our experience with *E. coli* suspensions we have found that a continuous stream of O_2 bubbles can be used for oxygenation without causing significant magnetic field inhomogeneities, provided the bubbles are small (≤2 mm in diameter) and the flow is regular.

The top spectrum of Fig. 6 shows the ^{31}P spectrum of freshly harvested succinate-grown cells prior to O_2 bubbling. The cell suspension medium contains succinate that is not metabolized in the absence of oxygen. The difference between the internal and external pHs observed prior to oxygenation is an artifact of sample preparation. When aerobically grown, midlog phase cells are harvested at ~4°C and the cultures are maintained aerobic during the cooling process prior to harvesting, the internal pH of recovered cells is usually between 7.3 and 7.5. Consequently, when the cells are resuspended in a medium whose pH is 6.4, as they have been in the experiment shown in

Fig. 6, an artificial pH gradient is created. On the other hand suspension in a pH 7.4 medium gives only a single P_i peak (see Fig. 5).

After the onset of oxygenation, NTP and sugar phosphate resonances appear and the P_i^{in} peak rapidly shifts downfield (Fig. 6). While oxygenation is maintained, the P_i^{in} resonance remains at -2.8 ppm, indicating an internal pH of 7.45. During the same period, P_i^{ex} moves upfield slowly, reflecting a change in pH^{ex} from 6.03 to 6.42 in ~15 minutes. The intensity of the P_i^{in} resonance decreases in the presence of O_2 as sugar phosphates and NTP are synthesized. When O_2 is turned off, however, it recovers its initial intensity within ~15 minutes. Furthermore, after the oxygenation is stopped the P_i^{in} resonance now shifts upfield, indicating a decrease in internal pH.

The time dependences of internal and external pH from Figs. 5 and 6 are shown in Fig. 7. Under anaerobic conditions (Fig. 7A), the cells generate and maintain a ΔpH of ~0.6 units for ~10 minutes after glucose addition as the external pH drifts from 7.1 to 6.2. This period corresponds to the glycolysis steady state characterized by constant phosphomonoester levels. When the glucose is exhausted (indicated by the leveling off of pH^{ex}) the ΔpH begins to collapse. Under aerobic conditions (Fig. 7B), the onset of oxygenation causes a rapid increase in internal pH to 7.55 concurrently with a slight decrease in external pH. During oxygen bubbling, the internal pH is clearly regulated and maintained at ~7.5, while the external pH drifts. The increase in pH^{ex} occurs as a consequence of net succinate consumption (Ugurbil *et al.*, to be published).

Because of the very high cell densities of the NMR samples, the O_2 levels in these suspensions are expected to be very low. In order to estimate the respiration rate under NMR conditions, steady-state O_2 levels were measured (Ogawa *et al.*, 1978) at different concentrations of cells using an O_2 electrode (Fig. 8). The suspensions contained succinate as the carbon source and were oxygenated by vigorous bubbling of air. Applying a simple enzymatic model of oxygen consumption by the cells (Ogawa *et al.*, 1978), one can write for the steady state

$$A([O_2]_S - [O_2]) = \frac{Q[E][O_2]}{[O_2] + K_m} \tag{3}$$

where $[E]$ is the *E. coli* concentration, Q the respiration rate of *E. coli* on succinate, K_m the Michaelis-Menton constant for the O_2 reducing enzyme, $[O_2]_S$ the O_2 concentration at the surface of the bubble, $[O_2]$ the bulk oxygen concentration, and A the rate of transfer to the bulk solution from the air bubbles. For $[O_2] \gg K_m$, $[O_2]$ will decrease linearly with increasing $[E]$, as observed for $[E] \leq 6.5 \times 10^9$ cells/ml. The linear portion

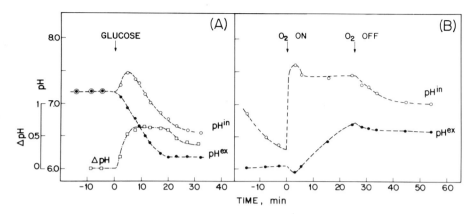

*Fig. 7.   Internal, external, and ΔpH in succinate-grown E.
coli cells in suspension: (A) under anaerobic conditions before
and after 25 mM glucose was added into the sample; (B) before,
during, and after oxygenation in the presence of 75 mM succinate.   The data were taken from the experiments shown in Figs.
5 and 6 for A and B, respectively.*

intercepts the abscissa at 8.3×10^9 cells/ml and the ordinate
at 208 μM.   From the nonlinear region of the $[O_2]$ vs. $[E]$ curve,
K_m is estimated to be ~0.7 μM.   If pure O_2 is used instead of
air, the estimated cell concentration at which $[O_2]$ equals K_m
will be ~10^{11} cells/ml, which is approximately one-fifth the
cell densities usually used for NMR measurement.   The following
experiment, however, demonstrates that this limitation does not
affect the internal pH values attained under NMR conditions.
In this experiment O_2 was bubbled continuously for 2 sec, followed by a period of non-bubbling, which was varied from 1 to
30 sec.   The internal pH values at steady state are plotted vs.
repetition rate in Fig. 9.   Clearly, the internal pH becomes
independent of the repetition rate when this rate exceeds ~0.1
sec^{-1}.   It is not totally surprising that a constant pH^{in} value
is attained by these cells even at somewhat reduced respiration
rates because it takes pH^{in} several minutes to decay after O_2
bubbling is discontinued (see Fig. 7).

C. ATPase Inhibition

It is clearly seen from these NMR experiments that *E. coli*
generate and maintain a transmembrane pH gradient when supplied
with a fermentable, ATP-generating substrate such as glucose or
with O_2 and a substrate of the electron transfer chain.   As dis-

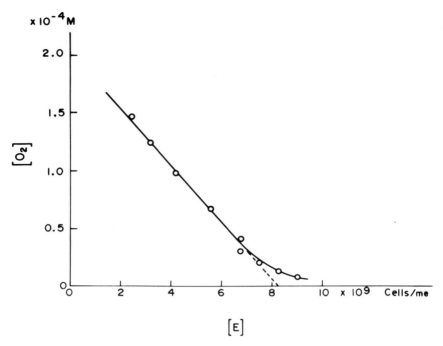

Fig. 8. Steady-state oxygen concentration vs. cell concen-
tration for succinate-grown cells respiring on succinate at
23°C. [O₂] was measured with an oxygen electrode and the cell
concentrations were determined by monitoring the absorbance at
650 nm. Oxygenation was accomplished by bubbling air at a flow
rate of 33 cm³/min. Suspensions were in M9 minimal medium, pH
7, supplemented with 5 mM succinate.

cussed previously, the chemiosmotic hypothesis postulates that
transmembrane proton gradients are coupled by the membrane-
bound ATPase to ATP synthesis and hydrolysis. Consequently,
when the ATPase is inhibited, glycolysing anaerobic cells should
not be capable of developing a ΔpH but should generate ATP.
Conversely under aerobic conditions these inhibited cells should
still be capable of developing a ΔpH, although their ability to
synthesize ATP should be drastically reduced.

1. DCCD

   Dicyclohexylcarbodiimide (DCCD) is a specific and irreversible inhibitor of bacterial and mitochondrial ATPases (Harold,
1972). DCCD inhibits oxidative phosphorylation in bacteria and
mitochondria and prevents growth in bacterial cultures. Fur-

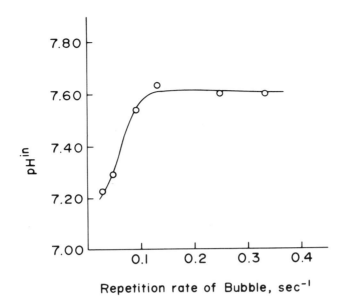

Fig. 9. The internal pH values determined from the position of the inorganic phosphate ^{31}P NMR peak as a function of the repetition rate of a 2-sec long oxygen bubbling period. During the bubbling period the gas flow was kept constant at a rate of 36 ml/minute. Cells were grown in M9 using succinate as the carbon source and were suspended for NMR at a concentration of ~3 × 10^{11} cells/ml in the growth medium.

thermore, when treated with DCCD, isolated and purified ATPase-membrane preparations or detergent-solubilized ATPase complexes exhibit little or no ATP hydrolysis activity (Harold *et al.,* 1969; Hare, 1975; Yoshida *et al.,* 1977). This potent inhibitor has been shown to react with the component of the ATPase that is associated with the cell membrane (Altendorf and Zitzman, 1975; Hare, 1975; Fillingame, 1975, 1976).

Figure 10 shows the ^{31}P NMR spectra of anaerobic glucose-grown *E. coli* cells at different times after glucose addition. This experiment serves as a control for the DCCD-inhibited cells (Fig. 11). The sequence of events observed in Fig. 10 after the introduction of glucose is very similar to those observed in anaerobic succinate-grown *E. coli* cells (Fig. 5), except that the catabolism of glucose is faster and FDP levels are much higher. The single P_i resonance broadens 2 minutes after glucose addition and by 4.5 minutes clearly splits into two distinct peaks reflecting the formation of a ΔpH. Internal P_i intensity decreases as the sugar phosphates and NTP are synthe-

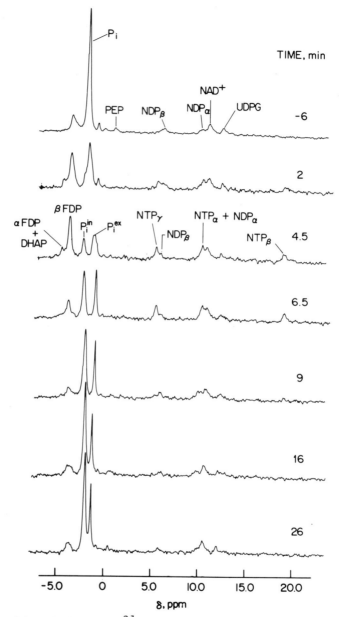

Fig. 10. 145.7 MHz ^{31}P NMR spectra of glucose-grown E. coli cells at 20°C. All experimental conditions are identical to those given for Fig. 1. Glucose was added to a final concentration of 25 mM in the suspension at time zero. Times given for each spectrum represents the middle of the 2 minute accumulation period referred to the time of glucose addition.

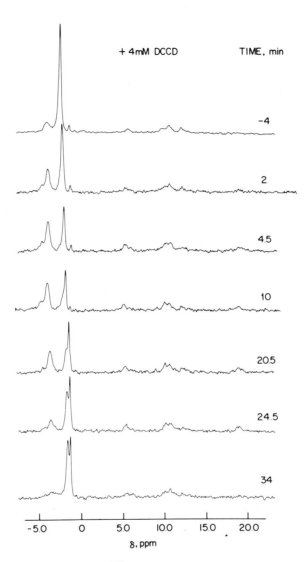

Fig. 11. 145.7 MHz ^{31}P NMR spectra of anaerobic E. coli cells treated with 4 mM DCCD. Cells used for this experiment were taken from the same batch used for the experiment shown in Fig. 10. All conditions were identical to those specified for Fig. 10, except 4 mM DCCD was added to the NMR sample after the sample was warmed up to 20°C, 20 minutes prior to glucose addition.

sized. At the same time the external P_i resonance shifts upfield as the acidic end products of glycolysis are formed. The
intracellular concentration of P_i is estimated to be ~9 m$\underline{M}$ when
it reaches its maximum at 4.5 minutes (Ugurbil *et al.*, 1978a).
When glucose starts running out, the external pH drift slows
down and the P_i^{ex} resonance becomes sharper. The NTP level,
however, remains high until the FDP pool is reduced to ~10% of
its maximum level. Finally, the NTP intensity and the ΔpH diminish with time.

When another sample from the same batch of cells is treated
with 4 m$\underline{M}$ DCCD, the ^{31}P NMR spectrum observed as a function of
time from glucose addition is shown in Fig. 11. Although the
phosphomonoesters and NTP are synthesized by the DCCD-treated
cells, the ΔpH formation is almost completely inhibited; only a
small downfield shoulder on the P_i resonance appears at 10
minutes.

The time dependence of the metabolite levels and of pH^{in},
pH^{ex}, and ΔpH in the presence of 0, 2, and 4 m$\underline{M}$ DCCD is plotted
in Fig. 12. In going from 0 to 4 m$\underline{M}$ DCCD, the [NTP]/[NDP] ratio
measured during the steady state of glycolysis increased from
~2 to ~3 even though DCCD also reduced the glycolysis rate. In
addition, the rate at which the [NTP]/[NDP] ratio decayed at the
termination of glycolysis (with concomitant increase in P_i
levels) was significantly reduced upon DCCD treatment. Both of
these observations are consistent with the inhibition of the
ATPase by DCCD and further indicate that the generation of ΔpH
by the ATPase is the major pathway for ATP hydrolysis.

As discussed above, DCCD treatment is not expected to inhibit ΔpH formation in *E. coli* respiring on succinate. This is,
in fact, observed as illustrated in Fig. 13. Both spectra shown
in Fig. 13 are the sum of eight consecutive 2 minute spectra of
oxygenated *E. coli* cells that displayed constant internal pH,
as well as constant phosphomonoester and nucleotide phosphate
levels. The untreated cells (Fig. 13, top) synthesized large
quantities of NTP (~7 m$\underline{M}$ intracellular) and we estimate the
[NTP]/[NDP] ratio to be $\geq$10. Upon treatment with 3 m$\underline{M}$ DCCD, the
NTP level is reduced approximately threefold (Fig. 13, bottom)
and the [NTP]/[NDP] ratio undergoes a drastic reduction to ~1.
A contribution from the NDP β-phosphate is now observable at
5.5 ppm slightly upfield of the NTP γ peak, whereas none was detectable in the absence of DCCD (Fig. 13).

The time dependences of the internal and external pHs for
the samples studied in Fig. 13 are shown in Fig. 14. The data
presented in Fig. 13 were the sum of spectra obtained from 4 to
20 minutes after the onset of oxygenation. It is seen that pH^{in}
level attained by the DCCD-inhibited cells is the same as in the
control, and furthermore both are constant during oxygenation
after the initial burst, although the substrate oxidation rates
differ by a factor of 2 as reflected by the pH^{ex} drift. In

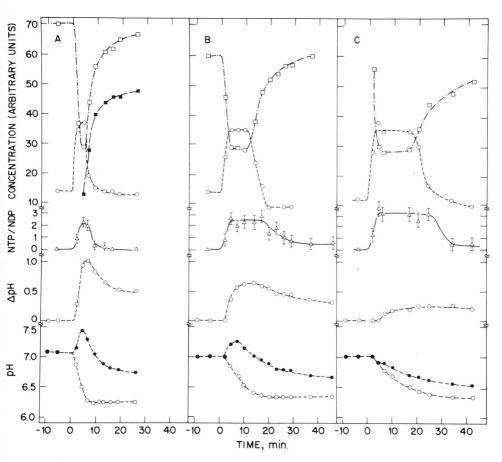

Fig. 12. Inorganic phosphate and total sugar phosphate concentrations, NTP/NDP ratio, internal pH, external pH, and ΔpH vs. time in anaerobic E. coli cells at 20°C. Data were obtained from the experiments of Figs. 10 and 11 and another experiment performed under conditions identical to the first two. The cells for all three experiments were obtained from one batch, which was divided into three after the final suspension under NMR conditions. Each point represents an average of 2 minutes. (A) Control, no DCCD; (B) 2 mM DCCD; (C) 4 mM DCCD. In the upper curves □ , total P_i; ● , internal P_i; ○ , sugar phosphates.

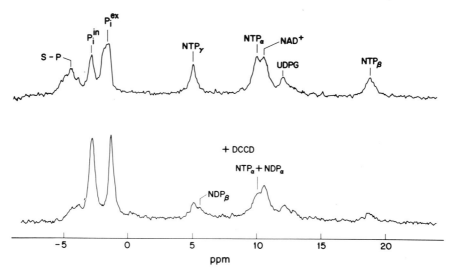

Fig. 13. The effect of DCCD on the ^{31}P NMR spectra of
succinate-grown E. coli cells respiring on succinate. Top
spectrum shows the untreated cells. In this experiment, spectra were obtained consecutively in 2 minute blocks (200 scans,
$90°$ pulses, 0.68 sec repetition time) before, during, and after
oxygenation. The spectra shown represent the sum of eight consecutive 2-minute spectra obtained during oxygenation. All
other conditions are identical to those of Fig. 6.

other words, whether or not they are inhibited by DCCD, intracellular pH of E. coli cells is regulated at ~7.5 and the ΔpH
maintained depends only upon the external pH. The broader P_i^{ex}
resonance shown in Fig. 13 (top) is a consequence of the faster
drift of the external pH values in the control, as can be seen
from Fig. 14.

2. ATPase⁻ Mutants

ATPase mutants provide another system for evaluating the
role played by the bacterial ATPase in generating proton gradients and ATP metabolism. Several well-characterized strains
in E. coli ATPase mutants exist (Simoni and Postma, 1975).
These mutants are selected on the basis of their ability to
grow on carbon sources like glucose (whose metabolism leads to
ATP synthesis by substrate level phosphorylation) but not on
pure electron-donating substrates such as succinate, which, however, they will oxidize. In short, in these mutants phosphorylation of ADP to ATP is uncoupled from substrate oxidation.

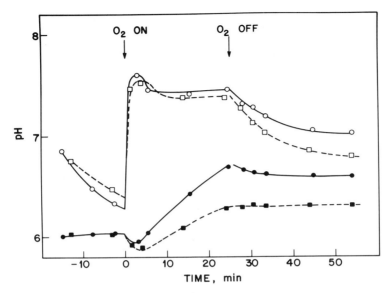

Fig. 14. The effect of DCCD on the internal and the external pH in succinate-grown cells at 20°C. The data were obtained from the individual 2 minute spectra of the experiment shown in Fig. 13. Internal pH and external pH are designated by open and closed symbols, respectively. Squares represent the DCCD treated cells (see Fig. 13 legend), and circles the control.

ATPase complexes isolated from some of these mutants, in particular in NR70 used here, also lack ATP hydrolysis activity, although this is not universally the case (Simoni and Postma, 1975).

The results we have obtained on the ATPase⁻ mutant NR70 and its wild-type parental strain NR7 are very similar to those observed in strain MRE 600 in the absence and presence of DCCD. When respiring on lactate, ATPase⁻ NR70 developed a ΔpH; however, no net ATP synthesis occurred. This is illustrated in Fig. 15. Under anaerobic conditions while metabolizing glucose NR70 displayed high levels of NTP but did not generate a ΔpH as shown in Fig. 16.

It should be emphasized that this mutant is characterized as being uncoupled and lacking ATPase activity. Consequently, the observation that it does not synthesize ATP during substrate oxidation is not surprising. However, its inability to generate ΔpH anaerobically in the presence of ATP is not *a priori* expected outside the context of the chemiosmotic hypothesis.

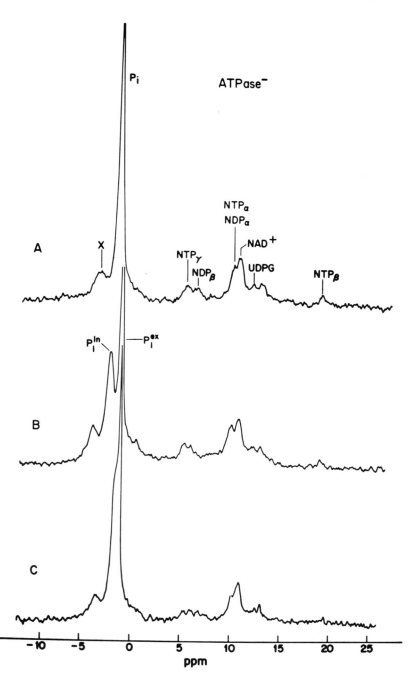

Fig. 15

In addition to being uncoupled, NR70 vesicles and to a
lesser extend whole cells have been reported to display a
marked defect in respiration driven transport (Rosen, 1973a,b).
Consistent with this observation, in our measurements, NR70
cells maintained a smaller ΔpH when respiring on lactate compared to coupled, wild-type strains NR7, AN180, and MRE 600.
Under aerobic conditions, all of the coupled strains were capable of maintaining a constant internal pH of ~7.5 in the external pH range 6 to 7. NR70 cells, however, were able to attain a pH^{in} of approximately 7.0 to 7.2 in the same external
pH range, thus effectively generating a smaller ΔpH.

D. The Effect of Uncouplers

Chemical uncouplers of oxidative phosphorylation dissociate respiration from phosphorylation even at very low concentrations. Typically uncouplers block phosphorylation but, unlike the ATPase inhibitor DCCD, they stimulate respiration.
These classes of chemicals include 2,4-dinitrophenol, tetra-
chlorosalicylanalide, decachloroborene, and carbonyl cyanide
p-trifluoromethoxyphenyl hydrazone (FCCP). Many uncouplers are
lipid-soluble acids, which will exist in substantial amounts at
physiological pH values in both the protonated and the anionic
forms. Within the framework of the chemiosmotic hypothesis,
uncouplers were proposed to act as proton conductors across the
cell membrane (Mitchell, 1966, 1968). The leakage of protons
mediated by the uncouplers would dissipate the proton gradient
and the energy required to synthesize ATP. Numerous studies
performed on mitochondria, bacteria, membrane vesicles of these
systems, and artificial lipid bilayers have been interpreted
to suggest that uncouplers specifically enhance proton diffusion

*Fig. 15. ^{31}P NMR spectra of E. coli ATPase$^-$ mutant, strain
NR70, before (A), during (B), and after (C) oxygenation in the
presence of 16 mM D-lactate at 20°C. Cells were grown aerobically at 37°C in M9 minimal media supplemented with glucose,
and were harvested and prepared for NMR under our standard conditions (see Fig. 1 legend) except the pellet collected at the
end of harvesting and washing was suspended in three times the
pellet volume of 10 mM Na_2HPO_4, 10 mM K H_2PO_4, 100 mM PIPES,
50 mM MES, 40 mM NaCl, pH 6.5. This density corresponds approximately to 1.3 × 10^{11} cells/ml. Oxygenation was accomplished by bubbling continuously at the rate of 20 cm^3/minute.
The data were collected in consecutive 2 minute blocks as described for Fig. 13. The spectra shown represent the sum of
four such 2 minute spectra.*

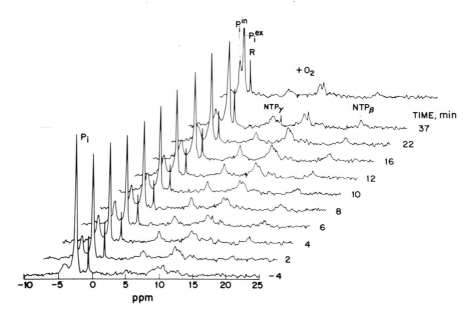

Fig. 16. ^{31}P *NMR spectra of E. coli ATPase$^-$ mutant, NR70, during anaerobic glycolysis on glucose and when oxygenated subsequently (top spectrum). Cells were suspended in the medium specified for Fig. 14 except that it lacked NaCl and its pH was adjusted to 7.5 with 1 $\underline{M}$ NaOH. 25 m$\underline{M}$ glucose was added at time zero. 40 minutes after glucose addition, O_2 was bubbled in the suspension. Each spectrum consists of 400 free induction decays, obtained with 45° pulses and 0.34 sec repetition time, at 20°C.*

across the hydrophobic membrane. The acceptance of this explanation, however, remains less than universal.

^{31}P NMR measurements made on anaerobic and aerobic *E. coli* cells in the presence of FCCP are in agreement with the mode of action suggested for uncouplers by the chemiosmotic hypothesis. FCCP prevents the formation of a ΔpH and the net synthesis of NTP by *E. coli* cells respiring on succinate (spectra not shown).

Under anaerobic conditions, in the presence of FCCP, glucose addition to *E. coli* suspension led to synthesis and subsequent consumption of FDP and DHAP (monitored by the ^{31}P resonances of these metabolites) and to net acid production (Fig. 17). However, a transmembrane proton gradient is clearly not generated and the NTP α, β, and γ phosphate resonances do not build up even though ATP production must be taking place because glycolysis is occurring. Consequently, hydrolysis of the ATP synthesized during glycolysis must be stimulated by FCCP.

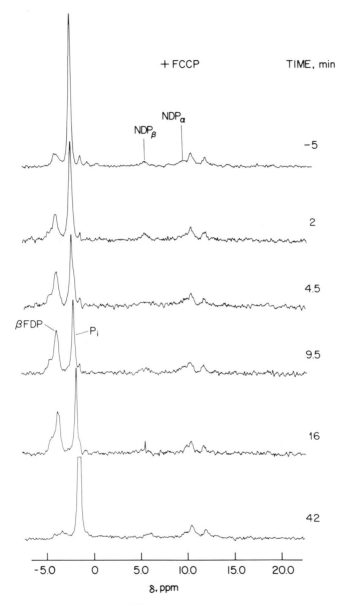

Fig. 17. 145.7 MHz ^{31}P NMR spectra of anaerobic E. coli cells (strain MRE 600) in suspension at 20°C in the presence of 75 μM FCCP. Experimental conditions were identical to those given for Fig. 1. FCCP was added 20 minutes prior to the addition of 25 mM glucose, which was added at time zero.

The effects of FCCP and DCCD on the NTP levels in anaerobic cells during glycolysis are illustrated in Fig. 18. Figure 18A and B shows that the NTP levels are approximately the same with or without DCCD, while Fig. 18C shows that FCCP destroys the NTP. If, prior to the addition of FCCP and glucose, *E. coli* are treated with 4 mM DCCD, a ΔpH is still not developed, but the NTP levels are now built up to the same levels seen in glycolysing cells treated with 4 mM DCCD alone (Fig. 18D). Clearly, FCCP effects must be mediated by the DCCD-sensitive mem-

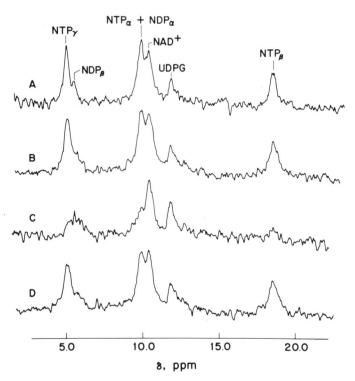

Fig. 18. Upfield region of 145.7 MHz NMR spectra of anaerobic E. coli cells: (A) Control, 400 scans; (B) 4 mM DCCD, 800 scans; (C) 75 μM, FCCP, 800 scans; (D) 75 μM FCCP and 4 mM DCCD, 800 scans. The spectra for A, B, and C were obtained by summing the individual 2 minute, 200 scan spectra of Figs. 10, 11 and 17. D was obtained from a different run performed under the conditions of Fig. 10. Control experiment for D gave identical results to that of Fig. 10. In each case the sums included those spectra which displayed the maximum amount of sugar phosphate intensity observed in that particular series.

brane ATPase. Furthermore, it can be concluded from the data
that the FCCP simultaneously stimulates the ATPase hydrolytic
activity and collapses the transmembrane proton gradient.

In summary, it is seen that substrate oxidation in aerobic
E. coli cells generates a ΔpH whose magnitude is regulated, effectively keeping the internal pH at ~7.5. How this regulation
occurs remains to be explained. Aerobically grown cells, when
given glucose under anaerobic conditions, generate a ΔpH of 0.6
to 1 units. The cytoplasmic pH under these conditions is not
maintained at 7.5 but changes with external pH. Succinate-
grown cells maintain a constant ΔpH of ~0.6 units during anaerobic glycolysis in the external pH range of 7.3 to 6.2.

The results presented on ΔpH generation and NTP production
by *E. coli* cells, and the effects of uncoupler FCCP and ATPase
inhibition on these quantities are in excellent qualitative
agreement with the chemiosmotic hypothesis. From the chemiosmotic point of view, however, these measurements are bioener-
getically incomplete in the absence of a determination of the
membrane potential, $\Delta\psi$. The ATPase free energy

$$\Delta G = \Delta G^0 + RT \log \frac{ATP}{[ADP] \; P_i^{in}} \tag{4}$$

which, as discussed in Section III, is postulated to be proportional to $\Delta\mu_{H^+}$ can be measured for *E. coli* from the data available from the NMR spectra, assuming the [NTP]/[NDP] ratio in
these cells is proportional to the [ATP]/[ADP] ratio. ΔG^0 for
ATP hydrolysis to ADP and P_i has been extensively studied and
tabulated as a function of pH and free Mg^{2+} concentrations
(Rosing and Slater, 1972). In our measurements, the internal
pH of these cells is determined simultaneously with the P_i^{in}
concentration and the [NTP]/[NDP] ratio. Assuming a free Mg^{2+}
concentration of 1 m$\underline{M}$ in anaerobic *E. coli* the value of
ΔG_{ATPase} was determined to be 10 kcal at its maximum, while in
the respiring cellular suspensions it was ≥12 kcal. If while
calculating ΔG_{ATPase} we assume a free Mg^{2+} concentration of
10 m$\underline{M}$, the values only change by 10%.

IV. RATE MEASUREMENTS *IN VIVO*

One of the areas where NMR has been extensively and very
profitably used is the determination of chemical reaction rates.
Consequently, it is not unexpected that the applications of high-
resolution NMR to cellular problems would include rate measure-

ments *in vivo*. The NMR techniques that have been used for rate determinations cover the time range from fractions of a second to minutes or longer. Simple measurements of resonance intensities or chemical shift changes with time provide kinetic information with a time resolution as short as ~1 minute in the case of *E. coli* cells. On the other hand, spin polarization transfer studies allow the determination of rates that can compete with spin-lattice relaxation rates of the nuclei participating in the reactions. In *E. coli*, observed ^{31}P spin-lattice relaxation times (T_1s) are fractions of a second. A third kind of measurement can be performed to determine the response of the system under study to repetitive perturbations. This type of measurement can be accomplished by synchronizing collection of free induction decays (FIDs) with the repetitive perturbation, and summing up those FIDs obtained in a given time slot following each perturbation. For example, if the perturbation is repeated n number of times, n FIDs each of which were recorded after a time interval of Δt following the perturbation can be summed up to measure the response of the system at time Δt. In *E. coli* suspensions the effects of oxygen pulsing upon the inorganic phosphate peaks have been followed by this method (Ogawa et al., 1978) while in muscles it has been used to follow the consequences of electrical stimulation (Gadian et al., 1978). For completeness we mention here another kind of rate measurement, which is unique in some ways to ^{13}C NMR, and is discussed in Section V. In these experiments the measured "scrambling" of a ^{13}C labeled substrate can be used to determine the *in vivo* rates of competing pathways.

In this section, we discuss examples of the first two kinds of measurements of enzymatic rates *in vivo*. In one case, we illustrate the effect of internal pH on the rate of glycolysis as monitored by consecutive ^{31}P NMR spectra obtained in 2 minute intervals (Ugurbil et al., 1978a). In the second case we show how the saturation transfer technique has been used to measure the apparent unimolecular rate constant for the ATPase catalysed synthesis of ATP from ADP and P_i under aerobic steady state conditions (Brown et al., 1977).

A. The Time Course of Glycolysis

^{31}P NMR measurements of glycolysis in *E. coli* illustrate the breadth and simultaneity of information that can be obtained about the time course of metabolite levels and internal pH. The main features of the glycolytic process observed in glucose-grown *E. coli* cells shown in Figs. 10-12 are that the addition of glucose results in a reduction of internal P_i with a simultaneous increase in FDP and NTP concentrations to their steady-state levels. Most of the P_i is stored in FDP during the steady state with smaller amounts in NTP and other intermediates. During the steady state there is a continuous drop of external pH, indicating the formation of acids. When the glucose is depleted, the rate of acid production is reduced, together with a rapid decrease in the level of FDP and the reappearance of P_i. The NTP level, however, remains high even after the level of FDP drops, indicating the continuing synthesis of NTP as the pool of phosphorylated intermediates is being exhausted.

A rough estimate of the rate of glycolysis was obtained by dividing the amount of added glucose by the duration of the steady state. A better estimate has been obtained by using the measured buffering capacity of the suspension and following the acidification of the medium, as expressed in the shift of the external P_i resonance. The rates calculated by this method agreed well with those determined from the length of the steady state (Ugurbil *et al.*, 1978a).

We have observed that glycolytic rates were drastically reduced in the presence of the ATPase inhibitor DCCD or the uncoupler FCCP (see Figs. 12 and 17). Moreover, the rate of glycolysis decreased during the course of these experiments although the levels of detectable glycolytic intermediates were constant. The control of the glycolytic rate is thought to be mediated mostly by the allosteric enzymes phosphofructokinase (PFK) and pyruvate kinase (PK) (Hess, 1973). Both of these enzymes are known to be affected by the levels of ADP, ATP, FDP, and PEP, while PFK in addition is controlled by other effectors. In our experiments DCCD increased the [NTP]/[NDP] ratio and FCCP decreased it, while both reduced the glycolytic rate. In both cases the FDP level was not different from the control, and the [PEP]/[ADP] ratio was so low that it could not drastically affect PFK activity (Blangy *et al.*, 1968). No correlation was found between measurable glycolytic rates and these metabolite levels in our experiments. However, to a first approximation it was possible to correlate the low rates of glycolysis with low internal pH.

In order to demonstrate the relationship between internal pH and the rate of glycolysis we have plotted in Fig. 19 both the initial rates of glycolysis as well as the rates measured at subsequent times vs. internal pH. In addition, Fig. 19 contains points obtained from DCCD- and FCCP-treated cells.

Figure 19 clearly indicates that under anaerobic conditions internal pH is an important parameter in determining the rate of glycolysis, although not necessarily the only one. Low intracellular pH was also implicated as the cause of decreased glycolytic rates in stationary phase *S. faecalis* cells that had been kept at 4°C for several days (Zarlengo and Abrams, 1962). On the assumption that aging had lowered the internal pH, Zarlengo and Abrams added ammonium ions, which raised the internal

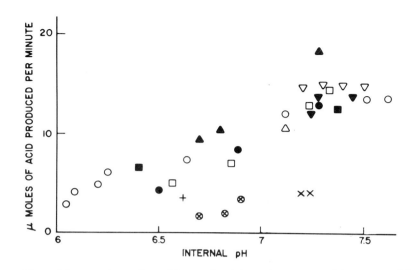

*Fig. 19. Rate of acid production in anaerobic E. coli cells
in the presence of glucose as a function of internal pH at 20°C.
This rate was calculated as described in the text using the external medium-buffer capacity measured in the presence of the
cells, and the external pH drift. Rates were obtained only at
those points where the intracellular sugar phosphate concentrations were at their maximum (i.e., during steady state). △,
Calculated from the experiment shown in Fig. 10; the cellular
suspension medium contained 100 mM PIPES, 50 mM MES, and 20 mM
P_i as buffers. 25 mM glucose was added to the NMR sample.
□, ⊗, obtained from experiments performed under identical
conditions to △ (same suspension medium) except 50 mM glucose
was added to the sample. ▽, ▼, Calculated from two different
experiments in which 50 mM glucose was added to sample but the
external medium buffer capacity was approximately doubled by
increasing the PIPES concentration to 200 mM. ▲, ■, ●,
Initial rates calculated from three different runs. ▲ was obtained from cells that were divided into three groups and resuspended in media with pH values of 7.4, 6.9, and 6.75. 25 mM
glucose was added after the NMR sample was allowed to sit at
20°C for ~20 minutes. ●, Identical to ▲ except the pH values
of the suspension media used were 7.4, 7.0, and 6.6. ■, Identical to ▲ except suspension media used had pH values 7.4 and
6.5. ×, ⊗, Rates obtained from cells treated with 2 and 4 mM
DCCD, respectively (Fig. 12B,C). +, Rate seen in the presence
of 75 μM FCCP (shown in Fig. 17). The control for the DCCD and
FCCP treated cells is △.*

pH. After this treatment the glycolytic rate increased, supporting their interpretation. In the present measurements we
have been able to refine this correlation and to determine the
pH profile of the glycolytic rate. It is not possible at
present to point out the specific reaction that is inhibited
by the lowering of internal pH. Most *E. coli* enzymes have high
pH optimum and thus can be affected. However, it has been
shown by *in vitro* experiments that PFK from several sources
shows a pH rate profile that is very similar to that shown in
Fig. 19 (Trivedi and Danforth, 1966; Ui, 1966). In the presence
of DCCD, inhibition of the glycolytic rate appears to be mediated by a factor other than pH because the two points at pH
7.20 and 7.24 coming from the sample with 2 mM DCCD deviate
from the remainder of the points in Fig. 19. A similar decrease in the glycolytic rates in the presence of DCCD has previously been observed in the strict anaerobe *S. faecalis*
(Harold *et al.*, 1969, 1970). Contrary to our results, uncouplers such as FCCP have been shown to stimulate glycolysis. For
example, Harold *et al.* (1970) had shown that acid production
rate in *S. faecalis* increases in the presence of FCCP. Unlike
our experiments, however, in the experiment reported by Harold
et al. (1970) the external pH and consequently the internal pH
was kept high enough so that glycolysis was probably not inhibited by pH.

B. ATPase Kinetics as Studied by Saturation Transfer

The basis of the saturation transfer technique is that an
individual spin will "remember" for a time of the order of its
T_1 any perturbation away from its equilibrium state (Forsen and
Hoffman, 1963; Gupta and Redfield, 1970; Glickson *et al.*, 1974).
Thus, if a spin on a molecule undergoing chemical exchange is
saturated, i.e., the population difference between the nuclear
Zeeman levels is eliminated by the application of a low-power
rf field at its resonant frequency, it will transmit that saturation to the molecular species with which it is exchanging.
By measuring the resultant reduction in magnetization at the
resonant frequency of the second species we can determine the
unidirectional exchange rate provided the T_1s are known.
For example, in the specific case of two species, A and B,
in exchange,

$$A \underset{k_{-1}}{\overset{k_1}{\rightleftharpoons}} B \tag{5}$$

the Block equation for the magnetization of B, M_z^B, is given by the expression

$$\frac{dM_z^B}{dt} = - \frac{M_z^B - M_{z0}^B}{T_1^B} + k_1 M_z^A - k_{-1} M_z^B \qquad (6)$$

where M_{z0}^B is the unperturbed equilibrium magnetization of B and T_1^B is the spin-lattice relaxation time of B in the absence of the chemical coupling to species A. If a continuous rf field at the resonant frequency of A is applied with sufficient power to reduce M_z^A to zero, then during the steady state when $(d/dt)M_z^B = 0$ the magnetization of B will be given by

$$M_z^B = \frac{1/T_1^B}{k_{-1} + 1/T_1^B} M_{z0}^B \qquad (7)$$

or the change in M_z^B, ΔM_z^B, by

$$\Delta M_z^B = \frac{k_{-1}}{k_{-1} + 1/T_1^B} M_{z0}^B \qquad (8)$$

Hence, as one might expect, the faster the exchange relative to T_1, the stronger the effect. We have, of course, assumed the system is in slow exchange with respect to the differences in the chemical shifts of A and B; otherwise, one would not see two separate resonances. Note that T_1^B is not equal to the relaxation time that would be measured in an ordinary T_1 measurement in the presence of chemical exchange with A; in the presence of chemical coupling one would see a complicated nonexponential decay characterized by T_1^A, T_1^B and the exchange rates. However, in the particular results discussed below we shall see that the inverses of the exchange rates are five- to tenfold longer than the T_1s and thus do not affect the apparent T_1s strongly.

Using saturation transfer, we were able to measure the unidirectional rate for ATPase-catalyzed synthesis of ATP from ADP and P_i in *E. coli*. The chemical and the NMR parameters of the measurement are illustrated schematically in Fig. 20.

The spectra in Fig. 21 show our results at $25^\circ C$ with aerobic *E. coli* cells that were respiring on endogeneous carbon sources during our measurements. Figure 21A shows the typical spectrum obtained under these conditions. The spectrum is very similar to those obtained with succinate-grown cells respiring

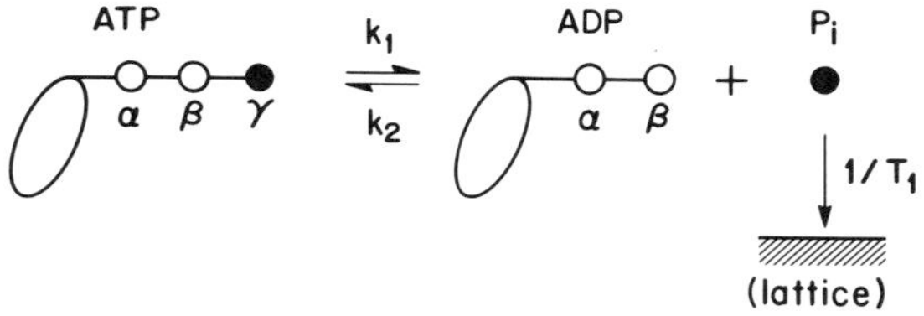

Fig. 20. Schematic representation of the fate of ATP γ phosphate spin in the presence of ATPase.

on succinate (see Figs. 6 and 13) except for the presence of the polyphosphate peak. Under these conditions, there are no changes in the intensities of the resonances for approximately 2 hours. Thus, the system is in a steady state. Figure 21B illustrates the spectrum of this sample obtained when the ATP position is irradiated, while in Fig. 21A the irradiating frequency was positioned halfway toward the P_i^{in} resonance (as indicated by the arrow). Figure 21C is the difference between the two. The spectra were obtained by switching the irradiating frequency between the two positions every 30 sec for a total of 30 minutes to average out any time-dependent changes during the accumulation.

As can be clearly seen in Fig. 21C the two spectra cancel perfectly except in two places, the NTP γ and the internal P_i. Note particularly that the external P_i shows no sign of magnetization transfer. The reduction of the P_i is 20% of its original value. Figure 22 shows the same experiment, on the same cells, after they were incubated for 10 minutes with DCCD in order to inhibit the ATPase. Here, there is no transfer to P_i^{in} within our signal-to-noise. This result clearly shows that the exchange that transfers the saturation to P_i^{in} spins is dominated by the ATPase-catalyzed reaction. Using a T_1 of 0.4 sec for the P_i^{in}, which was measured on a similar, DCCD-treated sample, we calculate an apparent unimolecular rate constant for the rate from P_i^{in} to ATP due to the DCCD-sensitive ATPase that is ≈ 0.8 sec^{-1}. The possible errors in our measurement are difficult to estimate beyond the contributions from the signal-to-noise in the difference spectrum. Probably the most severe is in the T_1 measurement. However we estimate that these errors could not change or vary the calculated unimolecular rate constant by more than a factor of 2.

We have also observed, albeit with worse signal-to-noise, the reverse reaction in which saturating the internal P_i results in a decrease in the ATP resonance. The poor signal-to-noise stems from the fact that only one-half of the nucleotide triphosphates in *E. coli* are ATP, and that the T_1 of ATP is only 0.2 sec. The measured reduction in the NTP γ peak was interpre-

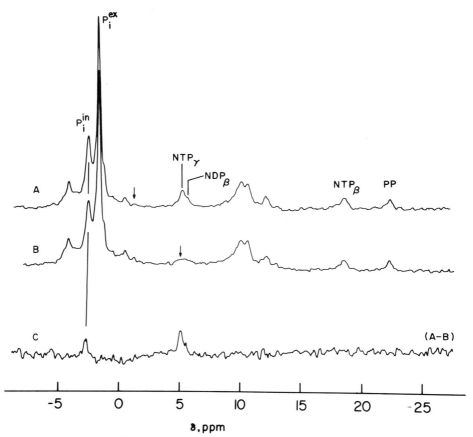

Fig. 21. ^{31}P NMR spectra of aerobic glucose-grown E. coli
at 25°C. The samples contained ~5 × 10¹¹ cells/ml. The arrows indicate the frequencies of the low-power pulses used in
B to saturate the NTP$_\gamma$ peak. The repetition time was 0.17 sec
and the pulse angle was 60°. The spectra consist of 4000 scans
each, taken in alternate 30-sec intervals. The peaks labeled
P_i^{in} and P_i^{ex} correspond to intracellular and extracellular inorganic phosphate, respectively. The peak at 22 ppm is polyphosphate. The peak identified as NTP$_\gamma$ consists of approximately
50% ATP and 50% nonadenine nucleotide triphosphates. The high-
field shoulder on this peak is due to the β phosphates of the
nucleoside diphosphates. The P_i^{in}-to-ATP ratio derived from
these intensities is 20 ± 5.

ted to give an apparent unimolecular rate constant of 10 sec^{-1}
for the ATP → P$_i$ + ADP reaction, but because of an even lower
signal-to-noise ratio here the possible errors are as much as

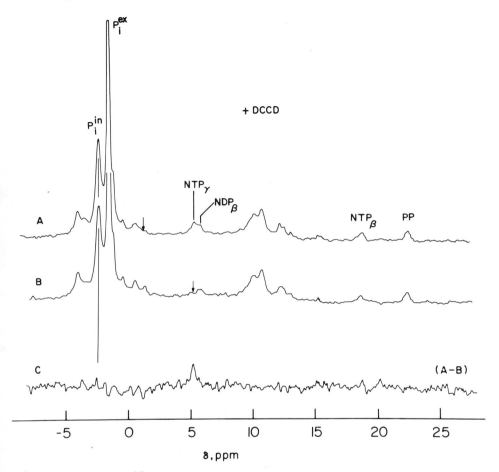

Fig. 22. ^{31}P NMR spectra of the sample in Fig. 21 taken 10 minutes after the addition of 1 m$\underline{M}$ DCCD. All other conditions are as in Fig. 21.

a factor of 3. Note that the ratio of these two rate constants is approximately equal to the observed ratio of P_i to ATP_γ intensities in Fig. 21.

It should be emphasized that the undirectional rate constant measured in this experiment describes the one-way flow across the ATPase and is different from net velocities measured, for example, by changes in the ATP pool as a function of time. Such measurements determine the difference between the two undirectional rates in opposing directions; while these undirectional rates could be very fast, the difference may in fact be very small. Note that in our samples after O_2 is

stopped (see Fig. 6) or, in anaerobic cells, after glucose and
FDP run out (Fig. 12) the NTP intensity decays with a time constant that is of the order of minutes, whereas the undirectional rate for the ATPase is of the order of ~ 1 sec^{-1}. This observation has interesting implications, provided that the unidirectional rates measured for the ATPase in the aerobic steady
state are applicable to the period when NTP levels are decaying.
If one includes the "high-energy" intermediate that drives the
ATP synthesis, the coupled ATPase reaction is given by the expression $X_1 + ADP + P_i \rightleftarrows ATP + X_2$. In the chemiosmotic hypothesis, X_1 and X_2 would be extracellular and intracellular protons.
If there were no other reactions coupled to ATP hydrolysis or
$X_1 \rightarrow X_2$ conversion, then the chemical potential or the activities of the species involved in this reaction will not change
when O_2 is turned off, or when glucose runs out. However, because of synthetic pathways, which require energy in the form
of ATP or small but finite "leaks" in the membrane, which collapse the proton gradient, the activities (to a first approximation concentrations for ATP, ADP, and P_i) will start to
change after oxygenation ceases or glucose is depleted. These
side reactions and/or leaks must determine the slow rate at
which ATP pool size decreases (which is in minutes) because
ATPase is capable of bringing the coupled reaction to equilibrium in a time of the order of seconds. It should be pointed
out that ATPase is a complicated enzyme that is probably regulated by numerous effectors; thus it is possible that the unidirectional rates measured during the aerobic state may not be
applicable when the ATP level is decreasing or $\Delta\mu_{H^+}$ is collaps-
ing. However, the strength of the present experiment is that
it shows how the ATPase rate can be measured *in vivo*, suggesting
how similar measurements made under different conditions might
allow us to understand the regulation of these rates.

V. ^{13}C NMR STUDIES

While ^{31}P NMR has been very valuable in studying cellular problems, it is of course restricted to measuring phosphorylated metabolites. In many ways the ^{31}P NMR measurements can be complemented by ^{13}C NMR on the same system.

Recently we have used ^{13}C-enriched glucose to monitor metabolism in anaerobic and aerobic *E. coli* suspensions. We followed the time course of glucose metabolism in well-resolved spectra obtained with 1 minute accumulations, and identified intermediates as well as end products. Using the methods we have developed to increase the ^{31}P signal to noise of cellular samples it was possible (Ugurbil et al., 1978b) to detect resonances of metabolic intermediates in addition to those of end products and substrates such as had been observed earlier (Eakin et al., 1972, 1975; Schaefer et al., 1975; Matwiyoff and Needham, 1972; Tanabe, 1973; Séquin and Scott, 1974; McInnes and Wright, 1975; McInnes et al., 1976; Smith et al., 1975; Kainosho et al., 1977). In our experiments we have also been able to determine the distribution of the ^{13}C labels among the different carbons of the intermediates and of the end products. This has allowed us to follow pathways in considerable detail *in vivo* and to evaluate the relative rates of alternative reactions. It is important to note that these NMR experiments uniquely determined *in vivo* the extent of labeling among intermediates that may have rapid-turnover rates, and thus avoid the uncertainties that may be encountered in extracting such metabolites.

Figure 23 shows a sequence of the 90.5 MHz ^{13}C NMR spectra of a suspension of anaerobic *E. coli* cells prepared as for the ^{31}P experiments and observed at the times indicated after glucose addition. Each spectrum consists of 200 free induction decays accumulated for 1 minute. The resonances have been assigned on the basis of chemical shift information (Gorin and Mazurek, 1975; Koerner et al., 1973; Horsley et al., 1970) and considerations of the known metabolic pathways. Prior to glucose addition, only the natural abundance ^{13}C resonances of the buffers (labeled B) used in the external medium are detectable (Fig. 23). After adding 1-^{13}C glucose one sees at 0.5 minutes the well-resolved α and β peaks of C-1 glucose as well as several weak resonances that stem from the natural abundance ^{13}C nuclei at the other positions of the two glucose anomers. The formation of intermediates and end products while glucose is being catabolized is monitored by the subsequent spectra. Specific assignments have been made to FDP, succinate (S), and the methyl groups of lactate (L), acetate (A), ethanol (E), alanine (ALA), and valine (VAL).

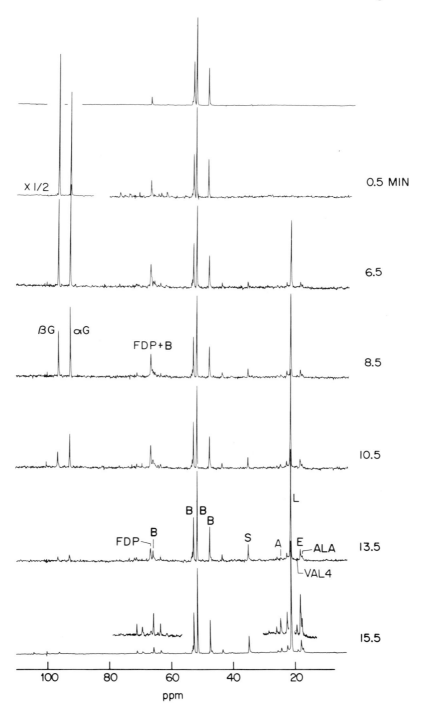

Fig. 23

Upon oxygenating the sample by bubbling O_2 through the suspension, the ^{13}C NMR spectrum evolves in time as shown in Fig. 24. Each of these spectra represents ~8 minutes of accumulation. Six of the new peaks that appear upon oxygenation have been assigned on the basis of their chemical shifts to the C-2, C-3, and C-4 carbons of glutamate and glutamine, as indicated in Fig. 24.

The intensities of the major resonances observed in Figs. 23 and 24 are illustrated as a function of time in Figs. 25 and 26 for the anaerobic and the aerobic series, respectively. The data for Fig. 26 were obtained from the experiments shown in Fig. 24. Figure 25 was prepared from an experiment done under conditions identical to that of Fig. 23 except that the MES buffer in the suspension medium was replaced with PIPES in order to avoid interference with the C-1 βFDP resonance. It is

Fig. 23. 90.52 MHz ^{13}C NMR spectra of anaerobic E. coli cells at 20°C as a function of time from ^{13}C-1 glucose addition. The chemical shift scale is with respect to tetramethylsilane (TMS). E. coli cells were grown in M9 minimal medium using glucose as the carbon source. Midlog cells were harvested as described in Ugurbil et al. (1978a). The pellet was suspended in a medium containing 10 mM Na_2PHO_4, 10 mM KH_2PO_4, 200 mM PIPES, and 50 mM MES (pH 7.5, adjusted with NaOH) at a density of ~5 × 10^{11} cells/ml. 1-^{13}C glucose (90% enriched) was added to a final concentration of 50 mM in the NMR sample, at time 0. Top spectrum (1600 scans) shows the natural abundance ^{13}C peaks (assigned to the PIPES and MES buffers) detectable in the suspension prior to glucose addition. All subsequent spectra, except the last one, represent 200 FIDs accumulated in 1 minute. The last spectrum consists of 1600 scans. The time given for each spectrum indicates the middle of the accumulation period, referred to glucose addition. Repetition time and pulse angle used were 0.34 sec and 45° respectively.

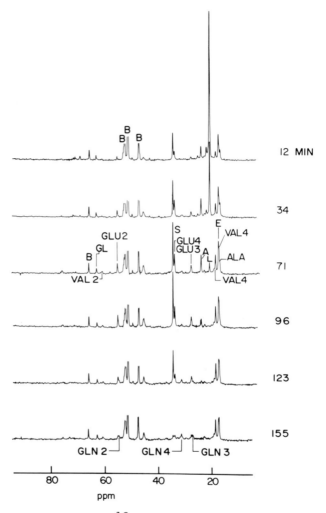

Fig. 24. 90.52 MHz ^{13} NMR spectra of aerobic E. coli cells
at 20°C as a function of time from oxygenation. The sample is
the same as in Fig. 23. 3 minutes after the bottom spectrum
in Fig. 23 was obtained, cells were oxygenated by bubbling O_2
through them continuously at the rate of 18 cm^3/min. Each
spectrum represents an average of 1600 FIDs obtained in ~8 minutes. Times given indicate the middle of the accumulation period from the start of oxygenation.

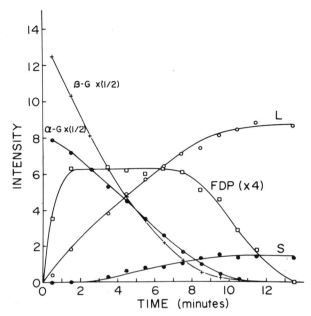

Fig. 25. Peak intensities of the major detectable ^{13}C resonances in anaerobic E. coli cells as a function of time from $1-^{13}C$ glucose addition. Peak intensities (heights) are given in arbitrary units. Data were obtained from an experiment performed under identical conditions to that of Fig. 23 except the resuspension medium contained 250 mM PIPES instead of 200 mM PIPES and 50 mM MES.

clearly seen that kinetics of substrate consumption and end product formation can be followed while simultaneously monitoring the time dependencies of intermediates.

These spectra show that the ^{13}C-enriched metabolites give *in vitro* spectra that are better resolved than corresponding ^{31}P NMR spectra and have slightly better signal-to-noise even though the smaller nuclear moment of the ^{13}C reduces its intensity by approximately a factor of four. The resolution is better because the ^{13}C full widths at half-height are ~15 Hz at 90.5 MHz (except for the βFDP peak, whose width is ~30 Hz) while at the same magnetic field and 145.7 MHz the ^{31}P widths range from 35 to 90 Hz. Furthermore, the range of ^{13}C chemical shifts in the abundant metabolites is ~100 ppm as opposed to ~20 ppm for ^{31}P in the corresponding phosphates. The relative ^{13}C intensities are increased two- or three-fold by the narrower lines. Presumably the nuclear Overhauser effect is contributing another factor of ~3, which would explain the observed intensities.

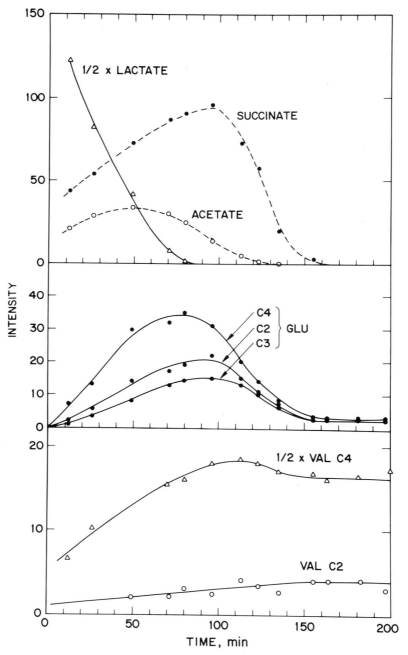

Fig. 26. Peak intensities of major resonances in aerobic
E. coli cells as a function of time from the onset of oxygenation. Data were obtained from the experiment shown in Fig. 24.

The observed end products of anaerobic glucose catabolism are consistent with the classification of *E. coli* as a "mixed-acid" bacteria. Glycolysis in strain MRE 600 under our conditions (see Figs. 23 and 25) produces mainly lactate, which presumably is derived directly from pyruvate by the action of lactate dehydrogenase. Succinate, ethanol, and acetate are produced to a much lesser extent. The relative distribution of the end products under our conditions is somewhat different from previously reported values (Doelle, 1975). However, the ratios of the end products formed may depend on environmental conditions and may even vary among different strains.

The two amino acids valine and alanine belong to the "pyruvate" class of amino acids, and can therefore be synthesized during glycolysis. Each valine molecule is expected to be produced from two pyruvate molecules by a mechanism that should label exclusively the two methyl (C4) carbons (Lehninger, 1970). One of the two methyl carbon resonances appears at 19.9 ppm and is labeled VAL4 (in Fig. 23). The second is discernible as a shoulder on the ethanol peak but is not well resolved in the anaerobic spectra. However, when the cells are aerated, the overall valine intensity increases and this second methyl carbon (also labeled VAL4) becomes more prominent (see Fig. 24).

During anaerobic glycolysis under our conditions, FDP is the most abundant intermediate of the Embden-Meyerhof-Parnas (EMP) pathway. It rises rapidly to a constant level, which is maintained for ~6 minutes. Its intracellular concentration during this "steady-state" period was estimated to be ~13 m$\underline{M}$ from the analogous ^{31}P NMR experiments discussed above. The other intermediate of the EMP pathway that we have positively identified in the intact cell spectra is glucose-6-phosphate (G6P) (intracellular concentration ~2 mM), which was observed in the *in vivo* ^{13}C NMR spectrum when [6-^{13}C] glucose was used as the glycolysis substrate (not shown). The chemical shift of the C-1 G6P is very similar to that of C-1 glucose and is not resolved from the intense C-1 glucose resonances in the intact cell spectrum.

The FDP region of the ^{13}C NMR spectra of glycolysing cells is illustrated with better signal-to-noise in Fig. 27. Six 1 minute spectra that displayed constant FDP intensity during glycolysis were added together to improve the signal-to-noise ratio. Peaks 1 and 3 are not yet identified, although the time course of peak 3 indicates that it is another intermediate. Peak 4 stems from an impurity in the ^{13}C-1 glucose used, while peak 2 is due to the natural abundance ^{13}C in the C-2 position of lactate. Note that the resonance located at 65.3 ppm has been assigned to the 6 carbon of FDP. Because of the importance of this assignment in considerations of isotopic "scrambling" it was checked both by addition of [1,6-^{13}C] FDP to a cellular extract and by enzymatic digestions of the FDP in the

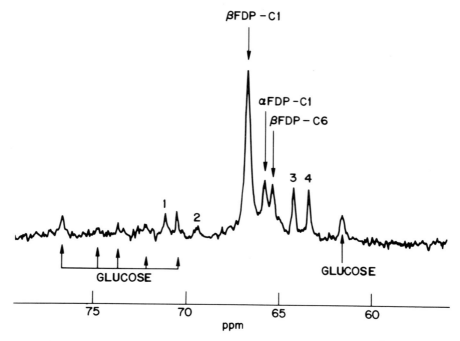

Fig. 27. 55 to 80 ppm region of the 90.52 MHz ^{13}C NMR
spectrum of glycolysing E. coli cells in the presence of 1-^{13}C
glucose. The spectrum was obtained from the same experiment
as Fig. 25 by adding together six 1 minute spectra that displayed constant FDP levels.

extract (Ugurbil *et al.*, 1978b). In similar experiments using
[6-^{13}C] glucose no sign of FDP enrichment at the 1 carbon position was observed to within the noise limits of the spectrum
(~5% of the C6 peak).

We have schematically indicated in Fig. 28 the relevant
steps involving the reactions of FDP. Clearly, if both aldolase and TPI are in equilibrium, a label introduced at either
C-1 or C-6 would be equally distributed between these two positions. We have observed that a label introduced at C-1 does
appear at C-6 while the converse does not happen. Since the
pathway proceeds through 3GAP it can be seen from Fig. 28 that
the $^{13}CH_2OPO_3$ position of 3GAP must contain the ^{13}C label 50%
of the time. Hence the weak intensity observed at C-6 of FDP
in Fig. 27 means that the upward reaction through aldolase is
at most one-third as fast as the downward reaction. We say at
most because the labeling of C-6 of FDP might have occurred
through the transaldolase shunt, a possibility that is presently
being evaluated.

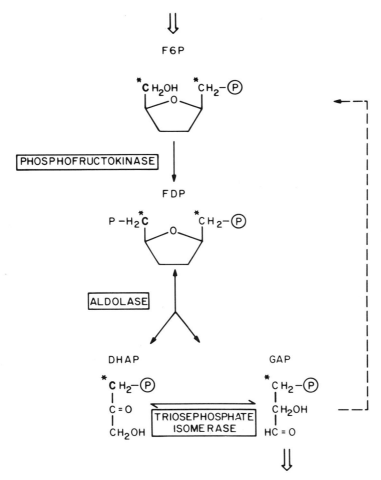

Fig. 28. Reactions catalyzed by the enzyme phosphofructo-kinase, fructose 1-6 diphosphate aldolase, and triose phosphate isomerase. Dashed line indicates transaldolase and/or trans-ketolase activity. Stars indicate all carbons that could be enriched with ^{13}C when $[1-^{13}C]$ or $[6-^{13}C]$ glucose are used as the substrate. Double arrows show the direction of flow during glucose catabolism under anaerobic conditions in MRE 600. It should be mentioned that although E. coli could synthesize gly-cerol from DHAP, this was not the case for MRE 600 under our conditions.

The major resonances detectable during oxygenation that were absent in the anaerobic spectrum are the C-4, C-3, and

C-2 carbons of glutamate. Glutamate is ascribed a central role in the synthesis of several amino acids. It itself is synthe sized by direct amination of α-ketoglutarate, a TCA cycle intermediate. Because the ^{13}C-H dipole-dipole interaction is expected to be the dominant T_1 relaxation mechanism (Kuthman *et al.*, 1970) GLU C-2 and C-3 should have similar T_1s, while for C-4 T_1 might be somewhat longer. Therefore, to a first approximation, the ratios of the peak intensities will reflect the actual distribution of the label between GLU C-2 and C-3 while underestimating its abundance in C-4. Note that the C-4 resonance (GLU4) is about twice as intense as the C-2 and C-3 peaks. It can be deduced from the known chemistry that a label entering the TCA cycle as the methyl carbon of acetyl-CoA will, in the first turn of the cycle, be incorporated exclusively into GLU C-4. In the second turn, the label will move into GLU C-2 and C-3, whose enrichment will be equal because of the symmetry of succinate. Alternatively, ^{13}C-enriched carbons could come directly into oxaloacetate from PEP via phosphoenol oxaloacetate (Doelle, 1975); these, however, will label GLU C-2 in their first turn in the TCA cycle. Thus, the spectra showing a stronger signal from GLU C-4 indicate that in our aerobic samples the label is preferentially entering the TCA cycle as acetyl-CoA rather than as oxaloacetate. Our observation of succinate under <u>anaerobic</u> conditions is presumably due to this latter pathway.

Intensities of the resonances in the present spectra cannot be used to measure accurately the absolute concentrations of the corresponding metabolites because our pulsing rate was faster than the expected spin-lattice relaxation rates. In addition, contributions of the nuclear Overhauser effect to the intensities may be different for each resonance. While these limitations can only be removed by future experiments, the time dependences of the present resonances can still be used to obtain important kinetic information. From Fig. 24, it can be seen that the utilization rate of α glucose increases in the first 1.5 minutes after glucose addition and then remains approximately constant for 7 minutes during the steady state characterized by the time-independent FDP concentration. The rate of β-glucose consumption, however, changes with time during the same period. The concentration dependences of the consumption rates for the α and β glucose anomers taken from the steady-state points between 1.5 and 8.5 minutes can be fitted to a Lineweaver-Burk plot for independent anomeric binding. The constant rate of α glucose utilization during the glycolytic steady state suggests that this anomer has saturated the enzyme(s) responsible for its consumption. The inverse of the β-glucose utilization rate, on the other hand, is linearly dependent on the inverse of its concentration and its plot yields an apparent K_m of 7 m<u>M</u>. On this model the α glucose specific site had a smaller K_m, i.e., ≤ 1 mM. The data, including the points outside the steady state range, can also be fitted satisfactorily to a model that assumes competition between α and β anomers for the same binding site but with the same V_{max}. This model yields a K_m of 3.8 and 2.5 mM for the α and β anomers, respectively.

The rapid utilization of glucose in *E. coli* is predominantly accomplished by the PEP-dependent phosphotransferase system that catalyzes the entry of glucose into the bacteria as G6P (Kundig *et al.*, 1964; Roseman, 1975, Kornberg, 1975). Genetic studies have shown the existence of two distinct PEP-dependent phosphotransferase systems for glucose translocation in *E. coli* cells. *E. coli* strains that contain both and only one (strains K2.1t and K2.1.22a, respectively) of these transport systems have been isolated. When we monitored glucose consumption in anaerobic suspensions of these two strains (kindly supplied by Dr. Jones Mortimer, University of Cambridge) we have observed the α and β glucose anomers always to be at equilibrium and consumed at the same rate in both strains. When ~85% α 1-^{13}C glucose was used, anomeric equilibrium was established within ~2 minutes in suspension containing ~10^{11} cells/ml. Another wild-type *E. coli* strain (NR7, the parental strain for the ATPase mutant NR70) displayed similar properties. Clearly, unlike MRE600, these three strains possess glucose epimerase activity capable of competing with the rate of translocation of the two anomers across the cell membrane.

VI. CONCLUSION

Many articles on NMR investigations of cells, including ours, have in the past concluded by suggesting that the NMR method has great potential applications to cells. In this review, we have tried to break with that tradition and have shown in the particular case of *E. coli* cells what has actually been achieved. In summary, these achievements are as follows.

With considerable completeness and with 1 to 2 minutes time resolution, we have followed the rate of glucose consumption and the subsequent formation of FDP while simultaneously determining the free energy of hydrolysis of ATP to ADP. Measurements of ΔpH showed it to be proportional to this free energy. The interpretation of these results was the generally accepted hypothesis that the ATP was hydrolyzed by a membrane bound ATPase that used the energy to translocate protons (outward), thereby creating a proton chemical potential across the membrane. This hypothesis was supported by the cellular response to DCCD, an inhibitor of ATPase, in which the ATP lasted longer

and the ΔpH was practically eliminated. An ATPase mutant be-
haved similarly to the DCCD-treated cells, but its inhibition
was complete so that under anaerobic conditions a transmembrane ΔpH was undetectable. The action of the uncoupler FCCP
was also observed clearly in the NMR spectra, and in the presence of the uncoupler both ΔpH and the ATP resonances were eliminated. Similar results on individual aspects of these observations have generally been obtained in *E. coli* and other bacteria using conventional chemical methods. However, these NMR
experiments have presented us with a panoply of simultaneous
and quantitative measurements of these pathways that avoid some
of the questions raised about other methods.

In addition, we have shown from the ^{13}C NMR experiments that
substrate consumption and end-product formation can be monitored
while intermediates of the metabolic pathway under investigation
(in this case, glycolysis) are observed. Furthermore, in the
event of label "scrambling," as observed in FDP during glycolysis, detailed information can be obtained about the kinetic
parameters of that section of the pathway. ^{13}C NMR also made
it possible to obtain information on anomeric distribution and
specificity; in particular we have observed that *E. coli* MRE
600 clearly utilized the α and β glucose anomers at different
rates and intracellular FDP is in anomeric equilibrium during
glycolysis at 20°C.

Finally, in the saturation transfer experiments, using ^{31}P
NMR, it has been possible to measure unidirectional rates dur-
ing steady state for the ADP + P$_i$ $\overset{\rightarrow}{\leftarrow}$ ATP reaction as catalyzed by
the membrane-bound ATPase.

These results demonstrate that NMR can be used to obtain information of varying complexity and sophistication on intact
cell suspension. At one level, NMR can serve as a noninvasive
spectroscopic tool for measuring the levels of metabolites *in
vivo*. It is also an accurate pH meter that can distinguish
among compartments of the different pHs and at the same time
provide information on the distribution of pH values in the
sample. Both ^{31}P and ^{13}C NMR are capable of providing kinetic
information *in vivo*, ranging from simple flow measurements to
unidirectional enzymatic rates, making possible studies aimed
at understanding metabolic rates and their control. All of the
measurements make NMR an exciting new method in cell biology.

REFERENCES

Altendorf, K. A., and Zitzman, W. (1975). *FEBS Lett. 59,* 268-
    272.
Blangy, D., Buc, H., and Monod, J. (1968). *J. Mol. Biol. 31,*
    13-35.

Brown, T. R., Ugurbil, K., and Shulman, R. G. (1977). *Proc. Nat. Acad. Sci. USA 74*, 5551-5553.

Burt, C. T., Glonek, T., and Barany, M. (1976). *J. Biol. Chem. 251*, 2584-2591.

Cox, G. B., and Gibson, F. (1974). *Biochem. Biophys. Acta 346*, 1-25.

Doelle, H. W. (1975). "Bacterial Metabolism." Academic Press, New York.

Eakin, R. T., Morgan, L. O., Gregg, C. T., and Matwiyoff, N. A. (1972). *FEBS Lett. 28*, 259-264.

Eakin, R. T., Morgan, L. O., and Matwiyoff, N. A. (1975). *Biochemistry 14*, 529-540.

Fillingame, R. H. (1975). *J. Bacteriol. 124*, 870-883.

Fillingame, R. H. (1976). *J. Biol. Chem. 251*, 6630-6637.

Forsen, S., and Hoffman, R. A. (1963). *J. Chem. Phys. 26*, 958-959.

Glickson, J. D., Dadok, J., and Marshall, R. G. (1974). *Biochemistry 13*, 11-14.

Gorin, P. A. J., and Mazurek, M. (1975). *Can. J. Chem. 53*, 1212-1223.

Gupta, R. K., and Redfield, A. G. (1970). *Science 169*, 1204-1205.

Hare, J. F. (1975). *Biochim. Biophys. Res. Commun. 66*, 1329-1337.

Harold, F. M. (1972). *Bacteriol. Rev. 36*, 172-230.

Harold, F. M. (1977). *Curr. Top. Bioenerg. 6*, 89-143.

Harold, F. M., Baarda, J. R., Baron, J., and Abrams, A. (1969). *J. Biol. Chem. 244*, 2261-2268.

Harold, F. M., Pavlasova, E., and Baarda, J. R. (1970). *Biochim. Biophys. Acta 196*, 235-244.

Henderson, T. O., Costello, A. J. R., and Omachi, A. (1974). *Proc. Nat. Acad. Sci. USA 71*, 2487-2490.

Hess, B. (1973). *Symp. Soc. Exp. Biol. 27*, 105-131.

Horsley, W., Sternlicht, H., and Cohen, J. S. (1970). *J. Am. Chem. Soc. 92*, 680.

Hoult, O. I., Busby, S. J. W., Gadian, D. G., Radda, G. K., Richards, R. E., and Seeley, P. J. (1974). *Nature 252*, 285-287.

Koerner, T. A. W., Cary, L. W., Bhacca, N. S., and Younathan, E. S. (1973). *Biochem. Biophys. Res. Commun. 51*, 543-550.

Kornberg, H. L. (1975). *Biochem. Soc. Trans., 558th Edinburgh Meeting 3*, 835-837.

Kundig, W., Ghosh, S., and Roseman, S. (1964). *Proc. Nat. Acad. Sci. USA 52*, 1067-1074.

Kuthman, K. F., Grant, D. M., and Harris, R. K. (1970). *J. Chem. Phys. 52*, 3439-3448.

Lehninger, A. L. (1970). "Biochemistry." Worth Publ. New York.

Lowry, O. H., Carter, J., Ward, J. B., and Glaser, L. (1971). *J. Biol. Chem. 246*, 6511-6521.

McInnes, A. G., and Wright, J. L. C. (1975). *Acc. Chem. Res. 9,* 313.

McInnes, A. G., Walter, J. A., Wright, J. L. C., and Wining, L. C. (1976). "Topics in Carbon-13 NMR Spectroscopy" (G. C. Levy, ed.), Vol. 2, p. 123. Wiley (Interscience), New York.

Mandel, M., and Westley, J. W. (1964). *Nature 203,* 302.

Matwiyoff, N. A., and Needham, T. E. (1972). *Biochem. Biophys. Res. Commun. 49,* 1158-1164.

Mitchell, P. (1961). *Nature 191,* 144-148.

Mitchell, P. (1966). *Biol. Rev. (Cambridge) 41,* 445-502.

Mitchell, P. (1968). "Chemiosmotic Coupling and Energy Trans-duction." Glynn Research Ltd., Bodmin U.K.

Moon, R. B., and Richards, J. H. (1973). *J. Biol. Chem. 248,* 7276-7278.

Navon, G., Ogawa, S., Shulman, R. G., and Yamane, T. (1977a). *Proc. Nat. Acad. Sci. USA 74,* 87-91.

Navon, G., Ogawa, S., Shulman, R. G., and Yamane, T. (1977b). *Proc. Nat. Acad. Sci. USA 74,* 888.

Navon, G., Navon, R., Shulman, R. G., and Yamane, T. (1978). *Proc. Nat. Acad. Sci. USA 75,* 891-895.

Ogawa, S., Shulman, R. G., Glynn, P., Yamane, T., and Navon, G. (1978). *Biochem. Biophys. Acta 502,* 45-50.

Padan, E., Zilberstein, D., and Rottenberg, H. (1976). *Eur. J. Biochem. 63,* 533-541.

Racker, E. (1970). "Membranes of Mitochondria and Chloroplasts." (E. Racker, ed.), pp. 127/171. Van Nostrand Reinhold Co., New York.

Ramos, S., Schuldiner, S., and Kaback, H. R. (1976). *Proc. Nat. Acad. Sci. USA 73,* 1892-1896.

Roberts, R. B., Abelson, P. H., Cowie, D. B., Bolton, E. T., and Britten, R. J. (1963). "Studies of Biosynthesis in *E. coli,*" p. 5, Publ. 607. Carnegie Institution of Washington, Washington, D.C.

Roseman, S. (1975). *Ciba Found. Symp. (New Ser.) 31,* 225-241.

Rosen, B. P. (1973a). *J. Bacteriol. 116,* 1124.

Rosen, B. P. (1973b). *Biochem. Biophys. Res. Commun. 53,* 1289-1296.

Rosing, J., and Slater, E. C. (1972). *Biochem. Biophys. Acta 267,* 275-290.

Rottenberg, H. (1975). *Bioenergetics 7,* 61-74.

Rottenberg, H. (1976). *FEBS Lett. 66,* 159-163.

Saez, M. J., and Lagunas, R. (1976). *Mol. Cell. Biochem. 13,* 73-78.

Salhany, J. M., Yamane, T., Shulman, R. G., and Ogawa, S. (1975). *Proc. Nat. Acad. Sci. USA 72,* 4466-4970.

Schaefer, J., Stejskal, E. O., and Beard, C. R. (1975). *Plant Physiol. 55,* 1048-1053.

Séquin, U., and Scott, A. I. (1974). *Science 186,* 101.

Simoni, R. D., and Postma, P. W. (1975). *Annu. Rev. Biochem.* *44*, 523-554.

Tanabe, M. (1973). *Biosynthesis 2*, 241.

Trivedi, B., and Danforth, W. H. (1966). *J. Biol. Chem. 241*, 4110-4114.

Ugurbil, K., Rottenberg, H., Glynn, P., and Shulman, R. G. (1978a). *Proc. Nat. Acad. Sci. USA 75*, 2244-2248.

Ugurbil, K., Brown, T. R., den Hollander, J. A., Glynn, P., and Shulman, R. G. (1978b). *Proc. Nat. Acad. Sci. USA 75*, 3742-3746.

Ui, M. (1966). *Biochim. Biophys. Acta 124*, 310-322.

Weibel, K. E., Mor, J. R., and Fiechter, A. (1974). *Anal. Biochem. 58*, 208-216.

West, I. C., and Mitchell, P. (1974). *FEBS Lett. 40*, 1-4.

Yoshida, M., Okamoto H., Sone, N., Hirata, H., and Kagawa, Y. (1977). *Proc. Nat. Acad. Sci. USA 74*, 936-940.

Zarlengo, M., and Abrams, A. (1962). *Biochim. Biophys. Acta 71*, 65-77.

SUBJECT INDEX

591